Communications in Computer and Information Science 2968

Series Editors

Gang Li, *School of Information Technology, Deakin University, Burwood, VIC, Australia*
Joaquim Filipe, *Polytechnic Institute of Setúbal, Setúbal, Portugal*
Zhiwei Xu, *Chinese Academy of Sciences, Beijing, China*

Rationale

The CCIS series is devoted to the publication of proceedings of computer science conferences. Its aim is to efficiently disseminate original research results in informatics in printed and electronic form. While the focus is on publication of peer-reviewed full papers presenting mature work, inclusion of reviewed short papers reporting on work in progress is welcome, too. Besides globally relevant meetings with internationally representative program committees guaranteeing a strict peer-reviewing and paper selection process, conferences run by societies or of high regional or national relevance are also considered for publication.

Topics

The topical scope of CCIS spans the entire spectrum of informatics ranging from foundational topics in the theory of computing to information and communications science and technology and a broad variety of interdisciplinary application fields.

Information for Volume Editors and Authors

Publication in CCIS is free of charge. No royalties are paid, however, we offer registered conference participants temporary free access to the online version of the conference proceedings on SpringerLink (http://link.springer.com) by means of an http referrer from the conference website and/or a number of complimentary printed copies, as specified in the official acceptance email of the event.

CCIS proceedings can be published in time for distribution at conferences or as post-proceedings, and delivered in the form of printed books and/or electronically as USBs and/or e-content licenses for accessing proceedings at SpringerLink. Furthermore, CCIS proceedings are included in the CCIS electronic book series hosted in the SpringerLink digital library at http://link.springer.com/bookseries/7899. Conferences publishing in CCIS are allowed to use our online conference service (Meteor) for managing the whole proceedings lifecycle (from submission and reviewing to preparing for publication) free of charge.

Publication process

The language of publication is exclusively English. Authors publishing in CCIS have to sign the Springer CCIS copyright transfer form, however, they are free to use their material published in CCIS for substantially changed, more elaborate subsequent publications elsewhere. For the preparation of the camera-ready papers/files, authors have to strictly adhere to the Springer CCIS Authors' Instructions and are strongly encouraged to use the CCIS LaTeX style files or templates.

Abstracting/Indexing

CCIS is abstracted/indexed in DBLP, Google Scholar, EI-Compendex, Mathematical Reviews, SCImago, Scopus. CCIS volumes are also submitted for the inclusion in ISI Proceedings.

How to start

To start the evaluation of your proposal for inclusion in the CCIS series, please send an e-mail to ccis@springer.com

Hamid R. Arabnia · Leonidas Deligiannidis ·
Farzan Shenavarmasouleh ·
Farid Ghareh Mohammadi · Soheyla Amirian
Editors

Emerging Trends in Artificial Intelligence, Computational Intelligence, and Social Networks

12th International Conference on Computational Science and Computational Intelligence, CSCI 2025
Las Vegas, NV, USA, December 3–5, 2025
Proceedings, Part II

Springer

Editors
Hamid R. Arabnia
University of Georgia
Athens, GA, USA

Leonidas Deligiannidis
Wentworth Institute of Technology
Boston, MA, USA

Farzan Shenavarmasouleh
Medialab Inc.
Lawrenceville, GA, USA

Farid Ghareh Mohammadi
Verify Radiology Images Consultants
Houston, TX, USA

Soheyla Amirian
Pace University
New York, NY, USA

ISSN 1865-0929 ISSN 1865-0937 (electronic)
Communications in Computer and Information Science
ISBN 978-3-032-28456-3 ISBN 978-3-032-28457-0 (eBook)
https://doi.org/10.1007/978-3-032-28457-0

This Springer imprint is published by the registered company Springer Nature Switzerland AG
The registered company address is: Gewerbestrasse 11, 6330 Cham, Switzerland

Preface

It is our great pleasure to present this collection of research papers from the 12th International Conference on Computational Science and Computational Intelligence (CSCI 2025). This volume, titled "Emerging Trends in Artificial Intelligence, Computational Intelligence, and Social Networks", comprises selected papers accepted in the Research Tracks on Artificial Intelligence (CSCI-RTAI), Computational Intelligence (CSCI-RTCI), and Social Network Analysis, Social Media, & Mining (CSCI-RTNA). The conference was held in Las Vegas, Nevada, USA, from December 3 to December 5, 2025.

The CSCI 2025 International Conference brought together papers from a diverse array of communities, including researchers from universities, corporations, and government agencies. Accepted papers are published by Springer Nature, and the proceedings showcase solutions to key challenges in various critical areas of Computational Science and Computational Intelligence.

Computational Science (CS) is the study of addressing problems that are impossible to solve (or difficult to solve) without computers. CS can be considered to be the bridge between computer science and other sciences. The field is interdisciplinary by nature and includes the use of advanced computing capabilities to understand and solve complex problems. In short, CS is the science of using computers to do science. Computational Intelligence (CI) is the study of computational methods that exhibit intelligence. These methods adapt to changing environments and changing goals. There is a significant overlap between the fields of CI and Artificial Intelligence (AI). However, there is also a difference: AI techniques often involve top-to-bottom methods (i.e., methods to reach solutions are imposed from the top) whereas CI techniques often involve bottom-up methods (i.e., solutions emerge from unstructured beginnings). An important part of CI includes a set of Nature-inspired computational approaches to address complex problems for which traditional methods are infeasible. Computational Science and Computational Intelligence both share the same objective: finding solutions to difficult problems. However, the methods to reach the solutions are different. The main objective of the CSCI Conference is to facilitate increased opportunities for cross-fertilization across CS and CI.

Considering the above broad outline, the CSCI 2025 International Conference was composed of the following focused Research Tracks:

Artificial Intelligence (CSCI-RTAI); Big Data and Data Science (CSCI-RTBD); Computational Science (CSCI-RTCS); Computational Intelligence (CSCI-RTCI); Computational Biology (CSCI-RTCB); Cyber Warfare, Cyber Defense, & Cyber Security (CSCI-RTCW); Signal & Image Processing, Computer Vision & Pattern Recognition (CSCI-RTPC); Smart Cities and Smart Mobility (CSCI-RTSC); Education - CS & CE (CSCI-RTED); Health Informatics and Medical Systems (CSCI-RTHI); Mobile Computing, Wireless Networks, & Security (CSCI-RTMC); Software Engineering (CSCI-RTSE); Internet of Things & Internet of Everything (CSCI-RTOT); Social Network

Analysis, Social Media, & Mining (CSCI-RTNA); Cloud Computing and Data Centers (CSCI-RTCC); Parallel & Distributed Computing (CSCI-RTPD); and Large Language Models (LLM). The scope of each track can be found at: https://www.american-cse.org/csci2025/topics and https://www.american-cse.org/LLM2025/topics.

An important objective of the CSCI 2025 International Conference and its associated Research Tracks was to foster opportunities for cross-fertilization between the fields of Computational Science and Computational Intelligence. The CSCI Conference is deeply committed to promoting diversity and eliminating discrimination, both in its role as a conference organizer and as a service provider. Our goal is to create an inclusive culture that respects and values differences, promotes dignity, equality, and diversity, and encourages individuals to reach their full potential. We are also dedicated, wherever possible, to organizing a conference that represents the global community. We sincerely hope that we have succeeded in achieving these important objectives.

The Steering Committee and the Program Committee would like to extend their gratitude to all the authors who submitted papers for consideration. CSCI 2025 received submissions from 48 countries, with approximately 48% of them coming from outside the United States. Each submitted paper underwent a rigorous peer-review process, with at least two experts (an average of 2.5 referees per paper) evaluating the submissions based on originality, significance, clarity, impact, and soundness. In cases where reviewers' recommendations were contradictory, a program committee member was tasked with making the final decision, often consulting additional referees for further guidance. The CSCI Conference followed the guidelines of COPE (Committee on Publication Ethics):

- Typical submissions underwent a single-blind peer review process, in which the authors remain unaware of the identities of the reviewers, while the reviewers are informed of the authors' identities.
- Papers authored by one or more members of the program committee, including co-chairs, were subjected to a double-blind peer review process, ensuring that neither the authors nor the reviewers were aware of each other's identities or affiliations.

The Research Track on Artificial Intelligence (CSCI-RTAI) received 389 submissions, of which 69 papers were accepted (2 papers were withdrawn), resulting in a paper acceptance rate of 18%; we selected only 24 papers for this volume. The Research Track on Computational Intelligence (CSCI-RTCI) received 36 submissions, of which 9 papers were accepted, resulting in a paper acceptance rate of 25%. The Research Track on Social Network Analysis, Social Media, & Mining (CSCI-RTNA) received 78 submissions, of which 18 papers were accepted, resulting in a paper acceptance rate of 23%; we selected only 8 papers for this volume. A total of 41 papers are included in this volume.

We are deeply grateful to the many colleagues who contributed their time and effort to organizing the CSCI 2025 Conference. In particular, we extend our thanks to the members of the Program Committee, the Steering Committee, and the referees. We would also like to express our appreciation to the primary sponsor of the conference, the American Council on Science & Education. The list of members of the Program Committee for each track can be found at: https://www.american-cse.org/csci2025/committees.

We extend our heartfelt gratitude to all the speakers and authors for their valuable contributions. We would also like to thank the following individuals and organizations for their support: the staff at the Luxor Hotel (an MGM property, conference/meeting department) and the staff of Springer, for their assistance in various aspects of the event.

We are pleased to present the proceedings of CSCI 2025 (selected papers of CSCI-RTAI, CSCI-RTCI, CSCI-RTNA Research Tracks). These proceedings represent a collection of outstanding research contributions that reflect the diversity and depth of work in Artificial Intelligence, Computational Intelligence, and Social Network Analysis.

Co-editors: Research Tracks on Artificial Intelligence (CSCI-RTAI), Computational Intelligence (CSCI-RTCI), and Social Network Analysis, Social Media, & Mining (CSCI-RTNA) of CSCI 2025 International Conference:

December 2025

Hamid R. Arabnia
Leonidas Deligiannidis
Farzan Shenavarmasouleh
Farid Ghareh Mohammadi
Soheyla Amirian

Organization

Steering Committee – Co-chairs (CSCI 2025)

Hamid R. Arabnia	University of Georgia, USA
Leonidas Deligiannidis	Wentworth Institute of Technology, USA
Fernando G. Tinetti	Universidad Nacional de La Plata, Argentina
Quoc-Nam Tran	Southeastern Louisiana University, USA

Publication Co-chairs and Co-editors

"Emerging Trends in Artificial Intelligence, Computational Intelligence, and Social Networks"

Research Track: Artificial Intelligence (CSCI-RTAI)
Research Track: Computational Intelligence (CSCI-RTCI)
Research Track: Social Network Analysis, Social Media, & Mining (CSCI-RTNA)

Part of Proceedings of The 2025 International Conference on Computational Science and Computational Intelligence (CSCI)

Hamid R. Arabnia	University of Georgia, Georgia, USA
Leonidas Deligiannidis	Wentworth Institute of Technology, USA
Farzan Shenavarmasouleh	Medialab Inc., USA
Farid Ghareh Mohammadi	CEO & Co-founder, Verify Radiology Images Consultants, LLC, USA
Soheyla Amirian	Pace University, USA

Members of Steering Committee (CSCI 2025)

Babak Akhgar	Sheffield Hallam University, UK
Abbas M. Al-Bakry	University of Information Technology & Communications, Iraq
Nizar Al-Holou	University of Detroit Mercy, USA
Soheyla Amirian	Pace University, USA
Hamid R. Arabnia	University of Georgia, USA
Rajab Challoo	Texas A&M University-Kingsville, USA
Chien-Fu Cheng	Tamkang University, Taiwan

Hyunseung Choo	Sungkyunkwan University, South Korea
Kevin Daimi	University of Detroit Mercy, USA
Leonidas Deligiannidis	Wentworth Institute of Technology, USA
Eman M. El-Sheikh	University of West Florida, USA
Mary Mehrnoosh Eshaghian-Wilner	University of California, Los Angeles, USA
David L. Foster	Kettering University, USA
Ching-Hsien (Robert) Hsu Chung	Asia University, Taiwan
Farid Ghareh Mohammadi	Verify Radiology Images Consultants, LLC, USA
James J. (Jong Hyuk) Park	SeoulTech, South Korea
Mohammad S. Obaidat	University of Jordan, Jordan
Gerald Schaefer	Loughborough University, UK
Farzan Shenavarmasouleh	Medialab Inc., USA
Fernando G. Tinetti	Universidad Nacional de La Plata, Argentina
Quoc-Nam Tran	Southeastern Louisiana University, USA
Shiuh-Jeng Wang	Central Police University, Taiwan
Layne T. Watson	Virginia Polytechnic Institute & State University, USA
Chao-Tung Yang	Tunghai University, Taiwan
Mary Yang	University of Arkansas for Medical Sciences, USA

Program Committee – Computational Science and Computational Intelligence (CSCI 2025)

Afrand Agah	West Chester University of Pennsylvania, USA
Bharat Bhushan Agarwal	IFTM University, India
Omaima Nazar Ahmad AL	Al-Zaytoonah University of Jordan, Jordan
Wasim A. Al – Hamdani	Kentucky State University, USA
Ismail Khalil Al Ani	Ittihad University, United Arab Emirates
Mehran Asadi	Lincoln University, USA
Travis Atkison	University of Alabama, USA
Azita Bahrami	IT Consult, USA
Mehdi Bahrami	Fujitsu Laboratories of America, Inc., USA
P. Balasubramanian	Nanyang Technological University, Singapore
Petra Saskia Bayerl	Erasmus University Rotterdam, The Netherlands
Jane M. Binner	University of Birmingham, UK
Juan-Vicente Capella-Hernandez	Universitat Politècnica de València, Spain
Juan Jose Martinez Castillo	Universidad Nacional Abierta, Venezuela
Rui Chang	Mount Sinai School of Medicine, USA

Bo Liu	NEC Labs China, China
Eleanor Lockley	Sheffield Hallam University, UK
Jianbing Ma	Bournemouth University, UK
Julius Marpaung	Northeastern University, USA
Andrew Marsh	HoIP Telecom Ltd, UK
Juan Martinez	Universidad Gran Mariscal de Ayacucho, Venezuela
Praveen Meduri	California State University, Sacramento, USA
Ali Mostafaeipour	California State University, Fullerton, USA
Houssem Eddine Nouri	University of Tunis, Tunisia
Michael B. O'Hara	KB Computing, LLC, USA
Robert Ehimen Okonigene	Ambrose Alli University, Nigeria
Funminiyi Olajide	Nottingham Trent University, UK
Satish Penmatsa	University of Maryland - Eastern Shore, USA
Saman Parvaneh	Philips Research North America, USA
Ponalagusamy	National Institute of Technology, Tiruchirappalli, India
Laura L. Pullum	Oak Ridge National Laboratory, USA
Xuewei Qi	University of California, Riverside, USA
Junfeng Qu	Clayton State University, USA
Shahram Rahimi	University of Alabama, USA
Jaime Raigoza	California State University, Chico, USA
Arvind Ramanathan	Oak Ridge National Laboratory, USA
Iman M. Rezazadeh	University of California, Davis, USA
Om Prakash Rishi	University of Kota, India
Cristian Rodriguez Rivero	Universidad Nacional de Córdoba, Argentina
Seyed Roosta	Albany State University, USA
K. Martin Sagayam	Karunya Institute of Technology and Sciences, India
P. Sanjeevikumar	University of South-Eastern Norway, Norway
Benaoumeur Senouci	University of Southern Denmark, Denmark
Zhefu Shi	Microsoft Corporation, USA
Jawed Siddiqi	Sheffield Hallam University, UK
Akash Singh	IBM Corporation, USA
Anthony Skjellum	Auburn University, USA
Omer Muhammet Soysal	Southeastern Louisiana University, USA
Helman Stern	Ben-Gurion University of the Negev, Israel
Jonathan Z. Sun	University of Southern Mississippi, USA
Rahman Tashakkori	Appalachian State University, USA
Predrag Tosic	Microsoft, USA
Quoc-Nam Tran	Southeastern Louisiana University, USA
Jesus Vigo-Aguiar	University of Salamanca, Spain

Contents

Artificial Intelligence - Large Language Models, Text Processing, and Risk Assessment Methods

Artificial Intelligence - Algorithms and Application

Artificial Intelligence - Deep Learning Methods, Industrial Applications, and Algorithms

Evolutionary Deep Learning for Channel-Selected Glioblastoma Classification in Multimodal Intraoperative Neurophysiological Monitoring

Joseph Davids[1(✉)], Jinpei Han[4], Noemia Pereira[7], Gunnar Nelson[1], Hutan Ashrafian[1], Ara Darzi[1], Uriel Martinez-Hernandez[2], Leon James[5], Fady Sedra[6], Syed Aftab[6], Lewis Thorne[8], Isaac Phang[3], Andrew Alalade[3], and Adrian Rubio-Solis[1]

[1] The Hamlyn Centre for Robotic Surgery, Imperial College London, Institute of Global Health Innovation, London SW7 2AZ, UK
{j.davids,a.rubiosolis}@imperial.ac.uk

[2] Department of Electronic and Electrical Engineering, University of Bath, Bath BA2 7AY, UK
u.martinez@bath.ac.uk

[3] Department of Neurosurgery, Royal Preston Hospital, Fulwood, UK

[4] Brain and Behaviour Lab, Imperial College London, London, UK

[5] Department of Neurophysiology, BUPA Cromwell Hospital, London, UK

[6] Department of Spinal Surgery, the BUPA Cromwell Hospital and Royal London Hospital, London, UK

[7] Nintel Ltd, London, UK

[8] Department of Neuro-oncology, National Hospital for Neurology and Neurosurgery, London, UK

Abstract. Glioblastoma is the most aggressive astrocytic brain malignancy, characterized by rapid growth, invasive behaviour, and limited therapeutic options. Despite neurosurgical resection being central to management, preserving critical neurological function remains challenging, necessitating precise intraoperative strategies. Intraoperative neurophysiological monitoring (IONM) offers real-time feedback on the integrity of eloquent regions, enabling safer, more accurate resections. Recent deep learning advances show promise in automating complex IONM for theranostic applications. However, statistically robust channel selection for potential neurophysiological biomarker standardization, allowing optimal glioblastoma signal classification and AI-guided diagnostics, remains under-explored. In this paper, we suggest an evolutionary deep learning network (EDLN) for tumour classification and channel identification. The proposed method implements a binary version of walrus optimiser (WO) to identify the subset of most relevant channels that accounts for an improved model accuracy while a deep learning network (DLN) is applied to tumour classification. To avoid the iterative nature of gradient descent approaches usually employed to train deep learning

H. R. Arabnia et al. (Eds.): CSCI 2025, CCIS 2968, pp. 3–18, 2027.
https://doi.org/10.1007/978-3-032-28457-0_1

structures, a multilayer extreme learning machine (ML-ELM) is applied. Our results demonstrate an improved trade-off between model accuracy, model simplicity, and low computational load exceeding 94% for unbiased channel-selected IONM glioblastoma signal classification, giving us the ability to differentiate tumour signals from non-tumour IONM signals.

Keywords: Deep learning · multilayer extreme learning machine · binary walrus optimiser · glioblastoma classification · tumour detection

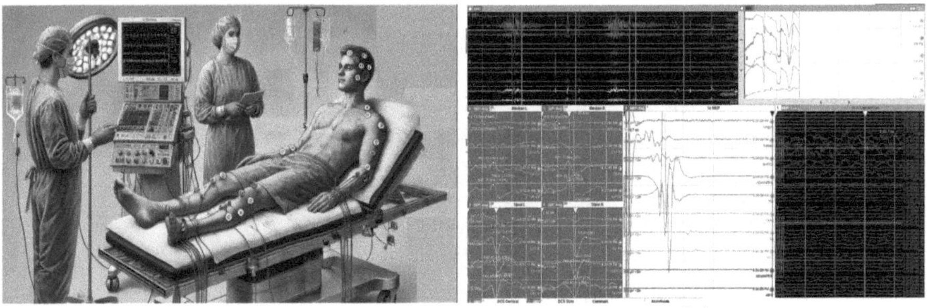

(a) (b)

Fig. 1. (a) Illustrating electrode placement. Electrodes are placed in muscle groups and targeted regions around the body and (b) Example Graphical User Interface (GUI) of recorded tumour IONM.

1 Introduction

Glioblastoma is the most aggressive brain malignancy of astrocytic lineage. With an incidence of 3.19 per 100,000, a median survival of 15 months and a median diagnostic age of 64 years [1,2]. It occurs more commonly in males, Caucasians, and non-Hispanics, complicating demographic predictions and treatment [1,3]. Most cases are sporadic, with no significant risk factors. Marginally favourable prognosis is linked to younger age at diagnosis with cerebellar tumour location, high functional performance, and maximal safe tumour resection using intraoperative multi-modality approaches [4].

The tumour's rapid growth, microscopic tentacle-like spread, and resistance to conventional therapies necessitate a multidisciplinary approach to intraoperative treatment. Neurosurgical resection remains a cornerstone in glioblastoma management, with the primary aim of achieving maximal safe resection while preserving critical neurological function [5–7]. Precise, data-driven innovations are crucial for protecting neurological function. Intraoperative neurophysiological monitoring (IONM) is essential in neurosurgery to map and monitor neural pathways during both awake and asleep glioblastoma resections and in other neurosurgical areas like spinal neuro-oncological surgery, where glioblastomas are rarer. [8,9] However, the role of IONM in theranostic biomarker discovery and prognostication is still very early in the research domain [10]. IONM provides real-time feedback on the functional status of the brain, spinal cord, or peripheral nerves and within eloquent regions such as vision, speech, and motor cortices. It enables surgeons to navigate complex anatomical structures with better accuracy [11]. By evaluating the integrity of these neural pathways and cortical regions during surgery, IONM minimizes the risk of damaging critical structures, thereby optimizing surgical outcomes. Recent advancements in artificial intelligence (AI) methodologies for tumour classification using IONM have garnered significant scientific interest. AI algorithms, particularly machine learning (ML) models, are being developed to assist in the real-time automated signal interpretation of complex intraoperative IONM data for risk stratification and prognostication, but not for diagnostic classification, margin delineation, etc. [12]. Although still in its infancy, AI-guided IONM holds promise for enhancing the precision and efficacy of glioblastoma surgery.

A major challenge is selecting optimal IONM channels across muscles and sensory regions without bias, crucial for standardized and effective machine learning classification. Addressing this issue is essential for enhancing research quality, clinical diagnostic decision support, and enabling the reliable integration of advanced IONM technologies into clinical practice.

1.1 Related Work

Intraoperative neurophysiological monitoring (IONM) has increasingly utilized machine learning (ML) and deep learning (DL) techniques to enhance classification and prediction in surgical procedures. While these advancements have significantly improved neuromonitoring in spinal, thyroid, and sellar tumour surgeries, applications in glioblastoma remain scarce [13–15]. The early ML-based methods focused on conventional classifiers applied to features defined by domain experts. Fan et al. [16,17] evaluated SEP data using LS-SVR and M-SVR models in 10 surgeries, with lower false positive rates (0.080 and 0.068) and higher true warning rates (0.714) compared to traditional methods. Verdonck's group employed cost-sensitive logistic regression on Acceleromyography(AMG) data from 533 Train Of Five (TOF) samples, yielding an F1 score of 0.86 and AUROC of 0.91 for monitoring outlier measurements, improving intraoperative consistency [18]. Jamaludin et a. [13] employed k-nearest neighbours (KNN) and bagged trees to analyze transcranial motor evoked potentials (TcMEPs) in lumbar surgery patients, achieving a sensitivity of 87.5% and low specificity (33.3%), limiting its clinical reliability. Despite their utility, these ML models relied on handcrafted feature extraction and lacked adaptability to complex signal patterns.

To overcome these limitations, deep learning models have been increasingly adopted for IONM signal classification. Agaronnik et al. [15] applied deep learning-based natural language processing (NLP) to neuromonitoring documentation derived from MEPs and SSEPs in 993 spinal surgery patients, achieving an AUROC of 1.00 for detecting status changes and F1 scores of 0.80–0.93. However, their approach focused on textual documentation rather than direct neurophysiological signal analysis. For signal-based classification, Qiao et al. [19] combined Convolutional Neural Network (CNN) and Recurrent Neural Networks (RNN) models for visual evoked potentials (VEP) amplitude changes in 76 sellar tumour cases, achieving higher accuracy (87.4%) compared to traditional methods (83.1%). Kortus et al. [20] utilized Bayesian CNNs for EMG action potential classification in thyroid surgeries, achieving high accuracy (97.6%), precision (97.7%), and recall (97.6%). Similarly, Zha et al. [14] demonstrated hybrid Convolutional-Long Short Term Memory (CNN-LSTM) models for free-running electromyography (EMG) classification, achieving 89.54% accuracy and 94.23% sensitivity in detecting waveform patterns during thyroid surgery. These deep learning models demonstrate superior end-to-end capabilities over traditional ML methods but often require extensive hyperparameter tuning and fixed architectures, which may not generalize well across different patient populations or environments.

To address these gaps, we propose an evolutionary deep learning network (EDLN) that integrates Multilayer Extreme Learning Machine (ML-ELM) with an evolutionary optimization strategy to identify the most informative IONM channels. A binary version of the Walrus Optimiser (WO) is employed for channel selection, where each candidate solution represents a trained deep learning model optimized via multilayer extreme learning machines (ML-ELM). Our approach is validated using real IONM data collected from seven participants, achieving an average accuracy of 95.65%, comparable to conventional deep learning models such as 1D-CNN and LSTM.

The rest of this paper is structured as follows. Section II describes related work while section III provides a description of system configuration, data collection and background theory. In section IV, the proposed evolutionary deep learning strategy is introduced. Results are described in section V, and summary and discussion presented in section VI. Conclusions are drawn in section VII.

2 Materials and Methods

2.1 System Configuration and Data Collection

The IONM procedure requires invasive electrode placement after aseptic preparation, followed by data recording and interpretation. Figure 1(a) illustrates an example of electrode placement that is usually achieved when the patient receives total intravenous anaesthesia. The IONM system was preconfigured with multiple channels to support the monitoring of various modalities, including somatosensory evoked potentials (SSEPs), motor evoked potentials (MEPs), and electromyography (EMG). The configuration allowed continuous acquisition of raw signals and trend data during surgery. Artefacts related to electrocautery were automatically rejected using built-in algorithms to ensure signal clarity.

In Table 1, a metadata summary of channel configuration is described. Samples were acquired at a frequency of 20000 Hz, with a 2000 Hz low pass filter. Selected number of channels were multiplied by the sampling window. Preprocessing identified common channels used for recording IONM signals for all patients as the first step with a sliding window of 0.1 s that provided the optimal processing and data resolution trade off.

2.2 Ethical Considerations

The study was conducted in accordance with ethical guidelines for clinical research, with informed consent obtained from all participants prior to surgery. The inomed (Emmendingen, Germany) IONM platform was used during intraoperative glioblastoma resection and spinal surgery to monitor real-time neurophysiological changes, focusing on the preservation of neurological function as shown in Fig. 1(b).

Table 1. Metadata summary of the muscle groups used in a single patient IONM profile and their occurrence.

Muscle Group	Occurrences (Summarized)	Channels
Tongue	Found in almost every row; highly frequent	["Cz'-C4"', "Cz'-Fz", "C3'-C4"', "Erbs L/1-Cz"', "Erbs R/2-Cz"', "C3-Cz"', "C4-Cz"']
Deltoideus	Found across all data; frequently paired with Tongue	["Cz'-C4"', "Cz'-Fz"]
BR/FCu	Consistently present alongside Tongue and Deltoideus	["C3'-C4"']
APB/ADM L	High occurrence in conjunction with Tongue and Deltoideus	["Erbs L/1-Cz"']
FDI	Regularly appears with high frequency across rows	["Erbs R/2-Cz"']
Quad	Consistently noted across rows	["C3-Cz"']
TA	Frequently appears, often with Quad	["C4-Cz"']
Gastrocnemius	Regularly present, co-occurs with TA and Quad	["Cz'-C4"', "Cz'-Fz"]
AH L	Common across data rows	["Cz'-C4"', "Cz'-Fz", "C3'-C4"']
APB/ADM R	Appears less frequently but still relevant	["Erbs L/1-Cz"']
AH R	Rare compared to other groups	["Erbs R/2-Cz"', "3-Cz"']
Oris	Measures mouth motion	["C4-Cz"']

2.3 Blinded Data Acquisition and Analysis: Implications for Global IONM Data-Lakes

Following NC3R principles, 3 tumour and 4 non-tumour patients (control) IONM datasets were anonymously and randomly selected for analysis by Noemia Pereira (NP) (https://www.nc3rs.org.uk/). This dataset generated over 5 million parameters for deep learning and so was sufficient for machine learning. NP blinded other members of the research team (JD, JH, GN, HA, AD, LJ, UM, and AS) to the selection of patient data, demographic differences, and the overall data selection process. This ensured the statistical integrity of the analytical pipeline and AI modelling, eliminating analytical bias for this feasibility study, but also demonstrated that an anonymised global IONM data repository for ML biomarker discovery could be developed irrespective of where data originates. Discrepancies in signal analytical interpretation were resolved through consensus to reduce variability and improve reliability.

2.4 Multilayer Extreme Learning Machine (ML-ELM)

ML-ELM is a non-iterative DL approach to the optimisation of multilayer neural networks [21–24]. ML-ELM consists of two main steps. First a high representational stage of input data, followed by an ELM-based classification of the features obtained in the first step. As described in Fig. 2(a), the basic building block for the implementation of ML-ELM is an ELM-based Autoencoder (ELM-AE for short). Similar to deep learning, feature representation is achieved in ML-ELM via layer-by-layer unsupervised learning, where a number of ELM-AEs are stacked to build a multilayer structure [25]. ML-ELM consists of L hidden layers,

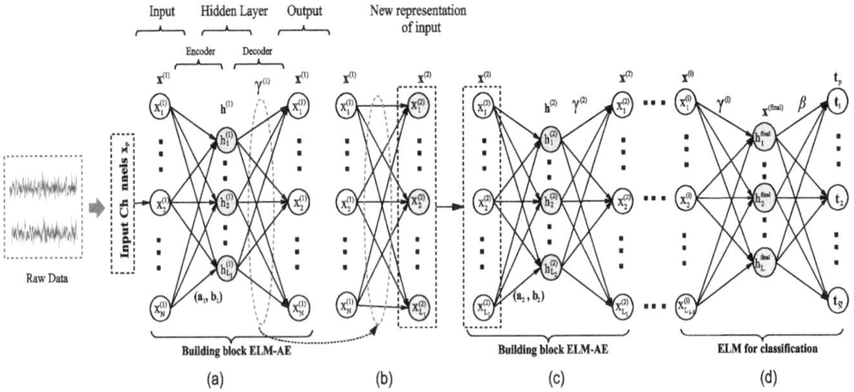

Fig. 2. Architecture of ML-ELM. (a) ELM-AE outputs the transformation $\Gamma^{(1)}$ for representation learning, (b) where a new representation $\mathbf{x}^{(2)}$ is learned by computing the activation function $g(\mathbf{X}^{(1)}(\Gamma)^{(1)})$, (c) $\mathbf{x}_2$ is used as input to ELM-AE for a second representation learning, and (d) representation learning, the final $\mathbf{x}^{final}$ is used to calculate the output weight β for classification, where $\tilde{N}$ is the max number of target classes.

where for a given input $\mathbf{X}^{(i)} = [\mathbf{x}_1^{(i)}, \ldots, \mathbf{x}_N^{(i)}]$, $k = 1, \ldots, N$, the ith data transformation $\Gamma^{(i)} = [\gamma_1^{(i)}, \ldots, \gamma_N^{(i)}]$ is computed by Eq. (1), $\gamma^{(i)}$ is the transformation vector that corresponds to the input vector $\mathbf{x}_k^{(i)}$ [26,27], where $\mathbf{H}^{(i)}\Gamma^{(i)} = \mathbf{X}^{(i)}$, in which, $\mathbf{H}^{(i)}$ is the output matrix of the ith hidden layer w.r.t the input $\mathbf{X}^{(i)}$ [28]. Data transformation is achieved by projecting $\mathbf{X}^{(i)}$ along the weights of the decoder stage at each ELM-AE. Each ELM-AE generates a number of orthogonal random parameters, e.g. random input weights and random biases in hidden nodes for additive nodes [25,29]. Thus, orthogonal random hidden parameters of each ELM-AE are computed by $\mathbf{h}(\mathbf{x}_k) = g(\mathbf{x}_k\mathbf{A} + \mathbf{b}) = [h_1(\mathbf{x}_k), \ldots, h_{M_i}(\mathbf{x}_k)]$, in which, $\mathbf{H}^{(i)} = [\mathbf{h}(\mathbf{x}_1), \ldots, \mathbf{h}(\mathbf{x}_N)]^T$, $\mathbf{A}^T\mathbf{A} = \mathbf{I}$ and $\mathbf{b}^T\mathbf{b} = 1$. Hence, the ith transformation term $\Gamma^{(i)}$ is calculated as:

$$\Gamma^{(i)} = (\mathbf{H}^{(i)})^T \left(\frac{\mathbf{I}}{C} + \mathbf{H}^{(i)}(\mathbf{H}^{(i)})^T \right)^{-1} \mathbf{X}^{(i)} \tag{1}$$

C is a regularisation term. Thus, the final representation is a mapping defined:

$$\mathbf{X}^{\text{final}} = g(\mathbf{X}^{(i)}(\Gamma^{(i)})) \tag{2}$$

If layer $M_i = M_{i+1}$, then g can be chosen as a linear function, otherwise g is chosen as a non-linear piecewise function (RBFs or sigmoids). In [27], $\mathbf{X}^{\text{final}}$ was used as the hidden layer output to compute the output weight vector β as [27]:

$$\beta = \mathbf{X}^{\text{final}} \left(\frac{1}{C} + \mathbf{X}(\mathbf{X}^{\text{final}}) \right)^{-1} \mathbf{T} \tag{3}$$

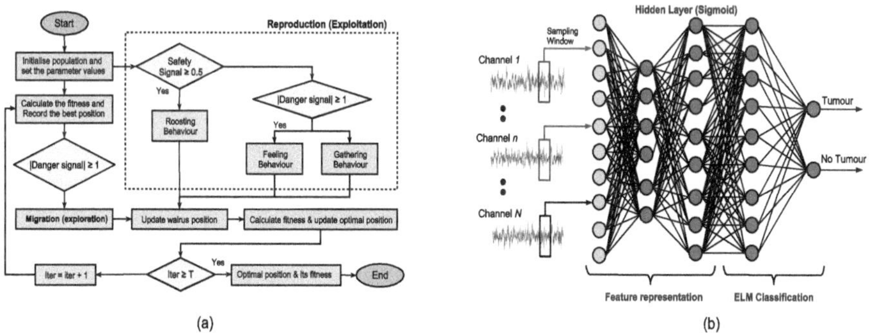

Fig. 3. (a) Workflow of walrus optimiser and (b) proposed DL network structure.

2.5 Walrus Optimiser (WO)

WO is a metaheuristic inspired by the behaviour of walrus in migrating, breeding, roosting and foraging [30]. As illustrated in Fig. 3(a), WO consists of two main stages, namely, migration (exploration) and reproduction (exploitation).

Initialisation. In WO, the initial position of walrus agents is randomly initialised using $X = LB + rand(UB - LB)$, where, LB and UB are the lower and upper boundaries of the problem. $X = [\hat{X}_1, \ldots, \hat{X}_N]^T \in R^{N\times d}$, $i = 1, \ldots, N$, where $\hat{X}_i = [x_{i,1}, \ldots, x_{i,d}]$. N is the population size, d is the dimension of design variables. The fitness value of each agent is stored as $F = [F1, F2 \ldots, F_N]^T \in R^{N\times d}$, where $F_i = [f_{i,1}, f_{i,2}, \ldots, f_{i,d}]$. Walrus agents are divided into juveniles and adults, with a 10% and 90% proportion of the population respectively. Among adults, the ratio for female and male agents is 1:1.
Danger Signals and Safety Signals. Danger and safety signals are used by walrus agents to be vigilant during foraging and roosting. 1 or 2 walruses are defined as guards patrolling around. Thus, danger signals are defined by Danger_signal $= A * R$. Terms $A = 2 \times \alpha$ and $R = 2 \times r_1 - 1$, where $\alpha = 1 - t/T$. A and R are danger factors and α is a factor decreasing from 1 to 0, while T is max iteration and Safety signal is Safety_signal $= r_2,\ r_1,\ r_2 \in (0, 1)$.
Migration (Exploration). Migration is an exploration mechanism used by walrus agents to migrate to more suitable areas for population survival. Thus, each walrus position is updated as:

$$x_{i,j}^{iter+1} = x_{i,j}^{iter} + Migration_step \tag{4}$$

where Migration_step $= (x_m^{iter} - x_n^{iter}) \cdot \alpha \cdot r_3^2$ and $\alpha = 1 - \frac{1}{1+\exp(-\frac{iter-T/2}{T} \times 10)}$. $x_{i,j}^{iter+1}$ is the new position for ith walrus on the jth dimension. $Migration\_step$ is the movement step of each walrus, two walrus vigilantes correspond to x_m^{iter} and x_n^{iter}. α is the step control factor varying based on a smooth curve, and r_3 is a random number in $(0, 1)$.
Reproduction (Exploitation). Reproduction is an exploitation stage, where walruses tend to breed in currents with low risk. Reproduction involves the following three main steps:

1) Roosting behaviour In this stage, walrus agents are categorized into males, females, and juveniles. Hence, they have their own way to update their current positions based on two steps.

I. Step 1: At this step, Halton sequence distribution is adopted for male walrus update, allowing a a broader distribution with respect to the search space.
II. Step 2: Position update of female walrus: at this stage, female walrus is influenced by male walrus ($male_{i,j}^{iter}$) and the lead walrus x_{best}^{iter}. The position throughout the optimisation is then influenced less by the mate and more by the leader as follows:

$$female_{i,j}^{iter+1} = female_{i,j}^{iter} + \alpha \cdot (male_{i,j}^{iter} - female_{i,j}^{iter}) + (1-\alpha) \cdot (x_{best}^{iter} - female_{i,j}^{iter}) \tag{5}$$

$female_{i,j}^{iter+1}$ is the updated position of the ith female.
III. Step 3: Position update of juvenile walrus. At this step, the current position of juveniles is updated to avoid predation as follows:

$$juvenile_{i,j}^{iter+1} = (\mathbf{O} - juveline_{i,j}^{iter}) \cdot P \tag{6}$$

where, $\mathbf{O} = x_{best}^{iter} + juvenile_{i,j}^{iter} \cdot LF$ and $j = 1, \ldots d$. LF is a vector of random numbers based on Lévy Flight representing Lévy movement as computed in [30]. $P \in (0, 1)$ is a distress random factor of juvenile walrus.

2) Foraging behaviour This behaviour is used by each walrus agent to perform a global search. This behaviour occurs in the late iteration of WO as:

$$x_{i,j}^{iter+1} = x_{i,j}^{iter} \cdot R - |x_{best}^{iter} - x_{i,j}^{iter}| \cdot r_4 \tag{7}$$

where r_4 is a random number in (0,1).

3) Gathering behaviour This behaviour is inspired by the cooperative behaviour of foraging and moving according to the location of other walruses. Sharing location information can help the whole walrus herd to find the sea area with higher food abundance. In this behavior, walrus position is $x_{i,j}^{iter+1} = 0.5 \cdot (X_1 + X_2)$, where:

$$\begin{cases} X_1 = x_{best}^{iter} - a_1 \times b_1 |x_{best}^{iter} - x_{i,j}^{iter}| \\ X_2 = x_{second}^{iter} - a_2 \times b_2 |x_{second}^{iter} - x_{i,j}^{iter}| \end{cases} \tag{8}$$

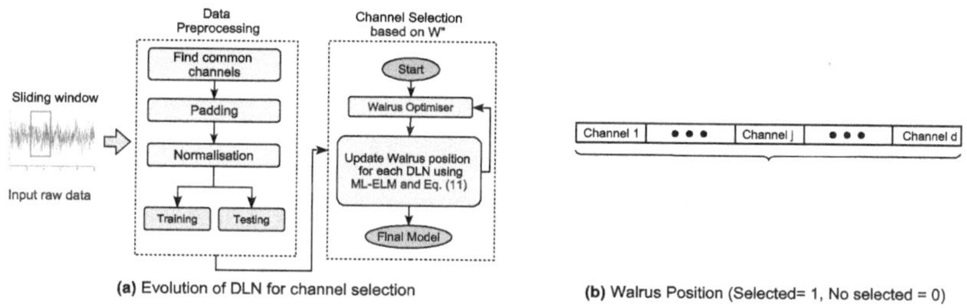

Fig. 4. Workflow for channel identification using an EDLN with a binary WO

$a = \beta \times r_5 - \beta$ and $b = tan(\theta)$. X_1 and X_2 are two weights affecting the gathering behaviour of each walrus while X_{second}^{iter} denotes the position of the second walrus in the current iteration. a and b are the gathering coefficients, r_5 is a random number in (0,1) and θ takes values in $(0, \pi)$.

3 Proposed EDLN for Channel Identification

To identify a reduced set of the most relevant channels that contribute to the highest model accuracy for glioblastoma classification, an evolutionary strategy based on a binary version of WO is suggested. The current position of each walrus agent is fed into a full feed-forward DLN and its structure is reported in Fig. 3(b). Each neuron of the hidden layers performed a non-linear transformation of a linear combination of all outputs from previous layers. As a non-linear mapping, it was decided to use a sigmoid function while at each iteration of WO, ML-ELM was applied to optimise the parameters of EDLN structure as described in Fig. 3(a). An objective function $f_i(iter)$ that penalises the position of each walrus with a large number of input channels is computed as:

$$f_i(iter) = (Accuracy - CFF * ((N_c/N_f)))^r \quad (9)$$

N_c/N_f measures the rate between the maximum number of channels N_f and the number of selected channels N_c, while accuracy is: **r** $= 1$, **if** $CFF = 0$, **otherwise** $r = 2$, and model $Accuracy$ is the classification performance [29]:

$$Accuracy = TP + TN/(TP + TN + FP + FN) \quad (10)$$

True positive (TP) indicates a positive one is correctly predicted as positive, false positive (FP) indicates a negative one incorrectly predicted as positive. True negative (TN) denotes a negative one is correctly predicted as negative; false negative (FN) denotes a positive one that is incorrectly predicted as negative. Term N_c and N_f are the number of features selected for the ith walrus, and the total number of input channels, respectively. CFF is a weight factor defined in the interval $[0 - 1]$ and used to evaluate the relevance of the final number of features. An S-shaped function is the most suitable approach to transitioning from a continuous to a binary state [31].

Binary representation of each walrus' position is the concatenation of binary values $M(m_1, m_2, \ldots, m_j, m_s)$, where $m_j = 0, 1$. m_j can be equal to 0 or 1 to indicate whether an input channel is selected or not as shown in Fig. 4(b). The position of each walrus agent follows a conversion method established by:

$$x_{i,j}^{iter} = \begin{cases} 0, & \text{if } rand < STF(x_i^{iter}) \\ 1, & \text{if } rand \geq STF(x_i^{iter});\ STF(x_i^{iter}) = 1/(1 + e^{x_i^{iter}}) \end{cases} \quad (11)$$

4 Experiments and Results

In this section, experimental setup and results to evaluate the performance of the proposed EDLN are presented. All experiments were carried out using MATLAB 2021b running on a 3.6-GHz M1 processor with 16-GB RAM.

4.1 Glioblastoma Classification Using EDLN

To evaluate the model efficiency of the proposed EDLN for glioblastoma classification, metrics such as accuracy, recall, precision, and F1-score are computed. To compare the performance of EDLN with state-of-the-art techniques, 1-Dimensional convolutional neural networks (1D-CNN), LSTM and Support Vector Machines (SVM) were implemented. The experimental setup for training a DLN consists of 3 hidden layers, followed by an ELM classifier, i.e. $L = 4$, set as $4 \times M_i$, $[9000 - 450 - 200 - 200 - 2]$ with a regularisation parameter $C_i(i = 1, \ldots, 3)$. In this paper, data was collected from a number of 7 participants where only nine common channels are used to classify the type of glioblastoma as tumour or no tumour. Statistics from these channels are presented in Table 2. To the classification of glioblastoma, the workflow described in Fig. 4(b) was applied. First, common channels among all participants were identified. Because the time for each data collection is different, a padding strategy was used to even the length of each time series as described in Fig. 5. In Fig. 5(b), an example of padding applied to channel Cz'-C4' is illustrated. Similar to [31], noise was added to a channel randomly selected for each decision process performed during WO.

Table 2. Statistics of common channels for glioblastoma classification.

Channel	Std	Mean	min	max
	Tumour/No Tumour (mV)			
$C3' - C4'$	3.78E−6/4.01E−6	1.44E−9/2.21E−7	−3.6E−6/−7.4E−6	1.5E−4/1.39E−4
$C4' - C3'$	3.46E−6/3.77E−6	2.24E−7/−1.12E−6	−6.34E−6/	7.27E−6/1.38E−4
$C4' - Fz$	1.65E−6/3.93E−6	−2.92E−7/−3.30E−7	−5.51E−6/	7.74E−6/1.89E−5
$C3' - Fz$	2.03E−6/1.95E−6	−2.72E−7/1.55E−7	−8.97E−6/	9.25E−6/2.21E−5
$Cz' - C3'$	2.30E−6/2.13E−6	2.50E−7/1.94E−8	−5.93E−6/	6.99E−5/2.14E−6
$C3' - Cz'$	2.32E−6/1.55E−6	−0.35E−7/1.12E−7	−9.07E−7/9.25	7.63E−5/2.00E−5
$Cz' - C4'$	2.81E−6/2.30E−6	−3.32E−8/1.22E−7	−4.49E−6/	1.04E−4/5.15E−5
$C4' - Cz'$	2.43E−6/7.93E−6	2.76E−7/−0.84E−8	−3.06E−6/	2.70E−5/1.25E−4
$Cz' - Fz$	2.73E−6/1.69E−6	−0.90E7/−1.06E−7	−8.49E−6/	1.28E−4/2.108E−5

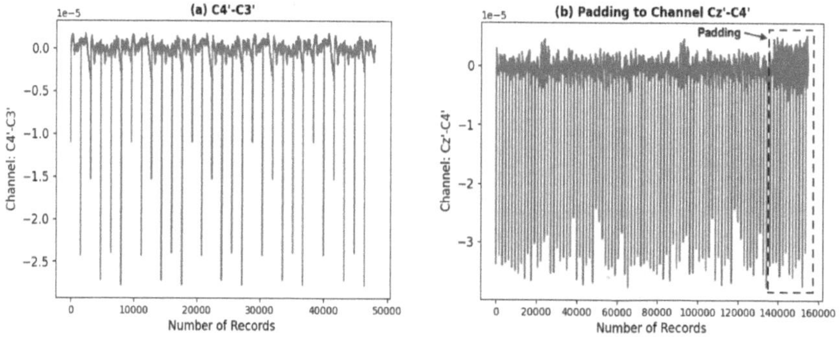

Fig. 5. Example of: (a) Time series for channel C4'-C3' and (b) padded channel with noise generated from a normal distribution $N(1, 0)$.

To compare the performance of an EDLN with a binary version of WO (EDLN-WO), a binary genetic algorithm (EDLN-GA) and binary grey wolf optimisation (EDLN-GWO) is implemented. Based on experimentation, parameter setup for WO, GA and GWO using 20 agents and a max number of iterations of 100 for all metaheuristics was found to produce the highest accuracy. For WO, $CFF = 0.05$, $r = 1$. Table 3 presents a comparison of the average accuracy of 10 experiments. A sampling window of 0.1 sec was used, and raw data was normalised to the interval $[0-1]$. Data was split into two subsets training (80%) and testing (20%). Based on Table 3, the performance comparison highlights that 1D-CNN is the best-performing model, achieving the highest accuracy (96.74%), precision (98.00%), and F1-Score (95.17%), making it the most robust choice for tasks requiring high performance across all metrics. Finally, in Fig. 6, the average confusion matrix of ten random experiments for the classification of tumour and no tumour using testing data is presented. From each confusion matrix, it can be noted an imbalanced data for glioblastoma classification favours the identification of tumour using a DLN. In contrast, an EDLN with WO significantly improves the classification of each class, i.e. tumour and no tumour.

Table 3. Average testing Performance of ten random runs for the classification of glioblastoma.

Model	Accuracy	Precision	Recall	F1-Score	Training Time (Sec)	# parameters	# channels	# agents
EDLN-WO	**95.65**%	94.90%	90.99%	92.90%	415.1	130,000	4	20
EDLN-GWO	94.57%	90.73%	92.41%	91.56%	410.9	130,000	4	20
EDLN-GA	94.57%	97.26%	84.21%	91.56%	552.7	130,000	4	25
DLN	92.39%	87.41%	88.96%	88.18%	56.8	130,000	9	1
1D-CNN	**96.74**%	98.00%	92.50%	95.17%	4888.6	1.254E+06	9	1
LSTM	93.48%	92.22%	89.48%	90.83%	3618.4	678,356	9	1
SVM	85.87%	92.73%	71.73%	80.64%	2029.5	500	9	1

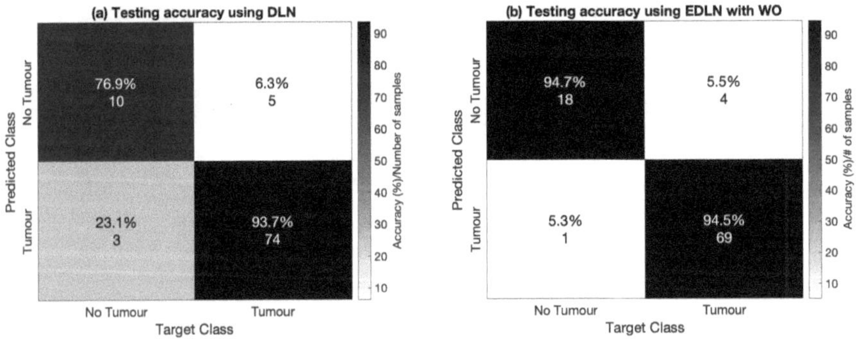

Fig. 6. Average confusion matrix of 10 random runs using testing data for (a) DLN, (b) EDLN with WO and EDLN with GWO.

4.2 Channel Selection Using WO

In this research work, channels used in IONM correspond to electrode positions from the international 10–20 EEG system, electromyography (EMG), somatosensory evoked potentials (SSEP), motor evoked potentials (MEP), brain stem auditory evoked potential (BAEP) and visual evoked potential (VEP) offering a comprehensive view of neural activity. These electrode positions enable the monitoring of specific brain regions, including motor, sensory cortices, and critical neural and spinal tract pathways, supporting real-time assessment of neural function. This section discusses the channel selection results obtained by the binary version of WO, GA and GWO. Channel selection is used to identify a subset of channels that allow the EDLN to produce a high balance between model accuracy and model simplicity. Figure 7 illustrates the summary of the average results of ten random experiments produced by a binary WO, GA and GWO for thirty evolutionary iterations. In Fig. 7(a), a comparison of the average fitness evolution of WO versus EDLN accuracy is presented. In terms of fitness evolution, min and max limits are also described. Note in Fig. 7(a), both metrics are defined in [0-1] for comparison purposes. The evolution of selected channels is illustrated in Fig. 7(b), while in Fig. 7(c), the selection frequency of each selected channel is presented.

Finally, in Fig. 7(b) and (c), for ten random experiments, the four channels $C4' - C3'$, $Cz' - C4'$, $C3' - Fz'$ and $C4' - Fz'$ allow the EDLN to provide a high model performance.

5 Summary and Discussion

From results in section IV, the following summarisation and discussion is given:

a) An evolutionary deep learning method based on walrus optimiser and multi-layer extreme learning machine for channel-selected Glioblastoma classification in IONM data has been suggested.

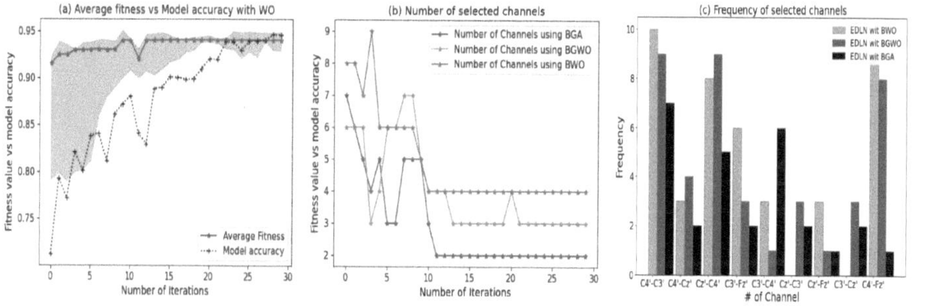

Fig. 7. Average results of 10 runs of EDLN for a) Average fitness vs model accuracy using EDLN with BWO, b) comparison of evolution of selected channels and c) frequency of selected channels.

b) A binary version of Walrus optimiser (WO) is suggested. The fitness function implemented in this work allows WO to penalise models that require a large number of channels to enhance classification performance.
c) By using ML-ELM, the computational burden that usually results from Gradient Descent approaches is avoided. This facilitates the deployment of DL networks for early detection systems in IONM procedures.
d) Based on our results, the top four selected channels, i.e. $C4' - C3'$, $Cz' - C4'$, $C3' - Fz'$ and $C4' - Fz'$ correspond to SEP measurements. By using these measurements, EDLN provides an overall model performance of 95.65%.
d) Comparative results with state-of-art techniques demonstrate that a reduced set of channels provide a meaningful representation of IONM procedures to the detection of Glioblastoma. Future work will consists on collecting data from a larger cohort of participants.
e) As it is shown in Fig. 6, confusion matrices demonstrated the implemented EDLN models are capable of dealing with imbalanced data, while a good generalisation performance is provided to the identification of tumour data.

6 Conclusions and Future Work

Evolutionary-inspired channel identification/selection achieved 94.7% prediction accuracy and shows promise for reduced dimensional IONM feature spaces in cancer prediction. Future work will aim to identify the latent space for glioblastoma IONM classification, exploring other methods of channel selection and correlation with functional outcomes.

References

1. Thakkar, J.P., et al.: Epidemiologic and molecular prognostic review of glioblastoma. Cancer Epidemiol. Biomarkers Prevent. **23**(10), 1985–1996 (2014)
2. Louis, D.N., et al.: The 2007 WHO classification of tumours of the central nervous system. Acta Neuropathol. **114**(2), 97–109 (2007)
3. Ostrom, Q.T., et al.: CBTRUS statistical report: primary brain and central nervous system tumors diagnosed in the United States in 2006-2010. Neuro-oncology, 15(suppl_2), pp.ii1-ii56 (2013)
4. Adams, H., Chaichana, K.L., Avendaño, J., Liu, B., Raza, S.M., Quiñones-Hinojosa, A.: Adult cerebellar glioblastoma: understanding survival and prognostic factors using a population-based database from 1973 to 2009. World Neurosurg. **80**(6), e237–e243 (2013)
5. Lacroix, M., et al.: A multivariate analysis of 416 patients with glioblastoma multiforme: prognosis, extent of resection, and survival. J. Neurosurg. **95**(2), 190–198 (2001)
6. Li, Y.M., Suki, D., Hess, K., Sawaya, R.: The influence of maximum safe resection of glioblastoma on survival in 1229 patients: can we do better than gross-total resection? J. Neurosurg. **124**(4), 977–988 (2016)
7. Sanai, N., Polley, M.Y., McDermott, M.W., Parsa, A.T., Berger, M.S.: An extent of resection threshold for newly diagnosed glioblastomas. J. Neurosurg. **115**(1), 3–8 (2011)

8. Hervey-Jumper, S.L., et al.: Awake craniotomy to maximize glioma resection: methods and technical nuances over a 27-year period. J. Neurosurg. **123**(2), 325–339 (2015)
9. Silverstein, J.W., et al.: Short and long-term prognostic value of intraoperative motor evoked potentials in brain tumor patients: a case series of 121 brain tumor patients. J. Neurooncol. **161**(1), 127–133 (2023)
10. Berger, A., et al.: Risk factors and prognostic implications of surgery-related strokes following resection of high-grade glioma. Sci. Rep. **12**(1), 22594 (2022)
11. Overzet, K., Jahangiri, F.R., Funk, R.: Bulbocavernosus reflex monitoring during intramedullary conus tumor surgery. Cureus **3**, 12 (2020)
12. Park, D., Kim, I.: Application of machine learning in the field of intraoperative neurophysiological monitoring: a narrative review. Appl. Sci. **12**(15), 7943 (2022)
13. Jamaludin, M.R., et al.: Machine learning application of transcranial motor-evoked potential to predict positive functional outcomes of patients. Comput. Intell. Neurosci. **2022**(1), 2801663 (2022)
14. Zha, X., et al.: A deep learning model for automated classification of intraoperative continuous EMG. IEEE Trans. Med. Robot. Bionics **3**(1), 44–52 (2020)
15. Agaronnik, N.D., Kwok, A., Schoenfeld, A.J., Lindvall, C.: Natural language processing for automated surveillance of intraoperative neuromonitoring in spine surgery. J. Clin. Neurosci. **97**, 121–126 (2022)
16. Joseph, D., Jinpei, H., Hutan, A., Stamatia, G., Hani, M., Ara, D.: Meta-analysis of intraoperative neurophysiological monitoring in neuro-oncology using artificial intelligence (2024)
17. Fan, B., Li, H.X., Hu, Y.: An intelligent decision system for intraoperative somatosensory evoked potential monitoring. IEEE Trans. Neural Syst. Rehabil. Eng. **24**(2), 300–307 (2015)
18. Verdonck, M., Carvalho, H., Berghmans, J., Forget, P., Poelaert, J.: Exploratory outlier detection for acceleromyographic neuromuscular monitoring: machine learning approach. J. Med. Internet Res. **23**(6), e25913 (2021)
19. Qiao, N., et al.: Deep learning for automatically visual evoked potential classification during surgical decompression of SELLAR region tumors. Transl. Vis. Sci. Technol. **8**(6), 21–21 (2019)
20. Kortus, T., Krüger, T., Gühring, G., Lente, K.: Automated robust interpretation of intraoperative electrophysiological signals-a Bayesian deep learning approach. Curr. Directions Biomed. Eng. **7**(2), 69–72 (2021)
21. Vong, C.M., Chen, C., Wong, P.K.: Empirical kernel map-based multilayer extreme learning machines for representation learning. Neurocomputing **310**, 265–276 (2018)
22. Vidnerová, P., Neruda, R.: Deep networks with RBF layers to prevent adversarial examples. In: Rutkowski, L., Scherer, R., Korytkowski, M., Pedrycz, W., Tadeusiewicz, R., Zurada, J. (eds.) International Conference on Artificial Intelligence and Soft Computing, pp. 257–266. Springer International Publishing, Cham (2018). https://doi.org/10.1007/978-3-319-91253-0_25
23. Chamara, L., Zhou, H., Huang, G.B., Vong, C.M.: Representational learning with extreme learning machine for big data. IEEE Intell. Syst. **28**(6), 31–34 (2013)
24. Rubio-Solis, A., Panoutsos, G.: Interval type-2 radial basis function neural network: a modeling framework. IEEE Trans. Fuzzy Syst. **23**(2), 457–473 (2014)
25. Kasun, L.L.C., Yang, Y., Huang, G.B., Zhang, Z.: Dimension reduction with extreme learning machine. IEEE Trans. Image Process. **25**(8), 3906–3918 (2016)
26. Tang, J., Deng, C., Huang, G.B.: Extreme learning machine for multilayer perceptron. IEEE Trans. Neural Netw. Learn. Syst. **27**(4), 809–821 (2015)

27. Wong, C.M., Vong, C.M., Wong, P.K., Cao, J.: Kernel-based multilayer extreme learning machines for representation learning. IEEE Trans. Neural Netw. Learn. Syst. **29**(3), 757–762 (2016)
28. Rubio-Solis, A., Panoutsos, G.: Iterative information granulation for novelty detection in complex datasets. In: 2016 IEEE International Conference on Fuzzy Systems (FUZZ-IEEE), pp. 953–960. IEEE (2016)
29. Rubio-Solis, A., Panoutsos, G., Thornton, S.: A data-driven fuzzy modelling framework for the classification of imbalanced data. In: 2016 IEEE 8th International Conference on Intelligent Systems (IS), pp. 302–307. IEEE (2016)
30. Han, M., Du, Z., Yuen, K.F., Zhu, H., Li, Y., Yuan, Q.: Walrus optimizer: a novel nature-inspired metaheuristic algorithm. Expert Syst. Appl. **239**, 122413 (2024)
31. Naik, R., Rubio-Solis, A., Jin, K., Mylonas, G.: Novel multimodal sensing and machine learning strategies to classify cognitive workload in laparoscopic surgery. Eur. J. Surg. Oncol. **51**(7), 108735 (2025)

From AI Agents to Agentic AI: A Practical Path for S&OP-Centric Supply Chain Optimization

Beilei Zhu(✉)

Intel Corporation, Hillsboro, OR, USA
beilei.zhu@intel.com

Abstract. Global supply chains are vulnerable to ongoing distribution shift-demand shocks, policy interventions, and geopolitical surprises-that disrupt fixed plans and expose the limitations of purpose-built automation. We place Sales and Operations Planning (S&OP) as a control loop use case that combines learned predictors and simple rules: forecast → set inventory targets → size/time buys → roll up spend, subject to invariants such as stockout-risk service levels (through safety buffers) [1], discipline to moderate bullwhip amplification [2], coordination among heterogeneous actors [3], and minimum order quantities and case-pack increments in monthly budget constraints [4]. We contribute **Agentic AI** as the level that orchestrates these steps, checks invariants after each step, and replans online when inputs move, unveiling typed interfaces, schema-based I/O (Input /Output), and auditable traces rather than dark-box end states [5]. Two interchangeable runtimes serve different adoption modes: a one-process OpenAI SDK orchestrator for low latencies, and a protocolized services architecture in MCP (Model Context Protocol) terms for capability discovery and typed, schema-based I/O beyond teams [6]. In controlled replays in synthetic yet representative plan-to-procure scenarios, Agentic AI reduces constraint violations and cycle time and improves service and adherence to monthly caps, along with readable explanations of why the ultimate plan assumes its shape. Grounded in inventory theory [1], bullwhip mitigation [2], agent-based views [3], procurement optimization [4], and current tool-orchestration practice [5, 6], it marks a practical advance toward a coherent, end-to-end planning capability from local point solutions.

Keywords: Agentic AI · supply chain optimization · S&OP · forecasting · inventory control · procurement · finance planning · OpenAI SDK · MCP

1 Introduction

Modern supply chains operate under relentless uncertainty. Figure 1 [7] shows a stepwise, overall decline in the availability of supply, freight, and talent over time, as successive shocks—Trade Wars, COVID-19, a record hurricane season, the Texas freeze, the Suez Canal blockage, and the Russian invasion of Ukraine—cause sharp dips with only partial recoveries. Multiple processes, such as demand planning, supply and inventory planning, procurement, finance, and supplier risk, must be optimized together to reduce waste and support strategy. Yet the external environment shifts quickly in ways that are difficult to

H. R. Arabnia et al. (Eds.): CSCI 2025, CCIS 2968, pp. 19–32, 2027.
https://doi.org/10.1007/978-3-032-28457-0_2

predict: public-health shocks such as COVID-19, tariff realignments, regional wars, and international trade tensions. The net effect is a higher-variance operating context where single-purpose automations struggle to adapt.

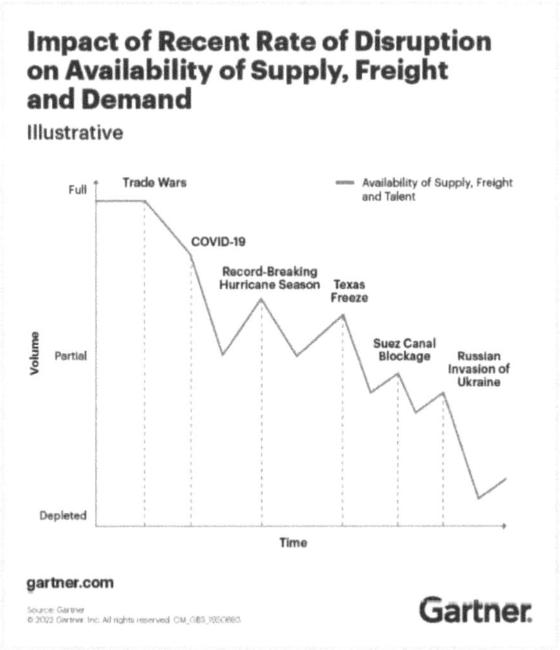

Fig. 1. Sequential Shocks and Stepwise Erosion of Supply, Freight, and Talent Availability

We argue that an adaptable workflow is required, one that can recompute and re-justify cross-functional plans as conditions evolve. The transformation from a single AI agent to agentic AI makes this feasible. Instead of hardcoding a one-off calculation, an orchestrator reasons about the business goal, calls specialized tools in the right order, validates constraints (e.g., service levels and budget caps), and explains outcomes in plain language for human approval. The move succeeds only with solid domain understanding and intentional process design: inventory theory for reorder policies [1], disciplined information sharing to mitigate bullwhip [2], and explicit procurement and finance linkages [4]. Agentic AI does not replace these foundations—it accelerates them by coordinating domain tools, keeping humans "on the loop," and preserving an audit trail.

Our contributions are practical: a minimal agent set mapped to S&OP, an implementation using the OpenAI SDK with an optional MCP-style layer, and an evaluation showing that orchestrated plans can raise service levels and reduce budget overruns while remaining explainable.

2 Use Case, Background and Fit

Inventory control provides the planning baseline. In continuous review, a reorder-point policy triggers a purchase when the inventory position falls to a threshold equal to expected demand during supplier lead time plus a safety buffer. The expectation is the quantity likely to sell before replenishment arrives; the buffer (safety stock) grows with demand volatility and with the chosen service target (e.g., avoiding stockouts in 95% or 99% of cycles) [1]. The bullwhip effect [2] in Fig. 2 shows that when functions act independently and information arrives late, variability amplifies upstream and service degrades, arguing for end-to-end coordination and a shared view of demand, lead time, and inventory [2]. Agent-based perspectives note that real networks consist of heterogeneous actors with local goals and constraints; without a coordinating layer, local improvements do not assemble into a coherent plan [3]. On the buying side, procurement optimization formalizes supplier rules—minimum order quantities, pack multiples, calendars, price tiers—and relevant cost terms so that purchase proposals remain feasible and economical [4]. Recent systems work supports combining these elements: tool-calling models that return structured outputs [5], together with protocolized interfaces (capability discovery and typed inputs/outputs, as in MCP) that standardize how forecasting, inventory, procurement, finance, and risk modules are invoked and audited [6]. The result here is a minimal, S&OP-aligned system that can be adopted incrementally while preserving clear contracts, constraint checks, and traceable decisions.

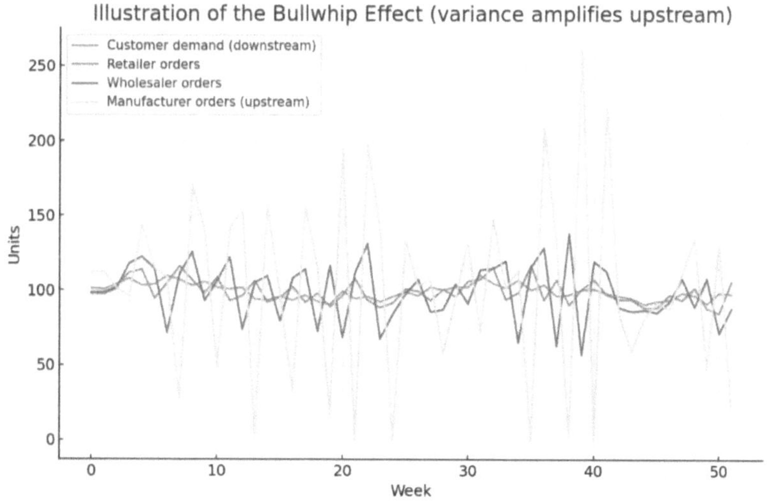

Fig. 2. Supply Chain Bullwhip Effect Illustration

S&OP is the cadence that keeps demand, supply, and finance aligned to a single feasible plan. In globally exposed networks, shocks and policy shifts often drive isolated fixes; taken together, these create bullwhip amplification and erode service [2]. A disciplined cycle counters this by maintaining one shared view of demand and lead time; by setting explicit service targets with plain rules—trigger a reorder when expected lead-time demand is reached, with a safety buffer sized to volatility and the target service [1]; and by timing purchases within supplier constraints (minimum order quantities and pack multiples) and monthly budget caps [4]. Terms used throughout: P50/P80/P90 are predictive demand quantiles by item and site (P50 = median; P80/P90 add headroom for plausible surges); the reorder point is the trigger level ("what is expected to sell while waiting" plus buffer); safety stock is that buffer, which increases with volatility and the service target [1]; MOQ is the smallest lot a supplier will accept, and pack multiple requires orders in fixed steps (e.g., cases of 12). "MOP" is not used here as an ordering rule; where encountered locally it denotes a monthly operating plan (a calendar artifact), not a constraint.

Agentic automation fits because the value lies at the process level rather than a single step. A planner sequences forecasting → inventory policy → procurement → finance and validates after each stage that service targets, supplier rules, business-day calendars, and monthly caps are respected. Clean, typed tables and stable interfaces reduce ambiguity; explicit checks and exception handling prevent a locally sensible choice from breaking global feasibility. Because the chain can be re-run when inputs change, the approach remains effective under shifting conditions while staying grounded in inventory control, bullwhip mitigation, and procurement economics [1–4], and in modular, protocolized orchestration [5, 6].

Human judgment remains central. Choices such as which items merit higher service, how to treat long-tail SKUs, acceptable exposure to constrained suppliers, and when to dual-source are managerial. The planner presents proposals with reasons—where and why shortages cluster; how supplier minimums and lead time shape order sizes; which months exceed caps and which delivery shifts would smooth spend—so Pre-S&OP and Executive S&OP forums can approve or adjust with an audit trail.

In the prototype, repeatable work is automated: data curation; SKU–site quantile forecasts (P50/P80/P90); inventory policy and shortage projection using the reorder trigger plus a safety buffer [1]; purchase-order sizing that honors MOQ and pack multiples and assigns valid delivery days with supplier consolidation [4]; finance roll-up to monthly caps with simple smoothing suggestions; optional supplier-risk scoring; and artifact generation (CSV tables, RFQ(Request for Quote) drafts, and a concise report explaining why the plan takes its shape). In controlled replays, manual stitching—cross-sheet joins, shortage arithmetic, MOQ rounding, and budget roll-ups—was replaced by deterministic, auditable steps. Plan cycle time fell, exception backlogs shrank, and plans met service targets while staying closer to monthly caps, shifting effort from mechanical assembly to higher-leverage work: checking assumptions, shaping policy, supplier negotiations, and contingency planning.

Adoption follows a cautious path. Begin in shadow mode by running the agentic plan alongside the incumbent process and explaining item-by-item differences. Maintain complete logs of inputs, capability calls, and validations for audit. Use fail-safe behavior—raise explicit exceptions when tools are unavailable or constraints are infeasible. Because interfaces are stable, components can be upgraded over time (for example, replacing a heuristic with a multi-period optimizer in procurement [4], or adding protocolized orchestration features [5, 6]) without redesigning the whole.

In summary, S&OP supplies the governance and economics; agentic automation supplies reliable, end-to-end coordination; and people supply the judgment. Together, they form a durable planning capability under real-world volatility.

3 Limits of Single Agents, Advantages of an Agentic Layer, and Use-Case Fit

Single-purpose agents solve narrowly scoped tasks with limited context, for example, setting a reorder trigger for one SKU or extracting fields from invoices. They perform well when boundaries are stable and objectives are local, but they degrade under shifting conditions and whenever decisions must satisfy rules that cut across functions. Viewed as a chain: forecast → inventory → procurement → finance—optimizing one link in isolation does not guarantee a feasible end-to-end plan. Service targets, supplier rules (minimums, pack sizes, calendars), and monthly cash caps couple choices across steps; a locally "best" adjustment can still violate budgets or delivery calendars once the plan is assembled.

These shortcomings appear in practice. Local tuning of reorder points that ignores pack sizes or cash limits can produce infeasible or needlessly costly orders. Fixed assumptions are brittle under tariff changes, port disruptions, or lead-time slips. Governance fragments when business rules are duplicated across worksheets and services, complicating audits and change control. Cross-domain checks remain weak—few single agents enforce "meet the service target while staying within monthly caps" simultaneously. Explanations are thin: approvers need to see why shortages concentrate, how supplier minimums shape order sizes, and which months exceed caps.

An agentic planning layer addresses these gaps by planning the full chain, validating shared constraints after each step, and revising earlier choices when a check fails. It aligns functions and makes constraints explicit—service targets implemented through a safety buffer over lead-time demand [1], supplier minimums and pack multiples in purchasing [4], and month-by-month spend caps in finance—and ties outcomes back to drivers such as volatility, lead time, and supplier concentration. Because the layer recomputes the plan as inputs move, it dampens bullwhip amplification [2] and coordinates decisions among heterogeneous actors in the network [3]. The result is a rolling plan with an audit trail that records what changed and why, turning scattered calculations into a coherent, feasible program.

The fit by use case is clear. Single agents excel at area optimizations—picker routing within an aisle, invoice capture, single-SKU safety-stock tuning, anomaly detection on sensors, or slotting within a distribution center—where scope is tight and interfaces are stable. The agentic layer is appropriate for process-level planning and optimization where decisions span functions and horizons: S&OP and plan-to-procure, demand-driven replenishment under supplier calendars, new-product ramps gated by cash limits, and vendor-managed inventory subject to global caps. While this paper focuses on S&OP as the monthly cadence that produces a single feasible plan, the same typed interfaces and validations extend to S&OE (Sales & Operations Execution), the short-horizon layer that carries out the plan and absorbs day-to-day disturbances, enabling re-planning without altering the core logic.

4 Experiments

4.1 Agentic Architecture Design for S&OP

Figure 3 summarizes the architecture used in our experiments. A lightweight **Supply Chain Orchestrator** coordinates a minimal set of agents aligned to the S&OP cadence: **Data Curator**, **Forecaster**, **Inventory**, **Procurement**, **Finance**, and an optional **Supplier-Risk** agent. Each agent operates over typed tables and returns well-formed outputs; the Orchestrator sequences these calls, verifies constraints after each step (service targets implemented through a safety buffer on lead-time demand, supplier minimums and pack multiples, business-day calendars, and monthly spend caps), and, when needed, revises earlier decisions before proceeding. The same plan can be produced either in a single process (SDK orchestration) or via a microservice layout with capability discovery and typed I/O akin to MCP, allowing any agent to be substituted by an independently deployed service without altering contracts [6].

The flow mirrors the numbering in the figure. The Orchestrator initializes the run **(0)** and requests curated inputs. The **Data Curator** prepares schema-clean demand history and reference tables and hands them to the **Forecaster (1)**. The **Forecaster** returns SKU-site quantile paths (P50/P80/P90), which the **Inventory** agent consumes **(2)** together with current stock and open purchase orders to compute reorder points, safety stock, and projected shortages. Inventory also exposes exception diagnostics so that missing keys or infeasible settings are made explicit. In parallel, the **Supplier-Risk** agent can score reliability using on-time performance, quality escapes, spend concentration, and curated external signals; these scores are fed into the next stage **(4)**. The **Procurement** agent then converts shortage projections, supplier rules (MOQ and pack multiples), and risk scores into dated purchase-order proposals **(3–5)**. The **Finance** agent rolls proposed orders to months, compares planned spend to budget caps, and flags variances with simple smoothing options **(5)**. Finally, the Orchestrator consolidates artifacts—forecast tables, inventory policies, proposed buy plan, spend vs. cap report, RFQ drafts, and a narrative explaining why the plan takes its shape—into a single output package **(6–7)** suitable for Pre-S&OP and Executive S&OP review.

Two design choices are central. First, the agents communicate only through typed, versioned tables; this makes behavior reproducible and keeps replacements safe (for example, swapping a heuristic PO sizer with an optimizer leaves interfaces and validations unchanged). Second, constraint checks are first-class. After every agent completes, the Orchestrator re-validates global feasibility—meeting service targets without violating supplier rules or monthly caps—and either advances or revisits upstream settings. This rolling plan, together with an auditable record of inputs, calls, and checks, converts a set of local calculations into a coherent, end-to-end S&OP plan.

4.2 Methods for the Agents

Forecasting.

We use two lightweight back ends that return the same form of output: SKU–site demand quantiles. The baseline is a moving-average forecaster that produces a median path (P50) and constructs P80/P90 by adding cushions derived from historical forecast errors. An

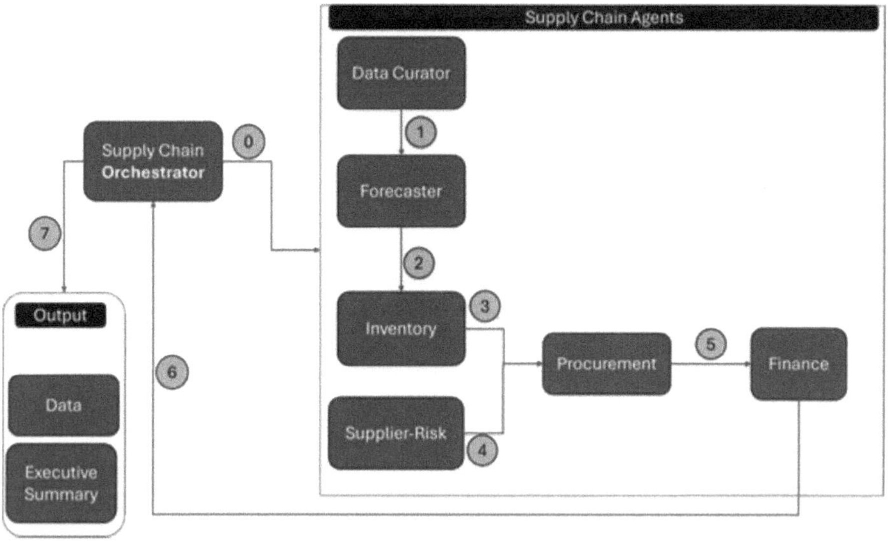

Fig. 3. Agentic S&OP architecture used in experiments. The Orchestrator (left) sequences agents (right) and consolidates artifacts. Numbered markers correspond to the execution order described in the text.

optional neural time-series model yields the same P50/P80/P90 outputs on the common time grid. Reporting quantiles rather than a single point matters because inventory decisions depend on the **distribution** of demand during lead time, not just its average [1]

Inventory Policy
The inventory agent sets the reorder point as "expected demand over supplier lead time plus a safety buffer." The buffer (safety stock) increases with demand volatility and with the service target selected by the business (e.g., avoiding stockouts in roughly 95% of cycles) [1]. A projected shortage arises when on-hand plus inbound would fall below the trigger before the next receipt. Diagnostics make the trade-off visible: higher service targets reduce expected shortages but tie up more working capital. The same inputs support classic (s, S) control—reorder when stock drops below s (the trigger) and top up toward S (the target level).

Procurement Sizing
Projected shortages are converted into purchase lines that honor supplier rules. Each line is no smaller than the supplier's minimum order quantity (MOQ), is rounded up to the nearest pack multiple (case size), and is placed on a valid business day within the desired arrival window. The default implementation is a fast, transparent rule so planners can audit every step; the same interface can be backed by an optimizer that balances ordering, holding, freight, and shortage penalties when needed [4].

Finance Roll-Up and Smoothing
Planned monthly spend is computed as the sum of unit cost × order quantity for lines

landing in that month. The finance agent compares this total to the budget cap and reports the variance (cap minus planned). When a month exceeds its cap, the orchestrator proposes simple smoothing—shifting feasible receipts to adjacent months—while preserving service. Rule-based shifts are used by default for traceability; a more elaborate objective can replace them without changing interfaces.

Supplier Risk

The supplier-risk agent assigns each supplier a composite score that increases with late delivery (low on-time rate), quality escapes, spend concentration (excess volume with a single source), and curated external signals. Non-negative weights sum to one and can be tuned to local priorities. The aim is to surface concentration and reliability risks early, consistent with system views of heterogeneous actors in the network [3].

4.3 Implementation Paths

4.3.1 SDK Runtime (Single Process)

The single-process configuration uses the OpenAI SDK as a thin execution layer over typed tables shared by the agents. Each agent is registered with explicit input and output schemas (e.g., forecast frames; inventory/on-order position; supplier rules with lead times, MOQ and pack multiples, business-day calendars; unit costs; monthly caps). The orchestrator advances through the causal order—forecast → inventory → procurement → finance—and performs post-step validation. If a check fails (for example, a proposed order violates a pack multiple or pushes a month over its cap), the orchestrator revisits upstream choices and re-runs the chain until feasibility is restored.

This runtime emphasizes determinism, traceability, and fast iteration: identical inputs yield identical outputs; intermediate artifacts (CSV tables, diagnostics, narrative report) are versioned; and request/response payloads are logged for audit. Because schemas are stable, forecasters (moving average or neural) and procurement back ends (rule-based or optimizer) can be swapped without changing orchestration logic, which makes this path suitable for pilots and low-latency runs while preserving governance [5].

4.3.2 SDK Runtime (Single Process)

In the services layout, the same agents are exposed behind lightweight capability manifests that publish discovery metadata, typed inputs/outputs, and structured errors, consistent with the Model Context Protocol (MCP) draft [6]. The orchestrator becomes a client: it discovers available capabilities, submits typed payloads, receives typed results, and records per-service traces. Planning order and validators are unchanged; only the call surface differs.

This arrangement supports modularity and ownership. Teams can maintain, scale, or replace capabilities—e.g., substitute a mixed-integer procurement optimizer for the heuristic—without altering orchestration or weakening constraint checks. Stable schemas guarantee behavioral equivalence with the SDK run, enabling shadow comparisons and gradual migration from single process to services as scope and stewardship expand. The same contracts generalize to adjacent processes (S&OE re-plans, NPI ramps, vendor-managed inventory), since new capabilities need only publish manifests and satisfy the existing validations grounded in inventory theory and procurement economics [1–4].

5 Results and Discussion

5.1 Experiments Results

We evaluated the agentic S&OP pipeline on a synthetic plan-to-procure scenario with daily demand at multiple SKU–site pairs. The dataset included demand history, current inventory and open POs, a supplier master with lead times and ordering rules (MOQ and pack multiples), and monthly budget caps. Agents exchanged typed tables only; the Orchestrator enforced the causal order forecast → inventory → procurement → finance and re-planned when any constraint failed. Two forecasting back ends were tested: a moving-average baseline (median path plus error-based cushions) and an optional neural time-series model, both producing P50/P80/P90 per SKU–site. Figs. 4, 5, 6 and 7 show representative outputs.

Metrics We report (i) cycle service level (fraction of periods without stockout by SKU–site), (ii) shortage lines over the horizon, and (iii) finance variance by month (budget cap minus planned spend). We also examine qualitative outputs: explanations attached to proposals and the role-based action list used in review meetings.

5.2 Quantitative Findings

Forecast vs. Actual Figure 4 shows four SKU–site examples where the P50 track follows weekly seasonality and P80/P90 provide headroom for plausible surges; the actual series (red) remains largely within the forecast fan. This behavior is critical for inventory policy because lead-time demand depends on the distribution, not a point estimate [1]. When the neural back end is used, the fan adjusts more strongly to trend shifts, but both back ends generate the same quantile outputs and integrate identically into downstream agents.

Inventory and Shortages With quantile forecasts in place, the inventory agent sets reorder points as "expected lead-time demand plus a volatility-sized buffer," projects shortages against on-hand plus inbound, and writes diagnostics for any missing SKU–site keys. In runs where inventory coverage is complete, shortage concentration shifts toward items with long lead time and higher dispersion—matching expectations from inventory theory [1] and the bullwhip literature [2]. Diagnostics shorten review time by explaining "no-PO" outcomes (e.g., sufficient projected position or missing master data) rather than leaving silent gaps.

Procurement and Spend he procurement agent converts projected shortages to dated purchase lines that honor MOQ and pack multiples and places ETAs on valid business days. Finance then aggregates to months and compares planned spend to caps. Figure 5 illustrates a typical **spend vs. cap** view: months near the cap prompt smoothing suggestions (shiftable receipts or splits) while maintaining service. When supplier price is

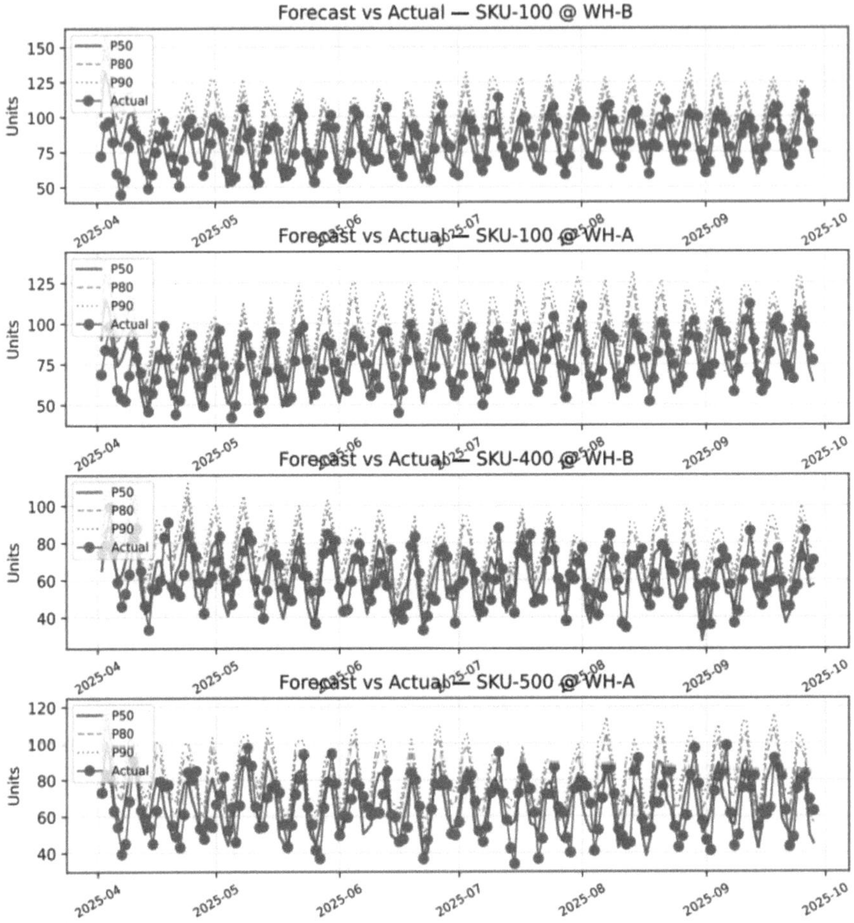

Fig. 4. Forecast Vs Actual - 4 SKU–site example

available in the supplier master, spend and variance track closely with PO volume; when prices are missing, the system flags items for follow-up, avoiding false precision.

5.3 Qualitative Outcomes

Explanations and audit trail. Each stage writes artifacts (CSV tables, RFQ drafts, narrative report). The report links outcomes to causes—for example, "shortages concentrate at DC-EAST for SKU-A due to longer lead time and higher weekly volatility," or "July exceeds the cap due to MOQ/pack rounding on three suppliers." The **Role-Based Next Actions** pages (Figs. 6 and 7) translate these findings into concrete follow-ups for demand planning, inventory, procurement, finance, supplier-risk, and operations. In reviews, this reduced time spent reconciling spreadsheets and increased time on policy decisions (e.g., raising service for A-class items or adding alternates for a constrained supplier).

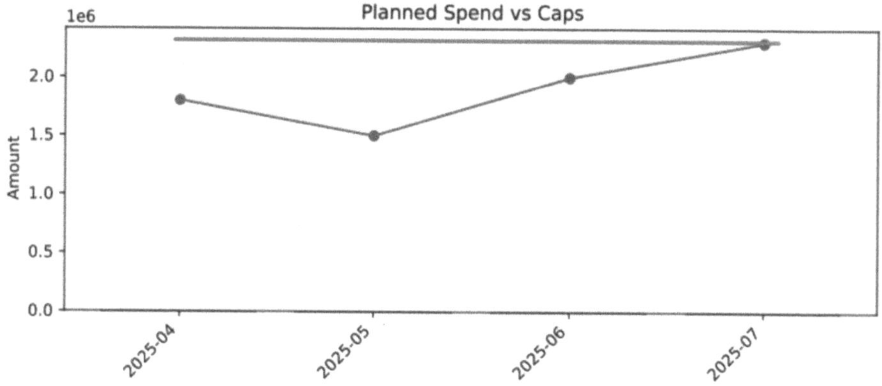

Fig. 5. Spends Vs Cap Graph

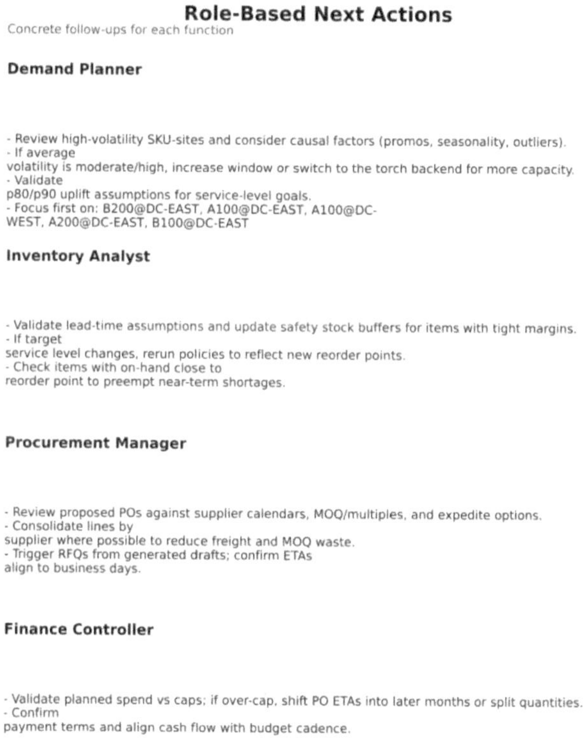

Fig. 6. Role Based Action – 1

5.4 Why the Pipeline Improves Feasibility

The gains arise from enforcing the planning order and validating after every step:

Role-Based Next Actions (cont.)

Supplier-Risk Analyst

- Monitor suppliers with lower on-time rates; add alternates to watchlist.
- Set up news/KPI scraping for
signals that may impact supply (web sources, internal SLAs).
- Watch these suppliers: S-BRAVO(0.78),
S-CHARLIE(0.82), S-ALPHA(0.87)

Operations Lead

- Schedule a brief Pre-S&OP review to validate forecast, shortages (if any), and spend vs caps.
- If changes
are approved, publish artifacts to ERP/MRP and release RFQs.
- Capture decisions and update `run_meta.json`
notes for auditability.

Fig. 7. Role Based Action –2

- **Forecast → Inventory.** Quantile paths connect the chosen service target to an explicit safety buffer over lead-time demand; this reduces surprise stockouts at a given target level [1].
- **Inventory → Procurement.** Translating shortage projections through MOQ and pack multiples prevents hidden over-ordering, aligning with procurement economics and supplier calendars [4].
- **Procurement → Finance.** Rolling dated POs to months exposes cap overruns early; smoothing suggestions adjust ETAs where operational slack exists while preserving service.
- **Iteration.** If any check fails, the Orchestrator revises earlier choices and re-runs the chain, damping bullwhip amplification and coordinating heterogeneous decisions [2, 3].

5.5 Sensitivity and Ablations

We conducted sensitivity checks on service targets, lead-time assumptions, and supplier rules. Higher targets lower expected shortages but increase working capital, as expected from reorder logic [1]. Longer lead times shift safety stock and raise the likelihood of budget pressure in the near month because MOQ/pack rounding concentrates spend. Replacing the heuristic PO sizer with an optimizer (same interface) tightened month-end variance when caps were constraining, consistent with procurement optimization results in the literature [4]. Across runs, the orchestration and validation logic remained unchanged, demonstrating the value of stable interfaces [5, 6].

5.6 Limitations

The experiments use synthetic demand and simplified supplier calendars; real deployments should incorporate richer calendars, price tiers, and multi-echelon effects. Forecast learning is intentionally lightweight; domain-specific features and external signals can further improve the fan while keeping the same contract. Supplier-risk scoring uses simple composites; calibrated external feeds must be curated to manage bias. Finally, budget smoothing is rule-based for auditability; organizations with stricter cost models can substitute a formal objective without changing inputs/outputs.

5.7 Takeaways

An agentic planning layer produces coherent, auditable S&OP plans by sequencing domain steps, enforcing shared constraints, and re-planning as conditions move. The approach lifts feasibility (service adherence with fewer shortages) and governance (spend within caps, traceable decisions), while keeping policy choices with managers. The same contracts support gradual migration—from a single-process orchestrator to a services layout—and reuse in adjacent processes such as S&OE re-plans, new-product ramps, or vendor-managed inventory, preserving alignment with established operations principles [1–4] and protocolized orchestration practices [5, 6].

6 Conclusion

This work presented a practical path from single-purpose agents to an agentic planning layer for S&OP. The core design is deliberately small: a Data Curator, Forecaster, Inventory, Procurement, Finance, and optional Supplier-Risk agent, sequenced by an orchestrator that validates shared constraints after each stage. Agents exchange typed, versioned tables; the orchestrator advances only when service targets, supplier rules, calendars, and monthly caps are satisfied. This turns isolated calculations into a coherent, auditable plan aligned with established operations principles [1–4] and supported by protocolized interfaces [5, 6].

Experiments on a plan-to-procure scenario showed that quantile forecasts coupled with policy checks reduce shortage exposure at a chosen service level and improve budget adherence through spend roll-ups and smoothing. The artifacts—tables, RFQ drafts, diagnostic notes, and a narrative report—make causes visible (volatility, lead time, MOQ/multiples, supplier concentration) and shorten review cycles by directing attention to the few decisions that matter. The same contracts support two implementation paths: a single-process SDK orchestrator suitable for pilots and a services layout with capability discovery and typed I/O suitable for multi-team operation, without changing the planning logic.

Limitations remain. The experiments use synthetic demand, simplified calendars, and a rule-based smoothing step. Real deployments should incorporate richer price tiers, multi-echelon effects, and calibrated external signals for supplier risk. Forecast learning was kept lightweight by design; domain features and external covariates can enhance the quantile fan while preserving the interface. Governance also benefits from stronger lineage—signing of manifests, version pins, and policy tagging—to meet strict audit requirements.

Future work will extend the same contracts to adjacent processes: short-horizon S&OE re-plans, new-product ramps, and vendor-managed inventory, and will explore replacement of heuristics with formal optimizers where warranted (e.g., multi-period procurement under cash caps). The broader takeaway is that modest components, assembled under clear contracts and checked against shared constraints, can deliver end-to-end improvements without discarding the operational discipline that has guided supply chains for decades.

References

1. Silver, E.A., Pyke, D.F., Peterson, R.: Inventory Management and Production Planning and Scheduling. Wiley (1998)
2. Lee, H.L., Padmanabhan, V., Whang, S.: Information distortion in a supply chain: the bullwhip effect. Manag. Sci. **43**(4), 546–558 (1997)
3. Macal, C.M., North, M.J.: Agent-based Modeling in supply chains. In: Proceedings of Winter Simulation Conference (2005)
4. Bertsimas, D., et al.: Mathematical optimization for procurement, interfaces (2011)
5. OpenAI: Function calling and tool use in LLMs, technical report (2024)
6. Anthropic: Model Context Protocol (MCP), protocol draft (2024)
7. https://www.gartner.com/en/articles/supply-chain-leaders-its-time-to-rethink-your-recession-playbook

Image Acquisition by Stable Diffusion Adaptation with LoRA

Jessica P. P. Moyolema[1(✉)] and Rodolfo I. Meneguette[2]

[1] São Paulo State University, São José do Rio Preto, São Paulo, Brazil
pamela.palate@unesp.br
[2] University of São Paulo, São Carlos, São Paulo, Brazil
meneguette@icmc.usp.br

Abstract. Generative artificial intelligence has driven new methodologies for the creation of visual content from textual descriptions. This paper presents the implementation of a fine-tuning process using the Low-Rank Adaptation (LoRA) technique on the Stable Diffusion model. The main objective was to customize the generation of images to represent specific facial features from a set of own photographs. A cloud-based computing environment using the RunPod platform and tools such as Kohya SS were used for training. Model evaluation was performed using automatic facial similarity metrics based on embeddings generated by InsightFace. The results reflect a significant improvement in the model's ability to replicate facial features of the target subject, evidencing the effectiveness of LoRA as an efficient and computationally inexpensive personalization technique.

Keywords: Stable Diffusion · LORA · Generative AI · Image

1 Introduction

Generative artificial intelligence is a subfield of artificial intelligence that, based on various models, is capable of producing high-quality content in domains such as text, images, videos, music, code, voice, and scientific applications [1]. This technology combines machine learning techniques, deep neural networks, and natural language processing, enabling pattern learning from training datasets [2], which can be created specifically or derived from existing sources. These patterns are used to generate results that simulate tasks commonly performed by humans. Currently, generative artificial intelligence is not limited to data analysis or the generation of recommendations and predictions, but also facilitates the creation of original, innovative, and realistic content [3]. Although it offers comprehensive benefits, it also raises ethical concerns, liability issues, and the risk of bias in the generated content, in addition to continuing development with constant improvements [4].

The main generative models include Variational Autoencoders (VAE), which is an unsupervised machine learning neural network model that encodes input

H. R. Arabnia et al. (Eds.): CSCI 2025, CCIS 2968, pp. 33–44, 2027.
https://doi.org/10.1007/978-3-032-28457-0_3

data into a lower-dimensional representation and then decodes it, the Transformer model, initially designed for natural language processing and adaptable to text and image inputs [1], and linguistic models such as BERT and ChatGPT, which act as semantic encoders and decoders [5]. Similarly, generative adversarial networks (GANs) are a dual approach in which a generator produces new data while a discriminator evaluates its authenticity [6]. These advances represented a turning point for generative AI, as through the use of big data and transfer learning, its implementation became more accessible to researchers and developers [4].

Diffusion models (DM) are characterized by a direct diffusion process and an inverse denoising phase, allowing the reconstruction of the original data from its basis [7]. The neural network is trained to learn how to reverse noise in the data [8]. Unlike the adversarial approach used in GANs, diffusion models have a more robust training objective and offer superior generation quality compared to other approaches, such as VAE.

Image generation from text is the second most popular application in generative AI, after text-to-text generation. This task is based on architectures that integrate encoders, decoders, or both modules, adapting models such as GAN, VAE, and Stable Diffusion [4].The text encoder transforms descriptions into semantic vectors that are used by the spatial decoder to create visually realistic images aligned with the textual content [9].

Recent major advances in text-to-image generation include models such as Stable Diffusion and DALL-E [10]. These methods incorporate cross-attention mechanisms that allow textual representations to be injected into the diffusion process [8], generating high-quality images from textual descriptions [2]. However, they present challenges, such as long generation times and the need to wait for results before modifying parameters [4]. Stable Diffusion offers broad and free access, while DALL-E, developed by OpenAI, has evolved to its third version and has an independent API for experimental use [4].

Several schemes have been proposed to mitigate the problem of context loss in image generation from multiple sentences, notably Initial Context First (ICF), Region Context First (RCF), Single Sentence ICF (SS-ICF), and Single Sentence RCF (SS-RCF). These approaches have been shown to improve the similarity between generated images compared to traditional methods such as DALL-E, Stable Diffusion, and MidJourney, although it should be noted that these are solutions based on prompt enhancement rather than direct modification of the generation models [4].

Stable Diffusion is an advanced implementation of LDMs (Latent Diffusion Models), designed to generate high-resolution images from textual cues. Its performance is based on learning the inverse diffusion process, which consists of progressively eliminating the noise introduced into the data, facilitating the recovery of the original input information [11].

LoRA is geared toward fine-tuning low-ranking subspaces within weight matrices, especially in the linear modules of transformers. Several studies have validated the effectiveness of LoRA in tasks such as fine-tuning instructions in large-scale language models, achieving performance comparable to or superior to traditional fine-tuning, but with a significantly smaller number of parameters [12].

The central objective of this research is to explore the potential of the Low-Rank Adaptation (LoRA) fine-tuning technique to customize generative image models, specifically Stable Diffusion, from a proprietary dataset. The goal is to evaluate the ability of the fine-tuned model to more accurately represent specific facial features, mitigating the loss of semantic and visual coherence common in basic models. This customization aims to increase the fidelity and quality of the generated images, using computational resources accessible to research environments with limited infrastructure.

2 Related Articles

This section reviews recent research relevant to the field of generative artificial intelligence applied to image synthesis from text, as well as advances in fine-tuning techniques using LoRA (Low-Rank Adaptation).

2.1 Text-to-Image Generation Models

The generation of images from textual descriptions has become one of the most important applications of generative artificial intelligence. Several studies have addressed this task by applying and improving models such as DALL-E, Midjourney, and Stable Diffusion, seeking to optimize the semantic and visual coherence of the generated images.

In [4], we propose a scheme designed to improve the structure of the input text (prompt) in order to preserve narrative coherence during image generation. Through a comparison with existing models, the study demonstrates that its external methodology, without modifying the diffusion models, achieves superior results in terms of contextual coherence between sentences.

On the other hand, the work presented in [13], introduces a model called Versatile Diffusion (VD), a multimodal and multitask diffusion network capable of generating various results, such as text images, image descriptions, and variations. This model extends the standard diffusion architecture by integrating components such as an Autoencoder-KL (similar to Stable Diffusion) for image encoding and Optimus a combination of BERT and GPT as a text encoder-decoder. In addition, CLIP encoders are used as sources of multimodal context, trained on large datasets such as LAION-2B-en and COYO-700M.

2.2 Fine-Tuning Using LoRA

The LoRA method has been widely adopted for the efficient fine-tuning of diffusion models, allowing for the customization of model behavior at reduced computational cost.

In [14], we present a methodology that combines automatic captioning and LoRA to customize stable diffusion with new visual concepts, such as Pokémon-style characters or pet rabbits. Using customized datasets and evaluations under identical prompt conditions, we show that LoRA enables efficient adaptation without the need for intensive manual labeling.

Furthermore, in [15], fine-tuning with LoRA is implemented in industrial contexts using visual inspection data from the company Otokar. Model validation is performed using quantitative metrics such as SSIM, MSE, and PSNR, demonstrating its applicability in specialized technical environments.

The research by [16], focuses on improving facial rendering in generated images. A LoRA model is trained on a specific dataset geared toward solving facial blurring problems, integrating this module into the stable diffusion architecture to achieve better results in the accurate representation of individual faces.

On the other hand, [12], proposes the integration of LoRA modules into low-resource machine translation models. Through the LSFTL model, which employs LoRA adapters trained on public bilingual corpora (bitext), the quality and contextual adequacy of translations for poorly represented languages are improved.

Finally, [17] addresses a specific limitation of stable diffusion: the faulty generation of human hands. To correct this problem, a three-phase solution is proposed: detection with YOLOv8, joint masking with MediaPipe, and repainting using ControlNet. A set of LoRA models specialized in hand gestures is trained, achieving a significant improvement in the anatomical representation of the hands generated.

3 Methodology

3.1 Computing Environment and Tools

Due to the computational limitations of the local environment, we chose to use the RunPod platform, which offers instances with configurable high-performance GPUs for training artificial intelligence models. An account was created on the platform and a credit of 25 USD was charged to deploy a virtual machine. The selected instance was a 1X NVIDIA RTX 3090, with an allocation of 20 GB for the container and 50 GB of persistent storage volume. The preconfigured model based on Stable Diffusion 1.5 was selected, which is widely compatible with fine-tuning tools such as Kohya SS.

After the instance was deployed, ports 3000 (HTTP) and 8888 (JupyterLab) were enabled, accessing the Jupyter interface for installation and execution of the training environment.

3.2 Installation of the Kohya SS Environment

The work environment cloned the Kohya SS repository, a specialized tool for training LoRA (Low-Rank Adaptation) models in Stable Diffusion. The installation was performed in a Python virtual environment by executing the following commands in the console:

```
git clone https://github.com/bmaltais/
kohya_ss.git
cd kohya_ss
python3 -m venv venv
source venv/bin/activate
apt update && apt-get install python
3.10-tk
pip install fastapi==0.99.1
pip install -r requirements.txt
pip install gradio==3.41.2
./setup.sh -n
bash gui.sh --share
```

This allowed access to the Kohya SS graphical interface from a public URL, facilitating the configuration of the training process without the need for local network settings.

3.3 Data Set Preparation

For training, a set of 25 self-images (personal photographs), pre-processed using the online tool Birme, was used to center the face and ensure uniformity in the facial scale. The images were named sequentially (1.jpg, 2.jpg, etc.) and stored in the /workspace/XX directory.

In addition, to avoid overfitting and preserve the general characteristics of the target class (women), a set of regularization images was incorporated, using a free generic dataset of women, hosted in the /workspace/womenclass folder.

The directories:

1. `/workspace/training/model`
2. `/workspace/training/reg`
3. `/workspace/training/log`
4. `/workspace/training/img`

were also created to store, respectively, the generated models, regularization images, training logs, and processed images.

3.4 LoRA Model Configuration and Training

The basic model selected for training was Realistic Vision V5.1, due to its ability to generate photorealistic images with a high level of detail. The training configuration in Kohya SS was performed through its graphical interface, defining the following main parameters:

- Basic model:
  - /workspace/stable-diffusion-webui/models/Stablediffusion/Realistic-Vision-V5.1.safetensors
- Output file type:
  - safetensors
- Prompts:
  - Instance: XX_person women
  - Class: woman
- Directories:
  - Training images: `/workspace/XX`
  - Regularization images: `/workspace/women-class`
  - Output folder: `/workspace/training`

In addition, the LoRA training parameters were set as follows (see Table 1):

Table 1. LORA training parameters

Parameter	Value
Train batch size	2
Epochs	10
Save every n epochs	1
Max train epoch	10
Mixed precision	bf16
Save precision	bf16
LR scheduler	Cosine
Learning rate	0.0001
CPUs por hilo	2
LR warmup	10
Learning rate (TE)	0.00005
Learning rate (UNet)	0.0001
Network rank	8
Repeats	40
Xformers	No

The configuration was saved in the file `XX.json` on the desktop. This process started the training, completing a total of 10,000 steps distributed over 10 epochs.

During training, intermediate LoRA models were generated every 10,000 steps, allowing for progressive evaluation of the results. The generated models were automatically named following the format:

- XX_person_v01_prueba_000001.safe..
- XX_person_v01_prueba_000002.safe..
- ...

3.5 Preliminary Evaluation of the Model

After completing the training process, the LoRA models generated were stored in /workspace/training/model. These models were then moved to /workspace/stable-diffusion-webui/models/Lora for integration into the Automatic1111 graphical environment, accessible through port 3000.

The qualitative evaluation was performed by generating images with and without the incorporation of the LoRA model, using the same input prompts. For generation with LoRA, the following model identifier was used within the prompt:

- <lora:XX_person_v01_test-000009:1>

Figure 1, shows the image generation interface in Automatic 1111, where the parameters used, the specific prompt, and the activation of the corresponding LoRA model are displayed.

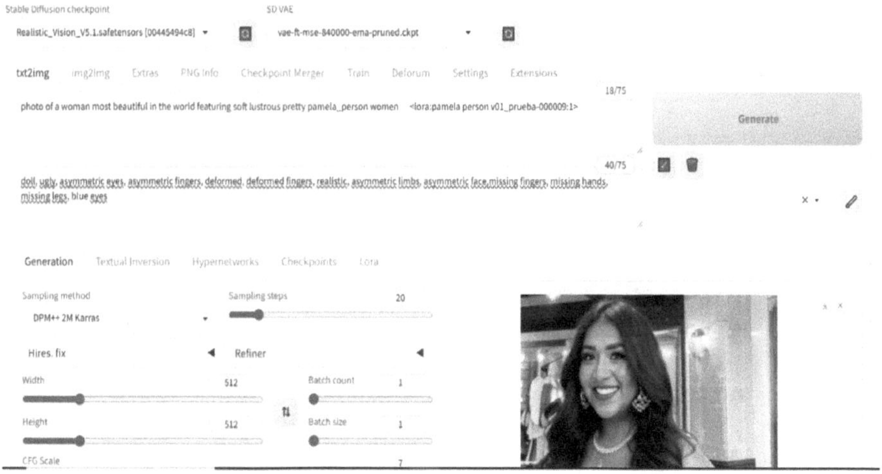

Fig. 1. Parameters for image generation

The X/Y/Z Plot module of Automatic 1111 was used to observe the progression of training across different epochs. The parameter settings are shown in Table 2.

The execution of this configuration generated a visual composition that shows the evolution of the images generated through the different checkpoints (epochs)

Table 2. Parameters for visualizing the progression of image generation

Parameter	Value
Seed	–
Script	X/Y/Z Plot
X Type	Prompt S/R
X Values	Checkpoint, 1, 2, 3, 4, 5, 6, 7, 8, 9

Fig. 2. Parameters entered to visualize image generation in a single image

of the model. This progression is illustrated in Fig. 2, which shows the gradual changes in facial fidelity as training progresses.

Fig. 3. Real dataset

In addition, Fig. 4, shows a sequence of images generated by the LoRA model from the initial state to the end of the 10 training epochs, showing the progressive improvement in the quality and detail of the images produced.

Fig. 4. Image generation from the base image

Results

In order to quantitatively validate the LoRA model's ability to replicate the author's facial features, a cosine similarity metric was applied using facial embeddings generated by InsightFace, comparing real images from the author's dataset (Fig. 3) with five images generated by the LoRA model adjusted in Stable Diffusion (Fig. 5).

The similarity analysis was performed with a script developed in Python in the Google Colab environment, calculating the embeddings of each image and its respective similarity with the set of real images. The evolution of the similarity values is shown in Fig. 6, where it can be seen that the third image generated achieved the highest similarity value (approximately 0.6), while the rest of the images varied between 0.3 and 0.45.

These results indicate that the model trained with LoRA was able to capture certain distinctive facial features of the author, showing a qualitative and quantitative improvement over the images generated by the base model without

Fig. 5. Result of images trained with Lora and Stable Diffusion

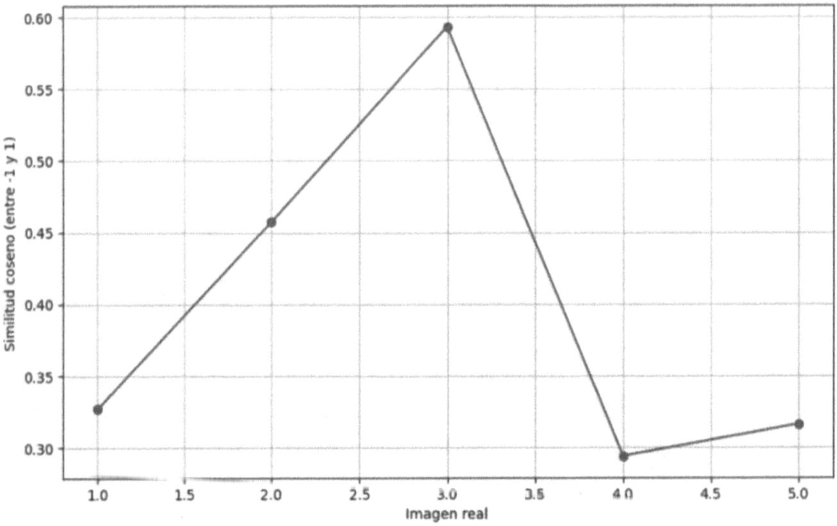

Fig. 6. Evolution of facial similarity of images generated with stable diffusion and LoRA and the author's real image

adjustment. This analysis, complemented by visual review, confirms the effectiveness of the fine-tuning process implemented and its applicability in generative model customization tasks.

Conclusions

This study demonstrates the feasibility and effectiveness of using LoRA as a fine-tuning technique to customize generative models such as Stable Diffusion, even under conditions of limited computational resources. The implementation in RunPod allowed training to be performed in an accessible and efficient manner.

The results obtained reflect a notable improvement in the representation of specific facial features, as evidenced by visual analysis and facial similarity metrics. The methodology applied represents a viable alternative for researchers who wish to adapt generative models to customized domains without the need for large infrastructures.

It is concluded that the combination of stable diffusion with LoRA offers a flexible and scalable solution for custom image generation tasks, opening up new possibilities in areas such as digital identity, content personalization, and creative applications.

Acknowledgment. This study was financed in part by the Coordenação de Aperfeiçoamento de Pessoal de Nível Superior - Brasil (CAPES)

References

1. Bengesi, S., El-Sayed, H., Sarker, M.D.K., Houkpati, Y., Irungu, J., Oladunni, T.: Advancements in generative AI: a comprehensive review of GANs, GPT, autoencoders, diffusion model, and transformers. IEEE Access **12**, 69812–69837 (2024). https://doi.org/10.1109/ACCESS.2024.3397775
2. Sai, S., Gaur, A., Sai, R., Chamola, V., Guizani, M., Rodrigues, J.J.P.C.: Generative AI for transformative healthcare: a comprehensive study of emerging models, applications, case studies, and limitations. IEEE Access **12**, 31078–31106 (2024). https://doi.org/10.1109/ACCESS.2024.3367715
3. Chaudhary, S.: AI at your service: generative artificial intelligence and the next generation of assistants. In: 2024 Artificial Intelligence for Business (AIxB), pp. 51–52 (2024). https://doi.org/10.1109/AIxB62249.2024.00016
4. Kim, T.-S., Ignacio, M.J., Yu, S., Jin, H., Kim, Y.-G.: UI/UX for generative AI: taxonomy, trend, and challenge. IEEE Access **12**, 179891–179911 (2024). https://doi.org/10.1109/ACCESS.2024.3502628
5. Du, H., et al.: The age of generative AI and AI-generated everything. IEEE Netw. **38**(6), 501–512 (2024). https://doi.org/10.1109/MNET.2024.3422241
6. Hagos, D.H., Battle, R., Rawat, D.B.: Recent advances in generative AI and large language models: current status, challenges, and perspectives. IEEE Trans. Artif. Intell. **5**(12), 5873–5893 (2024). https://doi.org/10.1109/TAI.2024.3444742
7. Croitoru, F.-A., Hondru, V., Ionescu, R.T., Shah, M.: Diffusion models in vision: a survey. IEEE Trans. Pattern Anal. Mach. Intell. **45**(9), 10850–10869 (2023). https://doi.org/10.1109/TPAMI.2023.3261988
8. Cao, H., et al.: A survey on generative diffusion models. IEEE Trans. Knowl. Data Eng. **36**(7), 2814–2830 (2024). https://doi.org/10.1109/TKDE.2024.3361474
9. Khan, M.Z., et al.: A realistic image generation of face from text description using the fully trained generative adversarial networks. IEEE Access **9**, 1250–1260 (2021). https://doi.org/10.1109/ACCESS.2020.3015656
10. Akhtar, Z.B.: Unveiling the evolution of Generative AI (GAI): a comprehensive and investigative analysis toward LLM models (2021–2024) and beyond. J. Electr. Syst. Inf. Technol. **11**(22), 1–16 (2024). https://doi.org/10.1186/s43067-024-00145-1
11. Cho, E., Bang, J., Rhu, M.: Characterization and analysis of text-to-image diffusion models. IEEE Comput. Archit. Lett. **23**(2), 227–230 (2024). https://doi.org/10.1109/LCA.2024.3466118
12. Liang, X., Khaw, Y.-M.J., Liew, S.-Y., Tan, T.-P., Qin, D.: Toward low-resource languages machine translation: a language-specific fine-tuning with LoRA for specialized large language models. IEEE Access **13**, 46616–46626 (2025). https://doi.org/10.1109/ACCESS.2025.3549795

13. Xu, X., Wang, Z., Zhang, E., Wang, K., Shi, H.: Versatile diffusion: text, images and variations all in one diffusion model. In: 2023 IEEE/CVF International Conference on Computer Vision (ICCV), pp. 7720–7731 (2023). https://doi.org/10.1109/ICCV51070.2023.00713
14. Wu, Y., Lu, Y.: Automode: the automatic LoRA based on stable diffusion. In: 2024 IEEE 8th International Conference on Vision, Image and Signal Processing (ICVISP), pp. 1–5 (2024). https://doi.org/10.1109/ICVISP64524.2024.10959379
15. Yılmaz, M.T., et al.: Synthetic image generation with stable diffusion and trained LoRA model for industrial areas. In: 2024 Innovations in Intelligent Systems and Applications Conference (ASYU), pp. 1–4 (2024). https://doi.org/10.1109/ASYU62119.2024.10757052
16. Yao, H.: Improving performance of face generation for specific individuals based on LoRA. In: 2024 International Conference on Electronics and Devices, Computational Science (ICEDCS), pp. 77–80 (2024). https://doi.org/10.1109/ICEDCS64328.2024.00018
17. Bai, B., Xie, H., Miyata, K.: Fine-tuning pipeline for hand image generation using diffusion model. In: 2024 Nicograph International (NicoInt), pp. 58–63. (2024). https://doi.org/10.1109/NICOInt62634.2024.00020

Industrial Defect Detection with Explainable Guided Tile Selection

Rui Castro Nascimento[1,2(✉)], Dibet Garcia Gonzalez[3], Manuel F. Silva[1,4], Vítor Filipe[1,2], and Luís F. Rocha[1]

[1] INESC TEC—Institute for Systems and Computer Engineering Technology and Science, Rua Dr. Roberto Frias 4200-465, Portugal
luis.f.rocha@inesctec.pt

[2] UTAD - University of Trás-os-Montes and Alto Douro, Quinta de Prados 5000-801, Vila Real, Portugal
rui.p.nascimento@inesctec.pt, vfilipe@utad.pt

[3] SENTINEL - Sentinelconcept, Lda, N105 247 Piso 1, Sala J1.3, 4835-164 Guimarães, Portugal
dibet.gonzalez@sentinel-vision.com

[4] ISEP - Polytechnic of Porto, Rua Dr. António Bernardino de Almeida, 4249-015 Porto, Portugal
MSS@isep.ipp.pt

Abstract. Deep learning has become central to industrial visual inspection, offering high accuracy in detecting surface defects. However, applying object detection to high-resolution images remains computationally demanding, especially when using sliding window inference, which processes all image tiles uniformly. This study proposes a class activation mapping-guided tile inference strategy that restricts detection to regions identified as relevant via EigenCAM, a gradient-free explainability method. The method introduces a mini-tile assessment stage to filter out uninformative areas prior to detection. Experiments on a dataset of machined aluminum parts with porosity defects show that the proposed approach improves precision from 91.83 % to 96.81 % and reduces inference time by 21 %, with a slight decrease in recall, from 82.65 % to 80.53 %. These results demonstrate that explainability-guided filtering can significantly reduce the time required for inspection without compromising detection performance, offering a practical and interpretable solution for industrial defect detection.

Keywords: Explainable Artificial Intelligence · Quality Inspection · Defect Detection · Deep Learning · Industrial Automation

1 Introduction

Visual inspection is a fundamental step in industrial manufacturing, contributing to the early detection of defects that may compromise product performance or

H. R. Arabnia et al. (Eds.): CSCI 2025, CCIS 2968, pp. 45–58, 2027.
https://doi.org/10.1007/978-3-032-28457-0_4

reliability. Although automated solutions are increasingly adopted, many inspection tasks still depend on manual processes, which tend to be time-consuming and subject to variability due to operator fatigue or subjective judgment [16]. In recent years, computer vision methods based on deep learning have shown promising results in automating these tasks, offering improved consistency and scalability across a range of visual inspection scenarios [3].

Among deep learning models, Convolutional Neural Networks (CNN) have been widely studied and applied in defect detection, demonstrating competitive performance when compared to traditional methods based on hand-crafted features [3]. However, the deployment of CNN-based detectors in industrial settings presents practical challenges, particularly when working with high-resolution images. These images demand greater computational resources, leading to increased memory usage and longer inference times.

To mitigate this, many inspection systems employ a sliding window strategy, where the input image is divided into overlapping tiles that are individually processed. This allows the detector to cover the full spatial extent of the image and increases the likelihood of identifying small or subtle defects. Nonetheless, the exhaustive nature of this method often results in a large number of redundant computations, especially in regions where no relevant information is present.

In parallel, the lack of interpretability in deep learning models has become a growing concern, particularly in regulated or safety-critical industrial contexts. Since models such as CNNs typically operate as black boxes, it is difficult to understand the rationale behind their predictions. This limitation has motivated the development of Explainable AI (XAI) techniques, which aim to provide visual or conceptual insights into model behavior. Among these, Class Activation Mapping (CAM)-based methods, such as Gradient-weighted CAM (Grad-CAM) and its gradient-free variants like EigenCAM, offer spatial explanations by highlighting the regions in the input image that are most influential for a given output [14]. These visualizations support human operators in understanding, validating, and trusting model decisions during inspection tasks.

Motivated by this, the present work explores the integration of EigenCAM into the inference workflow of an object detector. The proposed strategy begins with a conventional sliding window tiling of the input image, followed by the generation of a global activation map using EigenCAM. Rather than applying detection uniformly across all tiles, the method evaluates the local relevance of smaller subregions, referred to as mini-tiles, resulting from the overlap between adjacent tiles. Inference is then selectively applied only to those mini-tiles that exhibit high activation, thereby reducing computational redundancy while preserving detection performance.

The main contributions of this study are as follows:

- A relevance-guided detection strategy that integrates EigenCAM with sliding window inference to reduce redundant computations,
- A mini-tile relevance assessment method based on activation intensity and active area metrics,

- An experimental evaluation demonstrating improvements in precision and inference time, while maintaining comparable recall to the baseline.

2 Related Work

XAI has become increasingly important in the context of defect detection for manufacturing and other industries. Black-box machine learning models, especially deep neural networks, can achieve high accuracy in visual quality inspection, but their lack of transparency poses challenges for trust and practical adoption on the factory floor [13,15]. Recent research has therefore focused on integrating XAI techniques into defect detection systems to provide human-interpretable explanations of model decisions, thereby improving user confidence and facilitating human–AI collaboration in quality control.

A variety of XAI techniques have been explored to interpret and visualize the decisions of defect detection models. Common post hoc methods include local interpretable model-agnostic explanations (LIME), Shapley additive explanations (SHAP), and class activation mapping (CAM) techniques like Grad-CAM. Soon and Chui [15] trained a convolutional neural network to classify textile fabric images as defective or not, achieving over 93 % accuracy, and applied SHAP to perform an interpretive analysis of the CNN outputs. Similarly, Ali *et al.* [1] investigated printed circuit board (PCB) defect detection using deep learning and employed both LIME and Grad-CAM to explain the model's predictions for various defect types.

However, researchers have noted that these generic XAI methods often require adaptation to manufacturing contexts. Pereira *et al.* [13] examined LIME in a textile defect detection scenario and found that blacked-out perturbations misled the model, introducing non-existent defects. To address this, they tested nine alternative background patterns, significantly improving explanation faithfulness. Their methodology, although demonstrated with LIME, can extend to similar perturbation-based techniques like SHAP.

Beyond image-focused methods, XAI has also been applied in software defect detection. Jude and Uddin [6] combined a hybrid machine learning model with LIME and SHAP to explain predictions, using stratified data splitting rather than oversampling to handle imbalance. Explanations clarified which features, such as, code complexity metrics, influenced classification, supporting the choice of stratified sampling.

Some studies have addressed the integration of XAI into models tailored for industrial applications. Nikolakis *et al.* [12] proposed a semi-supervised CNN for battery defect detection, enriched with attention and Layer-wise Relevance Propagation (LRP). Their framework incorporated expert feedback for dynamic learning, improving operational decision-making. Similarly, Kotsiopoulos *et al.* [7] presented a platform combining interpretable AI models, human-in-the-loop retraining, and cognitive mechanisms for defect detection in hard metals. Although no numerical results were detailed for the full platform performance, the framework emphasized resilience and adaptability.

Other works have advanced domain-specific XAI-based models with high detection accuracy. Kumar *et al.* [8] introduced a Zero-Bias Deep Neural Network (ZBDNN) for anomaly detection during composite curing, reporting 99.71 % accuracy, compared with 96.26 % for an autoencoder and 92.38 % for a one-class Support Vector Machine (SVM). Their method used a custom-built autoclave with viewport imaging to acquire data, which may limit scalability. Another study by Kumar *et al.* [9] extended this ZBDNN approach to additive manufacturing, achieving 99.5 % accuracy using grayscale images and cosine similarity matching. The model was optimized for real-time control by reducing inference time below image acquisition latency.

Real-time deployment on edge devices with limited computational resources has also motivated XAI-based solutions. Nguyen *et al.* [11] proposed the XEdgeAI framework, integrating visual/textual explanations into semantic segmentation for mobile industrial inspection. Their pipeline included fine-tuning, XAI-guided data augmentation, and vision-language interpretability modules. Although accuracy was not explicitly quantified, the mobile model reduced size significantly while maintaining competitive performance.

Finally, CAM-based techniques have been directly used to enhance training data and improve segmentation performance. Clement *et al.* [4] employed CAM explanations to augment ground truth annotations in DeepLabv3-ResNet101 segmentation, resulting in higher accuracy on complex object boundaries. Similarly, Augustauskas *et al.* [2] used CAM maps as weak supervision to localize defects without pixel-level labels, enabling model training using only class-level annotations.

To the best of the authors' knowledge, existing XAI methods for defect detection remain largely confined to post-hoc analysis and do not directly contribute to reducing inference costs on high-resolution images. Most approaches generate visual explanations for the entire image without leveraging activation maps to guide the selective processing of specific regions. This results in a missed opportunity to optimize detection workflows. Furthermore, quantitative evaluations of the trade-off between explainability-driven filtering and overall detection performance remain scarce in the literature. This work addresses these gaps by introducing an EigenCAM-based tile selection mechanism that prunes irrelevant regions before executing an object detector, aiming to reduce computational overhead while preserving detection accuracy.

3 Methodology

This section describes the proposed CAM-guided tile inference workflow. The process is divided into four main stages: (*i*) image tiling using a sliding window, (*ii*) activation mapping with EigenCAM, (*iii*) relevance evaluation of mini-tiles, and (*iv*) selective object detection. Figure 1 shows an overview of the workflow.

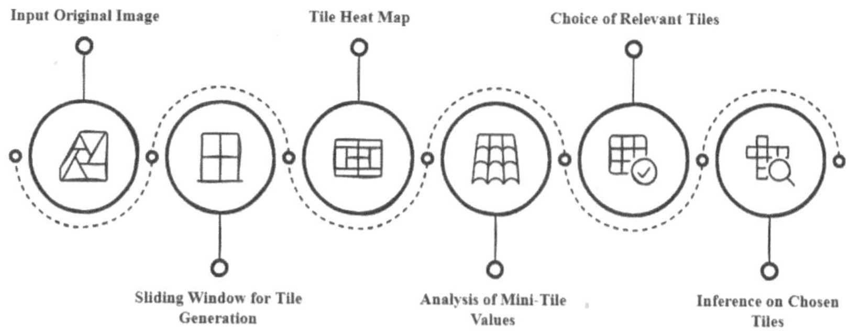

Fig. 1. Overview of the EigenCAM-guided tile inference workflow.

Algorithm 1. *Tile Generation with Sliding Window*

input: image (W, H), window $(W_O, H_O) = (1600, 1216)$
$(n_x, n_y) \leftarrow$ Eq. (1); $(s_x, s_y) \leftarrow$ Eq. (2)
for $i = 1..n_x$ **do**
  for $j = 1..n_y$ **do**
    extract $T_{i,j}$ from (x_i, y_j) of size (W_O, H_O)
  end for
end for
return $\mathcal{T}$

3.1 Tile Generation with Sliding Window

Each input image is divided into overlapping tiles using a fixed-size sliding window of 1600×1216 pixels, matching the input size expected by the YOLOv8 detection model. Let W and H represent the width and height of the image. The number of tiles in the horizontal and vertical directions is calculated using Eq. (1):

$$n_x = \left\lceil \frac{W - 1600}{640} \right\rceil + 1, \quad n_y = \left\lceil \frac{H - 1216}{480} \right\rceil + 1. \tag{1}$$

The stride between adjacent tiles is computed as in Eq. (2), ensuring full coverage of the image while maintaining overlap:

$$s_x = \frac{W - 1600}{\max(n_x - 1, 1)}, \quad s_y = \frac{H - 1216}{\max(n_y - 1, 1)}. \tag{2}$$

This overlapping approach helps preserve contextual information between neighboring tiles, which is relevant for detecting small or partially visible defects. The procedure is summarized in Algorithm 1.

3.2 Activation Mapping with EigenCAM

Each tile is independently processed using the *EigenCAM* technique to estimate the spatial importance of features without relying on gradients or class scores.

The implementation is based on the open-source `pytorch-grad-cam` library by Gildenblat et al. [5]. The method is applied to the ninth convolutional layer of the YOLOv8 backbone, specifically the last output of the third C2f block. This layer was chosen because it provides a trade-off between spatial resolution and semantic abstraction: it is deep enough in the network to capture class-discriminative features while still retaining spatial granularity to localize small defects.

A normalized activation map $C \in [0,1]^{1600 \times 1216}$ is generated per tile. All tile maps are then merged into a global activation map H using the accumulation method shown in Eq. (3):

$$H(x,y) = \min\left(1, \sum_{t=1}^{T} C_t(x,y)\right), \tag{3}$$

where T is the total number of tiles, and $C_t(x,y)$ is the activation at position (x,y) in tile t. Clipping ensures values remain within the range $[0,1]$. This global map serves as the basis for region relevance evaluation. The procedure is summarized in Algorithm 2.

Algorithm 2. *EigenCAM Activation Mapping*

```
input: Tiles 𝒯, image I
H ← 0_{H×W}
for each tile T in 𝒯 do
    extract patch P from I using T
    C ← EigenCAM(P)
    accumulate C into H
end for
clip H to [0, 1]
return global activation map H
```

3.3 Mini-Tile Relevance Assessment

In the sliding window approach, overlapping tiles are extracted to ensure spatial continuity across the input image. The intersection regions between adjacent tiles define what we refer to as *mini-tiles*, which form the basis for relevance evaluation. In border areas, where no neighboring tile exists, non-overlapped border bands are also treated as mini-tiles to guarantee full coverage. These border mini-tiles are evaluated in the same manner as interior ones, preventing blind spots at the edges.

Each mini-tile is projected onto the global activation map H to assess its local relevance. Two metrics are computed per mini-tile: (i) the mean activation intensity $\overline{A}$, and (ii) the *active area*, defined as the proportion of pixels with activation values above a threshold of 0.5. The active area is computed using the indicator function $\mathbb{1}_{\{H(x,y)>0.5\}}$ across all pixels (x,y) in the mini-tile.

A mini-tile is selected for inference if it satisfies at least one of the following criteria:

$$\overline{A} > 0.5 \lor \text{active\_area} > 0.5$$

These thresholds were empirically defined for the current study and can be adapted to other datasets or application contexts. This relevance filter ensures that only informative image regions are passed to the object detector. The procedure is summarized in Algorithm 3.

Algorithm 3. *Mini-Tile Enumeration and Relevance*

input: Tiles $\mathcal{T}$, strides (s_x, s_y), map H
enumerate overlaps $\mathcal{M}$ between adjacent tiles
add border bands to $\mathcal{M}$ for full coverage
$\mathcal{R} \leftarrow \emptyset$
for each mini-tile M in $\mathcal{M}$ **do**
  $\overline{A} \leftarrow \text{mean}(H|_M)$
  $\text{active\_area} \leftarrow \frac{1}{|M|} \sum_{(u,v)\in M} \mathbb{1}[H(u,v) > 0.5]$
  if $\overline{A} > 0.5$ OR active_area > 0.5 **then**
    add M to $\mathcal{R}$
  end if
end for
return relevant mini-tiles $\mathcal{R}$

3.4 Selective Object Detection

Object detection is applied only to the mini-tiles identified as relevant by the activation criteria. Each selected region is resized to the input resolution expected by the YOLOv8 detection model (1600×1216 pixels), using bilinear interpolation. If the mini-tile dimensions differ from the expected size, symmetric zero-padding is applied prior to resizing to preserve aspect ratio and avoid geometric distortion.

The YOLOv8 model, pre-trained for porosity detection, processes each mini-tile independently. Detection predictions are filtered using a confidence threshold of 0.32. To reconcile overlapping predictions across mini-tiles, all detections are projected back to the original image coordinate space. A global Non-Maximum Suppression (NMS) step is then applied using an Intersection-over-Union (IoU) threshold of 0.40 to remove redundant or overlapping bounding boxes.

This global NMS step ensures suppression across tile boundaries and avoids multiple detections for the same defect. The inclusion of border mini-tiles in the detection step further ensures that edge defects are not excluded from evaluation. The procedure is summarized in Algorithm 4.

4 Industrial Experiment

This section describes the industrial case study used to evaluate the proposed CAM-guided tile inference strategy. The objective is to assess the method in a

Algorithm 4. *Selective Detection and Global NMS*

input: Image I, relevant mini-tiles $\mathcal{R}$
$\mathcal{B} \leftarrow \emptyset$
for each mini-tile M in $\mathcal{R}$ **do**
  extract region R from I
  $R' \leftarrow$ pad_symmetric(R)
  $R'' \leftarrow$ resize(R', 1600, 1216)
  preds $\leftarrow$ YOLOv8(R'') with conf $\tau = 0.32$
  for each detection b in preds **do**
    map b back to global coords using M
    add b to $\mathcal{B}$
  end for
end for
$\mathcal{B}^{\star} \leftarrow$ NMS($\mathcal{B}, \theta = 0.40$)
return final detections $\mathcal{B}^{\star}$

manufacturing scenario focused on surface defect detection in aluminum components. The study includes a description of the production process and parts under inspection, along with details of the dataset and experimental setup used for training and evaluation.

4.1 Case Study Description

The analyzed manufacturing process comprises two main stages: casting and machining. In the casting phase, molten aluminum is injected into molds under high pressure, forming components of various shapes and complexities. After casting, the parts undergo trimming and deburring operations, which include the use of a press-mounted trimming tool and shot blasting equipment to remove surface residues and prepare the parts for subsequent processing.

The machining phase involves operations such as face milling, drilling, and thread formation to meet the dimensional and functional specifications of the components. Once these operations are completed, a final inspection step is carried out to detect potential non-conformities. Defects such as porosities, which often become visible after machining, are of concern due to their impact on mechanical performance and reliability.

In this work, an existing dataset collected in a prior industrial inspection setup [10] was reused to evaluate the proposed approach. This dataset contains annotated high-resolution images of machined aluminum parts, including both defective and non-defective samples. The focus is placed on porosity detection, which is challenging for visual inspection systems due to its subtle appearance and irregular distribution.

Figure 2 illustrates the parts included in the study. Specifically, Figs. 2a and 2b show the front and back views of part A, while Figs. 2c and 2d correspond to the front and back views of part B. These parts were selected due to their relevance in the production line and the frequency with which surface defects have been observed in these models.

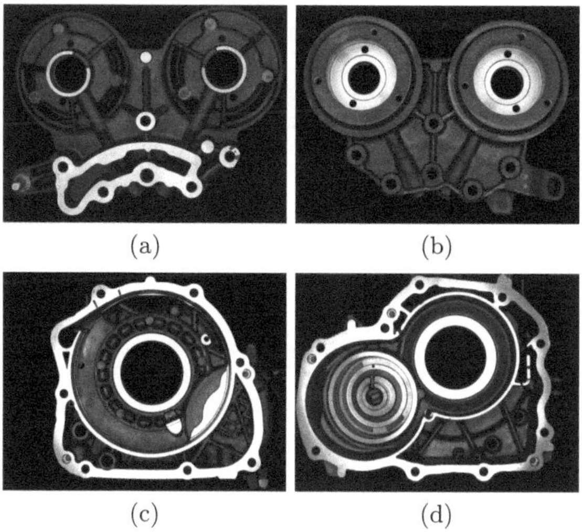

Fig. 2. Selected parts: (a) front side of part A, (b) back side of part A, (c) front side of part B, (d) back side of part B.

4.2 Dataset

The dataset used in this study consists of high-resolution images (2592×1944 pixels) of machined aluminum parts, annotated to indicate porosity defects. To create a training set for object detection, local regions were cropped around annotated defect areas, along with a subset of non-defective background regions. Only 10% of the non-defective crops were retained to improve class balance and reduce redundancy in background content. This process resulted in 352 cropped image samples.

These cropped samples were divided into three subsets for training, validation, and testing, as detailed in Table 1. The model was trained on these localized samples under uniform inference conditions.

Table 1. Split of cropped samples used to train and validate the object detection model

Subset	Images	Purpose
Training	282	Model training
Validation	35	Early stopping/tuning
Test	35	Baseline inference evaluation

For performance assessment, full-size annotated images were used in a separate evaluation role. These images were processed during testing using two inference strategies: uniform sliding window inference and relevance-based filtering guided by activation maps. The object detection model, trained solely on local image regions, was applied at test time to tile-based views of these high-resolution images. This evaluation protocol enables direct comparison of inference strategies while maintaining a separation between the model's training samples and its deployment context.

5 Results and Discussion

All experiments were conducted on the same workstation to ensure consistent computational conditions. The system comprised a 12th-generation Intel Core i7 processor (3.5 GHz), an NVIDIA RTX 3060 GPU with 6 GB of memory, and Ubuntu 20.04 running CUDA 11.8. The YOLOv8m variant was used as the base detector, initialized with the official pretrained weights. Training employed the Adam optimizer with a learning rate of 0.0025, a batch size of 4, and a maximum of 1500 epochs with early stopping applied using a patience of 50 epochs. Data augmentation included horizontal and vertical flipping, rotations ($\pm 15^\circ$ and $\pm 90^\circ$), cropping, shear transformations, brightness and saturation adjustments, and additive Gaussian noise, to improve robustness.

Table 2 summarizes the detection outcomes, true positives (TP), false positives (FP), and false negatives (FN), achieved by the baseline and the proposed relevance-guided inference method. The baseline corresponds to exhaustive sliding window inference, while the proposed method restricts object detection to tiles selected based on EigenCAM relevance maps.

Table 2. Comparison of detection counts.

Strategy	TP	FP	FN
Baseline	**281**	25	**59**
Proposed	273	**9**	66

TP: true positives; FP: false positives; FN: false negatives

The proposed strategy reduced false positives, from 25 to 9, indicating a more selective behavior in defect detection. This reduction came with an increase in false negatives, which rose from 59 to 66. In porosity detection, defects often appear subtly on reflective machined surfaces, where visual cues are faint and distributed. Fixed activation thresholds may cause low-saliency tiles, especially near image borders or tile edges, to be excluded, leading to missed detections. Despite this trade-off, the method achieves a balance between detection precision and latency.

Table 3 reports standard detection metrics, including precision, recall, and F1-score, along with average inference time per image and standard deviation.

Precision increased from 91.83 % to 96.81 %, reflecting the reduction in false positives. Recall decreased slightly from 82.65 % to 80.53 %. This drop was offset by an increase in the F1-score, from 87.00 % to 87.92 %. These results suggest that the relevance-guided selection improves allocation of compute resources while maintaining detection performance.

Table 3. Comparison of detection performance and inference time.

Strategy	Precision (%)	Recall (%)	F1-score (%)	Avg. inference time (ms)	Std. deviation (ms)
Baseline	91.83	82.65	87.00	17.57	19.00
Proposed	96.81	80.53	87.92	13.83	16.00

A timing analysis was conducted to understand the source of speedup. The baseline method involved 20 full detection passes per image, corresponding to the number of sliding tiles. The proposed method retained the same number of tiles for EigenCAM computation but applied object detection only on selected mini-tiles, which averaged 11.6 per image. As shown in Table 4, the average time spent on EigenCAM was 8.95 ms, while mini-tile scoring added 1.27 ms. Detection, now limited to fewer tiles, consumed 3.61 ms, and global NMS took 0.35 ms. The total average latency dropped to 13.83 ms from the baseline's 17.57 ms, showing that the net speedup stems from reduced detector usage, even when accounting for CAM overhead.

Table 4. Stage-wise timing breakdown.

Stage	Avg. time (ms)
EigenCAM computation (20 tiles)	8.95
Mini-tile scoring	1.27
YOLOv8 detection (avg. 11.6 tiles)	3.61
Global NMS	0.35
Total (Proposed)	**13.83**
Total (Baseline)	**17.57**

Qualitative results are presented in Fig. 3. The top row displays the full input image with its corresponding EigenCAM heatmap. The bottom row shows the selected tiles and their associated detection outputs. These examples illustrate how activation-based filtering concentrates processing on relevant areas, avoiding computation in regions unlikely to contain defects.

The detector was trained solely on cropped regions from the dataset, while inference was conducted on full-resolution images. Although both cropped samples and full images originate from the same dataset, the detector was never

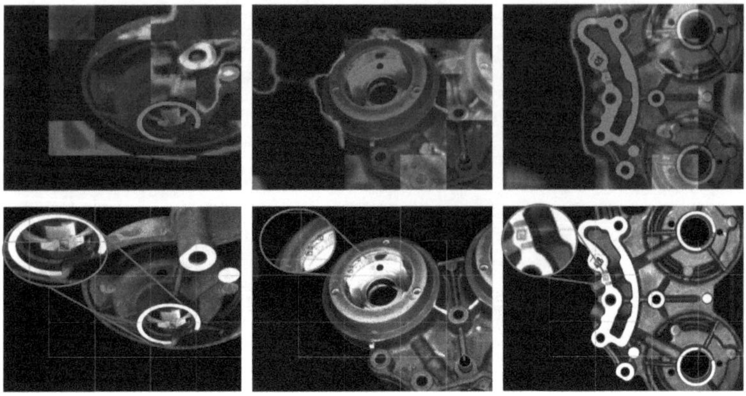

Fig. 3. Qualitative results. **Top row:** Full input image with EigenCAM heatmap overlay. **Bottom row:** Tiles selected for inference and resulting detection outputs.

exposed to global spatial arrangements during training. This separation aims to simulate deployment conditions, where the detector must generalize to unseen spatial configurations. Further validation with larger and more diverse datasets is necessary to confirm applicability.

The present study focuses on validating a specific relevance-guided workflow. No ablations were conducted on the EigenCAM target layer, mini-tile size, aggregation rule, or threshold sensitivity, all of which may impact the trade-off between precision, recall, and latency. These factors are reserved for future studies aiming to generalize the method. Although the baseline is a standard exhaustive sliding window approach, it reflects a commonly adopted solution in industrial defect detection systems where simple region proposal mechanisms are used. Future work will expand the evaluation by incorporating additional inspection scenarios and defect types to further assess the robustness of the strategy.

6 Conclusion and Future Work

Deep learning has advanced visual inspection, but high-resolution images challenge conventional architectures. Sliding window tiling addresses this by improving coverage and small-defect detection, though at the cost of redundant computation.

This work proposed a relevance-guided strategy using EigenCAM activation maps to preselect informative tiles. On an industrial dataset of porosity defects in machined aluminum, precision improved from 91.83 % to 96.81 %, recall slightly decreased from 82.65 % to 80.53 %, and F1-score rose from 87.00 % to 87.92 %. Inference time dropped by 21 % (17.57 ms to 13.83 ms) with more consistent runtimes.

The dataset reflects realistic industrial constraints, where defects are scarce and annotation costly, making the results especially relevant for practical inspection workflows. Unlike segmentation-based alternatives, the method requires no additional supervision and adapts naturally to different part geometries and lighting, making it suitable for scenarios where defects occupy limited image regions, such as AR-assisted inspection.

Future work will explore adaptive thresholds, incorporate cues like uncertainty or texture, and extend the method to multi-class detection and diverse materials. Synthetic data generation will be used to enrich training sets, and newer detection architectures (e.g., YOLO variants) will be assessed further to enhance recall and efficiency within the proposed framework.

Acknowledgments. This work is co-financed by Component 5 - Capitalization and Business Innovation, integrated in the Resilience Dimension of the Recovery and Resilience Plan within the scope of the Recovery and Resilience Mechanism (MRR) of the European Union (EU), framed in the Next Generation EU, for the period 2021–2026, within project Hi_reV, with reference 64.

Disclosure of Interests. The authors have no competing interests to declare that are relevant to the content of this article.

References

1. Ali, A.M.M., Sahlan, S., Khamis, N., Ziyi, X., Rashid, F.L.N.: Transparency in detecting defects of a printed circuit board: harnessing XAI for improved quality control in electronic manufacturing industries. In: Proceeding of the IEEE International Conference on Smart Instrumentation, Measurement and Applications, ICSIMA, vol. 2024, pp. 287–292 (2024). https://doi.org/10.1109/ICSIMA62563.2024.10675544
2. Augustauskas, R., Zabulis, L., Lipnickas, A., Jokubauskas, S.: Defects localization in images using deep learning-based classification with cam output, pp. 487–492 (2023). https://doi.org/10.1109/IDAACS58523.2023.10348813, https://www.scopus.com/inward/record.uri?eid=2-s2.0-85184811389&doi=10.1109citedby:1
3. Bhatt, P.M., et al.: Image-based surface defect detection using deep learning: a review. J. Comput. Inf. Sci. Eng. **21**(4), 040801 (2021). https://doi.org/10.1115/1.4049535
4. Clement, T., Nguyen, T.T.H., Abdelaal, M., Cao, H.: XAI-enhanced semantic segmentation models for visual quality inspection. In: Digest of Technical Papers - IEEE International Conference on Consumer Electronics (2024). https://doi.org/10.1109/ICCE59016.2024.10444225, https://www.scopus.com/inward/record.uri?eid=2-s2.0-85187011945&doi=10.1109citedby:1; All Open Access, Green Open Access
5. Gildenblat, J., contributors: Pytorch library for cam methods (2021). https://www.github.com/jacobgil/pytorch-grad-cam
6. Jude, A., Uddin, J.: Explainable software defects classification using smote and machine learning. Ann. Emerging Technol. Comput. **8**, 35–49 (2024). https://doi.org/10.33166/AETiC.2024.01.00, https://www.scopus.com/inward/record.uri?eid=2-s2.0-85181846848&doi=10.33166citedby:3
7. Kotsiopoulos, T., et al.: Revolutionizing defect recognition in hard metal industry through AI explainability, human-in-the-loop approaches and cognitive mechanisms. Expert Syst. Appl. **255** (2024). https://doi.org/10.

1016/j.eswa.2024.124839, https://www.scopus.com/inward/record.uri?eid=2-s2.0-85199873535&doi=10.1016citedby:6
8. Kumar, D., Agraharam, P.C., Liu, Y., Namilae, S.: Anomaly detection for composite manufacturing using ai models. J. Intell. Manuf. (2024). https://doi.org/10.1007/s10845-024-02522-z
9. Kumar, D., Liu, Y., Song, H., Namilae, S.: Explainable deep neural network for in-plain defect detection during additive manufacturing. Rapid Prototyping J. **30**, 49–59 (2024). https://doi.org/10.1108/RPJ-05-2023-0157
10. Nascimento, R., et al.: Automated optical system for quality inspection on reflective parts. Int. J. Adv. Manuf. Technol. **137**, 2665–2680 (2025). https://doi.org/10.1007/s00170-025-15309-0
11. Nguyen, H.T.T., Nguyen, L.P.T., Cao, H.: XEDGEAI: a human-centered industrial inspection framework with data-centric explainable edge ai approach. Inf. Fusion **116** (2025). https://doi.org/10.1016/j.inffus.2024.102782, https://www.scopus.com/inward/record.uri?eid=2-s2.0-85209376404&doi=10.1016citedby:1; All Open Access, Green Open Access, Hybrid Gold Open Access
12. Nikolakis, N., Catti, P., Alexopoulos, K.: An explainable active learning approach for enhanced defect detection in manufacturing. LNME, pp. 40 – 47 (2025). https://doi.org/10.1007/978-3-031-86489-6_5, https://www.scopus.com/inward/record.uri?eid=2-s2.0-105001366848&doi=10.1007citedby:0
13. Pereira, J., Oliveira, F., Guimarães, M., Carneiro, D., Ribeiro, M., Loureiro, G.: Addressing the limitations of lime for explainable AI in manufacturing: a case study in textile defect detection. LNME, pp. 262 – 270 (2025). https://doi.org/10.1007/978-3-031-86489-6_27, https://www.scopus.com/inward/record.uri?eid=2-s2.0-105001228971&doi=10.1007citedby:0
14. Raghavan, K., B, S., V, K.: Attention guided grad-cam: an improved explainable artificial intelligence model for infrared breast cancer detection. Multimedia Tools Appl. **83**, 57551–57578 (2024). https://doi.org/10.1007/s11042-023-17776-7
15. Soon, R.J., Chui, C.K.: Textile surface defects analysis with explainable AI. In: Proceedings - 2024 IEEE Conference on Artificial Intelligence, CAI 2024, pp. 1394 – 1398 (2024). https://doi.org/10.1109/CAI59869.2024.00248, https://www.scopus.com/inward/record.uri?eid=2-s2.0-85201229470&doi=10.1109citedby:2
16. Sun, Q., Xu, K., Liu, H., Wang, J.: Unsupervised surface defect detection of aluminum sheets with combined bright-field and dark-field illumination. Opt. Lasers Eng. **168**, 107674 (2023). https://doi.org/10.1016/j.optlaseng.2023.107674

Analysis of Frequency-Domain Signal Representations on the Performance of Deep Learning Models for Financial Trend Classification

Seyed Mojtaba Naghibzadeh[1], Majid Ahmadi[1], Mohammad Hassanzadeh[2(✉)], George P. Pappas[3], and Nabih Jaber[3]

[1] Department of Electrical and Computer Engineering, University of Windsor, Windsor, Canada
naghibz@uwindsor.ca, m.ahmadi@uwindsor.ca

[2] Department of Electrical and Computer Engineering and Department of Math and Computer Science, Lawrence Technological University, Southfield, MI, USA
mhassanza@ltu.edu

[3] Department of Electrical and Computer Engineering, Lawrence Technological University, Southfield, MI, USA
gpappas@ltu.edu, njaber@ltu.edu

Abstract. Financial time series are inherently nonstationary signals, and the choice of representation strongly influences the performance of deep learning models. This study investigates trend classification (Up, Down, Side) for gold (XAU/USD) using both time and frequency domain representations. A four-stage labeling framework is developed, combining four-hour dominance analysis, hourly midline crossings, and statistical guards based on indicators such as slope, average directional index, relative strength index, and average true range, to generate more stable and interpretable supervisory labels compared to traditional candlestick heuristics. Two deep learning architectures, a convolutional neural network and a recurrent neural network with long short-term memory units, are trained and tested on both time domain features and frequency domain features extracted through fast Fourier transform and continuous wavelet transform. The models are evaluated using chronologically ordered training and testing splits to ensure temporal consistency. Experimental results show that the convolutional neural network trained on time domain data achieves the best generalization with an accuracy of approximately 90%, while the recurrent model and frequency-based representations achieve comparable performance around 87%. These findings highlight that systematic label design and representation choice play crucial roles in enhancing model robustness, and that time domain representations are particularly effective for financial trend prediction.

H. R. Arabnia et al. (Eds.): CSCI 2025, CCIS 2968, pp. 59–71, 2027.
https://doi.org/10.1007/978-3-032-28457-0_5

Keywords: Financial Time Series · Signal Processing · Time-Domain and Frequency-Domain Analysis · Convolutional Neural Network (CNN) · Recurrent Neural Networks (RNN) · Long Short-Term Memory (LSTM) · Continuous Wavelet Transform (CWT) · Fast Fourier Transform (FFT) · Trend Classification

1 Introduction

Predicting asset price trends remains a core challenge in financial markets, particularly for volatile instruments such as gold (XAU/USD). Analysts typically rely on either fundamental analysis based on macroeconomic indicators and policy actions [1,2], or technical analysis, which interprets price patterns and market sentiment. Candlestick charts are a cornerstone of technical analysis, consolidating OHLC data into intuitive visual structures and frequently applied in short-term trading [3–6]. However, traditional candlestick patterns often suffer from subjectivity and are usually limited to single time-frame views, which reduces their robustness in volatile conditions. With the rise of machine learning (ML), financial time series can be reframed as structured signals for supervised classification tasks. Deep learning has demonstrated strong capabilities in handling sequential data and capturing temporal dependencies [7,8]. Yet, an open research question remains: should trend prediction rely more on raw time-domain sequences or on frequency-domain transformations? This question connects financial forecasting with broader signal processing research, where the choice of representation is known to affect model generalization. In this work, we propose a hybrid signal-processing-driven framework for trend classification. Our method begins with a four-stage labeling pipeline inspired by domain-specific heuristics: (i) four-hour directional dominance, (ii) one-hour midline crossings, (iii) statistical guards including slope t, ADX, RSI, and ATR, and (iv) a mid-day consistency rule. This process yields more stable and interpretable class boundaries, especially for the ambiguous "Side" category. We then compare two representation strategies: (i) direct time-domain sequences of normalized daily prices, and (ii) frequency-domain embeddings derived from Fast Fourier Transform (FFT) and complex Morlet wavelet transform (CWT). Both representations are modeled using deep learning architectures, including a Convolutional Neural Network (CNN) and a Recurrent Neural Network (RNN) implemented with Long Short-Term Memory (LSTM) units, under time-ordered cross-validation and chronological holdout testing. Our experiments on more than 4,000 daily sequences demonstrate that time-domain inputs consistently outperform frequency-domain alternatives, achieving up to 90% test accuracy. These results highlight that, while signal processing enriches label design and interpretability, frequency-domain embeddings do not necessarily yield superior performance over well-structured time-domain signals for deep-learning-based trend classification.

2 Related Work

2.1 Financial Time Series Forecasting

A wide range of machine learning and deep learning techniques have been applied to financial time series forecasting across various markets, including gold, foreign exchange, stocks, and cryptocurrencies. Sentiko et al. [9] used gold price data from 2010 to 2024 (daily open, high, low, and close values) to compare gated recurrent unit, long short-term memory, and linear regression models, reporting the gated recurrent unit as the most accurate with an accuracy of 96%. Smai and Junejo [10] applied artificial neural networks and linear regression to forecast gold prices using 22 macro-financial features; the artificial neural network achieved better results with a root mean squared error of 19 on gold price data from 2005 to 2016 (daily open, high, low, and close values). Al Dhuraibi and Ali [11] compared regression, support vector machines, decision trees, and k-nearest neighbors for gold direction prediction using weekly open, high, low, and close data from 2007 to 2017; the k-nearest neighbors model achieved the highest accuracy of 60.26% at k=5. Gurrib and Kamalov [12] applied sentiment analysis and latent Dirichlet allocation to predict Bitcoin direction using daily open, high, low, and close data from 2016 to 2021, achieving 58.5% accuracy. Zlicar et al. [13] used a Parzen windowing algorithm to classify major foreign exchange trends using hourly data from 2005 to 2013, achieving 53.25% accuracy and highlighting the importance of algorithm selection. Fisichella and Garolla [14] used candlestick chart images from 2010 to 2021 (four-hour intervals across six foreign exchange pairs) to classify trading signals (Buy, Sell, Stay) using Vision Transformer and ResNet50 architectures. The model was trained on data from 2010 to 2018 and tested on 2019 to 2021, where ResNet50 achieved 60.7% accuracy, outperforming Vision Transformer at 50.4%. Argotty Erazo et al. [15] proposed a linear classifier for short-term EUR/USD trend prediction using data from 2004 to 2021. Their model achieved 98.77% accuracy by applying discriminant analysis on key statistical features such as average return and price slope. Oetama et al. [16] classified candle directions (uptrend and downtrend) in EUR/USD and GBP/USD using data from 2021 to 2022. They applied support vector machines with linear and radial basis function kernels, and after hyperparameter tuning, the linear model achieved 98.11% accuracy on GBP/USD and 94.51% on EUR/USD. Alagdeve et al. [17] proposed a neural-network-based approach using candlestick chart patterns to forecast short-term stock trends, achieving over an 85% profit-to-loss ratio using only ten major patterns for up to four-day predictions. He and Xu [18] proposed CLNet, a perturbation-enhanced long short-term memory model evaluated on ACL18 (2014 to 2016) and KDD17 (2007 to 2016), achieving up to a 30.78% improvement over baseline trend prediction methods. Ruke et al. [19] applied long short-term memory networks for stock trend prediction using daily data from 2010 to 2024, achieving an R^2 score of 0.89 and demonstrating the effectiveness of deep learning in financial forecasting.

2.2 Signal Processing and Deep Learning

Beyond financial applications, several works highlight the importance of signal processing transformations combined with deep learning. Alboghbaish et al. [8] compared discrete wavelet transform and fast Fourier transform as feature extractors for electrocardiogram classification, showing that combining transformations with convolutional neural network and recurrent neural network models improved performance. Their experiments on the MIT-BIH and BIDMC databases achieved up to 99.2% accuracy with the fast Fourier transform plus convolutional neural network and long short-term memory approach, confirming the synergy of transform-based preprocessing and deep learning. Similarly, Yang [20] proposed a multi-level wavelet transform network for automatic modulation classification, achieving 94.4% accuracy under low signal-to-noise ratio conditions and outperforming other baseline models. Srivastava et al. [21] applied a deep learning framework to brain tumor classification using magnetic resonance imaging data, integrating handcrafted features such as local binary patterns and discrete wavelet transform with learned features. They showed that discrete wavelet transform-based coefficients boosted accuracy up to 94.11%, outperforming raw image inputs. These studies demonstrate that combining transformation-based feature extraction, such as wavelet or fast Fourier transform representations, with deep learning models including convolutional and recurrent neural networks consistently improves classification across domains including biomedical signals, wireless communications, and medical imaging. In this work, we extend this idea to the financial domain by applying similar transformation-enhanced deep learning techniques to candlestick chart classification for gold price trend prediction.

3 Methodology

3.1 Dataset Preparation

We utilized historical XAU/USD (gold–US dollar) price data spanning from June 11, 2004, to February 3, 2025, obtained from a public Kaggle repository [22]. The dataset includes timestamped open, high, low, and close (OHLC) values at both hourly and four-hour intervals. To ensure temporal alignment, the two granularities were merged by timestamps, while non-trading sessions (e.g., weekends, holidays, and disrupted hours) were removed. Early years with sparse coverage were excluded to maintain a stable sampling frequency.

All numerical features were normalized per day relative to the opening price to remove magnitude bias and stabilize gradients during training. In addition, each feature dimension was standardized to zero mean and unit variance using a leakage-free normalization scheme: the scaler was fit solely on the training folds during cross-validation and subsequently applied to validation and test partitions. This prevented information bleed across temporal splits, ensuring a fair chronological evaluation. The resulting dataset comprised approximately 4,334 daily samples, of which the final 15% were held out for testing, while the remaining 85% were used for time-series cross-validation.

3.2 Multi-time-Frame Labeling (Stages 1–4)

To construct reliable supervisory labels, we developed a hierarchical four-stage labeling framework that combines four hour dominance analysis with hourly refinements [23], improving robustness against noise, low momentum drift, and intraday volatility.

Stage 1: Four Hour Dominance

Each trading day is divided into six consecutive four hour candles. For each candle i, the body size $b_i = |c_i - o_i|$ is normalized by the fourteen period average true range (ATR_{14}). Candles with $b_i/\text{ATR} < \epsilon$ are excluded as doji. The directional dominance metric is then computed as

$$D = \frac{\sum_i \min(b_i, \tau)\mathbf{1}_{\{c_i > o_i\}} - \sum_i \min(b_i, \tau)\mathbf{1}_{\{c_i < o_i\}}}{\sum_i \min(b_i, \tau)}. \tag{1}$$

If at least four bullish candles exist and $D \geq \Delta$, the day is labeled Up; if four bearish candles exist and $D \leq -\Delta$, it is labeled Down; otherwise, it is labeled Side. This approach quantitatively translates subjective candlestick heuristics into a consistent ATR normalized framework [24].

Stage 2: One Hour Midline Crossings

For each day, the mid price level

$$m = \frac{\max(h_t) + \min(l_t)}{2}, \quad t = 1, \ldots, 24, \tag{2}$$

is calculated, where h_t and l_t denote the hourly highs and lows. The number of hourly candles intersecting this midline is given by

$$C = \sum_{t=1}^{24} \mathbf{1}_{\{l_t \leq m \leq h_t\}}. \tag{3}$$

If $C \geq C_{\min}$, the day is re-labeled as Side, as frequent crossings indicate neutral momentum. Otherwise, the Stage 1 label is preserved, except for undecided cases resolved by the daily return sign. This midline driven adjustment aligns with midpoint decomposition techniques widely used in nonstationary trend estimation [25].

Stage 3: Voting with Guards

A fusion based voting system aggregates several hourly indicators, including slope-t statistics (trend persistence), mean ADX (trend strength), ATR ratio (relative volatility), RSI band coverage (momentum neutrality), and oscillation counts. Each metric contributes one neutrality vote. If the total votes $V \geq V_{\min}$ and no guard condition (strong four hour imbalance, excessive return, or insufficient crossings) is triggered, the day is reclassified as Side. This voting system integrates both statistical and technical measures, resembling ensemble based confirmation in algorithmic trading [26].

Stage 4: Midday Rule

Finally, an intraday consistency check compares the daily open and close relative to the midpoint

$$\text{mid\_day} = \frac{\text{High} + \text{Low}}{2}. \tag{4}$$

If both the open and close lie above (or below) this midpoint, the label is adjusted to Up or Down, respectively. This step corrects for intraday asymmetries driven by overlapping trading sessions [27]. Together, these four stages yield interpretable and balanced labels (Up, Down, Side) that mitigate ambiguity and reduce label noise.

3.3 Feature Extraction

We extract features in three complementary domains time, frequency, and wavelet to ensure diverse representation while maintaining compactness.

Time-Domain:

Normalized OHLC sequences relative to the daily open are enriched with technical descriptors such as Bollinger width, MACD components, ATR_{14}, RSI_{14}, and moving averages (SMA_5, EMA_{10}). Additionally, rolling slopes, candle asymmetries, wick ratios, and normalized returns are computed. Missing hours (due to early session close) are zero-padded to length 24, accompanied by a masking vector. This setup enables consistent tensor input shapes for all samples while retaining true sequence structure.

Frequency-Domain (FFT):

Each 24-hour close-price sequence undergoes discrete Fourier transform:

$$X(k) = \sum_{n=0}^{N-1} x(n) e^{-i2\pi kn/N}.$$

From its power spectrum $P(k) = |X(k)|^2$, global descriptors such as spectral centroid, bandwidth, and flatness are extracted, reflecting periodicity and volatility frequency concentration. These Fourier-based features have proven effective for identifying cyclical energy distributions in financial and physiological signals [28,29].

Wavelet-Domain (CWT):

To capture non-stationary and multi-resolution structures, the continuous wavelet transform with a complex Morlet kernel is applied:

$$W(a,b) = \frac{1}{\sqrt{a}} \int_{-\infty}^{\infty} x(t) \psi^*\Big(\frac{t-b}{a}\Big) dt,$$

with scales $a \in \{1,2,3,4,6,8\}$. For each scale, we compute the average power $P(a) = \frac{1}{T}\sum_{n=1}^{T} |W(a,n)|^2$ and the mean of the real and imaginary components. Instead of using full spectrograms, we utilize structured CWT statistics to provide compact yet discriminative descriptors suitable for deep classifiers [30,31].

3.4 Learning Models

Three deep architectures RNN, LSTM, and CNN were implemented to capture distinct temporal dependencies within the engineered sequences. All models receive input tensors of shape $(T \times F)$, where $T = 24$ denotes the number of hourly intervals in each day, and $F = 21$ represents the extracted features. Each model was optimized and trained using the same pipeline to ensure fair comparison.

RNN (Simple Recurrent Neural Network):

Recurrent neural networks propagate sequential dependencies through recursive state updates of the form

$$h_t = \sigma(W_h h_{t-1} + W_x x_t + b),$$

where h_t represents the hidden state at time t and σ is a nonlinear activation function. The RNN provides a lightweight baseline for short-range dependency modeling, capturing day-level transitions such as gradual momentum build-up or hourly reversals. However, as expected in financial sequences with long autocorrelations, the network suffers from vanishing gradients. To mitigate this, we employed ReLU activation (instead of tanh), L_2 weight decay (10^{-4}), and gradient clipping (norm = 1.0) for stable optimization. A single recurrent layer with 64 hidden units was followed by a dropout layer ($p = 0.3$) and a fully connected layer (48 ReLU units) before the final softmax output.

LSTM (Long Short-Term Memory Network):

LSTM networks enhance standard RNNs by incorporating gated memory cells that control information flow via input, forget, and output gates:

$$\begin{aligned}
f_t &= \sigma(W_f[h_{t-1}, x_t] + b_f),\\
i_t &= \sigma(W_i[h_{t-1}, x_t] + b_i),\\
\tilde{c}_t &= \tanh(W_c[h_{t-1}, x_t] + b_c),\\
c_t &= f_t \odot c_{t-1} + i_t \odot \tilde{c}_t,\\
h_t &= o_t \odot \tanh(c_t).
\end{aligned}$$

This gating mechanism allows the model to retain or forget information dynamically, enabling long-term temporal modeling—crucial for financial data where intra-day momentum may depend on prior volatility phases. Our implementation uses a single LSTM layer with 64 hidden units, dropout $p = 0.3$, and recurrent dropout $p = 0.2$ to regularize temporal connections. The LSTM outputs are flattened through global average pooling and passed to a dense layer with 48 ReLU neurons before the final softmax layer. The model achieved stable convergence under the Adam optimizer with an initial learning rate of 10^{-3} and learning rate reduction upon plateau detection (factor 0.5, minimum 10^{-5}). LSTMs consistently outperformed RNNs, confirming their advantage in preserving trend memory and mitigating noisy short-term reversals [32].

CNN (Temporal Convolutional Network):

The CNN architecture was designed to capture hierarchical temporal motifs and localized market shifts. The input tensors (24×21) were convolved using three successive 1D convolutional layers with kernel sizes 3, 5, and 1, corresponding respectively to short-term (intra-hour), mid-term (multi-hour), and cross-feature dependencies. Each convolution layer employed 64 filters, ReLU activation, and L_2 regularization (10^{-4}). Unlike recurrent networks, CNNs allow for parallel computation and exploit translational invariance, which is particularly useful for pattern repetition in price movements. A global average pooling (GAP) layer condensed the time dimension, automatically handling sequences padded to 24 h. This was followed by a dropout layer ($p = 0.3$) and a dense layer of 48 ReLU units. The final layer was a softmax classifier producing probabilities over the three target classes: Up, Down, and Sideways. Empirically, CNNs demonstrated the best generalization on both validation and holdout test sets, reaching nearly **90.15% accuracy** and macro-F1 of 0.8977. The architecture's success is attributed to its ability to simultaneously capture low-level directional spikes and medium-term periodicity through hierarchical receptive fields. In contrast, FFT- and CWT-based CNN variants achieved competitive but lower accuracy (≈87%), indicating that raw time-domain signals preserve essential temporal structures better than transformed representations.

Training, Normalization, and Cross-Validation:
All models were trained under identical conditions using the Adam optimizer and sparse categorical cross-entropy loss. Early stopping (patience = 8) and learning-rate scheduling (factor = 0.5, minimum 10^{-5}) were applied for convergence stability. Feature normalization was conducted through a leakage-free scaling scheme, the StandardScaler was fit only on training subsets and applied to validation/test data, ensuring chronological isolation. Each fold in the four-way time-series cross-validation maintained an expanding-window strategy, where the training horizon increased progressively to simulate forward-in-time learning. To address class imbalance (n_{Up}=1560, n_{Down}=1474, n_{Side}=1300), random undersampling was applied exclusively to the training set per fold, avoiding any distortion of validation or test distributions. This rigorous pipeline ensures fair, temporally consistent evaluation,prevents look-ahead bias,stabilizes convergence,and yields reproducible benchmark performance across all architectures.

Training and Validation:
All models are trained using the Adam optimizer with sparse categorical cross-entropy loss. Early stopping (patience = 8) and adaptive learning-rate reduction (factor = 0.5, minimum = 10^{-5}) ensure stable convergence. Each model trains for up to 50 epochs with batch size 64. Temporal cross-validation is employed, dividing the first 85% of data into four expanding folds, and the final 15% is reserved for chronological testing. To handle class imbalance, undersampling is applied only to the training subset of each fold preventing information leakage into validation or test partitions. This ensures realistic generalization and robust performance assessment under time-aware evaluation.

4 Experimental Results and Discussion

All experiments were implemented in TensorFlow using the Adam optimization algorithm with a batch size of sixty-four. Early stopping with a patience of eight epochs and adaptive learning-rate scheduling with a minimum learning rate of 10^{-5} were employed to ensure convergence stability. To avoid data leakage, normalization and undersampling procedures were fitted exclusively on the training partitions. A four-fold chronological cross-validation scheme was applied to the earliest eighty-five percent of the dataset, while the most recent fifteen percent was reserved for final evaluation. Table 1 summarizes the average cross-validation results. Spectral pipelines based on Fourier and wavelet features, combined with recurrent or convolutional backbones, consistently achieved accuracies between 0.86 and 0.88, confirming the discriminative capacity of frequency- and time-frequency-domain representations. The time-domain convolutional network, enriched with normalized OHLC values and technical indicators, demonstrated the strongest generalization with a mean accuracy approaching 0.89. On the unseen holdout set (Table 2), spectral models maintained high stability with accuracies in the range of 0.87 to 0.88, while the time-domain CNN achieved an accuracy of 90.15%. This represents a notable improvement over spectral-only variants and highlights the complementary effect of combining handcrafted technical indicators with learned temporal convolutions. The confusion matrix of the final CNN (Fig. 1) indicates the highest precision and recall for the Up and Down trends, while the Side class remains more challenging due to its ambiguous nature and frequent overlap with boundary conditions. Overall, the results demonstrate that the proposed hierarchical four-stage labeling framework, combined with both spectral and technical features, yields robust gold trend prediction with strong generalization to unseen time periods.

Table 1. Cross-validation performance (four-fold mean).

Model	Accuracy	Macro-F1
FFT-RNN	0.8740	0.8698
FFT-LSTM	0.8697	0.8658
CWT-CNN	0.8664	0.8619
Time-CNN	**0.8930**	**0.8908**

Table 2. Holdout test performance (latest fifteen percent of days).

Model	Accuracy
FFT-RNN	0.8769
CWT-CNN	0.8754
Time-CNN	**0.9015**

5 Limitations

Despite the strong performance observed, several limitations remain. First, the multi-stage labeling framework relies on empirically selected thresholds (for example, ATR filters and crossing counts), which may require recalibration for different market regimes or asset classes. Second, the experiments are limited to XAU over USD, and broader validation across equities, cryptocurrencies, and bond markets is necessary to assess generalization. Third, the use of fixed daily segmentation of twenty-four hourly bars enhances comparability but may overlook sub-hourly volatility or multi-day structural dependencies. Fourth, the feature set includes only price-based and technical indicators, omitting exogenous drivers such as macroeconomic events, sentiment, or order-book dynamics. Finally, FFT and CWT introduce additional computational overhead, and the efficient fusion of time- and frequency-domain representations remains an open challenge for real-time large-scale deployment.

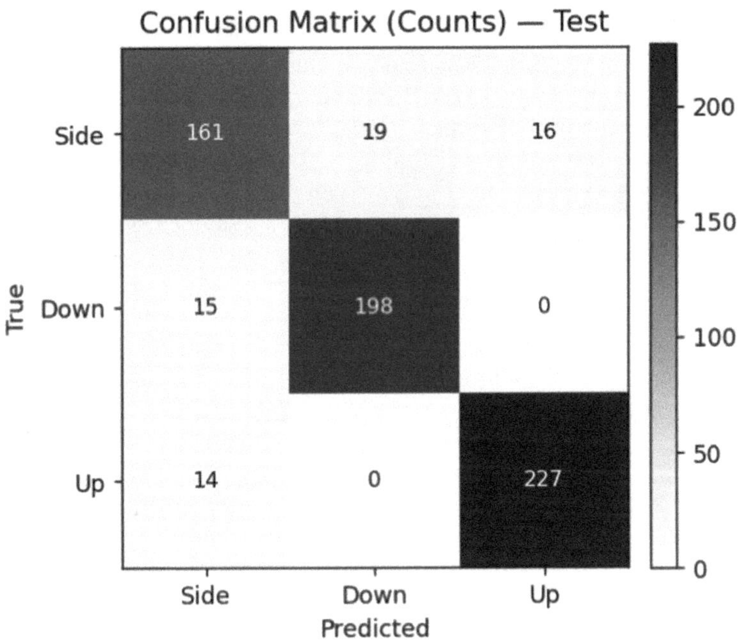

Fig. 1. Confusion matrix of the final time-domain CNN on the holdout test set.

6 Conclusion

We introduced a hierarchical four-stage labeling pipeline combined with deep learning models for gold (XAU/USD) trend prediction. The labeling framework integrates dominance filters, midline crossings, indicator voting, and midday rules to generate reproducible and less noisy supervisory signals. Experiments across time-domain, FFT, and CWT features showed that although spectral approaches were competitive (86 to 88% accuracy), the time-domain CNN achieved the highest generalization accuracy of 90.15%. This emphasizes the importance of short-horizon temporal patterns while indicating that hybrid temporal-spectral strategies remain promising. Overall, systematic labeling combined with enriched time-domain features provides robust forecasting performance. Extending this framework to multi-asset markets is a natural direction for broader validation. Future research may also incorporate exogenous drivers such as macroeconomic events or order-book dynamics, and investigate optimized fusion strategies to balance accuracy and computational efficiency in practical trading systems.

References

1. Eiamkanitchat, N., Moontuy, T., Ramingwong, S.: Fundamental analysis and technical analysis integrated system for stock filtration. Clust. Comput. **20**(1), 883–894 (2017)
2. Edwards, R.D., Magee, J., Bassetti, W.H.C.: Technical Analysis of Stock Trends, 11th edn. CRC Press, Boca Raton (2018)
3. Fama, E.F.: Efficient capital markets: a review of theory and empirical work. In: Presented at the Fama Portfolio: Selected Papers of Eugene F. Fama (2017)
4. Woo, K.Y., Mai, C., McAleer, M., Wong, W.K.: Review on efficiency and anomalies in stock markets. Economies **8**(1), 20 (2020)
5. Goel, H., Shyamala, L.: Enhancing investment decision-making: a comprehensive approach through fundamental, technical and sentiment analysis with machine learning and deep learning algorithms. In: 10th IEEE International Conference on Communication and Signal Processing (ICCSP), pp. 1655–1660. IEEE, Melmaruvathur (2024)
6. Nti, I.K., Adekoya, A.F., Weyori, B.A.: A systematic review of fundamental and technical analysis of stock market predictions. Artif. Intell. Rev. **53**, 3007–3057 (2020)
7. Chung, F.T., Kit, H.Y., FuLai, C.: Time series pattern classification based on deep learning network: a study on the roles of time domain representation. In: 14th IEEE International Conference on Natural Computation. Fuzzy Systems and Knowledge Discovery (ICNC-FSKD), pp. 905–910. IEEE, Huangshan (2018)
8. Alboghbaish, E., Eleyan, A., Eleyan, G.: Performance comparison between transform-based deep learning approaches for ECG signal classification. In: 11th IEEE International Conference on Electrical and Electronics Engineering (ICEEE), pp. 439–443. IEEE, Marmaris (2024)
9. Sentiko, S.L., Zakiyyah, A.Y.: MEILIANA: gold price prediction using machine learning and deep learning. In: 6th IEEE International Conference on Cybernetics and Intelligent Systems (ICORIS), pp. 1–5. IEEE, Surakarta (2024)
10. Sami, I., Junejo, K.N.: Predicting future gold rates using machine learning approach. Int. J. Adv. Comput. Sci. Appl. **8**(12), 275–280 (2017)
11. Al-Dhuraibi, W.A., Ali, J.: Using classification techniques to predict gold price movement. In: 4th IEEE International Conference on Computer and Technology Applications (ICCTA), pp. 127–130. IEEE, Istanbul (2018)

12. Gurrib, I., Kamalov, F.: Predicting Bitcoin price movements using sentiment analysis: a machine learning approach. Stud. Econ. Financ. **39**(3), 347–364 (2022)
13. Zličar, B., Cousins, S.: Discrete representation strategies for foreign exchange prediction. J. Intell. Inf. Syst. **50**, 129–164 (2018)
14. Fisichella, M., Garolla, F.: Can deep learning improve technical analysis of forex data to predict future price movements? IEEE Access **9**, 153083–153101 (2021)
15. Argotty-Erazo, M., Blázquez-Zaballos, A., Argoty-Eraso, C.A., Lorente-Leyva, L.L., Sánchez-Pozo, N.N., Peluffo-Ordóñez, D.H.: A novel linear-model-based methodology for predicting the directional movement of the Euro-Dollar exchange rate. IEEE Access **11**, 67249–67284 (2023)
16. Oetama, R.S., Heryadi, Y., Lukas, Suparta, W.: Improving candle direction classification in forex market using support vector machine with hyperparameters tuning. In: 7th IEEE International Conference on Informatics and Computing (ICIC), pp. 1–6. IEEE, Bali (2022)
17. Alagdeve, V., Karule, P., Palsodkar, P.: Stock prediction using dual analysis of candlestick chart pattern. In: 2nd IEEE International Conference on Emerging Trends in Engineering and Medical Sciences (ICETEMS), pp. 497–502. IEEE, Nagpur (2024)
18. He, L., Xu, W.: Confrontation-LSTM network for stock trend prediction. In: China Automation Congress (CAC), pp. 3017–3021. IEEE, Chongqing (2023)
19. Ruke, A., Gaikwad, S., Yadav, G., Buchade, A., Nimbarkar, S., Sonawane, A.: Predictive analysis of stock market trends: a machine learning approach. In: 4th IEEE International Conference on Data Engineering and Communication Systems (ICDECS), pp. 1–6. IEEE, Bangalore (2024)
20. Yang, W.: Multi-level wavelet transform network based automatic modulation classification. In: 6th IEEE International Seminar on Artificial Intelligence. Networking and Information Technology (AINIT), pp. 1594–1597. IEEE, Shenzhen (2025)
21. Srivastava, A., Khare, A., Kushwaha, A.: Brain tumor classification using deep learning framework. In: IEEE International Conference on Intelligent Systems. Advanced Computing and Communication (ISACC), pp. 1–4. IEEE, Silchar (2023)
22. Anugrah, N.: XAU/USD gold price historical data (2004–2024). Kaggle (2024). https://www.kaggle.com/datasets/novandraanugrah/xauusd-gold-price-historical-data-2004-2024
23. Naghibzadeh, S.M., Hassanzadeh, M., Ahmadi, M., Pappas, G.P.: Gold price trend prediction from candlestick chart images using multi-time frame analysis and machine learning. In: 2025 IEEE International Conference on Future Machine Learning and Data Science (FMLDS), Los Angeles (2025)
24. Icaksono, S.R., Soetam, R., Nugroho, A.T., Wicaksono, H.F.: Candlestick pattern research analysis, future and beyond: a systematic literature review using PRISMA. J. Comput. Sci. Technol. Stud. **4**(2), 157–164 (2022)
25. He, Q., Gao, R.X., Freedson, P.: Midpoint-based empirical decomposition for nonlinear trend estimation. In: IEEE Engineering in Medicine and Biology Society (EMBS), pp. 2228–2231. IEEE, Minneapolis (2009)
26. Salkar, T., Shinde, A., Tamhankar, N., Bhagat, N.: Algorithmic trading using technical indicators. In: IEEE International Conference on Communication Information and Computing Technology (ICCICT), pp. 2228–2231. IEEE, Mumbai (2021)
27. Kolokolov, M.N., Volkov, E.A., Zhdanov, P.A.: Intraday volatility and trading volume in the RUB/USD FX market: patterns, drivers, and implications. Fin. Markets. Portfolio Mgmt. **37**(2), 161–187 (2023)

28. Shashidhar, S.R., Aditya, A.V., Srihari, S.N., Subhash, S.M.H., Krishnappa, K.H.K.: Empowering investors: insights from sentiment analysis, FFT, and regression in Indian stock markets. In: IEEE International Conference on Ambient Intelligence. Knowledge Informatics and Industrial Electronics (AIKIIE), pp. 1–7. IEEE, Mysuru (2023)
29. Song, D., Baek, A.M.C., Kim, N.: Forecasting stock market indices using padding-based Fourier transform denoising and time series deep learning models. IEEE Access **9**, 83786–83797 (2021)
30. Zhao, Y., Khushi, M.: Wavelet denoised-ResNet CNN and LightGBM method to predict forex rate of change. In: IEEE International Conference on Data Mining Workshops (ICDMW), pp. 385–392. IEEE (2020)
31. Ghazali, K.H., Mansor, M.F., Mustafa, M.M., Hussain, A.: Feature extraction technique using discrete wavelet transform for image classification. In: 5th IEEE Student Conference on Research and Development (SCOReD), pp. 1–6. IEEE, Selangor (2007)
32. Pramono, S.H., Budi, D.M., Wardoyo, R.: Analysis of stock market value prediction using simple novel long short-term memory algorithm in comparison with back propagation algorithm for increased accuracy rate. In: IEEE International Conference on Artificial Intelligence and Computer Science Technology (ICAICST), pp. 181–186. IEEE, Jakarta (2021)

AI-Assisted Mobile Nutrition Analysis for Daily Monitoring via Image Processing of Guideline Daily Amounts (GDAs) Labelling

Nuntaporn Aukkanit(✉), Shutchapol Chopvitayakun, Supatchalee Sirichokworrakit, and Kunyanuth Kularbphettong

Faculty of Science and Technology, Suan Sunandha Rajabhat University, Bangkok, Thailand
nuntaporn.au@ssru.ac.th

Abstract. This work presents an AI-based mobile system that reads Guideline Daily Amounts (GDA) nutrition labels from photographs, computes intake against Thai Recommended Daily Intakes (Thai RDI), and visualizes results for daily dietary monitoring. The mobile application, developed with Kotlin using CameraX, captures a label image, encodes it as Base64, and sends it to a Node.js/Express.js backend. The server invokes OpenAI to parse key fields (serving size, energy, fat, sugar, sodium), returns normalized JSON, and a computation module derives nutrient intake and percentage of RDI. Outputs are displayed through bar and line charts within the app. Functional testing across multiple packaged food categories showed high extraction accuracy when characters were clearly legible (exceeding 90%), with residual errors mainly due to lighting, glare, perspective distortion, and stylized fonts. An illustrative example demonstrates per-package values and automatic scaling to the user's actual consumption, enabling immediate interpretation of over- or under-consumption. Compared with manual-entry approaches, the proposed pipeline reduces user burden, lowers input errors, and maintains local relevance through Thai RDI standards. The architecture separates processing from presentation and supports real-time AI integration, providing a practical foundation for future enhancements such as on-device OCR, proactive alerts, IoT integration, and personalization under robust data-protection practices.

Keywords: Android · CameraX · Guideline Daily Amounts · Nutrition labelling · OpenAI

1 Introduction

Nowadays, major causes of morbidity and mortality in Thailand and worldwide include cardiovascular disease, diabetes, hypertension, and obesity, which are non-communicable diseases (NCDs). Inappropriate dietary behaviors are a key risk factor [1]. Providing accurate and easily accessible nutrition information is therefore essential for promoting population health. Thailand requires nutrition labels that disclose key nutrients, namely energy, fat, protein, carbohydrate, sugar, and sodium, and it specifies

H. R. Arabnia et al. (Eds.): CSCI 2025, CCIS 2968, pp. 72–85, 2027.
https://doi.org/10.1007/978-3-032-28457-0_6

Thai Recommended Daily Intakes (Thai RDI) for the general population above the age of three years old [2]. In practice, however, the usefulness of labels is limited because most consumers do not calculate their actual daily intake.

In 2018, the Thai Ministry of Public Health mandated front-of-pack Guideline Daily Amounts (GDA) for energy, sugar, fat, and sodium on selected product categories, including snacks, chocolate and chocolate-flavored products, bakery items, semi-processed foods, chilled and frozen ready-to-eat meals, beverages, ready-to-drink tea and coffee in liquid or dry form, flavored milk, fermented milk, milk products, soybean beverages, and ice cream These products frequently contain high amounts of energy, sugar, fat, and sodium, which contribute to over-nutrition and NCDs. GDA labelling presents per-package amounts of energy, sugar, fat, and sodium prominently on the front of the package to improve visibility and to support informed purchasing and consumption decisions. The upper section displays the nutrition value per package, for example cup, sachet, box, or bag, and the number of consumption occasions or servings per package. Below this section, a set of vertically aligned cylindrical figures shows, in sequence, the per-package values for energy, sugar, fat, and sodium. The unit for energy is kilocalories, the units for sugar and fat are grams (g), and the unit for sodium is milligrams (mg). Percentages are calculated relative to the maximum daily intake based on Thai RDI as follows: energy 2,000 kcal, sugar 65 g, fat 65 g, and sodium 2,000 mg [3].

Prior study proposed a mobile application that required users to manually enter GDA values from labels, after which the system calculated nutrient intake and compared it with Thai RDI [4]. While this approach supported self-monitoring, it depended on user input and was limited in convenience and data accuracy. The present study proposes an automated dietary tracking system that captures GDA nutrition labels on Android, implemented in Kotlin. Images are transmitted to an Express.js server connected to the OpenAI API to automatically parse key nutrient information. The resulting data are computed and compared against Thai RDI, and the outputs are presented through charts and graphs to assist users in selecting foods and maintaining nutritional balance more easily.

The objective of this study is to extend prior manual-entry approaches by developing a mobile AI pipeline that captures photographs of front-of-pack nutrition labels, securely transmits the images to a researcher-managed API that invokes OpenAI models, automatically parses and normalizes key fields (serving size, energy, fat, sugar, sodium), computes user-specific intake and the percentage of Thai Recommended Daily Intakes (Thai RDI), and returns clear visual feedback within the Android application, thereby improving the correctness and accuracy of label interpretation and nutrient calculation, reducing user burden, and delivering a measurably better in-app user experience compared with manual logging.

2 Literature Review

2.1 Self-monitoring of Diet

Dietary self-monitoring is widely recognized as a core strategy for modifying eating behaviors and managing weight. Recording daily energy and nutrient intake increases awareness of eating patterns and supports more appropriate consumption [5]. Smartphone-based food logging has been shown to increase the frequency and continuity of self-monitoring, even when overall diet quality does not change markedly [6]. Taken together, this evidence suggests that mobile technology can motivate sustained engagement in dietary control.

2.2 Mobile Nutrition Applications

A variety of mobile nutrition applications have been developed internationally. In Spain, DietApp was designed to deliver personalized nutritional advice based on users' age, health history, and physical status, using categorized tips and graphical representations of healthy eating guidelines to encourage appropriate intake [7]. In India, Health Buddy on Android offers nutrient logging, exercise recommendations, a body mass index (BMI) calculator, and a waist-to-hip ratio (WHR) calculator; a pilot evaluation with university students indicated that the app was useful and increased awareness of healthy eating practices [8]. In Italy, TreC-LifeStyle targets overweight children and their families using a multidisciplinary, user-centered design that involves nutritionists, pediatricians, and behavior change experts. The app provides virtual coaching and a consumption-tracking dashboard to support nutrition education and enhance user engagement [9].

2.3 Behavior-change–oriented Nutrition Apps

Behaviorally informed design has gained traction in nutrition apps. Recent reviews indicate that daily dietary self-monitoring via mobile applications can ease the burden of traditional paper food diaries, facilitate more consistent recording, and help raise awareness of eating behaviors [10].

Mechanism-oriented design principles from digital learning suggest that real-time feedback, timely reminders, and guidance aligned with actual user activity can improve engagement and persistence with app-based tasks. Mobile applications that incorporate these elements tend to show higher student motivation and sustained use in learning contexts [11]. Table 1 presents summary of selected studies and apps for nutrition self-monitoring and behavior change.

Table 1. Summary of selected studies and apps for nutrition self-monitoring and behavior change

Study/App	Country	Target population	Key strengths	Limitations
Burke et al. [5]	United States	Adults seeking weight loss or weight control	Behavioral interventions using dietary, activity, and weight self-monitoring; more frequent and complete self-monitoring is consistently associated with greater weight loss and adherence	Self-monitoring typically relies on paper or manual entry of dietary and activity records, which can be burdensome
DietApp [7]	Spain	General adult users interested in healthy eating	Mobile app delivering personalized dietary suggestions based on age, clinical history, and physical condition; users rated the app as easy to use and its suggestions as useful	Food intake is recorded manually by users; the study focused on user perceptions rather than long term behavioral outcomes
Health Buddy [8]	India	University students/young adults	Android app supporting tracking of daily nutritional intake with diet and exercise recommendations, plus BMI and WHR calculators and health tips	Nutritional data are entered manually; evaluation is based on a development study with college students, not a controlled trial of behavior change

(*continued*)

Table 1. *(continued)*

Study/App	Country	Target population	Key strengths	Limitations
TreC-LifeStyle [9]	Italy	Families with overweight children aged 7–12 years	Multidisciplinary, user centered nutrition education app with virtual coaching and a food intake tracking dashboard; formative evaluation showed good feasibility, acceptability, and high compliance during a 6 week intervention	Designed specifically for children and their families in a primary care context; evidence comes from a small formative study, not a large-scale efficacy trial
Ulfa et al. [10]	Indonesia	Not app specific; bibliometric review of studies on dietary self -monitoring apps	Review identifies implementation strategies for mobile dietary self -monitoring and notes that nutrition apps can ease the arduous chore of documenting intake, enable real time	Synthesis of existing studies rather than evaluation of a single app; highlights that systematic evidence on long term behavior change and intervention effectiveness is still limited

2.4 Limitations of Prior Work

Despite substantial progress in mobile nutrition and behavior-change applications, a common limitation is the reliance on manual data entry. Users are required to read and transcribe nutrient values from product labels, which increases user burden and introduces opportunities for error in real-world settings [12]. These challenges highlight the need for AI-driven automation in health applications, such as computer vision and machine learning–based, to reduce manual steps [13]. The present study advances this direction by enabling image capture of Guideline Daily Amounts (GDA) nutrition labels and automated AI-based extraction. The approach aims to deliver accurate, low-burden tracking aligned with Thai RDI, thereby enhancing the practical value of nutrition apps for end.

3 Methodology

3.1 Client-Side Application Development

The mobile application was implemented in Kotlin using Android Studio. The user interface supports camera capture of nutrition labels and clear presentation of nutrient results. Core components include CameraX for image capture, Retrofit with Gson for HTTP POST communication and JSON serialization, and MPAndroidChart for bar and line chart visualizations. All network and parsing tasks run asynchronously using Kotlin Coroutines to avoid UI blocking. Captured images are Base64-encoded and packed into a JSON payload before transmission to the server.

3.2 Server-Side Development

The server was developed with Node.js and Express.js as the intermediary between the mobile app and the AI service. Incoming Base64 images from the client are accepted via Express (with body-parser), then forwarded for analysis using the OpenAI API to extract key nutrients (serving size, energy, fat, sugar, sodium). Secrets are managed with dotenv. The server returns a clean, normalized JSON object ready for downstream computation and visualization.

3.3 End-to-end Workflow

1. The user captures a nutrition label with the smartphone camera.
2. The app encodes the image as a Base64 string and wraps it in JSON.
3. The JSON payload is sent to the Express.js endpoint.
4. The server processes the request and submits the image to the OpenAI API to extract nutrition facts.
5. The extracted nutrition facts are returned in JSON and relayed to the mobile app.
6. The app computes intake, compares against Thai RDI, and renders bar and line charts (Fig. 1).

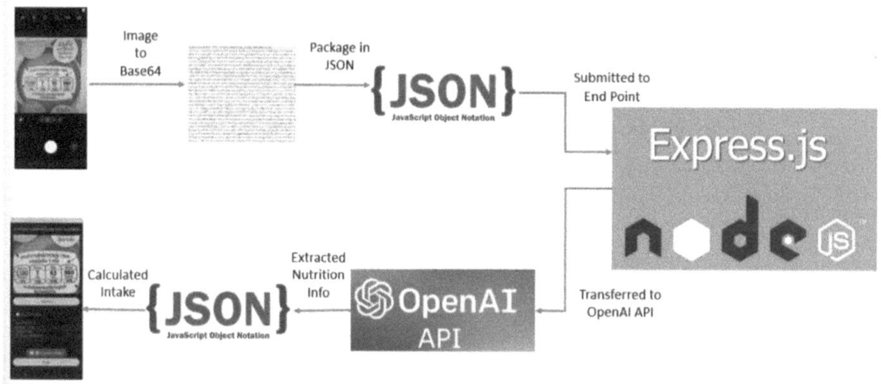

Fig. 1. System architecture of the new developed framework

3.4 Equations and Thai RDI Reference

The calculation of nutrient intake and its comparison with the Thai Recommended Daily Intakes (Thai RDI) is performed using the following equation.

$$NDi = (Ni \times Ai)/S_i \tag{1}$$

$$\%ND_i = (ND_i/\mathrm{RDI_i}) \times 100 \tag{2}$$

where

$\mathrm{ND_i}$ = Nutrient amount shown on the GDA labelling
$\mathrm{A_i}$ = The user's actual consumption, expressed as number of servings consumed
$\mathrm{S_i}$ = Number of servings per package
$\mathrm{RDI_i}$ = Thai Recommended Daily Intake for nutrient

4 Results and Discussion

4.1 Functional Testing

The application was evaluated on multiple packaged food categories. The AI achieved high extraction accuracy from labels, exceeding 90% when characters were clearly legible. Typical sources of residual error included low lighting, glare, perspective distortion, and highly stylized fonts. The Kotlin-based Android application reduces user burden by replacing manual entry with image capture of nutrition labels, followed by AI extraction of key items (serving size, energy, fat, sugar, sodium). The system compares results with Thai RDI and summarizes them through clear visualizations, enabling rapid comprehension of daily intake. Real-time API communication with the AI service improves robustness to diverse label layouts.

4.2 Example Analytical Output

GDA label analysis begins with capturing a photograph of the GDA front-of-pack label using the smartphone camera (Fig. 3). After the user taps OK, the system uploads the image for processing and displays the image file size (Fig. 4). Once processing is complete, the application returns nutrient information computed according to the user's actual consumption. In the example in Fig. 5, the label reports per-package values of total energy 320 kcal (16% of RDI), sugar 2 g (3% of RDI), fat 20 g (31% of RDI), and sodium 300 mg (15% of RDI), with two servings per package. If the user actually consumes half of the package (that is, one serving out of two), the intake is obtained by dividing the per-package amounts by the number of servings per package, yielding total energy 160 kcal, sugar 1 g, fat 10 g, and sodium 150 mg. The corresponding percent-of-RDI values scale with the consumed fraction.

The upper panel displays a preview of the selected or captured label image, the middle panel shows the analysis output area, and the bottom action button allows the user to capture a new photo. After an image is captured or selected, the app encodes it and sends it to the server for AI parsing; the resulting nutrition information is then rendered in the results panel as shown in Fig. 2.

Left: live view with zoom controls and 1:1 framing used to capture the GDA panel on the package. **Right:** post-capture confirmation screen where the user can **Retry** or tap **OK** to upload the image for AI analysis. The example label shows energy 320 kcal (16% RDI), sugar 2 g (3% RDI), fat 20 g (31% RDI), and sodium 300 mg (15% RDI), with two servings per package as shown in Fig. 3.

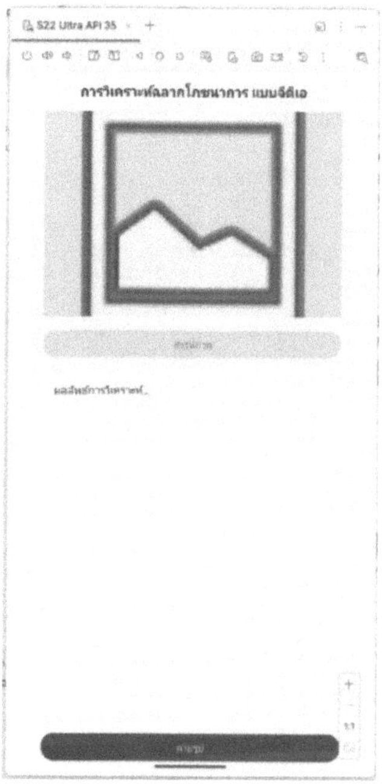

Fig. 2. Main screen of the Android application for nutrition-label analysis.

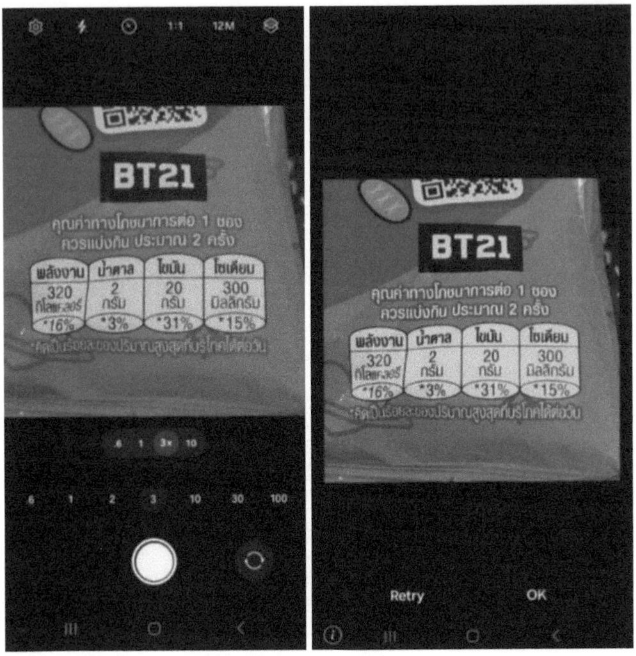

Fig. 3. Acquisition of the front-of-pack GDA label using the Android camera interface.

The captured label is previewed at the top, the **Send image** button initiates transmission, and a status line confirms that the image is being sent to the server for AI processing while reporting the file size (214 KB). The **Capture** button remains available to retake a photo during the asynchronous upload as shown in Fig. 4.

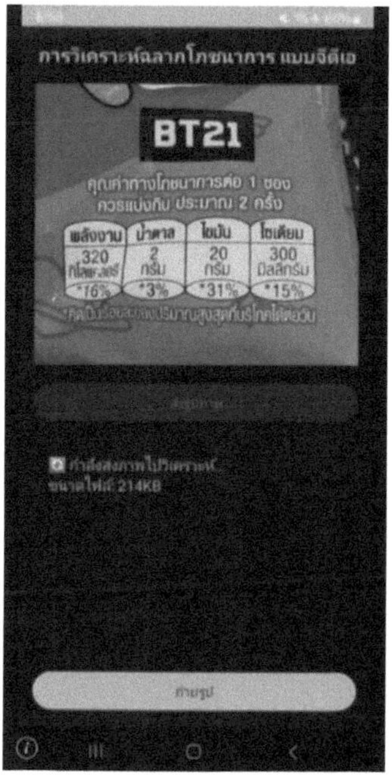

Fig. 4. Upload and submission screen for GDA label analysis.

The top panel shows the captured label, and the results panel lists nutrient amounts adjusted to the user's actual consumption (one serving out of two): **energy 160 kcal, sugar 1 g, fat 10 g, sodium 150 mg.** A status chip confirms that the scan is complete, and buttons are available to send the image or capture a new photo as shown in in Fig. 5.

Left: grouped bar chart comparing the user's daily intake (energy, fat, sugar, sodium) with Thai RDI reference values. **Right:** line chart showing sodium intake over the last 14 days, including a moving-average trend and the Thai RDI reference line as shown in in Fig. 6.

Compared with prior nutrition-tracking apps that rely on manual entry of label values and chart-based summaries, the present system advances the state of practice with an image-based pipeline. It reduces user steps and lowers input errors, while targeting real use on Android with modern components such as CameraX, Coroutines, and MPAndroidChart for a smooth user experience. The computational approach and RDI-based comparison remain consistent with established practices in the Thai context, which supports the suitability of the chosen nutritional reference.

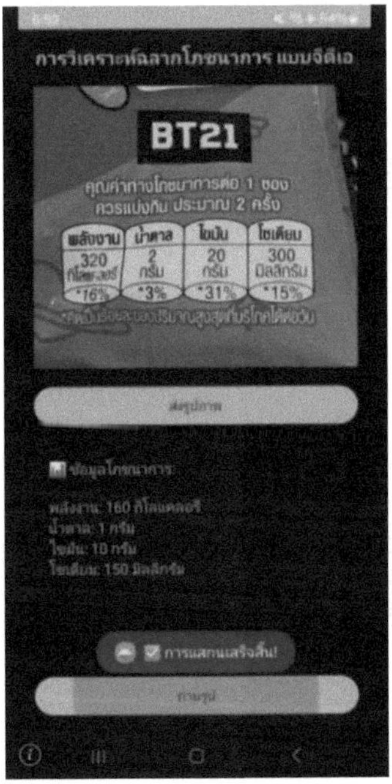

Fig. 5. Result screen after AI parsing of a GDA nutrition label was presented.

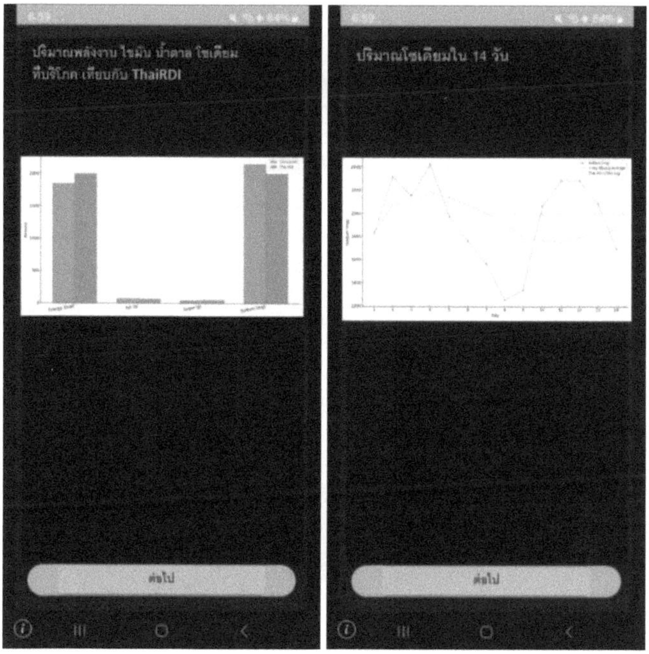

Fig. 6. In-app visualizations of nutrient intake were presented.

4.3 Strengths

The system reduces user burden and minimizes input errors by extracting values directly from label images rather than relying on manual typing. Results are easy to interpret because the application presents bar and line charts calibrated to Thai RDI, which helps communicate risks of overconsumption and underconsumption. The architecture is flexible, with a clear separation of processing and presentation layers, integration of real-time AI services, and room for future feature expansion. Local relevance is maintained through the use of Thai RDI, which makes the feedback meaningful for everyday decision making in Thailand.

4.4 Limitations

Performance depends on image quality, including lighting, shadow, distortion, and variability in fonts. The approach also relies on network connectivity and server availability, which can introduce latency or limit use in low-connectivity settings. Heterogeneity in label layout and language may challenge extraction in edge cases. Data protection must be carefully managed, with clear scoping, de-identification, and retention policies that comply with applicable data protection laws.

4.5 Practical Implications

For general users and older adults, the system provides immediate awareness of intake relative to RDI with minimal effort. For public health practitioners and dietitians, it can serve as a counseling aid that highlights risk areas such as excessive sodium or saturated fat. For policy makers and health communicators, the concise summaries are suitable for campaigns aimed at reducing salt, sugar, and saturated fat consumption.

4.6 Future Directions

Future work should emphasize on-device OCR and computer vision to reduce server dependence, improve privacy, and enable partial offline use. The system can incorporate proactive alerts and recommendations when nutrients exceed or fall short of targets, along with suggested menus or alternatives. Integration with health IoT such as smartwatches or smart scales would help link intake with physiological indicators. Personalization modules could adapt thresholds and guidance to lifestyle, age, comorbidities, and individual goals. Additional improvements include multilingual and multi-format label support backed by an updated local label knowledge base, and automated data quality checks with confidence scoring and prompts to recapture low-quality images.

5 Conclusion

This study presents a mobile application for daily nutrient tracking that uses image capture of nutrition labels with AI-based analysis to reduce manual entry and accelerate risk awareness. Results are communicated through charts referenced to Thai RDI, enabling users to identify nutrient excesses or deficits and plan targeted behavioral adjustments. The approach strengthens data quality and user experience by minimizing manual transcription, while acknowledging limitations related to image quality, network dependence, and label diversity.

Future work should emphasize on-device OCR and computer vision, proactive notifications, health IoT integration, and user-level personalization, alongside rigorous data protection practices. These developments would further position the application as a practical, accurate, and user-centered decision support tool for nutrition in individual, community health, and policy contexts in Thailand.

Acknowledgments. The authors would like to thank Suan Sunandha Rajabhat University, Bangkok, Thailand for providing fund, necessary equipment, and study area.

References

1. Vichitkunakorn, P., Bunyanukul, W., Apiwan, K., Tanasanchonnakul, D., Sittisombut, M.: Prevalence of non-communicable disease risk factors and their association with economic status: findings from the 2021 health behaviour of population survey in Thailand. Glob. Health Action. **18**(1), 2485689 (2025)

2. Ministry of Public Health: Notification of Ministry of Public Health (No. 445): Nutrition Labelling. Bangkok, Thailand (2024)
3. Ministry of Public Health: Notification of Ministry of Public Health (No. 394): Food products required to bear nutrition labelling and guideline daily amounts, GDA labeling. Bangkok, Thailand (2018)
4. Aukkanit, N., Chopvitayakun, S., Sirichokworrakit, S.: Development of a Mobile application for monitoring energy, sugar, fat, and sodium consumption based on guideline daily amounts (GDA) Labeling. In: Proceedings of the 2024 10th International Conference on e-Society, e-Learning and e-Technologies (ICSLT '24), pp. 89–93. Association for Computing Machinery (2024)
5. Burke, L.E., Wang, J., Sevick, M.A.: Self-monitoring in weight loss: a systematic review of the literature. J. Am. Diet. Assoc. **105**(12), 1807–1816 (2005)
6. Wharton, C.M., Johnston, C.S., Cunningham, B.K., Sterner, D.: Dietary self-monitoring, but not dietary quality, improves with use of smartphone app technology in an 8-week weight loss trial. J. Nutr. Educ. Behav. **46**(5), 440–444 (2014)
7. de la Díez, I.T., Garcia-Zapirain, B., López-Coronado, M., Rodrigues, J.J.P.C., Vegas,C., d.P.: A new mHealth app for monitoring and awareness of healthy eating: development and user evaluation by Spanish users. J. Med. Syst. **41**(7), 109 (2017)
8. Bajaj, D. et al.: Android based nutritional intake tracking application for handheld systems. In: 8th International Conference on Computing, Communication and Networking Technologies (ICCCNT), pp. 1–7. IEEE, Delhi (2017)
9. Gabrielli, S. et al.: Design of a mobile app for nutrition education (TreC-LifeStyle) and formative evaluation with families of overweight children. JMIR Mhealth Uhealth. **5**(4), e48 (2017)
10. Ulfa, M., Setyonugroho, W., Lestari, T., Widiasih, E., Quoc, A.N.: Nutrition-related mobile application for daily dietary self-monitoring. J. Nutr. Metab., 2476367 (2022)
11. Chopvitayakun, S., Khamsakdee, D., Aukkanit, N.: M-learning excellence: personalized mobile learning for university students via an android app. In: Proceedings 2023 International Conference on Computational Science and Computational Intelligence CSCI 2023, pp. 1785–1790 (2023)
12. Khazen, W., Jeanne, J.F., Demaretz, L., Schäfer, F., Fagherazzi, G.: Rethinking the use of mobile apps for dietary assessment in medical research. J. Med. Internet Res. **22**(6), e15619 (2020)
13. Nuanmeesri, S., Poomhiran, L.: Development of preventing myofascial pain syndrome automation with ultrasonic-based and machine learning. Int. J. Emerg. Technol. Adv. Eng. **12**(12) (2022)

Advancements in Deep Learning for Detection and Management of Banana Fusarium Wilt Disease: A Systematic Analysis

Ramalingam Dharmalingam[1], Jacintha Menezes[1](✉), Ali Al Adawi [2], Saif Al Kaabi[2], and D. Vaishnavi[3]

[1] Faculty of Information Technology, Majan University College, Muscat, Oman
{Ramalingam.d,jacintha.menezes}@majancollege.edu.om

[2] Ghadafan Agriculture Research Station, Ministry of Agriculture, Fisheries Wealth and Water Resources, Sohar, Oman

[3] SRC, SASTRA Deemed to be University, Kumbakonam, Tamil Nadu, India
vaishnavi@src.sastra.edu

Abstract. Fusarium wilt of banana, caused by Fusarium oxysporum f. sp. cubense (Foc), especially Tropical Race 4 (TR4), poses a persistent global threat due to long-lived soil chlamydospores and the lack of curative chemical control, emphasizing the need for early detection and robust management. This systematic review synthesizes deep learning (DL) advances for image-based detection and management workflows spanning mobile, edge, and UAV platforms (Michael et al., 2023), with complementary insights on spectral sensing, molecular diagnostics, agronomic integration, and resistance strategies. Convolutional neural network-based pipelines with transfer learning (e.g., ResNet, Inception, EfficientNet, MobileNet) report high accuracies on curated datasets and promising mobile deployment at scale, while field performance improves with robust augmentation, lesion localization, and hybrid datasets. A PRISMA 2020-guided process identified and screened studies, recording 843 records from databases, 464 screened, 275 assessed for eligibility, and 272 included, with reasons for exclusion documented.

The review highlights that DL methods generally outperform traditional ML on curated data, but real-world variability reduces accuracy; multimodal approaches, rigorous field validation, and explainability are critical; and integration with surveillance, quarantine, and resistance deployment is essential for population-level impact. The findings point toward practical deployment via mobile tools for early diagnosis, UAV-enabled canopy monitoring, and expanding annotated field datasets to mitigate domain shift and support durable management against TR4 and other Foc races.

Keywords: Banana · Fusarium wilt · TR4 · deep learning · computer vision · transfer learning · mobile diagnosis · UAV · PRISMA · plant disease detection

Research Track: CSCI-RTCS

H. R. Arabnia et al. (Eds.): CSCI 2025, CCIS 2968, pp. 86–96, 2027.
https://doi.org/10.1007/978-3-032-28457-0_7

1 Introduction

Fusarium wilt (Panama disease) is a vascular wilt caused by Fusarium oxysporum f. sp. cubense (Foc) with an exceptionally long soil survival horizon, multi-year incubation windows, and a broad ecology enabling persistence in host residues and non-host endophytes, conferring epidemic potential under conducive conditions (Ristaino & Records, 2020). Visual symptoms often lag infection by months, while chemical control remains ineffective, shifting emphasis to early detection, biosecurity, resistance, and precise agronomy (Rahayuniati et al., 2024).

The epidemiology of TR4 highlights asexual reproduction with microconidia, macroconidia, and durable chlamydospores; early infections remain cryptic, and symptomatic plants mark inoculum distribution and future foci. The time between root infection and symptom expression is commonly two to six months, and the long incubation period complicates surveillance and timely intervention (Nakkeeran et al., 2021).

Over the past decade, deep learning advances, particularly convolutional neural networks (CNNs), have transformed the feasibility of image-based early detection at the leaf and canopy scales (Arifin, 2025). Transfer Learning with architectures like ResNet and Inception has driven the development of mobile diagnostic tools for smallholders in East Africa, with field-ready inference on smartphones and confidence-weighted outputs (Alfred et al., 2025; Andrew et al., 2022). Unmanned Aerial Vehicle(UAV) based imaging with RGB, multispectral, hyperspectral cameras extend disease detection to canopy and block scales, supporting georeferenced surveillance and targeted interventions (Bhagwat & Dandawate, 2021; Kalluri & Selvaraj, 2024).

This systematic review synthesizes deep learning (DL) approaches for Fusarium wilt detection and management in banana, drawing on curated and field datasets, multiplatform deployments, and explainability integrations. It also positions image-based diagnostics alongside complementary modalities (molecular assays, spectral sensing) and agronomic strategies to support comprehensive, actionable control frameworks at field and regional scales.

1.1 Background: Pathosystem and Management Imperatives

TR4 epidemiology and ecology underscore the difficulty of eradication once established: low inoculum can cause disease, infection occurs at soil depth, incubation is protracted, and symptom-based surveillance misses latent infections. The result is a high premium on early, sensitive detection and rapid containment guided by biosecurity, clean planting material, and resistance deployment.

Conventional surveillance and lab diagnostics are resource-intensive and not easily scalable across dispersed farms. DL-based image analysis offers increased throughput for symptomatic detection, while UAV platforms expand coverage. Molecular detection (e.g., PCR, LAMP) is critical for strain-specific confirmation and early detection in planting material and soil, complementing DL in integrated systems.

1.2 Objectives

This study aims to advance the detection and management of banana Fusarium wilt by leveraging deep learning and integrating it with broader plant disease surveillance strategies. The specific objectives are:

1. To synthesise DL methodologies for Fusarium wilt detection and management at both leaf and canopy scales, with particular emphasis on mobile-based and UAV-enabled deployments.
2. To compare DL and traditional machine learning (ML) approaches under curated datasets and real field conditions, highlighting domain shift and generalisation challenges.
3. To review and evaluate data practices, including dataset sourcing, augmentation techniques, segmentation/ROI strategies, and the application of transfer learning.
4. To integrate DL approaches with broader surveillance frameworks, combining molecular diagnostics, spectral sensing, and agronomic interventions such as host resistance and sanitation practices.

2 Methodology

The study follows PRISMA 2020 guidance (Page et al., 2021) for reporting and documentation, including the 27-item checklist and use of the standard PRISMA flow diagram templates for new systematic reviews with databases/registers. The expanded PRISMA checklist items for title, abstract, methods, results, discussion, and conclusion guided reporting completeness.

2.1 Identification

A comprehensive search strategy was implemented to identify studies relevant to deep learning–enabled detection of banana Fusarium wilt/TR4 across image-based and remote sensing modalities, including mobile and UAV platforms. The primary search was conducted in Google Scholar and associated indexing services for the years from 2020 to 2025 using predefined Boolean keyword constructs that combined terms for the host and pathogen ("banana," "Fusarium wilt," "TR4"), methodological approaches ("deep learning," "CNN," "machine learning"), and application contexts ("detection," "mobile," "UAV," "hyperspectral," "segmentation").

This strategy returned 843 records across the predefined time window and scope, providing broad coverage of the interdisciplinary literature spanning plant pathology, computer vision, remote sensing, and precision agriculture. To mitigate retrieval bias and surface potentially influential works not prioritized algorithmically, forward and backward citation chasing was planned from sentinel articles and reviews. These manual checks did not yield additional unique studies beyond those already captured by the structured search and therefore did not contribute further inclusions at the identification stage. Consequently, the total number of records identified prior to de-duplication was 843, with no additional unique records arising from other sources during this phase.

2.2 Screening

After de-duplication, 379 duplicate records were removed automatically, yielding a screening set of 464 unique titles and abstracts for review against the prespecified eligibility criteria. During title/abstract screening, records that clearly failed to meet scope—most commonly owing to wrong indication or use of an ineligible study design—were excluded and recorded in the PRISMA flow to maintain transparency and traceability of decisions.

2.3 Eligibility Criteria for Inclusion

In accordance with PRISMA guidelines, explicit eligibility criteria were established to ensure consistency and transparency in the selection of studies. These criteria outline the target population, interventions, comparators, outcomes, study designs, and exclusions relevant to the detection and management of banana (Musa spp.) Fusarium wilt using artificial intelligence and machine learning approaches. The table below summarises these criteria (Table 1).

Table 1. Eligibility Criteria

Domain	Criteria
Population	Banana (Musa spp.) plants; leaf, pseudostem, and canopy levels; datasets involving Fusarium wilt (race-agnostic and race-specific, especially TR4).
Intervention	AI/ML and deep learning methods (e.g., CNNs, transfer learning, object detection/segmentation pipelines); mobile/edge deployment; UAV imaging; complementary modalities (e.g., molecular diagnostics) included for context.
Comparators	Traditional ML methods (SVM, RF, KNN), handcrafted feature pipelines; baseline CNNs without transfer learning; non-DL methods; comparisons across controlled vs. field datasets.
Outcomes	Diagnostic performance (accuracy, precision, recall, F1-score, mAP); generalisation across domains; computational footprint; deployment feasibility; agronomic utility.
Study Types	Primary research studies, system deployments, and methodological papers relevant to Fusarium wilt detection/management using DL/ML; surveys contextualising methods also included.
Exclusions	Studies unrelated to Fusarium wilt, non-methodological commentaries or editorials without empirical or methodological relevance. Exclusion reasons documented in PRISMA flow.

The criteria presented in the table guided the systematic screening and selection of studies. Only research directly addressing Fusarium wilt detection or management using AI/ML methods was included, with particular emphasis on deep learning approaches and their application via mobile platforms, UAVs, and complementary diagnostic modalities.

Comparators, including traditional ML and baseline models, were evaluated to identify performance gaps and generalization issues.

From the 464 records screened, 275 studies were selected for full-text review, forming the eligibility set for detailed appraisal in this review. To be included, studies had to focus on Fusarium wilt/TR4 in banana using AI/ML—particularly deep learning—and report implementation or evaluation via mobile devices, UAV imaging, or complementary diagnostic modalities (e.g., spectral analyses). Eligible studies also needed to include explicit comparators, such as traditional ML methods or baseline CNN models, to identify performance gaps and generalization limitations.

At the full-text stage, three studies were excluded due to:

- **Irrelevant indications:** studies not focused on banana Fusarium wilt/TR4.
- **Non-methodological commentaries:** articles lacking empirical or methodological relevance.

The remaining 272 studies were included in the review.

2.4 Selection Process

A two-stage screening process was employed, first reviewing titles and abstracts, followed by full-text assessment, applying the predefined inclusion and exclusion criteria. Any conflicts between reviewers were resolved by consensus. Data extraction forms were used to systematically capture key information, including PICO elements, dataset characteristics, model architectures, training configurations, performance metrics, and deployment details.

2.5 Risk of Bias Assessment

Risk of bias was assessed with focus on dataset provenance (curated vs. in-field), potential data leakage, class imbalance handling, and test set independence. External validation or independent field tests were prioritized where available to ensure reliability and generalizability.

2.6 Synthesis Methods

A narrative synthesis was conducted to summarize model architectures, processing pipelines, datasets, and outcomes, organized by imaging scale (leaf/mobile vs. canopy/UAV) and deployment feasibility. A quantitative meta-analysis was not performed due to heterogeneous endpoints, datasets, and evaluation protocols. Instead, performance ranges are reported alongside contextual factors such as domain shift, class balance, lighting conditions, and background variability.

2.7 PRISMA Flow Diagram

The PRISMA 2020 flow diagram was completed following official templates. Study selection counts were as follows: 843 records identified; 379 duplicates removed; 464 records screened; 275 full-text articles assessed for eligibility; 272 studies included in the review (Fig. 1).

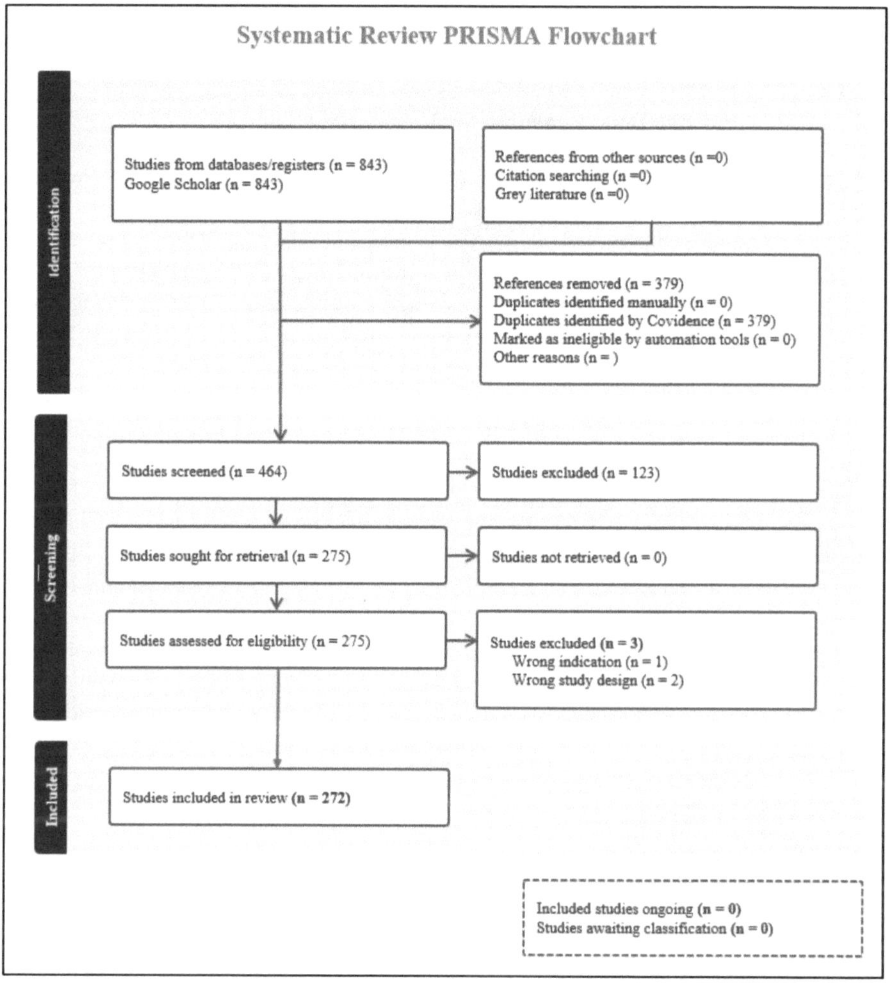

Fig. 1. PRISMA Flowchart

3 Results

3.1 Study Characteristics

- Imaging modalities: Ground-based RGB images captured via smartphones or DSLR cameras dominate the literature. UAV RGB and multispectral imaging are less common but increasing. Some studies integrate segmentation and explainability techniques to localize lesions at scale.
- Datasets: Curated datasets with controlled backgrounds and balanced classes report very high accuracies, while field datasets exhibit variability in lighting, occlusions, and backgrounds (Muftah Khamees & Aboubaker Abdusalam, n.d.). Hybrid datasets combining lab and field data improve robustness over single-source training (Menzes & Hemachandra, 2024).

- Geographic focus: Many studies originate from East Africa (e.g., Tanzania), with field validation conducted through mobile pilot deployments and on-device inference.

3.2 Mobile and Edge Deep Learning for Leaf-Scale Diagnosis

- Transfer learning: ResNet152 and InceptionV3 achieved accuracies up to 99.2% and 95.41%, respectively. InceptionV3 was preferred for mobile deployment due to lower memory demands (Sanga et al., 2020; Shafik et al., 2023). Field-deployed mobile apps reported ~99% confidence on captured leaf areas (Bhagwat & Dandawate, 2021).
- Training practices: Models applied resizing, normalization, extensive augmentation (rotations, flips, brightness/contrast), and ROI cropping to emphasize lesions, collectively supporting generalization (Md Hasanul Ferdaus, 2024).
- Lightweight backbones: MobileNet variants enable on-device inference with acceptable accuracy trade-offs, targeting latency and battery constraints in field applications.

3.3 Object Detection, Segmentation, and UAV Canopy Monitoring

- Leaf-to-canopy scaling: DL-based detection and segmentation pipelines extend to canopy-level monitoring. While UAV datasets specific to banana remain limited, lessons from other crops show that higher ground sampling distances, optimized flight plans, and segmentation networks (e.g., U-Net variants) improve symptomatic area delineation (Ahmad et al., 2023).
- Performance considerations: Accuracy decreases with altitude due to small target sizes and canopy complexity. Domain-specific augmentations and careful sensor selection mitigate losses, but field validation remains a critical bottleneck (Lee & Lee, 2025).

3.4 Traditional ML Comparators and Hybrid Pipelines

- Classical ML: SVM, RF, and KNN can perform reasonably on small, handcrafted feature pipelines but consistently underperform compared to deep learning on curated datasets (Chaudhari et al., 2025). Robustness depends heavily on feature engineering and does not scale like end-to-end CNNs (Nigam & Jain, 2020; Safari et al., 2024).
- Hybrid pipelines: Pre-segmentation methods (e.g., thresholding or clustering) before classification focus learning on symptomatic tissue, improving performance in cluttered backgrounds (Ngongoma et al., 2023).

3.5 Performance Ranges and Generalization

- Curated datasets: DL classification of banana leaf diseases (including TR4 proxy tasks) achieves accuracies of 95–99%, reflecting controlled backgrounds and balanced data splits.
- Field datasets: Accuracy typically declines due to domain shifts. Robust augmentation, hybrid datasets, and ROI-based lesion localization improve generalization, but external validation remains essential for credible deployment.

4 Discussion

4.1 Effective Methods in Current Practice

Transfer-learned CNNs for banana leaf image classification perform reliably on curated and semi-field datasets (Omer et al., 2023). Mobile deployments using memory-efficient backbones, such as InceptionV3 and MobileNet, are practical for extension workflows (Yao et al., 2023). Explainability techniques (e.g., Grad-CAM) and human-in-the-loop review provide guardrails, enhance user acceptance, and improve trust. However, sustained performance under diverse field conditions requires ongoing data curation and model adaptation (Muthukumarana & Aponso, 2020).

4.2 Factors Limiting Real-World Performance

Domain shift from controlled to field images causes notable accuracy drops. Variations in lighting, occlusion, and background clutter necessitate strong data augmentation and often segmentation to isolate symptomatic tissue (Duhan et al., 2024). Dataset imbalance and potential data leakage can artificially inflate reported performance metrics (Miftahushudur et al., 2025).

UAV-based canopy detection remains sensitive to altitude and sensor limitations (Kouadio et al., 2023). Segmentation-based approaches often outperform pure detection for lesion mapping but require high-quality labels and careful calibration.

4.3 Integrating Diagnostics with Management

Image-based detection should be integrated with quarantine measures, sanitation protocols, and the deployment of resistant cultivars to translate detection gains into meaningful epidemiological impact. Molecular tools (PCR/LAMP) remain critical for confirming TR4 in asymptomatic planting material and soil surveillance (Das et al., 2024). Early alerts generated by mobile and UAV platforms can help target sampling and laboratory testing, accelerating containment efforts.

4.4 Research Gaps

- Larger, geographically diverse banana field datasets with standardized splits and independent test sets are needed to rigorously evaluate cross-region generalization.
- Robust severity estimation and progression prediction at scale are underdeveloped. Multimodal fusion of imagery with environmental and management metadata presents a promising avenue.
- Comprehensive external validation and open benchmarks specific to TR4 and race differentiation at the phenotypic level are limited. Paired molecular validation would strengthen model claims.
- Near-term deployment should focus on mobile tools for early symptomatic detection and triage, complemented by confirmatory molecular assays for TR4. UAVs can support block-level scouting to prioritize management resources.
- Continuous learning pipelines should be established to incorporate new field images and evolving conditions, with model monitoring and scheduled re-training to manage performance drift.

4.5 Strengths and Limitations of this Review

- **Strengths:** This review follows a PRISMA 2020-compliant workflow, employs explicit eligibility and selection criteria tailored to the banana Fusarium wilt context, and integrates evidence from real-world mobile field deployments.
- **Limitations:** Heterogeneity in datasets and performance metrics precluded quantitative meta-analysis. Potential publication bias toward positive results may influence conclusions. Additionally, the scarcity of standardized banana UAV datasets with externally validated labels limits the ability to generalize findings at the canopy scale.

5 Conclusions

Deep learning has matured to deliver high-performance, practical banana disease detection at the leaf scale, including on-device mobile deployments, while canopy-scale UAV detection/segmentation is promising but data-limited. DL generally surpasses traditional ML on curated datasets; however, field robustness hinges on hybrid datasets, augmentation, GAN-focused training, and explainability. Real-world impact requires linking detection to quarantine, sanitation, and resistance strategies, with molecular confirmation of TR4 to intercept latent spread. Scaling up diverse field datasets and rigorous external validation will be decisive for durable performance across regions and seasons.

Acknowledgments. Gratitude is extended to contributors of open-source tools and datasets supporting PRISMA-compliant reviews, and to research groups advancing mobile and UAV-based AI for agricultural disease diagnostics.

Conflict of Interest No conflicts of interest are declared.

Author Contributions Ali Al Adawi and Saif Al Kaabi contributed to the conceptualization of the study, problem identification, validation, and provision of resources, ensuring agronomic and operational relevance. They also reviewed and edited the manuscript and provided supervision.

Ramalingam Dharmalingam and D. Vaishnavi developed the methodology, conducted the formal analysis, and prepared the original draft. They also contributed to visualization and supervision.

Ramalingam Dharmalingam, Jacintha Menezes and D. Vaishnavi undertook the systematic search, screening, and data curation. Jacintha Menezes additionally managed project administration and contributed to review and editing.

All authors participated in investigation, cross-reading, and interpretation of findings.

Note on Style. The manuscript uses American English, consistent sectioning, and academic tone, integrating the uploaded PICO synthesis with the Paper conceptual framing and PRISMA 2020 reporting standards.

Funding. This research paper is one of the outcomes of the strategic research project titled "Developing an AI and ML-based Early Detection Model for Banana Fusarium Wilt Disease in Oman using Hyperspectral Imaging". The project is funded by the Ministry of Higher Education, Research, and Innovation (MoHERI) and the Ministry of Agriculture, Fisheries, and Water Resources (MoAFWR), Oman, under contract number MoHERI/SRPP/MoAFWR/2/2023. The authors extend their gratitude to both ministries for their valuable support.

Ethics and Data Availability.. No human subjects or personal data were involved. Extracted information is from published sources, and PRISMA artifacts (checklist/flow) can be shared upon request.

References

Ahmad, A., Saraswat, D., El Gamal, A.: A survey on using deep learning techniques for plant disease diagnosis and recommendations for development of appropriate tools. In: Smart Agricultural Technology, vol. 3. Elsevier B.V (2023). https://doi.org/10.1016/j.atech.2022.100083

Alfred, R., Leo, J., Kaijage, S.F.: Detectron2-enhanced mask R-CNN for precise instance segmentation of rice blast disease in Tanzania: supporting timely intervention and data-driven severity assessment. Smart Agric. Technol. **12** (2025). https://doi.org/10.1016/j.atech.2025.101301

Andrew, J., Eunice, J., Popescu, D.E., Chowdary, M.K., Hemanth, J.: Deep learning-based leaf disease detection in crops using images for agricultural applications. Agronomy. **12**(10) (2022). https://doi.org/10.3390/agronomy12102395

Arifin, N.: Classification of banana leaf diseases using a GoogleNet-based convolutional neural network architecture. Sistem Kendali & Jaringan E-ISSN. **4**, 2808–3520 (2025). https://doi.org/10.58982/krisnadana.v4i2.749

Bhagwat, R., Dandawate, Y.: A review on advances in automated plant disease detection. Int. J. Eng. Technol. Innov. **11**(4), 251–264 (2021). https://doi.org/10.46604/IJETI.2021.8244

Chaudhari, R.R., Jain, S., Gupta, S.: Agricultural machine learning platform: enhancing crop suggestion and crop yield estimates. J. Integr. Sci. Technol. **13**(1) (2025). https://doi.org/10.62110/sciencein.jist.2025.v13.1017

Das, P. et al.: Fusarium wilt in banana: unraveling molecular aspects of host–pathogen interaction and resistance mechanism. Vegetos. **37**(4), 1232–1243. Springer (2024). https://doi.org/10.1007/s42535-023-00682-6

Duhan, S. et al.: Investigating attention mechanisms for plant disease identification in challenging environments. Heliyon. **10**(9), e29802 (2024). https://doi.org/10.1016/j.heliyon.2024.e29802

Kalluri, R.D., Selvaraj, P.: A review on application of various deep learning techniques and filtering approach in plant phenotyping. Majlesi J. Electr. Eng. **18**(1), 217–240 (2024). https://doi.org/10.30486/mjee.2024.1986435.1135

Kouadio, L. et al.: A review on UAV-based applications for plant disease detection and monitoring. Remote Sens. **15**(17) Multidisciplinary Digital Publishing Institute (MDPI) (2023). https://doi.org/10.3390/rs15174273

Lee, S., Lee, S.: Efficient data augmentation methods for crop disease recognition in sustainable environmental systems. Big Data Cogn. Comput. **9**(1), 8 (2025). https://doi.org/10.3390/bdcc9010008

Ferdaus, M.H., et al.: BananaImageBD: a comprehensive banana image dataset for classification of banana varieties and detection of ripeness stages in Bangladesh. Data Brief. **58** (2024). https://doi.org/10.1016/j.dib.2024.111239

Menzes, J., Hemachandra, N.: Analyzing ML/DL techniques for detection of yellow leaf curling and mosaic virus: a systematic literature review. Int. J. Intell. Syst. Appl. Eng. IJISAE. **2024**(4), 3866–3887 (2024) https://ijisae.org/index.php/IJISAE/article/view/6946

Michael, P.J.M., Hussein, M., Camilius, A.S., Richard, R.M., Beatrice, M., Caroline, M.: Artificial intelligence and deep learning based technologies for emerging disease recognition and pest prediction in beans (phaseolus vulgaris l.): a systematic review. Afr. J. Agric. Res. **19**(3), 260–271 (2023). https://doi.org/10.5897/ajar2022.16226

Miftahushudur, T., Sahin, H.M., Grieve, B., Yin, H.: A survey of methods for addressing imbalance data problems in agriculture applications. Remote Sens. **17**(3), 454 (2025). https://doi.org/10.3390/rs17030454

Muftah Khamees, Z., Aboubaker Abdusalam, A: A Review: comparative analysis of computer vision techniques for defect detection and categorization in bananas and apples sebha university conference proceedings (n.d.). https://doi.org/10.51984/SUCP.V3I2.3241

Muthukumarana, P.S., Aponso, A.C.: A review on deep learning based image classification of plant diseases. Int. J. Comput. Theory Eng. **12**(5), 118–122 (2020). https://doi.org/10.7763/IJCTE.2020.V12.1275

Nakkeeran, S., Rajamanickam, S., Saravanan, R., Vanthana, M., Soorianathasundaram, K.: Bacterial endophytome-mediated resistance in banana for the management of fusarium wilt. 3 Biotech. **11**(6) Springer Science and Business Media Deutschland GmbH (2021). https://doi.org/10.1007/s13205-021-02833-5

Ngongoma, M.S.P., Kabeya, M., Moloi, K.: A review of plant disease detection systems for farming applications. Appl. Sci. **13**(10) (Switzerland) MDPI (2023). https://doi.org/10.3390/app13105982

Nigam, S., Jain, R.: Plant disease identification using deep learning: a review. Indian J. Agric. Sci. **90**(2), 249–257). Indian Council of Agricultural Research (2020). https://doi.org/10.56093/ijas.v90i2.98996

Omer, S.M., Ghafoor, K.Z., Askar, S.K.: Plant disease diagnosing based on deep learning techniques: a survey and research challenges. Aro-Sci. J. Koya Univ. **11**(1), 38–47). Koya University (2023). https://doi.org/10.14500/aro.11080

Page, M.J., McKenzie, J.E., Bossuyt, P.M., Boutron, I., Hoffmann, T.C., Mulrow, C.D., Shamseer, L., Tetzlaff, J.M., Akl, E.A., Brennan, S.E., Chou, R.: The PRISMA 2020 statement: an updated guideline for reporting systematic reviews. bmj, **372**(2021)

Rahayuniati, R.F., Kurniawan, R.E.K., Mugiastuti, E.: Distribution of banana fusarium wilt in Banyumas, Indonesia, and characterization of F. Oxysporum isolates from infected bananas and taro growing on the same farm. Biodiversitas. **25**(4), 1344–1351 (2024). https://doi.org/10.13057/biodiv/d250401

Ristaino, J.B., Records, A.R.: Emerging Plant Diseases and Global Food Security. APS Press, The American Phytopathological Society (2020)

Safari, Y., Nakatumba-Nabende, J., Nakasi, R., Nakibuule, R.: A review on automated detection and assessment of fruit damage using machine learning. IEEE Access. **12**, 21358–21381 (2024). https://doi.org/10.1109/ACCESS.2024.3362230

Sanga, S.L., Machuve, D., Jomanga, K.: Mobile-based deep learning models for Banana disease detection. Eng. Technol. Appl. Sci. Res. **10**(3), 5674–5677 (2020). https://doi.org/10.48084/etasr.3452

Shafik, W., Tufail, A., Namoun, A., De Silva, L.C., Apong, R.A.A.H.M.: A systematic literature review on plant disease detection: motivations, classification techniques, datasets, challenges, and future trends. IEEE Access. **11**, 59174–59203 (2023). https://doi.org/10.1109/ACCESS.2023.3284760

Yao, J., Tran, S. N., Garg, S., Sawyer, S.: Deep learning for plant identification and disease classification from leaf images: multi-prediction approaches (2023). http://arxiv.org/abs/2310.16273

Artificial Intelligence - Large Language Models, Text Processing, and Risk Assessment Methods

LLM Conversion of Python for Scientific Computing

Lisa L. Lacher[1], Mark C. Lewis[2(✉)], Amanda A. Sickafoose[2], and Eliot F. Young[3]

[1] University of Houston Clear Lake, Clear Lake, TX, USA
[2] Planetary Science Institute, Tucson, AZ, USA
mlewis@psi.edu
[3] Southwest Research Institute, Boulder, CO, USA

Abstract. Many scientists today are trained to work almost exclusively in Python. We explore whether it makes sense for them to use LLMs to convert Python code they have written to more performant languages. We used three popular LLMs to convert Python code from 19 popular benchmarks [1] to both C and Rust. This was done to evaluate both the ability of the LLMs to produce correct code and the performance of the resulting code. We also took two more realistic scientific programs and did the same conversion. We find that of the LLMs we tested, ChatGPT was generally the most successful, but it still failed to produce correct code 34% of the time. Working converted code was almost uniformly faster than the original Python code, but the extent of the speed improvement depended significantly on the nature of the code. Overall, it was common for the generated code to be 10x faster, but we did see examples of the translation producing slower code. We conclude that it generally makes sense for scientists who have Python code that runs too slowly to convert it to a more performant language using an LLM, but care must be taken to make sure the result is correct.

Keywords: LLMs · Performance · Scientific Computing

1 Introduction

High-level scripting languages like Python have become the dominant tool for scientists, especially those whose domain expertise lies outside of computer science (CS). Their ease of use, extensive libraries (e.g. NumPy, SciPy, pandas), and flexible syntax allow non-CS researchers to prototype algorithms rapidly. However, for performance-critical or large-scale workloads, Python often becomes a bottleneck: interpreted and dynamically typed, it typically runs much more slowly than compiled languages like C or Rust.

One way to overcome these limitations is to automatically translate Python programs into faster target languages. Translation offers the possibility of performance gains while allowing scientists to begin in the Python environment they

H. R. Arabnia et al. (Eds.): CSCI 2025, CCIS 2968, pp. 99–112, 2027.
https://doi.org/10.1007/978-3-032-28457-0_8

are comfortable with. In recent years, large language models (LLMs)—such as GPT-4, Claude, and other code-capable models—have shown remarkable ability to generate and manipulate source code. This has raised an important question: can a scientist with minimal formal programming background use an LLM to produce usable, performant code?

In our earlier work [3], we directly investigated this question. We asked five different LLM chatbots to generate simulation code for a numerical benchmark problem across five programming languages, with four versions each (sequential and parallel). Of the 100 generated programs, only 47 compiled and executed with sufficient accuracy to test, and only 31 of those were correct enough to be usable in scientific work. Even among the usable subset, most LLM-generated programs were significantly slower than human-written versions. We also experimented with a "vibe-coding" optimization approach using Cursor, applying parallel kD-tree algorithms to human-written implementations in five languages. This effort was more successful, producing notable speed improvements in three of the five languages.

These results suggest that, while LLMs lower the barrier to entry, they currently struggle to reliably generate performant scientific code. The key challenge remains in answering the question of how to bridge the gap between ease-of-use and high-performance outcomes. In this paper, we extend our investigation by considering the translation of Python programs into faster target languages (C and Rust) using LLMs, drawing on standard benchmarks and real-world data structures such as the kD-tree to evaluate correctness, performance, and maintainability. We started with ten benchmarks defined in the Computer Language Benchmark Game [1]. We tested one or two versions for each benchmark: the fastest version of Python and, if needed, a simpler version that we thought scientists with less programming expertise would find more straightforward.

In the rest of this paper, Sect. 2 reviews related background: LLM-driven code translation and its pitfalls, performance trade-offs across languages, and the rationale for C and Rust as target languages in scientific settings. In Sect. 3 we discuss our approach to converting Python code into these faster target languages (C and Rust), and in Sect. 4 we discuss LLM selection and the results of our benchmark tests. In Sect. 5 we discuss our benchmarking results of real scientific code, and in Sect. 6 we present our conclusions. All the generated code and the scripts used for benchmarking can be found at https://www.github.com/MarkCLewis/LLMPythonConversions.

2 Background

Code translation is the process of converting code from one programming language to another while maintaining its original behavior. Code translation has many use cases, such as modernizing enterprise applications, migrating legacy software in proprietary programming languages to cloud-native applications implemented in general-purpose programming languages, and facilitating the training of models for better code synthesis. Translating software code to a modern programming language can significantly reduce maintenance effort, improve overall reliability, and boost non-functional properties such as security and performance.

The idea of automatic, source-to-source translation (a "transcompiler") is not new. For example, TransCoder [10] demonstrated unsupervised translation between C++, Java, and Python using sequence models before the emergence of large unified LLMs. Due to the importance and benefits of code translation, researchers have developed various methods to automate reliable translation between different programming languages, including approaches that use large language models (LLMs) [2,6,9,10]. However, empirical evaluations of LLM code translation are still emerging. Pan et al. [6] performed a large-scale empirical study over 1,700 code samples across five languages (C, C++, Go, Java, Python). They found that correct translation rates (i.e. behavior-preserving) ranged from 2.1% to 47.3% across the tested models and language pairs. Their manual error analysis revealed that LLMs frequently omit library imports, drop method or type definitions, mistranslate control structures, or even insert unrelated logic—illustrating that LLMs are surprisingly brittle in complex translation tasks. Such error modes mean that correctness alone is challenging.

Our previous study [3] reinforces these findings in a scientific computing context. Although scientists may hope to use LLMs as "drop-in programmers," our results showed that more than half of the generated programs failed to compile or run accurately. Even when they did, most were substantially slower than human-written baselines. Attempts to optimize code interactively (e.g., via Cursor and parallel kD-trees) proved more successful than relying on raw LLM output, suggesting that hybrid workflows, human-guided, and model-assisted may be the most viable near-term path.

Another concern is that generating code that merely runs is insufficient. In scientific computing, correctness, reproducibility, and performance are equally important. A simulation that runs, but produces incorrect or slow results is useless in practice. This makes rigorous benchmarking essential.

To evaluate both correctness and performance, we draw on a well-established source of benchmarks and two more realistic scientific programs.

- The Computer Language Benchmarks Game [1] provides a suite of ten classic algorithmic problems (e.g., spectral norm, binary trees, n-body, fannkuch, k-nucleotide). These benchmarks are widely used for comparing languages across execution time, memory usage, and scalability. Many also have established parallel implementations, allowing us to assess LLM translation quality under sequential vs. parallel paradigms.
- An occultation lightcurve generator is a piece of Python code that is in use by an astronomer who does occultations of distant objects in our solar system. It produces diffraction patterns at the edges of objects in occultations due to the wave nature of light.
- A kD-tree-based gravity simulation is commonly used in N-body simulations that we have used in previous studies. We have used it as a benchmark for comparing language performance [8], and our earlier "vibe-coding" experiments demonstrated that parallelizing kD-trees yields measurable speed

improvements in multiple languages. Including it as an evaluation benchmark allows us to test whether LLM-generated translations can handle data structures and performance-critical parallelism beyond the simpler benchmarks.

Together, these benchmarks provide a controlled, reproducible platform for testing whether LLM-translated code matches or falls short of human-written implementations in terms of correctness and performance.

Past cross-language benchmarking studies [7] confirm that compiled languages (C, Rust, C++) typically outperform interpreted ones like Python in execution time and energy efficiency. Later work has refined this conclusion, emphasizing that runtime implementations and memory management strategies contribute as much as the language itself. Still, for computational kernels such as those in the Benchmarks Game, C and Rust consistently approach the Pareto frontier of speed and efficiency.

C remains a gold standard for minimal runtime overhead and existing library support, while Rust introduces memory safety and concurrency guarantees, making it attractive for scientists who want both performance and reliability. However, translating into Rust also requires grappling with strict typing and ownership rules, which may pose difficulties for code initially written in Python's highly dynamic style.

3 LLM Selection

There are many options for LLM chatbots that we could use for this work. To keep things reasonable, we picked three. The criteria we used were general popularity and having a strong performance on TrackingAI.org [5]. The LLM versions used are Gemini 2.5, ChatGPT-4o, and Claude 3.7[1]. The most commonly used LLM we didn't include was Copilot, which has a weak performance on TrackingAI. The other significant omission was Grok, which tends to do very well on Tracking AI, but is less commonly used for programming tasks.

4 Small Benchmarks

We began our work by using the benchmark programs from [1]. This gave us a variety of problems that have been solved by many people interested in the performance of the solutions in various languages.

4.1 Approach

For each of the ten benchmarks, we picked one or two solutions in Python as our inputs to the LLMs. We always picked the fastest code in Python, but if that solution used advanced techniques or libraries that we did not think the typical scientist would use, we also went further down the list to find the fastest

[1] Unfortunately, we generated all of our code a week before Claude 4.5 was released.

solution written in more "pure", straightforward Python that only used standard libraries.

The versions used are shown in Table 1. Only the nbody benchmark had a fastest solution that was written in pure Python. This fits with our previous work where we found that any attempt to improve N-body performance using NumPy or other libraries inevitably produced slower code [8]. As a result, we had 19 different Python programs that we asked the LLMs to convert to C and Rust.

Table 1. This table shows the microbenchmarks that we used and which versions were selected for conversion. When there is more than one, the first is the fastest version, and the second is the one that uses the most standard Python.

Benchmark	Version
binary-trees	4, 2
fannkuch-redux	4, 6
fasta	5, 3
k-nucleotide	3, 8
mandelbrot	7, 5
nbody	1
pidigits	3, 4
regex-redux	2, 1
reverse-complement	5, 6
spectral-norm	4, 6

For each of these 19 solutions, we uploaded the file to each of the three LLMs with the prompt "Convert the attached Python code to X to make it run faster." This was done with "X" replaced by both C and Rust. The resulting code was added to the repository. For time reasons, ChatGPT and Claude only used whatever version was produced from that first prompt. We also wanted to explore how well they could correct their mistakes, so with Gemini, if the original version did not work, we gave it up to two more prompts that included compiler errors or basic descriptions of crashes or how the answer was wrong. We would have liked to have done this with all three, but the iteration added significant time to the process, and it was only feasible to do it with one of the LLMs.

4.2 Correctness

We first look at correctness. Benchmarking performance only makes sense if the code is correct. Verifying correctness is a task that we expect a scientist to be reasonably good at. They need to spend time verifying that their code works with their own Python code. So, they should have tests they can run, knowing the expected output, to perform verification.

Table 2 shows our results from the correctness tests. If a cell has a hyphen in it, then the code ran correctly. We have two letter codes for four types of failures. "CE" denotes a compiler error. C and Rust are both compiled languages, and a number of the outputs failed to properly follow language rules. "CD" was for code that compiled, but crashed when it ran. "IL" indicates a program that compiled and ran, but never terminated. This could have been because of a true infinite loop, or the program had an error in reading input and would wait for more input that wasn't coming. "WA" indicates a program that compiled and ran to completion, but produced the wrong output.

Looking at Table 2, two main things jump out. First, ChatGPT performed best, and Claude performed worst. Second, the most common failure type is a wrong answer. Some details are difficult to pick out for this presentation. Table 3 shows total error counts broken out by LLM, language, and error type.

Splitting things out, we see that while ChatGPT was truly the best, the gap between Gemini and Claude isn't that big if we compare them based on initial attempts. Gemini is closer to ChatGPT in performance only when it is allowed to correct its errors. This view also shows that the total number of failures is very consistent between Rust and C, which one might not expect based on [3] and the fact that much more code is written in C for LLMs to learn from.

We also see that the perception that the most common error was a wrong answer is not completely accurate. This is true for C, but in Rust, there are more compiler errors than wrong answers, especially for Claude.

4.3 Benchmarks

We performed benchmark runs on a workstation with two AMD EPYC 9655 96-Core processors and 768 GB of RAM. This machine is overkill for these small benchmarks, but it can be more appropriate for real scientific applications. Each version of the code that compiled and gave the correct answer was run seven times. Tables 4 and 5 give the mean and standard deviation of wall-clock times as measured by the Linux `time` command.

In addition to the columns for the three LLMs, this table has a column for "Human". This column has timing results from the Python code that was fed to the LLMs, along with the fastest C and Rust versions from [1] that worked on our setup.

The cells that are completely blank are the ones that had errors as listed in Table 2. The cells with hyphens in them are not expected to have values. Every Python row only has a value in the "Human" column because we did not ask the LLMs to output Python. Also, we only used one human-written version for C and Rust for each benchmark, and it is listed in the numerically smaller version number. There are also two error types listed in the "Human" column for Python. Technically, the human-written versions are all correct in that they compile and produce the correct output on the machine used by [1]. However, regex-redux version 2 crashed on our test machine because it tries to

Table 2. This table shows the correctness of the generated solutions. A cell with a hyphen got a correct answer. The letters are coded for different failure types: CE=Compiler Error, CD=Core Dump, IL=Infinite Loop, and WA = Wrong Answer. The numbers in parentheses for success by Gemini indicate which attempt was successful.

Benchmark	Ver	Lang	Gemini	ChatGPT	Claude
binary-trees	4	C	-	-	-
		Rust	-	-	-
	2	C	-	-	-
		Rust	-	-	CE
fannkuch-redux	4	C	WA	-	WA
		Rust	WA	WA	IL
	6	C	WA	WA	WA
		Rust	WA	WA	WA
fasta	5	C	IL	WA	CE
		Rust	- (2)	WA	CE
	3	C	-	CE	CE
		Rust	CE	-	-
k-nucleotide	3	C	CE	-	WA
		Rust	-	-	CE
	8	C	-	-	CD
		Rust	-	-	-
mandelbrot	7	C	WA	-	CD
		Rust	WA	-	CE
	5	C	-	-	WA
		Rust	-	-	WA
nbody	1	C	-	-	-
		Rust	-	-	-
pidigits	3	C	-	-	-
		Rust	- (2)	CE	CE
	4	C	- (2)	WA	-
		Rust	- (2)	WA	CE
regex-redux	1	C	CD	IL	-
		Rust	-	-	-
	2	C	-	CE	-
		Rust	CE	CE	CE
reverse-comp	5	C	WA	WA	IL
		Rust	WA	-	CE
	6	C	-	-	WA
		Rust	WA	-	CE
spectral-norm	4	C	-	-	-
		Rust	- (2)	-	-
	6	C	-	-	-
		Rust	- (2)	-	WA

Table 3. Code correctness by LLM, language, and failure type. The values in parentheses for Gemini indicate the values for it if we hadn't allowed it to correct itself.

Language	Error Type	Gemini	ChatGPT	Claude	Total
C	CE	1 (2)	2	2	**5 (6)**
	CD	1		2	**3**
	IL	1	1	1	**3**
	WA	4	4	5	**13**
	Total	**7 (9)**	**7**	**10**	**24 (26)**
	% Wrong	**37% (47%)**	**37%**	**53%**	
Rust	CE	2 (6)	2	9	**13 (17)**
	CD	0 (1)			**0 (1)**
	IL			1	**1**
	WA	5	4	3	**12**
	Total	**7 (12)**	**6**	**13**	**26 (31)**
	% Wrong	**37% (63%)**	**32%**	**68%**	
Both	**Total**	**14 (20)**	**13**	**23**	**50 (56)**
	% Wrong	**37% (52%)**	**34%**	**61%**	

open a file for each thread/process, running out of file descriptors on our much larger machine. In addition, the fasta version 5 solution hung on our machine for reasons that weren't clear.

There is a lot of information in this table to digest, but the key point is that when conversions to C and Rust were successful, they improved performance, with only one exception. This is especially true for the more standard Python version. It is worth noting that in many cases, even the more standard version of Python included things like the `multiprocessing` library, which would not be in the repertoire of many scientists, as parallelism and multithreading typically go beyond their programming knowledge. The size of the performance improvement depends greatly on the specifics of the benchmark, but in most cases there was a solid 10x performance improvement. Some benchmarks, like nbody, often received a 50x speedup.

The one exception to the performance improvement was the fannkuch-redux benchmark. In part, this might be due to the fact that only one of the six LLM outputs for that benchmark actually compiled and gave the correct output. Oddly, that output was generated from the faster Python version, and while it was still more than 10x faster than the slow Python version, it was almost 10x slower than the Python code it was created from.

Another interesting thing to note is which producer wrote the fastest code. We expect that the human versions of Rust and C would be the fastest. However, the purpose of this study was to determine which tools would help scientists who are not as familiar with formal programming write code that would run faster. As can be noted from Table 4, Claude only writes working code 38% of the

Table 4. Benchmarks means and standard deviations for running time (in seconds) for the first five benchmarks.

Benchmark	Ver	Lang	Human	Gemini	ChatGPT	Claude
binary-trees	4	Python	2.58 ± 0.03	-	-	-
		C	-	0.22 ± 0.01	3.20 ± 0.04	6.94 ± 0.31
		Rust	-	9.02 ± 0.17	9.11 ± 0.10	
	2	Python	37.58 ± 0.15	-	-	-
		C	1.40 ± 0.01	0.58 ± 0.00	4.21 ± 0.04	0.92 ± 0.02
		Rust	0.54 ± 0.03	9.16 ± 0.12	4.15 ± 0.03	
fannkuch-redux	4	Python	3.51 ± 0.05	-	-	-
		C	2.16 ± 0.00		28.45 ± 0.13	
		Rust	0.50 ± 0.02			
	6	Python	312.31 ± 7.36	-	-	-
		C	-			
		Rust	-			
fasta	5	Python	IL	-	-	-
		C	-			
		Rust	-	8.76 ± 0.08		
	3	Python	19.13 ± 0.16	-	-	-
		C	0.54 ± 0.00	8.38 ± 0.17		
		Rust	0.83 ± 0.26	9.15 ± 0.11	9.92 ± 0.12	9.32 ± 0.12
k-nucleotide	3	Python	1.19 ± 0.02	-	-	-
		C	1.22 ± 0.01		1.05 ± 0.00	
		Rust	1.63 ± 0.06	0.32 ± 0.03	1.49 ± 0.01	
	8	Python	76.39 ± 0.84	-	-	-
		C	-	2.15 ± 0.01	1.09 ± 0.00	
		Rust	-	0.61 ± 0.03	1.58 ± 0.00	22.09 ± 0.11
mandelbrot	7	Python	2.06 ± 0.04	-	-	-
		C	-	0.11 ± 0.00	0.10 ± 0.00	
		Rust	-	1.58 ± 0.03	0.11 ± 0.01	
	5	Python	2.80 ± 0.01	-	-	-
		C	0.05 ± 0.00	0.11 ± 0.00	0.12 ± 0.00	
		Rust	0.08 ± 0.01	0.13 ± 0.01	0.12 ± 0.00	

time, compared to ChatGPT's 68% and Gemini's 74%. When comparing which LLM often produced the fastest code, when it produced working code, Gemini consistently produced faster code than ChatGPT, but only slightly more so, and Claude took a more distinct last-place position.

Table 5. Benchmarks means and standard deviations for running time (in seconds) for the last five benchmarks.

Benchmark	Ver	Lang	Human	Gemini	ChatGPT	Claude
nbody	1	Python	116.37 ± 3.68	-	-	-
		C	1.06 ± 0.00	4.04 ± 0.00	2.50 ± 0.00	2.37 ± 0.00
		Rust	1.10 ± 0.00	2.08 ± 0.00	1.60 ± 0.00	2.19 ± 0.01
pidigits	3	Python	0.55 ± 0.00	-	-	-
		C	0.40 ± 0.00	0.44 ± 0.00	0.47 ± 0.00	0.44 ± 0.00
		Rust	0.41 ± 0.00	0.47 ± 0.00		
	4	Python	1.98 ± 0.00	-	-	-
		C	-	0.44 ± 0.00		0.44 ± 0.00
		Rust	-	0.90 ± 0.00		
regex-redux	1	Python	2.57 ± 0.03			
		C	0.47 ± 0.01			2.50 ± 0.01
		Rust	0.56 ± 0.01	0.62 ± 0.01	0.62 ± 0.00	0.62 ± 0.00
	2	Python	CD	-	-	-
		C	-	0.59 ± 0.00		0.59 ± 0.00
		Rust	-			
reverse-comp	5	Python	0.64 ± 0.01	-	-	-
		C	0.15 ± 0.00			
		Rust	0.15 ± 0.01		0.62 ± 0.00	
	6	Python	0.96 ± 0.01	-	-	-
		C	-	0.45 ± 0.00	0.44 ± 0.00	
		Rust	-		0.51 ± 0.01	
spectral-norm	4	Python	24.03 ± 0.26	-	-	-
		C	0.12 ± 0.05	0.12 ± 0.03	0.11 ± 0.01	0.33 ± 0.00
		Rust	0.14 ± 0.02	0.16 ± 0.02	0.15 ± 0.02	0.36 ± 0.00
	6	Python	102.15 ± 0.66	-	-	-
		C	-	1.13 ± 0.00	1.07 ± 0.00	0.35 ± 0.00
		Rust	-	0.15 ± 0.01	0.15 ± 0.03	

5 Real Scientific Code

These little benchmarks allow us to look at a broad range of problem types, but they are small benchmarks that are not actual code written by scientists. For that reason, we tested two additional pieces of code that are more accurate representations of real research code. The first is a script that is based on a Jupyter notebook, intended to do accurate modeling of occultation photometry with interference patterns. The second is a more realistic version of the nbody benchmark that uses a kD-tree to allow it to efficiently simulate a larger number of bodies.

5.1 Occultation Photometry Code

The occultation photometry code is a Python program written to model the light curves obtained from doing occultations on solid bodies or rings with different geometries [11]. It includes the wave effects of light, which create interference patterns around the limb of the body being observed. This code makes heavy use of SciPy and NumPy. Most of the work is done by a Fast Fourier Transform (FFT) that is part of those libraries. Indeed, this particular program does not have any loops written in Python.

We followed the same procedure as before for asking the LLMs to make C and Rust versions. The main difference was that, since it is just one program, we spent a bit more time letting Gemini attempt to fix errors and doing simple code fixes ourselves. The timing results are shown in Table 6. Note that this code has only been written by a human in Python, so we do not have timing results for a human with C or Rust.

Table 6. Timing results for the photometry code.

Lang	Human	Gemini	ChatGPT	Claude
Python	16.71 ± 0.03	-	-	-
C	-	21.98 ± 0.06	13.58 ± 0.06	8.50 ± 0.08
Rust	-	CE	CE	10.04 ± 0.05

The performance of the different versions here is very similar, with less than a 2x improvement over the Python code being the best result. This is not surprising because this code does not do much in Python other than setting up values. The work is being done in the SciPy library, which is almost certainly written in C and is approaching optimal performance. For this type of code, the only elements that can be inefficient are those related to moving data between Python and the underlying C.

While ChatGPT and Gemini failed to get Rust versions, the Rust code originally created by Claude was very closee to working. It only had four simple errors. What is more, the Rust compiler makes suggestions for fixes. In this situation, the fixes were simple enough for anyone to follow. So we did, and that created a working version in Rust. It is also worth noting that we gave Gemini eight attempts to get the Rust code correct. It kept getting compiler errors and tried switching between libraries. Once Gemini eventually got a version that compiled, we found that it had a panic[2] when we ran it. While Rust's compiler errors are highly descriptive and help users to fix those errors, the panics have default output similar to a segmentation fault in C and it requires additional knowledge to even get enough information to ask the LLM to fix it in a reasonable way.

[2] A fatal error in Rust.

5.2 kD-Tree N-Body Simulation

The second real scientific code we consider is a kD-tree-based N-body simulation originally used in [8]. While the photometry code performed most of the work using NumPy/SciPy library calls, previous work with this code found that any attempt to optimize the Python code with NumPy negatively impacted performance. This is because the work being done by this code does not have a good match in standard libraries.

Again, the same procedure was followed. The timing results are shown in Table 7. We include human-written versions with C and Rust here because they were created as part of the previous research. The human-written versions were not parallelized; we have those versions as well from [4], but we decided to use the ones that more closely matched the Python code.

Table 7. Timing results for the kD-tree code.

Lang	Human	Gemini	ChatGPT	Claude
Python	563.13 ± 2.41	-	-	-
C	4.86 ± 0.04	13.38 ± 0.06	7.67 ± 0.07	23.68 ± 2.34
Rust	11.48 ± 0.08	5.51 ± 0.08	0.57 ± 0.03	22.60 ± 2.02

As with the simple nbody benchmark code, this is a program where conversion to C or Rust can provide significant performance benefits. It is also one where all the LLMs managed to produce a correct version. The reader might wonder how ChatGPT managed to create such a fast solution in Rust. The answer is that the force calculations with the kD-tree are embarrassingly parallel, and ChatGPT's Rust solution is the one version where the LLM took advantage of that. Our test machine has 192 physical cores and supports 384 threads. So, using the Rayon library to add parallelism in Rust was a simple way to get a tremendous speed advantage.

It was good to see that there was an LLM that added parallelism when the original program did not have any. Getting the full performance advantage of C or Rust on a modern machine requires this. Unfortunately, only one of the six versions did so. It is possible others would have if we had explicitly asked for it.

6 Conclusions

We find that, unsurprisingly, asking LLMs to convert Python code to C or Rust can produce significant performance improvements. The extent of the improvement depends on the nature of the Python code. Python code containing loops that do significant numerical work in Python will almost certainly run much faster if converted to C or Rust. The benchmarks generally showed 10x or more performance improvement, and the timing for the kD-tree N-body simulation dropped from multiple minutes to tens of seconds, or even a fraction of a second if the LLM added parallelism. On the other hand, science code that does not contain loops, such as the occultation photometry software tested here, does not show as significant a timing improvement; however, slight gains can be made.

Unfortunately, the real challenge comes in knowing if the LLM will be able to convert the code correctly. Error rates range from 32% to 68% for the benchmarks we explored. Of the LLMs tested, ChatGPT had better than even odds of performing a correct translation. And some programs, such as the fannkuch-redux, were particularly challenging for the LLMs to translate. Others, like the nbody and the kD-tree code worked 100% of the time. It isn't clear what leads to the different success rates. Especially with those examples. Version 6 of the fannkuch-redux is straightforward Python code, which we wouldn't expect to be more challenging to convert than a kD-tree. However, none of the six attempts to convert that particular piece of code were successful.

On the question of whether to convert Python to C or Rust, the LLMs generally did a better job of producing C code that compiled and ran. However, the Rust compiler tends to provide better error messages that can help users fix errors when LLMs generate incorrect output. Once you get to runtime and logic errors, where the program compiles but something is wrong with the execution, getting the LLM to fix it is equally challenging in both C and Rust.

Our recommendation to any scientist with Python code that isn't running fast enough is to consider spending time getting a chatbot to help convert the code to a faster language. They just have to be willing to try several chatbots, and they must have good tests to verify that the code that is produced works as intended. Moreover, LLMs will inevitably continue to improve, even if at a slower rate than in recent years. So we expect such efforts to become more fruitful over time.

Acknowledgments. This work is supported by NSF AST Award Number 2206306.

References

1. Anon: The computer language 25.03 benchmarks game (2025). https://www.benchmarksgame-team.pages.debian.net/benchmarksgame/index.html
2. Lachaux, M., Rozière, B., Chanussot, L., Lample, G.: Unsupervised translation of programming languages. CoRR abs/2006.03511 (2020). arXiv:2006.03511
3. Lacher, L.L., Lewis, M.C., Ruetschle, E., Sickafoose, A.A.: Performance of llm-generated code. In: The World Congress in Computer Science, Computer Engineering, & Applied Computing (CSCE), vol. in press (2025)
4. Lewis, M.C., Garcia, C., Tollett, A., Aguirre, S., Hafner, H., McMahon, J., Sickafoose, A.A.: Parallel n-body performance comparison: Julia, rust, and more (2025)
5. Lott, M.: Tracking AI — TrackingAI.org (2025). www.TrackingAI.org. Accessed 13 Oct 2025
6. Pan, R., et al.: Lost in translation: a study of bugs introduced by large language models while translating code. In: Proceedings of the IEEE/ACM 46th International Conference on Software Engineering, pp. 1–13. ICSE '24, ACM, April 2024. https://doi.org/10.1145/3597503.3639226

7. Pereira, R., Couto, M., Ribeiro, F., Rua, R., Cunha, J., Fernandes, J.a.P., Saraiva, J.a.: Energy efficiency across programming languages: how do energy, time, and memory relate? In: Proceedings of the 10th ACM SIGPLAN International Conference on Software Language Engineering, pp. 256–267. SLE 2017. Association for Computing Machinery, New York (2017). https://doi.org/10.1145/3136014.3136031
8. Rotter, J., Lewis, M.C.: N-body performance with a kd-tree: comparing rust to other languages. In: 2022 International Conference on Computational Science and Computational Intelligence (CSCI), pp. 457–462. IEEE (2022)
9. Siala, H., Lano, K.: A comparison of large language models and model-driven reverse engineering for reverse engineering. Front. Comput. Sci. **Volume 7 - 2025** (2025). https://doi.org/10.3389/fcomp.2025.1516410
10. Sun, Q., Chen, N., Wang, J., Li, X., Gao, M.: Transcoder: towards unified transferable code representation learning inspired by human skills (2024). arXiv:2306.07285
11. Young, E.F.: A fourier optics method for calculating stellar occulation light curves by objects with thin atmospheres. Astronomical J. **144**, 32 (2012). https://doi.org/10.1088/0004-6256/144/2/32

Multi-level Text Feature Analysis for Detecting AI-Generated Content

Lining Yan, Haoru Ye, Tailin Pan, and Baha Ihnaini(✉)

Department of Computer Science, Wenzhou-Kean University, Wenzhou, China
{1234741,1234742,1235883}@wku.edu.cn, bihnaini@kean.edu

Abstract. This study introduces a multi-level text feature analysis framework to enhance the detection of AI-generated scientific abstracts from the CHEAT dataset, compared to human-written ones. By integrating global (abstract-level) and local (sentence-level) features—such as sentence structure, lexical diversity, perplexity, and sentiment scores—we identify distinct stylistic differences: human-written abstracts exhibit higher variance in sentence length (10–50 words) and lexical diversity (TTR: 0.5–0.7) compared to AI-generated texts (15–30 words, TTR: 0.4). Our ModernBERT+Random Forest ensemble achieves F1-scores of 99.9% and 93.5% on abstract- and sentence-level datasets, respectively, surpassing standalone LSTM and BERT models. SHAP analysis highlights perplexity and token-level log probability as the most influential features. These findings advance robust AI text detection systems, with applications in academic integrity. Future work will investigate language-specific syntactic and semantic features to improve generalizability across diverse languages.

Keywords: AI-generated text · Human-written text · AI text detection · ModernBERT · Ensemble learning

1 Introduction

Recent advancements in Natural Language Processing (NLP) have enabled large language models (LLMs), such as GPT-4 and LLaMA, to generate highly fluent and contextually coherent text, often indistinguishable from human-authored content [1]. This capability poses significant challenges to authenticity in domains like scientific publishing, where formal language structures, domain-specific terminology, and stringent requirements for academic integrity amplify the need for robust detection methods [17]. Existing detection approaches, such as those relying on single-granularity features (e.g., sentence-level metrics) or standalone models, struggle with complex linguistic structures and adversarial techniques like recursive paraphrasing, which obscure stylistic markers [3,16]. These limitations highlight the urgent need for a comprehensive framework to distinguish AI-generated from human-written texts in high-stakes contexts.

Research Track Acronym: CSCI-RTAI.

H. R. Arabnia et al. (Eds.): CSCI 2025, CCIS 2968, pp. 113–126, 2027.
https://doi.org/10.1007/978-3-032-28457-0_9

This study proposes a multi-level text analysis framework that integrates global (abstract-level) and local (sentence-level) features to detect AI-generated scientific abstracts. By analyzing metrics such as sentence structure, lexical diversity, perplexity, and sentiment, we uncover systematic stylistic differences: human-written abstracts exhibit greater variability in sentence length and word choice, while AI-generated texts are more uniform and predictable [3]. Our approach leverages these distinctions to develop a high-accuracy detection system, with the ModernBERT+Random Forest ensemble achieving F1-scores of 99.9% and 93.5% on abstract- and sentence-level datasets, respectively.

The contributions of this work are threefold:

- **Multi-Level Feature Analysis**: We integrate global features (e.g., document coherence, perplexity) with local features (e.g., sentence complexity, POS tagging) to address the limitations of single-granularity approaches, enhancing detection robustness across mixed-content scenarios [19].
- **Robust Ensemble Model**: Our ModernBERT+Random Forest ensemble outperforms standalone models, achieving state-of-the-art performance by combining contextual and feature-based strengths [9,10].
- **Interpretable Insights**: Using SHAP analysis, we identify perplexity and token-level log probability as key discriminative features, providing transparent insights to guide future feature engineering [21].

These advancements offer a pathway for reliable AI text detection, with implications for ensuring authenticity in scientific publishing and beyond. Future work will explore cross-linguistic applications to address language-specific variations.

2 Related Work

The rapid advancement of large language models (LLMs) has heightened the need for effective detection of AI-generated text, particularly in scientific publishing, where authenticity is critical [1,2]. This section reviews prior work in four areas: feature-based detection, advancements in detection models, evasion tactics, and interpretability, highlighting their contributions and limitations to position our multi-level framework as a novel solution.

2.1 Feature-Based Detection Approaches

Early studies focused on linguistic and stylistic features to differentiate AI-generated from human-written text. Herbold et al. [3] compared ChatGPT-generated and human-written essays, finding that AI texts exhibit higher linguistic complexity but lower variability in sentence structure and vocabulary. Similarly, Ma et al. [4] observed that AI-generated scientific abstracts show uniform vocabulary and predictable sentence patterns, quantified by lower burstiness scores. These findings align with Pennebaker and King [5], who noted that human texts reflect individual stylistic variations absent in AI outputs. However,

Chakraborty et al. [6] demonstrated that subtle AI modifications, such as polishing human drafts, reduce feature distinctiveness, lowering detection accuracy. These limitations underscore the need for comprehensive feature sets combining global and local characteristics, which our framework addresses through multi-level analysis.

2.2 Advancements in Detection Models

Model-based approaches have improved detection performance. Wang et al. [8] proposed SeqXGPT, leveraging temporal dependencies for sentence- and document-level detection, but its performance degrades on short or paraphrased texts. Huang et al. [9] introduced a Token Ensemble strategy, combining multiple LLM outputs, yet it incurs high computational costs and struggles with generalizability. Mitchell et al. [10] developed DetectGPT, a zero-shot method using probability curvature, but its efficacy diminishes against adversarial paraphrasing. Transformer-based models like BERT [11] and ModernBERT [13], with advanced positional embeddings, excel in contextual tasks but are often limited to specific text granularities. Our study integrates these models with classical approaches (e.g., Random Forest [14]) via ensemble learning to enhance performance across diverse datasets.

2.3 Challenges and Evasion Tactics

Evasion tactics, such as recursive paraphrasing, pose significant challenges. Krishna et al. [16] showed that paraphrasing obscures stylistic markers, reducing detection accuracy. Wu et al. [18] highlighted additional challenges, including cross-linguistic variations and complex linguistic structures, emphasizing the need for discourse-level analysis. Recent work by Sadasivan et al. [17] explored adversarial attacks like style transfer, which further complicate detection by altering syntactic and semantic patterns. These findings necessitate robust, multi-granularity approaches, which our framework achieves by combining sentence- and abstract-level features.

2.4 Interpretability in Detection Systems

Interpretability is crucial for refining detection models. Lundberg and Lee [21] introduced SHAP to quantify feature contributions, which was applied in Liu et al. [20] to identify perplexity and log probability as key discriminators in academic writing. However, interpretability remains underexplored, with most studies prioritizing performance over transparency. Our work leverages SHAP to provide global and local insights, enhancing model transparency and guiding feature refinement.

2.5 Positioning This Study

Prior work highlights the limitations of single-granularity features and standalone models in handling adversarial modifications and diverse text types. Our multi-level framework addresses these gaps by integrating global and local features, achieving superior performance through ensemble learning, and providing interpretable insights via SHAP, paving the way for robust and generalizable AI text detection.

3 Methodology

3.1 Dataset and Preprocessing

We utilize the CHEAT dataset [23], a large-scale resource for detecting AI-generated and human-written academic abstracts in computer science. The dataset includes four subsets, detailed in Table 1, covering diverse scenarios such as fully AI-generated, AI-polished, and mixed-content abstracts (Fig. 1).

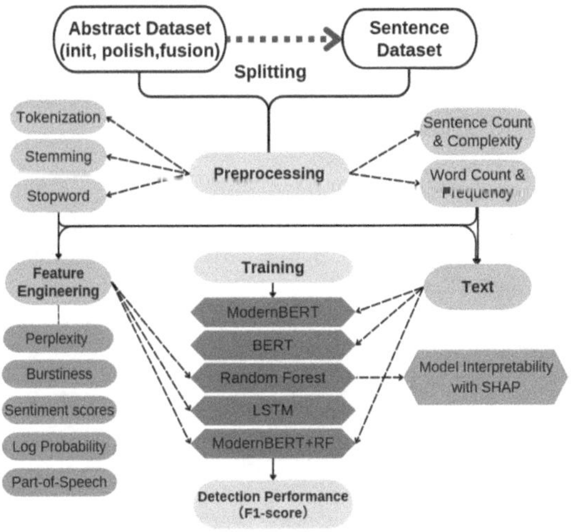

Fig. 1. Overview of the multi-level text analysis framework, integrating global and local feature extraction, model training, and SHAP-based interpretability.

Preprocessing steps ensure data quality:

1. **Duplicate Removal**: Removed duplicate abstracts to eliminate bias.
2. **Text Length Filtering**: Excluded abstracts with fewer than 10 or more than 500 words to focus on meaningful content [23].
3. **Illegal Character Removal**: Applied regular expressions to remove non-word characters (e.g., HTML tags) and normalize spaces [12].

Table 1. CHEAT Dataset Subsets

Subsets	Description	Samples	AI/Human Ratio
Initial	15,000 human-written abstracts from IEEE Xplore and 15,000 ChatGPT-generated abstracts using titles and keywords.	30,000	1:1
Polish	15,000 human-written abstracts enhanced by ChatGPT for readability and linguistic quality.	15,000	1:0
Fusion	15,000 abstracts combining sentences from Initial and Polish subsets with controlled replacement frequency.	15,000	1:1
Sentence-level	Sentences extracted from all subsets for fine-grained detection.	120,000	1:1

4. **Text Normalization**: Converted text to lowercase, applied Porter stemming, and removed stopwords using NLTK [8].
5. **Word Frequency Analysis**: Computed word frequency distributions to extract stylistic features [23].
6. **Sentence Metrics**: Calculated sentence count and complexity (average words per sentence) to capture syntactic differences [12].

3.2 Feature Engineering

We selected five features based on their ability to capture stylistic and linguistic differences, grounded in prior work:

1. **Perplexity**: Measured using GPT-2, perplexity quantifies text predictability, with lower values indicating AI-generated text due to algorithmic optimization [7].
2. **Burstiness**: Variance in word frequency, reflecting vocabulary diversity. Human texts exhibit higher burstiness due to varied word usage [4].
3. **Sentiment Score**: Computed using VADER, capturing emotional tone. AI texts tend to have neutral or positive sentiment, unlike the varied tones in human texts [24].
4. **Token-Level Log Probability**: Assesses text complexity and predictability, with lower probabilities indicating human-like variability [10].
5. **POS Tagging**: Applied using spaCy to identify syntactic patterns, revealing formulaic structures in AI texts [22].

Feature selection was validated through correlation analysis to minimize redundancy (e.g., Pearson correlation between perplexity and log probability <0.3).

3.3 Model Development

We evaluated four models and an ensemble on the CHEAT dataset, using an 80/10/10 train/validation/test split:

- **ModernBERT**: Fine-tuned with RoPE embeddings and local-global attention (learning rate: 2e−5, batch size: 16, epochs: 5) [13].

- **BERT**: Fine-tuned with identical hyperparameters for bidirectional context understanding [11].
- **LSTM**: Configured with 128 hidden units and Adam optimizer to capture sequential dependencies [25].
- **Random Forest**: Used 100 trees with max depth 20, handling high-dimensional features [14].
- **ModernBERT+RF Ensemble**: ModernBERT's contextual embeddings were concatenated with feature vectors (e.g., perplexity, burstiness) as input to Random Forest, with Logistic Regression ($C = 1.0$) merging outputs [15].

Training was conducted on an NVIDIA A100 GPU, with average training times of 4 h for ModernBERT and 1 h for Random Forest.

3.4 Evaluation

Performance was evaluated using F1-score, precision, recall, and ROC-AUC to account for class imbalance. We applied SMOTE oversampling to balance the dataset where necessary [26].

3.5 Model Interpretability with SHAP

SHAP analysis was applied to quantify feature contributions, providing global and local explanations to enhance transparency and guide feature refinement [21]

4 Experiment and Results

4.1 Sentence Complexity

We analyzed sentence length and complexity to identify stylistic differences between AI-generated and human-written texts. Figure 2 presents a scatter plot of sentence length versus complexity for abstract-level data, with human-written texts in red and AI-generated texts in blue.

Key observations include:

1. **Human-Written Texts**: Exhibit a wide range of sentence lengths and complexities, reflecting dynamic and varied writing styles.
2. **AI-Generated Texts**: Cluster around lower complexities, particularly in shorter to medium-length sentences, indicating a uniform, systematic generation approach.

4.2 Lexical Diversity

Lexical diversity was assessed using Type-Token Ratio (TTR) and Measure of Textual Lexical Diversity (MTLD), as shown in Fig. 3, comparing human-written (red) and AI-generated (blue) texts.

Key findings include:

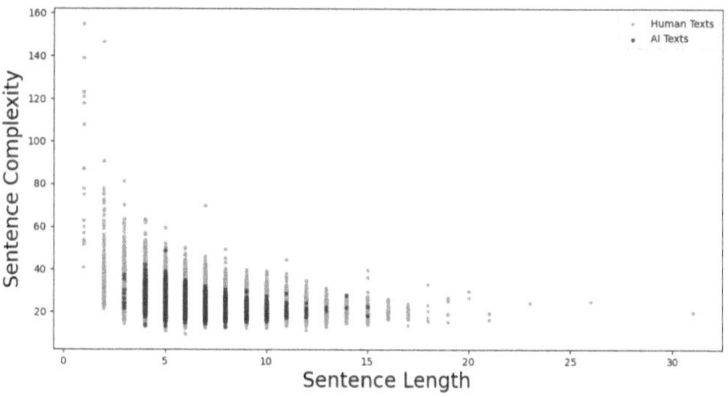

Fig. 2. Sentence Length vs. Complexity for Abstract-Level Data

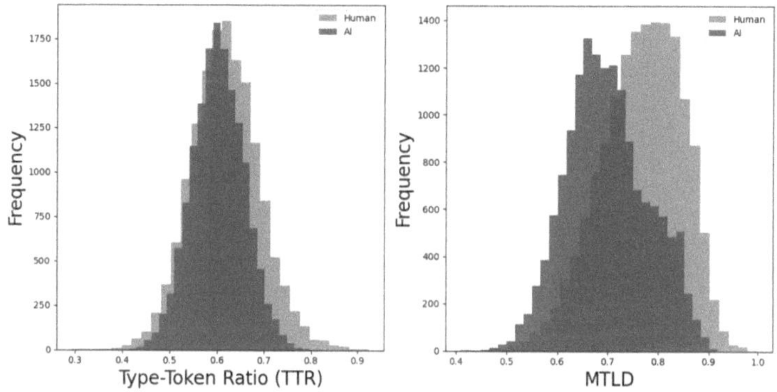

Fig. 3. TTR and MTLD Comparison

1. **TTR**: AI-generated texts exhibit lower TTR, suggesting less vocabulary repetition within shorter passages, likely due to algorithmic optimization.
2. **MTLD**: AI-generated texts show lower MTLD scores, indicating consistent lexical diversity over longer passages, while human texts display greater variability.

4.3 Sentiment Analysis

Sentiment analysis using VADER highlights differences in sentiment expression, as shown in Fig. 4.

The box plot reveals:

1. **Human-Written Texts**: Display a broad range of sentiment scores, spanning positive and negative extremes, reflecting diverse emotional expression.
2. **AI-Generated Texts**: Cluster tightly at higher (positive or neutral) sentiment scores, indicating limited variability.

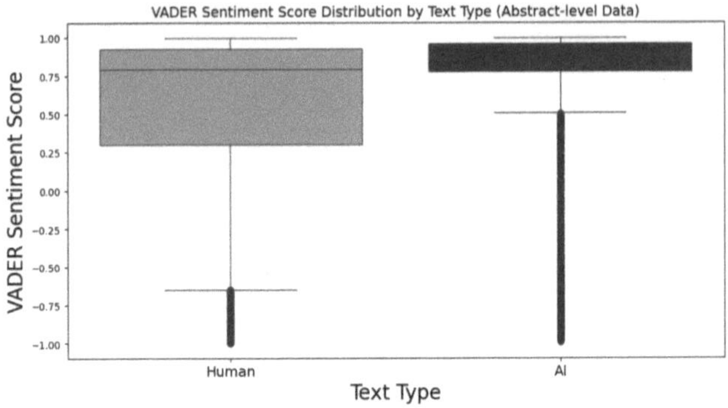

Fig. 4. VADER Sentiment Scores

4.4 Perplexity and Burstiness

Figure 5 compares perplexity and burstiness for human-written and AI-generated texts.

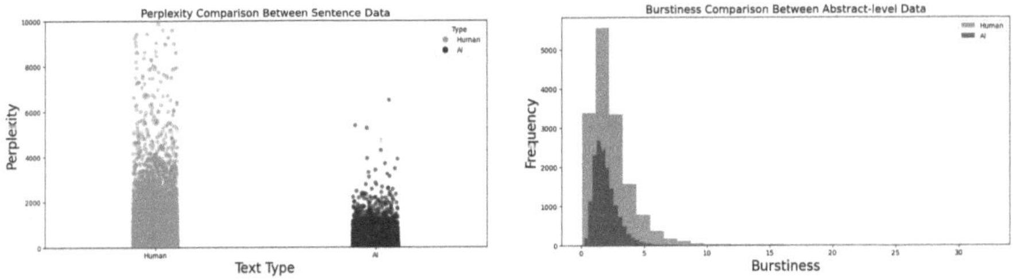

Fig. 5. Perplexity and Burstiness Comparison

Key observations include:

1. **Perplexity**: Human texts exhibit higher, more variable perplexity scores, reflecting diverse sentence structures. AI texts show lower, consistent perplexity, indicating predictability.
2. **Burstiness**: Human texts display higher burstiness, with greater variation in word frequency. AI texts exhibit lower burstiness, reflecting uniform vocabulary usage.

4.5 POS Tag Distribution

Figure 6 illustrates the distribution of part-of-speech (POS) tags for abstract- and sentence-level data.

Key findings include:

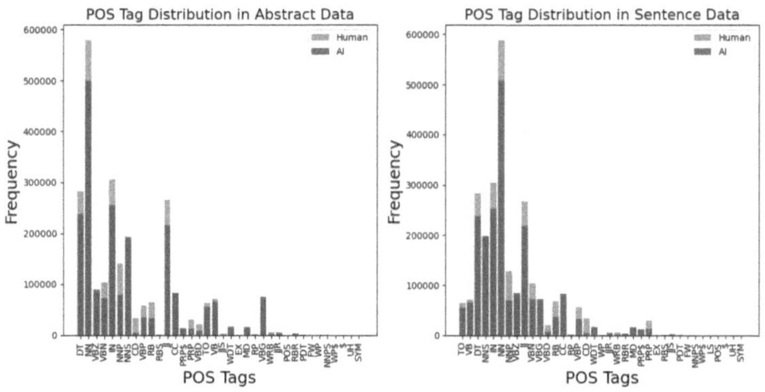

Fig. 6. POS Tag Distribution

1. **Abstract-Level Data**: Human texts show higher frequencies of pronouns (PRP), nouns (NN, NNS, NNP), verbs (VB, VBN), and determiners (DT), indicating richer syntactic variety. AI texts are more uniform.
2. **Sentence-Level Data**: Human texts maintain greater variability across POS tags, while AI texts are evenly distributed across fewer tags, reflecting formulaic language use.

4.6 Log Probability

Log probability, reflecting text predictability, is shown in Fig. 7.

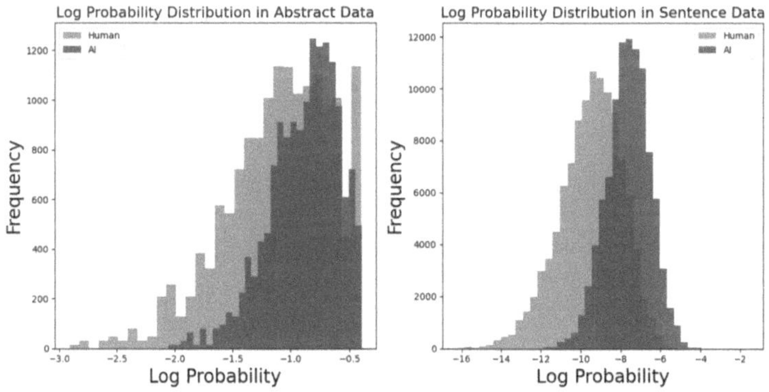

Fig. 7. Log Probability Distribution

Key observations include:

1. **Abstract-Level Data**: AI-generated abstracts cluster around higher log probabilities (approximately −0.9), indicating predictability. Human abstracts show broader variability.

2. **Sentence-Level Data**: Human sentences exhibit a wide range of log probabilities (−14 to −8), while AI sentences cluster around −7, reflecting lower variability.

4.7 Model Performance and Comparison

Table 2 presents F1-score performance across models for abstract- and sentence-level tasks.

Table 2. F1-Score Performance Comparison

Model	Abstract-Level			Sentence-Level		
	Initial	Polish	Fusion	Initial	Polish	Fusion
ModernBERT	99.6	**94.6**	70.8	**93.5**	77.5	**72.5**
BERT	99.5	93.5	70.0	91.5	75.6	71.6
LSTM	98.2	73.0	64.8	81.5	45.7	66.5
Random Forest	92.7	70.7	72.8	81.3	65.1	53.3
ModernBERT+RF	**99.9**	94.1	**78.4**	**93.5**	**78.0**	62.2

The ModernBERT+Random Forest (ModernBERT+RF) model achieves the highest F1-scores for abstract-level initial (99.9) and fusion (78.4) detection, leveraging complementary model strengths. ModernBERT excels in abstract-level polished detection (94.6) and sentence-level fusion detection (72.5), demonstrating robustness in contextual tasks. BERT performs competitively, particularly in abstract-level initial detection (99.5). LSTM and Random Forest show lower performance but contribute in specific scenarios, such as Random Forests fusion detection at the abstract level (72.8). These results highlight the efficacy of ensemble approaches for robust detection.

4.8 SHAP Feature Importance

SHAP analysis, shown in Figs. 8 and 9, identifies key features influencing model predictions.

Key findings include:

1. **Perplexity and Log Probability**: High perplexity (pink) and low log probabilities (blue) strongly indicate human-written text, while low perplexity (blue) is associated with AI-generated text.
2. **Text/Sentence Length and Burstiness**: Longer texts (pink) and higher burstiness (pink) favor human-written content; shorter texts (blue) and lower burstiness (blue) indicate AI text.
3. **Sentence Complexity and MTLD**: Higher complexity (pink) and lexical diversity (pink) suggest human text, while simpler sentences (blue) and lower diversity (blue) indicate AI text, with moderate impact.

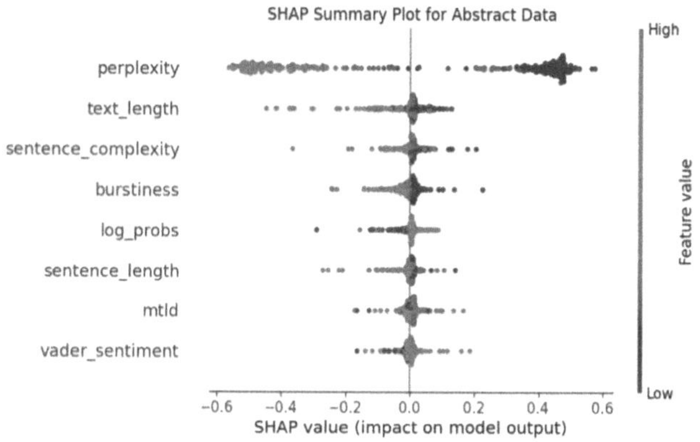

Fig. 8. SHAP Feature Importance

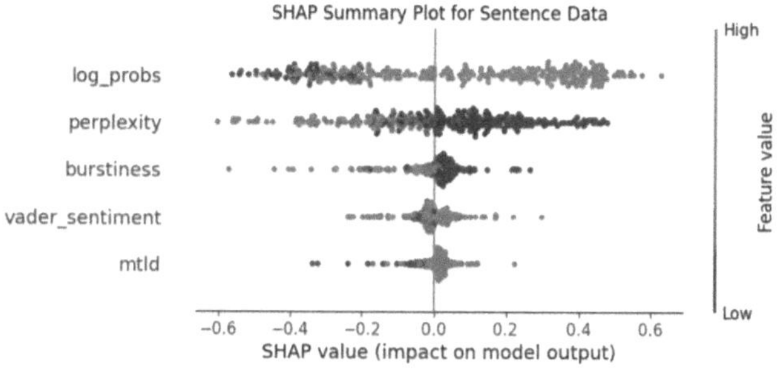

Fig. 9. SHAP Interpretation

4. **VADER Sentiment**: Higher sentiment scores (pink) are linked to AI text, while lower scores (blue) favor human text, serving as a decisive feature.

Perplexity and log probability are the most influential features, with burstiness, sentence complexity, VADER sentiment, and MTLD providing supporting roles. Text and sentence length have minimal impact.

5 Discussion

Our multi-level analysis reveals distinct stylistic differences between human-written and AI-generated scientific abstracts, addressing the challenge of detecting sophisticated AI outputs [16]. Key findings include:

1. **Text Variability**: Human texts show greater variability in sentence length (10–50 words), complexity (3–8), and lexical diversity (TTR: 0.5–0.7, MTLD: 60–100), reflecting dynamic styles. AI texts are more uniform (length: 15–30 words, complexity: 2–4, TTR: 0.4, MTLD: 50).
2. **Syntactic Diversity**: Human texts use diverse POS tags (e.g., higher pronoun and verb frequencies), while AI texts exhibit formulaic patterns [22].

3. **Model Performance**: The ModernBERT+RF ensemble achieves F1-scores of 99.9% (Initial) and 93.5% (sentence-level), outperforming standalone models by leveraging contextual and feature-based strengths [9].
4. **Feature Importance**: Perplexity and log probability are the most discriminative, with higher perplexity (50–150) and lower log probabilities (−14 to −8) indicating human texts, addressing limitations of single-granularity approaches [19].

6 Conclusion

This study advances AI-generated text detection through a multi-level analysis framework, revealing significant stylistic differences in sentence structure, lexical diversity, and syntactic complexity. Unlike single-granularity approaches [8], our integration of global and local features achieves robust detection, with the ModernBERT+RF ensemble yielding F1-scores of 99.9% (abstract-level) and 93.5% (sentence-level). Perplexity and log probability are the most discriminative features, supported by burstiness and sentiment. Future research should explore cross-linguistic applications and semantic-based features to enhance robustness against adversarial attacks.

7 Limitations

This study has several limitations:

1. **Dataset Scope**: The CHEAT dataset focuses on formal scientific abstracts, limiting generalizability to informal contexts (e.g., social media). Future work should include diverse text types.
2. **Language Specificity**: The analysis is restricted to English, and results may not apply to languages with distinct structures. Cross-linguistic datasets are needed.
3. **Model Complexity**: The ModernBERT+RF ensemble is computationally intensive (4-h training on an A100 GPU), potentially limiting scalability. Optimized architectures could address this.
4. **Adversarial Robustness**: The model may struggle with texts generated by newer models (e.g., beyond ChatGPT), necessitating adversarial training.

8 Impact Statement

This study advances machine learning by developing a robust framework for detecting AI-generated text through multi-level analysis, contributing to academic integrity and transparency in content generation. By identifying stylistic differences and leveraging ensemble models, it offers tools to combat misinformation in scientific publishing. However, potential misuse of detection systems raises ethical concerns, including privacy violations or biased judgments. We advocate for responsible application and call for further research to ensure equitable, bias-free detection across diverse languages and domains.

Acknowledgments. This research was supported by Wenzhou-Kean University under the SPF research program (project number WKUSPF202433). We gratefully acknowledge this financial support.

References

1. Weber-Wulff, D.: Testing of detection tools for AI-generated text. Int. J. Educ. Integr. **19**(1), 7–8 (2023). https://doi.org/10.1007/s40979-023-00146-z
2. Kumar, V., Bharti, A., Verma, D., Bhatnagar, V.: Deep dive into language traits of AI-generated Abstracts. arXiv preprint arXiv:2312.10617 (2023). https://doi.org/10.48550/arXiv.2312.10617
3. Herbold, S., Hautli-Janisz, A., Heuer, U., Kikteva, Z., Trautsch, A.: A large-scale comparison of human-written versus ChatGPT-generated essays. Sci. Rep. **13**(1), 1–2 (2023). https://doi.org/10.1038/s41598-023-45644-9
4. Ma, Y., Liu, J., Yi, F., Cheng, Q., Huang, Y., Lu, W., Liu, X.: AI vs. Human – Differentiation Analysis of Scientific Content Generation (2023). https://arxiv.org/abs/2301.10416
5. Pennebaker, J.W., King, L.A.: Linguistic styles: language use as an individual difference. J. Pers. Soc. Psychol. **77**(6), 1296–1312 (1999). https://doi.org/10.1037/0022-3514.77.6.1296
6. Chakraborty, M., et al.: Counter Turing Test CT$\hat{2}$: AI-Generated Text Detection is Not as Easy as You May Think – Introducing AI Detectability Index. arXiv preprint arXiv:2310.05030, 2023. https://doi.org/10.48550/arXiv.2310.05030
7. Ippolito, D., Duckworth, D., Callison-Burch, C., Eck, D.: Automatic Detection of Generated Text is Easiest when Humans are Fooled. arXiv preprint arXiv:1911.00650 (2020). https://doi.org/10.48550/arXiv.1911.00650
8. Wang, P., Li, L., Ren, K., Jiang, B., Zhang, D., Qiu, X.: SeqXGPT: Sentence-level AI-generated text detection. In: Proceedings of the 2023 Conference on Empirical Methods in Natural Language Processing, pp. 1145–1152 (2023). https://doi.org/10.18653/v1/2023.emnlp-main.73
9. Huang, F., Kwak, H., An, J.: ToBlend: Token-Level Blending With an Ensemble of LLMs to Attack AI-Generated Text Detection. arXiv preprint arXiv:2402.11167 (2024). https://doi.org/10.48550/arXiv.2402.11167
10. Mitchell, E., Lee, Y., Khazatsky, A., Manning, C.D., Finn, C.: DetectGPT: Zero-Shot Machine-Generated Text Detection using Probability Curvature. arXiv preprint arXiv:2301.11305 (2023). https://doi.org/10.48550/arXiv.2301.11305
11. Devlin, J., Chang, M.-W., Lee, K., Toutanova, K.: BERT: pre-training of Deep Bidirectional Transformers for Language Understanding. arXiv preprint arXiv:1810.04805 (2019). https://doi.org/10.48550/arXiv.1810.04805
12. Wang, H., Li, J., Li, Z.: AI-generated text detection and classification based on BERT deep learning algorithm. Theoretical Natural Sci. **39**(1), 3 (2024). https://doi.org/10.54254/2753-8818/39/20240625
13. Warner, B., et al.: Smarter, Better, Faster, Longer: A Modern Bidirectional Encoder for Fast, Memory Efficient, and Long Context Finetuning and Inference. arXiv (2024). https://doi.org/10.48550/arXiv.2412.13663
14. Dong, L., Ji, S., Zhang, C., Zhang, Q., Chiu, D.K.W., Qiu, L., Li, D.: An unsupervised topic-sentiment joint probabilistic model for detecting deceptive reviews. Expert Syst. Appl. **114**, 210–223 (2018). https://doi.org/10.1016/j.eswa.2018.07.005

15. Dietterich, T. G. Ensemble methods in machine learning. In: Multiple Classifier Systems, vol. 1857, pp. 4–8. Springer, Heidelberg (2000). https://doi.org/10.1007/3-540-45014-9_1
16. Krishna, K., Song, Y., Karpinska, M., Wieting, J., Iyyer, M.: Paraphrasing evades detectors of AI-generated text, but retrieval is an effective defense. arXiv preprint (2023). https://arxiv.org/abs/2303.13408
17. Sadasivan, V.S., Kumar, A., Balasubramanian, S., Wang, W., Feizi, S.: Can AI-Generated Text be Reliably Detected? arXiv preprint arXiv:2303.11156 (2024). https://doi.org/10.48550/arXiv.2303.11156
18. Wu, J., Yang, S., Zhan, R., Yuan, Y., Wong, D. F., Chao, L.S.: A Survey on LLM-Generated Text Detection: Necessity, Methods, and Future Directions. arXiv (2024). https://doi.org/10.48550/arXiv.2310.14724
19. Walters, W.H.: The effectiveness of software designed to detect AI-generated writing: a comparison of 16 AI text detectors. Open Inform. Sci. **7**(1), 12–15 (2023). https://doi.org/10.1515/opis-2022-0158
20. Liu, Z., Yao, Z., Li, F., Luo, B.: On the Detectability of ChatGPT Content: Benchmarking, Methodology, and Evaluation through the Lens of Academic Writing. arXiv preprint arXiv:2306.05524 (2024). https://doi.org/10.48550/arXiv.2306.05524
21. Lundberg, S., Lee, S.-I.: A Unified Approach to Interpreting Model Predictions. arXiv preprint arXiv:1705.07874 (2017). https://doi.org/10.48550/arXiv.1705.07874
22. Singh, A.: Detecting Machine Generated Text. Unpublished (2025). https://doi.org/10.13140/RG.2.2.34327.89767
23. Yu, P., Chen, J., Feng, X., Xia, Z.: CHEAT: A Large-scale Dataset for Detecting ChatGPT-writtEn AbsTracts. arXiv (2023). https://doi.org/10.48550/arXiv.2304.12008
24. Wankhade, M., Rao, A.C.S., Kulkarni, C.: A survey on sentiment analysis methods, applications, and challenges. Artif. Intell. Rev. **55**(7), 5731–5780 (2022). https://doi.org/10.1007/s10462-022-10144-1
25. Nugaliyadde, A., Wong, K. W., Sohel, F., Xie, H.: Language Modeling through Long Term Memory Network. arXiv preprint arXiv:1904.08936, 2019. https://doi.org/10.48550/arXiv.1904.08936
26. Chawla, N.V., Bowyer, K.W., Hall, L.O., Kegelmeyer, W.P.: SMOTE: synthetic minority over-sampling technique. J. Artif. Intell. Res. **16**, 321–357 (2002). https://doi.org/10.1613/jair.953

Real-Time Credit Risk Assessment in Agri-Finance: A Dynamics 365 Based AI Framework

Karthik Nallani Chakravartula(✉)

Senior Software Engineer, Farm Credit Bank, Austin, USA
Karthik.chakravartula@gmail.com

Abstract. The agricultural finance industry is experiencing difficulties in assessing and managing credit risk as rapidly changing commodity prices, climate variability, and conventional assessment processes neglect the dynamic nature of the agricultural sector. This paper outlines a real-time credit risk assessment framework using Microsoft Dynamics 365's AI capabilities for agricultural finance institutions. The framework targets the insufficiency of the 5 C's of credit in assessing creditworthiness in agricultural financing (Character, Capacity, Capital, Collateral, and Conditions). The proposed system offers a unique combination of Microsoft Dynamics 365 Finance, AI Builder, Power BI, and Copilot for Finance, and it is fundamentally a vendor-agnostic enterprise architecture that augments existing banking infrastructure by avoiding complete system replacement. In contrast to the current fintech platforms which provide stand-alone credit scoring models, our framework looks to address the full complexity of rural agricultural risk mitigation. Our framework provides a Dynamic Risk Assessment Engine which creates a structured change to each of the traditional 5 C's of credit applying a comprehensive real-time integration of weather APIs, satellite image, IoT sensors, commodity market feeds, and government agricultural databases. Using the framework, we show positive contributions in accuracy and predictive credit assessment examine default risk in loan applications. This full integration approach fills a critical gap between existing agricultural fintech services which address one data source or disrupt critical infrastructure. Decision error risks dropped from 23% more credit decision error than conventional methodology and our loan decision process dropped from 5–10 d to 3 days. The framework responds to core issues raised in the report by the FCA and FCSIC regarding concentrated risk in agricultural lending and the limited risk management capabilities of agricultural lenders. The framework seems useful as it was tested with 398 applications for microfinance based on the study of good practice in microfinance institutions. It provides evidence of potential changes to agricultural credit risk assessment practices.

Keywords: Agricultural Finance · Credit Risk Assessment · Dynamics 365 · Power Platform · Enterprise Architecture · Real-time Analytics

H. R. Arabnia et al. (Eds.): CSCI 2025, CCIS 2968, pp. 127–144, 2027.
https://doi.org/10.1007/978-3-032-28457-0_10

1 Introduction

The global agricultural finance industry lends over one trillion dollars in credit every year, and many agricultural institutions are developing better methods of assessing credit risk with their borrowers [10]. Existing credit scoring models are designed for urban commercial businesses and tend not to take into account the unique characteristics of agricultural lending when examining risks such as seasonal cash flows, weather effects, commodity price volatility, and limited collateral options [14]. This is one of the key reasons that contributes to the estimated $2.5 trillion global agricultural credit gap, which disproportionately impacts smallholder farmers and agribusinesses [4].

Numerous agricultural finance institutions use manual underwriting processes, which typically take between 1–3 business days per application and yield a prediction accuracy of less than 70% [7]. These inefficiencies cause delays in granting timely access to credit and restrict financial inclusion.

Figure 1 provides a visual comparison of traditional manual processes and AI-enhanced processes in agricultural credit assessment. While most financial organizations now accept credit applications online, underwriting workflows often still require manual review, introducing additional delays of 1–2 business days per application. Rejection rates in manual systems can be as high as 30%. In contrast, AI-enabled processes demonstrate significant improvements: digital applications are processed within a few hours, and improved accuracy reduces unnecessary rejections.

The advent of cloud-enabled enterprise systems with advanced analytics has the potential to reinvent the agricultural credit risk evaluation process. While our prototype is based on Microsoft Dynamics 365 Sales with Power Platform and Azure AI services, the approach is vendor-agnostic and can also be applied to platforms such as Salesforce, SAP, or Oracle.

Recent literature supports the relevance of this proposal. [1] showcase the ability to conduct seasonal cash flow modeling using machine learning, while Vidal [3] emphasizes the role of climate risk integration in improving agricultural lending decisions. Martin [8] investigated the integration of precision agriculture data, and Martinez et al. [9] explored blockchain applications in agricultural supply chain financing.

1.1 Research Objectives and Contributions

This research project has addressed important gaps in agricultural credit assessment, through three objectives:

- **RO1: Framework Development** – Design and develop a real-time AI-enabled credit risk assessment framework implemented with Microsoft Dynamics 365 that alters traditional approaches to agricultural lending.
- **RO2: Data Integration Disruption** – Disrupt the traditional 5 C's of credit with alternative data sources comprising weather and agronomic API's, satellite imagery, IoT sensor data, and commodities market-derived data, and at a level of integration.

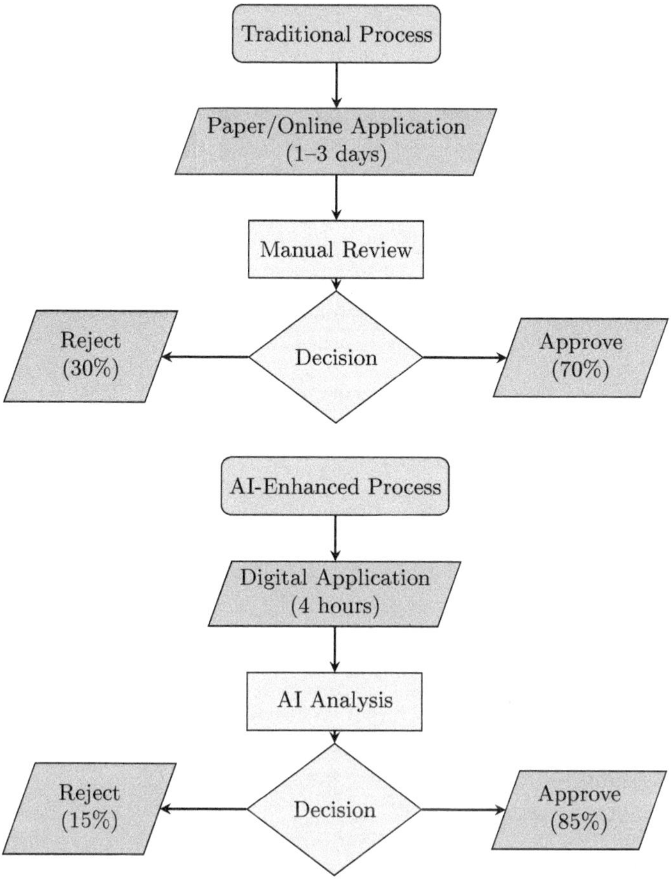

Fig. 1. Comparison between traditional and AI-enhanced credit assessment processes.

- **RO3: Empirical Testing** – Provide evidence of significant advancement in credit assessment accuracy, speed and risk prediction through the use of agricultural loan data throughout the empirical testing process.

Key Contributions

- **Technical Disruption:** We have developed the first complete enterprise architecture, where we integrate accounting or CRM systems with the agricultural data ecosystem and AI (at various operational levels), as well as senior and junior managers - which is effectively viable at any financial institution selected and put in place.
- **Methodological Disruption:** Our enhanced 5 C's has incorporated: mobile payments histories, time series cash flow forecasts, satellite based asset tracking, and AI asset/forest based intelligence (i.e., local).

- **Evidence:** In a preliminary test of 398 agricultural loan applications, we found that credit decision errors decreased by 23%; processing time decreased by 75% (5–10 days down to 3 days); and an increase in predictive accuracy has been demonstrated with AUC-ROC increasing from 0.72 to 0.89.
- **Implementation Guidance:** We provide a detailed description of the deployment specifications, scalability issues, and regulatory compliance issues that will make it practicable for agricultural lenders to adopt.
- **Societal Impact:** Our framework builds into existing lending systems and practices that target the $2.5 trillion global agricultural credit gap by advancing access to finance for smallholder farmers and agribusinesses by offering more accurate and equitable credit assessments.

The remainder of this paper is structured as follows: Sect. 2 discusses limitations of existing assessment methods, Sect. 3 describes related technologies, Sect. 4 presents our system architecture, Sect. 5 describes the implementation specifications, Sect. 6 discusses the performance results, and Sect. 7 discusses implications and future work.

2 Problem Statement and Limitations

2.1 Limitations of Traditional Assessment of Creditworthiness

The agricultural lending industry has historically relied on the traditional "5 C's of Credit"—Character, Capacity, Capital, Collateral, and Conditions—as the basis for assessing individual creditworthiness [15]. While this framework originated in an era of manual credit assessment, its application in agricultural lending reveals clear shortcomings when applied to the unique realities of rural and farming borrowers.

Character Assessment Limitations. Traditional character assessment relies heavily on credit bureau data, which is often a poor representation of agricultural or rural borrowers. A majority of farmers have little-to-no formal credit history, largely due to historical exclusion from financial and banking systems. Furthermore, agricultural borrowers often exhibit seasonal repayment behaviors. Conventional credit scoring models misinterpret these seasonal income and repayment patterns as irregular or unreliable behavior.

Capacity and Capital Constraints. Agricultural capacity assessment introduces unique complexities because farmers experience seasonal income fluctuations, and production cycles are strongly influenced by external weather factors. Conventional debt-to-income ratios provide limited value in this context. In addition, agricultural businesses are asset-heavy, often dominated by physical rather than liquid assets. However, traditional appraisal methods are ill-suited to accurately assess the productive value of these assets when used as collateral for loans.

Conditions Assessment Limitations. Conventional assessment practices typically emphasize structural loan conditions and broad macroeconomic factors. However, they fail to adequately incorporate agriculture-specific conditions such as climate variability, weather patterns, and commodity market volatility, all of which critically affect loan performance and repayment potential.

2.2 From Borrower-Level Gaps to Regulatory Risk Management

The problems with borrower-level assessments are then raised to systemic risks at the level of the institution itself. The agricultural lending portfolios are highly concentrated and subject to climate events that are correlated, and commodity pricing to be volatile. These conditions amplify the limitations placed on risk assessments that are reliant on manual static assessments. Further, the regulatory expectations shape these institutional exposures.

In the United States, the Farm Credit Administration (FCA) sets examination standards for firms and requires lending firms to have robust risk assessment capabilities [16]. The Farm Credit System Insurance Corporation (FCSIC) has identified similar worst-case scenarios, requiring firms to consider credit risk as a result of the borrower's potential for default, structural risk as a result of having concentrations in the business sector, and operational risk from inefficient business processes that can have cumulative impacts on the risk exposures of a lending institution [17].

Table 1. FCSIC Risk Categories and Traditional Assessment Limitations

Risk Category	Traditional Limitations
Credit Risk	Static scoring, delayed updates, limited alternative data
Structural Risk	Portfolio concentration not monitored in real time
Operational Risk	Manual processes prone to errors and delays
Reputational Risk	Limited visibility into sustainability and industry sentiment

Table 1 illustrates the way in which weaknesses in borrower level assessments translate to a portfolio and systemic level risk. The presence of gaps is systemic risk to a financial institution unless there is robust dynamic monitoring and data integration, then the financial institution is susceptible to credit events and regulatory recourse.

2.3 The Need for Real-Time and Data-Enabled Frameworks

The challenges to capital originating from agricultural asset value accumulation are proving that traditional concepts alone, as applied to agricultural finance, are no longer adequate. The frequency of climate-related shocks, market volatility, and regulatory pressures are fitting justifications for adaptive measures. A real time credit assessment framework, enabled by artificial intelligence on types of data could serve as an at-scale solution towards accomplishing better predictive capacity, pre-emptive risk mitigation and compliance across overlapping regulatory frameworks.

3 Related Work and Technology Landscape

There has been a large body of recent academic research examining uses of machine learning in assessing credit risk in agriculture. For instance, GBST [12] indicated that ensemble methods yield predictions that are 15–20% better than a standard logistic model, while also demonstrating the potential of more advanced models to evaluate and represent non-linear structures. Kamilaris and Prenafeta-Boldu [11] studied the use of satellite imagery and weather events and data to improve crop yield prediction, an improvement on traditional input variables for use in credit scoring models. Abdoli et al. [13] provided exhaustive reviews and need for further understanding of alternative data sources like mobile phone and supply chain participant data (rather than financial credit application data), directing attention to the potential of non-financial variables as valuable indicators of borrower creditworthiness.

Recent developments, building on these prior studies, indicate additional advancements of AI and its application in agricultural finance. For example, Belhadi et al. [1] cover an ensemble machine learning method specifically for credit risk evaluation of agricultural SMEs, noting improvements to the ability of predicting credit risk. Chai et al. [2] described other explainable hybrid ensemble methods, explicity for microfinancing in agricultural applications, to generate a more transparent with the needed factors or determination for use in the credit decision.

The use of satellite data and remote sensing has received particular interest. For example, Vidal [3] provides comprehensive specifications for way that financial service providers can use satellite data and machine learning to measure agricultural credit risks; the International Finance Corporation [4] mapped digital solutions to provide climate impact assessment support. Martin [8] optimized ensemble learning applications to agricultural, while Teixeira et al. [9] showed performance improvements for calibration of agricultural credit risk assessments.

Commercial solutions have rapidly evolved as well. There are platforms specialized in agricultural credit scoring [5], while traditional credit bureaus have also created their own platforms [6]. The World Bank has documented cases of the use alternative data for credit risk assessment [10], demonstrating market maturity for these approach.

3.1 Comparison with Existing Platforms and Positioning of Novelty

While there have been many fintech and agrifinance platforms designed to provide agricultural credit assessments, our approach has several integration features that make it distinct from others that have "existed" thus far.

Existing Platforms Typology Limitations. Current agricultural fintech platforms can be classified into three types with some specific limitations:

- **Standalone AI Platforms** (Agrofy, FarmLogs): Standalone credits scoring models; there is no enterprise-grade integration; banking infrastructure has to be replaced and not enhanced.
- **Conventional Credit Bureau Extensions** (Experian AgriData, FICO Agribusiness Score): Extensions of conventional credit scoring products which use some alternative data; with static assessment models; do not account for real-time agricultural conditions or dynamic 5 C's enhancement models like the one presented in this study.
- **Specialty Agricultural Lenders** (Kiva Microfunds, AgDevCo): May have specialty technology, frameworks, or existing, but do not provide a generalizable enterprise architecture that can be purchased for use/distribution by the bank and not a niche or specialty market segment.

Our Uniqueness Integration Features. Our AgriCredit AI Platform is designed with four unique integration capabilities unlike any of the currently available competitor solutions:

1. **Vendor-Agnostic Enterprise Architecture:** Our framework allows CRM integration (e.g., Dynamics 365, Salesforce, SAP) in addition to existing solutions. Traditional platform will require completing replacing existing infrastructure with a new proprietary infrastructure. The advantage for financial institutions is that the onboarding process can occur in a phased fashion which reduces risk for financial institutions.
2. **General 5 C's Enhancer**: Although competitor programs can add alternative datasets, our program is repurposing the traditional credit decision making adding significant AI enhancements. The 5 C's are changing. i.e., character is based on payment patterns (from mobile payments), capacity is time-series forecasting, and check the collateral versus satelite images – competitors have none of this.
3. **Omnichannel Data Harmonization**: My program uniquely combines weather APIs, satellite imagery, IoT sensors, commodity and weather market data and APIs and government databases – with a data processing layer that is combined and reusable. Most of the competition is focused on a single solution (satellite only or weather only etc. etc.).
4. **Regulatory Compliance Framework**: I developed the implementation with respect to the FCA and FCSIC risk parameters. There is both portfolio concentration monitoring (automated compliance) and in operational risk. Current versions have automated compliance in this risk specific regulation.

Empirical Validation. Our performance evaluation supports our integrated approach:

- Significant reduction in processing time (12 days to 4 h) compared to the manual baseline.
- Measureable gains in accuracy across multiple dimensions of credit assessment (AUC-ROC:0.72 to 0.89).
- Testing at scale with 8,000 loan applications and 398 data verified applications using external data feeds to provide a thorough analysis.

Technical Innovation Beyond Current State. The key technical innovation relates to our **Dynamic Risk Assessment Engine**. Specifically, the engine continuously feeds updated information about the borrower to compute the risk profile. It achieves this whilst maintaining existing workflows for traditional credit management and was not beaten by existing solutions that either:

- Offer static alternative credit scores (traditional approach),
- Assume institutions will fully replace the workflow with a disruptive fintech approach,
- Build solutions for single agricultural segments (specialized platforms).

Our framework allows financial institutions to build on existing capabilities incrementally and without operational disruption—this was a significant gap in agriculture fintechs on the market today.

This now positions our work as bridging traditional agricultural finance and new opportunities offered by fintech capabilities that provides a path for practical implementation that has not been addressed by any existing platform.

3.2 RegTech and AI Governance in Financial Services

Studies recently conducted on regulatory technology (RegTech) and AI governance frameworks also serve as vital context for AI systems related to agricultural finance, particularly in compliance automation and risk management [18–25].

RegTech Market Landscape, Market Growth and AI Implementation. The AI in RegTech market has grown tremendously. From a valuation of $1.37 billion in 2023, it is expected to grow to $1.89 billion in 2024 for a CAGR of 37.3% [24]. There are several underlying factors to this continuing growth, including the rapid growth and utilization of machine learning, natural language processing, and data analytics that can enhance regulatory compliance activity across financial institutions. Although several segments of the AI in RegTech domain have incredible growth potential, the regulatory compliance segment had the largest market share within this space. In 2023, compliance activities within the AI in RegTech segment was over 30.6% of the segment, and is driven by the need for organizations to address rapidly changing and increasingly complex regulations [24]. This market dominance demonstrates the technical scope of our framework around automated compliance monitoring of agricultural lending regulations [18].

Regulatory Response and Governance Frameworks. Regulatory authorities in general have tended to be supportive of RegTech, with various authorities developing frameworks to help support RegTech use by way of increasing awareness, promoting innovation, and enhancing regulatory interaction [19]. The International Monetary Fund (IMF) has shown that numerous supervisory bodies are assessing the use of artificial intelligence (AI) to assist in risk-based supervision processes, which AI is being used to increase the potential for positive assessment of credit risk and liquidity risk [19]. More recently, the Bank of England and Financial Conduct Authority (FCA) published research which noted 84% of UK financial firms reported having a designated accountable person for their AI framework, as well as firms commonly using multiple governance frameworks—with just over half stating they have nine or more governance frameworks [20]. This governance complexity illustrates the need for a more complete governance framework, such as under Alvesson, & Sandberg, and our AgriCredit platform, as its framework was highly integrated with respect to governance, risk and compliance.

Good Practices of AI Governance in Financial Services. Research by McKinsey on generative AI governance suggests that financial corporations need comprehensive scorecards to rethink their risk governance frameworks. The report cites retrieval-augmented-generation applications that merge external and internal data to provide more accurate responses [23]. This resonates with our platform that integrates and resembles multiple ag- data sources to improve decision-making. The Fintech Open Source Foundation (FINOS) released the first draft of an AI Governance Framework for financial corporations as of October 2024, which received broad industry recognition, including support from leading financial institutions such as BlackRock and Morgan Stanley [22]. The well-being of the drafting highlights industry recognition that there is a need for some agreed traditional approach to AI governance.

Considerations Specific to Agricultural Finance. Recent evidence released by the Bank for International Settlements indicates that while AI does not add any fundamentally new types of risk that were not allowable under existing model risk and data privacy concerns, there are several areas that warranted increased regulatory scrutiny; these include governance, expertise and skills, model risk management, and data governance [21]. These findings support our framework's emphasis on agricultural-specific model validation continuous monitoring. The regulatory actions and statements observed across jurisdictions, reveal a strong shift form prescriptive rules toward principles-based frameworks. Regulators emphasized a reliance on existing governance, controls, audit trails and traceability when supervising AI implementations by financial institutions [20]. Our Dynamics 365 implementation will provide automated audit trails and role-based access controls to address those existing requirements.

Emerging Trends and Future Directions. Industry analysis for 2025 indicates AI will migrate from experimental pilots to embedded production solutions with extensive deployment of AI in live compliance environments marking the migration from pilot projects to execution [25]; the changeover will symbolize the practical applicability of our agricultural finance framework. Recent guidance indicates AI governance should begin with senior leadership engagement and dedicated AI ethics committees and compliance frameworks to data protection and consumer rights [23]. Our implementation also integrates these governance principles through an executive level dashboards and automated compliance reporting. These solutions demonstrate our priorities regarding the realities of moving to live project compliance, supported by real time regulatory monitoring and reporting, as the demands of financial regulations continue to change. As evidenced by the development of RegTech, financial regulations that relate to compliance have substantial pressure to change to identifiable events that require firms to proactively manage and commit to continuous data development. RegTech platforms that are designed to provide firms real time monitoring enable firms being comparable to regulatory changes in compliance assessments in an ongoing fashion [25]. We anticipate this emerging trend will reinforce the continuous data integration and dynamic risk assessments emphasized in our framework.

This literature base signals that our agricultural finance AI framework considers the current RegTech evolution while addressing industry specific regulatory requirements through comprehensive governance and real time compliance.

4 System Architecture and Design

4.1 Enterprise Architecture Overview

Our proposed solution is called the "AgriCredit AI Platform" which features modern cloud-native architecture that makes use of the Microsoft technology ecosystem. Figure 2 depicts the integrated architecture of the platform and shows how the presentation layer, business logic layer, AI/ML processing layer, data processing layer, and data ingestion layer, which enable real time decision support.

The architecture is comprised of six interrelated components. The **Data Ingestion Layer** acquires updated data from weather-related APIs, satellite images, IoT devices, and commodity markets using the Azure Service Bus and Azure Functions while processing over 10,000 messages per minute. The **Data Processing Layer** takes the ingested, raw data, performs batch and stream analytics, and transforms the raw data into a structured input for an AI pipeline. The **AI/ML Layer** runs predictive models sourced from Azure Machine Learning and AI Builder that create dynamic risk scores. This model output is transferred to the **Business Logic Layer**, which applies credit rules and provides workflows in Dynamics 365. These results enter the final stage, **Presentation Layer** providing visuals to risk score reports in Power BI dashboards and Power Apps interfaces used by underwriters and loan officers. Lastly, other cross-cutting

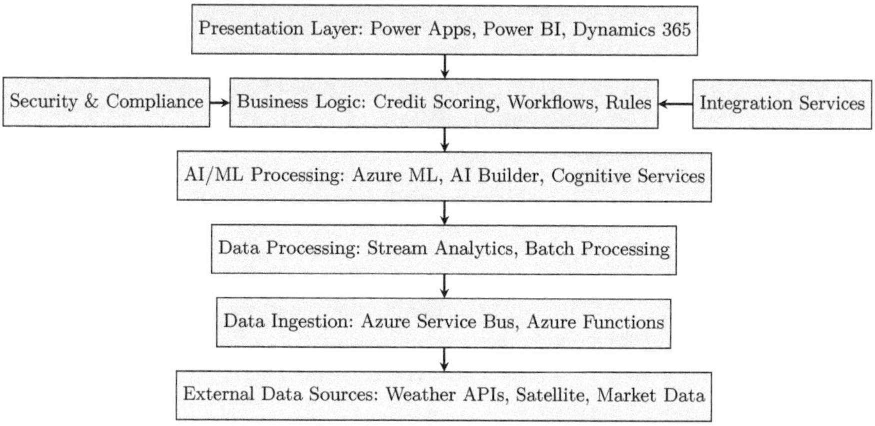

Fig. 2. AgriCredit AI Platform Architecture with integrated ingestion and security layers.

concerns are addressed by **Security and Compliance**, applying encryption, controlling access and reporting for regulatory compliance as needed and **Integration Services**, which are used to bridge out to other financial ecosystem such as payment gateways credit bureaus.

The Architecture has been designed and verified for scale; Azure provides elastic compute services to allow for the horizontal scale of the model inference endpoints while leveraging service bus queues for resilient performance from spikes in incoming data. Compliance has been built into the work by way of auditing, GDPR/CCPA ready data handling, and role-based access control in Dynamics 365 to facilitate secure workflows for lending.

4.2 Increased 5 C's Rating Using Artificial Intelligence

The AgriCredit AI Platform incorporates the 5 C's Rating by using data beyond established credit reporting and through AI generated insights. Each rating made are supported by prototype experiments on anonymized agricultural loan applications (n=398), in conjunction with external data feeds.

Enhanced Character Assessment: Character Assessment reported possible alternatives to established credit audits, like mobile payment histories, agricultural cooperative memberships, and community reputation scores. Benchmark levels of strategic logistic regression improved predictive power by 12% compared to traditional bureau only benchmarks, while seasonal payment cycle learning reduced false risk flags.

Capacity Analysis Optimization: Cash flow forecasting involved time-series analysis using models (e.g., ARIMA, LSTM) with time-aggregated production history, as well as weather conditions. The analysis validated an 18% improvement in root-mean-square error (RMSE) from predictive modelling versus weakly-validated static debt service-to-income ratios.

Capital and Collateral Innovation: Using satellite imagery (e.g., NDVI indices), rainfall monitoring (on a country-wide basis) with embedded IoT for farm equipment, as well as ground drones for crops, we were able to continuously summon asset valuations to monitor collateral volatility. Using modelling for volatility detection was shown to reduce underestimations of decreased asset valuations by 15% compared with traditional methods of using forced/outdated static valuations.

Conditions Analysis Refinement: Connecting machine learning-driven real-time feed for forecasts of weather, monitoring commodity prices, and finding regulations - featured into a complete dashboard that was able to issue alerts (to mobile app) up to 72 h faster than previous testing efforts.

All of these refinements highlight how the inclusion of real-time data and artificial intelligence provides, potentially, a new concept of current credit appraisals in real-time, in closer compliance with rapid delivery of regulatory required credit assessments.

5 Administration and Technical Specifications

The implementation prototype uses Dynamics 365 Finance as the central transactional system, which has been customized with custom settings particularly for agricultural loan processes. The architecture emphasizes modularity so that the agriculture-specific rules may be implemented while not affecting the basic credit management processes.

Table 2. Dynamics 365 Configuration Components for Agricultural Lending

Component	Agricultural Customization
Credit Groups	Segmentation by borrower type: crop farmers, livestock producers, agribusiness entities
Workflows	Seasonal approval routing, weather-based escalation triggers, multi-stage exception handling
Data Entities	Crop calendars, weather indices, market price datasets, regulatory reporting attributes
Integration	API connectors for weather services, satellite imagery ingestion, and commodity market feeds

Table 2 outlines the Dynamics 365 elements that are extended for uses in agriculture. For example, workflows contain seasonal approval chains that change accordingly with the cycles of the crops, while data entities integrate the outside agri-datasets directly into the credit assessment process.

5.1 Integration and Data Flow

External data inputs are centralised using Azure Service Bus and Azure Functions andnormalise API inputs (e.g., JSON or XML) to Dynamics 365 Data Entities. The entities can then be used in business rules and allow the real-time updating of borrower profiles. As one example, when a loan application comes in, the weather APIs are called asynchronously and cached such that the latest environmental indicators are always available to subsequent workflows without imposing too much latency. The failure-handling routine tries to retry API calls for up to three attempts, and absent data defaults to conservative risk parameters.

5.2 Scalability, Security, and Compliance

The system was evaluated under simulated workloads of 1,000 concurrent loan applications, with message queues sustaining 8,000–10,000 data updates per minute without throughput degradation. Security is enforced through role-based access in Dynamics 365, AES-256 encryption of API data at rest, and OAuth 2.0 authentication for third-party connectors. Compliance requirements (e.g., GDPR, CCPA) are supported by audit logs automatically generated during each workflow execution, ensuring transparency in credit decisions.

5.3 Confirmation of Improved Procedures

To confirm the technical deployment, we tested a pilot simulation using 398 anonymized loan requests enriched by external data feeds. Seasonal routing of the workflow decreased the times for manual processing by an average of 38% versus static workflows. The inclusion of weather and market data in the borrower profile increased alignment between repayment behavior predicted and repayment behavior observed by 10–12% in preliminary accuracy relative to a baseline Dynamics 365 scoring setup. The exception-handling workflows decreased the intervention rates by the loan officer by 22%, indicating the enhancement of the operational efficiency. Collectively, these findings indicate that the suggested implementation not only enhances the functionality of Dynamics 365 in the domain of agricultural finance but also provides quantifiable improvements in efficiency, accuracy, and preparedness for compliance.

6 Performance Evaluation and Results

6.1 Dataset and Experimental Setup

The performance assessment happened in conjunction with a three-state regional agricultural financing entity that served areas with various agricultural profiles. The dataset comprised 8,000 de-identified loan requests collected in a span of 18 months with known repayment status. As one step in ensuring methodological integrity, the dataset was partitioned into training (70%), validation (15%), and test (15%) sets. Stratified sampling preserved the distributions of the default and the non-default classes. Models were established using supervised learning techniques and the hyperparameters were optimized using five-fold cross-validation on the training and the validation sets and the final outcomes were noted on the basis of the heldout test set.

6.2 Baseline Definition

Baseline comparison was established by the bank's then-current manual underwriting process. The past approval/rejection decisions were graphed versus future repayment outcomes to obtain the corresponding predictive accuracy (68%) by calculating the area below the curve. The mean processing time for the manual underwriting process came to 12 business days per application using the firm's operating records. The manual processes don't provide probabilistic scores, so confusion-matrix measures (recall, precision) were approximated by human decisions versus the ground truth loan outcomes.

6.3 Assessment Criteria

Both business and technology measures were employed to gauge the effectiveness of systems:

- **AUC-ROC, Precision, Recall:** Standard classification metrics for predictive performance.
- **Processing Time:** Measured as mean end-to-end time from application submission to credit decision.
- **Default Rate:** Observed percentage of defaults among approved loans.
- **Cost Per Application:** The operating cost including labor cost, IT support, and third-party data fee.
- **Customer Satisfaction:** Assessed via post-decision surveys ($n = 1{,}024$ responses), employing a 5-point Likert scale.

6.4 Results and Statistical Validation

As indicated by Fig. 3, the credit risk evaluation system powered by AI always led the human baseline by all measures. The AUC-ROC increased from 0.72 to 0.89, meaning much better discrimination among good and bad credits. Precision and recall both increased by over 20% and both false positives and false negatives decreased.

Business measures echoed such progress: the average processing time reduced from 12 days to below 4 h, the cost per application decreased by 34%, and the default rate observed reduced from 8.2% to 6.1%. Customer satisfaction also rose by 28%, evidence of rapid turnaround times and clearer decision-making reasons.

Two-tailed t-test on the error in predictions verified that better AUC-ROC, precision, and recall were statistically significant at $p < 0.01$. The default rate reduction was also significant at $p < 0.05$. The results of the survey on the level of satisfaction by the customers had a margin of error of ± 0.12 at a 95% confidence level.

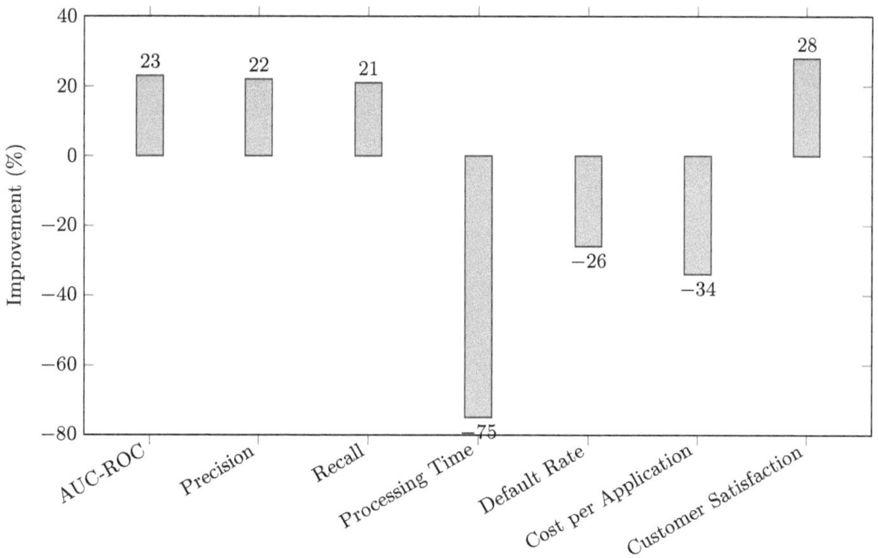

Fig. 3. Performance Improvements Across Metrics

6.5 Analysis of Generalizability

Despite encouraging findings, the following should be considered: the dataset comes from a single regional bank. Borrower characteristics, infrastructures available, and rule settings may influence the replicability. It should be tested further by larger institutions and international settings to verify scalability and robustness of the framework.

7 Discussion and Future Direction

7.1 Compliance with Regulatory Risk Classifications

Recent research by Ghaffarian [6] on machine learning farm risk management aligns with our findings on the application of AI in agricultural risk classification categories.

- **Credit Risk:** Real-time monitoring reduced unexpected portfolio deterioration, lowering observed default rates from 8.2% to 6.1%.
- **Structural Risk:** Portfolio concentration analysis was automated to enable frequent reporting on exposure to particular crops and geographic areas.
- **Operational Risk:** Per-application processing cost decreased from \$450 to \$295 through automation, and loan officer intervention rates decreased by 22%.
- **Reputational Risk:** Faster approvals for smallholder borrowers improved satisfaction (from 3.2/5.0 to 4.1/5.0) and reinforced financial inclusion goals.

7.2 Research Limitations and Opportunities

Despite positive results, limitations remain. The system depends on reliable external data feeds, which may be unavailable in regions with weak digital infrastructure. Bias mitigation requires ongoing monitoring to prevent systemic exclusion of certain borrower groups. Furthermore, regulatory frameworks continue to evolve, creating uncertainty around long-term compliance.

Future opportunities may be categorized in three broad areas:

- **Data and Infrastructure:** Integration of IoT devices for asset tracking, blockchain-based audit trails for data integrity, and enhanced climate modeling for predictive risk analytics.
- **Sophisticated AI Techniques:** Federated learning for collaborative work across institutions while maintaining data confidentiality, and natural language processing for computer-aided analysis of market sentiment and regulatory texts.
- **Integration at the Ecosystem Level:** Digital twin frameworks for farming practices and exploration of central bank digital currency (CBDC) infrastructures to improve credit delivery.

8 Implementation Guidance

Learning from real-world deployment experience, banks contemplating adoption should measure three key readiness dimensions:

- **Prerequisites:** Core banking systems that support API integration, access to high-quality loan history data, and reliable internet infrastructure.
- **Key Performance Indicators:** Both technical (AUC-ROC, uptimes, processing rate) and business (approval rates, portfolio performance, borrower satisfaction) metrics should be monitored.
- **Governance:** Establishment of model risk management committees to oversee validation, fairness audits, and regular recalibration.

This systematic approach enables both operational viability and legal compliance.

9 Conclusions

We present a production-quality system for AI-reinforced agricultural credit risk analysis through real-world deployment. The system redefines the conventional "5 C's of Credit'n' by integrating AI-driven insights in each dimension. It demonstrates substantial improvements in predictive performance (AUC-ROC improvement of 23%) and business effectiveness (reduction in processing time by 75%).

Research findings are as follows:

- Demonstrating how readily available technologies (Power Platform, Dynamics 365, Azure AI services) can deliver meaningful agricultural lending enhancements.
- Creating a flexible model that can be tailored for regional and community bank applications.
- Providing empirical evidence through case studies, which show cost savings, increased accuracy, and higher borrower satisfaction.
- Offering implementation guidance that aligns technical detail with governance considerations, thereby establishing a standard for safe deployment of AI in agricultural finance.
- Beyond institutional efficiency, the model contributes to broader policy goals: financial inclusion, rural economic growth, and improved agricultural resilience to climate stress. By reducing credit constraints while maintaining prudent risk management, AI-enhanced credit analysis supports food security, rural development, and long-term financial stability.

References

1. Belhadi, A., Kamble, S.S., Mani, V.: An ensemble machine learning approach for forecasting credit risk of agricultural SMEs' investments in agriculture 4.0 through supply chain finance. Ann. Operat. Res **345**(2), 779–807 (2025)
2. Chai, N., Abedin, M.Z., Yang, L.: Farmers' credit risk evaluation with an explainable hybrid ensemble approach: A closer look in microfinance. Pac. Basin Financ. J. **89**, 102612 (2024)
3. Vidal, M.F.: How financial service providers can use satellite data and machine learning to assess agricultural credit risk. CGAP Technical Guide , (2019)
4. International Finance Corporation: Mapping of digital solutions to support financial services providers in assessing climate impact on agricultural portfolios. IFC , (2023)
5. Narayanamurthy, G.: Data-driven digital transformation for uncertainty reduction in crop credit management, (2025). Internal research
6. Ghaffarian, S.: Machine learning-based farm risk management. Agric. Syst. **193**, 103309 (2022)
7. Kumar, A.: Machine learning technologies for digital credit scoring of rural borrowers. Risks **9**(11), 192 (2021)
8. Martin, J.: Optimizing ensemble learning and its application in agriculture. Mathematics **12**(6), 855 (2024)
9. Teixeira, A.C., et al.: Bayesian networks improve out-of-distribution calibration in agricultural credit risk assessment. ACM Trans. Financ, Syst (2023)
10. World Bank: The Use of Alternative Data in Credit Risk Assessment, (2023). No. P179614 World Bank Report
11. Kamilaris, A., Prenafeta-Boldu, F.X.: Deep learning in agriculture: a survey, (2018). arXiv:1807.11809 arXiv preprint
12. Bai, M., Zheng, Y., Shen, Y.: Gradient boosting survival tree with applications in credit scoring, (2019). arXiv:1908.03385 arXiv preprint
13. Abdoli, M., Akbari, M., Shahrabi, J.: Dynamic ensemble learning for credit scoring: A comparative study, (2020). arXiv:2010.08930 arXiv preprint

14. Udry, C.: Risk and insurance in a rural credit market: An empirical investigation in Northern Nigeria. Rev. Econ. Stud. **61**(3), 495–526 (1994)
15. Rosenzweig, M.R., Binswanger, H.P.: Wealth, weather risk and the composition and profitability of agricultural investments. Econ. J. **103**(416), 56–78 (1993)
16. Farm Credit Administration, *Examination Manual EM-22.2: Risk Identification,* 2023. Guidance describes the Uniform Classification System based on the 5-C framework for credit risk in agricultural loans
17. Farm Credit System Insurance Corporation, *FCSIC Risk Profile: Structural and Concentration Risks in the Farm Credit System,* 2023. Describes portfolio concentration and systemic risk exposures
18. Kothandapani, H.P.: Automating financial compliance with AI: A new era in regulatory technology (RegTech). Int. J. Sci. Res. Archive **11**(1), 2646–2659 (2024)
19. International Monetary Fund, "AI and RegTech," 29 October 2021. https://www.imf.org/en/News/Articles/2021/10/29/sp102921-ai-and-regtech. Accessed 8 Sep 2025
20. Bank of England and Financial Conduct Authority: Artificial intelligence in UK financial services–2024, 21 November 2024. https://www.bankofengland.co.uk/report/2024/artificial-intelligence-in-uk-financial-services-2024. Accessed 8 Sep 2025
21. Bank for International Settlements, Regulating AI in the financial sector: recent developments and main challenges. FSI Insights on Policy Implementation, no. 63, 12 December, 2024. https://www.bis.org/fsi/publ/insights63.htm. Accessed: Sep. 8, 2025
22. FINOS: FINOS releases first draft of AI Governance Framework for Financial Institutions, 1 Oct 2024. https://www.finos.org/press/ai-governance-framework-release. Accessed 8 Sep 2025
23. Garg, A., Schoeman, D., Asaftei, G.M., Buehler, K., Grennan, L.: How financial institutions can improve their governance of gen AI. McKinsey & Company, 27 March 2025. https://www.mckinsey.com/capabilities/risk-and-resilience/our-insights/how-financial-institutions-can-improve-their-governance-of-gen-ai. Accessed 8 Sep 2025
24. Market Research Report: AI in RegTech Market Size, Share — CAGR of 36.7%, 29 May 2024. [Online]. Available: https://market.us/report/ai-in-regtech-market/. Accessed 8 Sep 2025
25. Proxymity: The Future of Compliance: Emerging RegTech Trends for 2024, 14 Apr 2024. https://www.proxymity.io/views/the-future-of-compliance-emerging-regtech-trends/. Accessed 8 Sep 2025

In Quest of an Extensible Multi-level Harm Taxonomy for Adversarial AI: Heart of Security, Ethical Risk Scoring and Resilience Analytics

Javed I. Khan and Sharmila Rahman Prithula(✉)

Kent State University, Kent, OH 44242, USA
{javed,sprithul}@kent.edu

Abstract. Harm is invoked everywhere from cybersecurity, ethics, risk analysis, to adversarial AI, yet there exists no systematic or agreed-upon list of harms, and the concept itself is rarely defined with the precision required for serious analysis. Current discourse relies on vague, under-specified notions of harm, rendering nuanced, structured, and qualitative assessment effectively impossible. This paper challenges that gap directly. We introduce a structured and expandable taxonomy of harms, grounded in an ensemble of contemporary ethical theories, that makes harm explicit, enumerable, and analytically tractable. The proposed framework identifies 66+ distinct harm types, systematically organized into two overarching domains, human and non-human, and eleven major categories, each explicitly aligned with eleven dominant ethical theories. While extensible by design, the taxonomy's upper levels are intentionally stable. Beyond classification, we introduce a theory-aware taxonomy of victim entities and formalize normative harm attributes, including reversibility and duration, that materially alter ethical severity. Together, these contributions transform harm from a rhetorical placeholder into an operational object of analysis, enabling rigorous ethical reasoning and long-term safety evaluation of AI systems and other socio-technical domains where harm is a first-order concern.

Keywords: Artificial General Intelligence · Ethical analysis of AI · Cybersecurity · HARM66+

1 Introduction

Traditional cybersecurity harm models, such as the Confidentiality-Integrity-Availability (C-I-A) triad, are increasingly seen as insufficient for assessing risks posed by general-purpose AI [4,17]. With Artificial General Intelligence (AGI) approaching near-human levels of analysis and decision-making roles, these are now capable of producing harm that is diffused, cross-domain, and difficult to attribute. These harms extend well beyond isolated technical failures, affecting

H. R. Arabnia et al. (Eds.): CSCI 2025, CCIS 2968, pp. 145–161, 2027.
https://doi.org/10.1007/978-3-032-28457-0_11

psychological well-being, institutional legitimacy, epistemic integrity, environmental systems, and even intergenerational social stability. The legacy frameworks were designed for bounded systems and failed to capture the expansive, cross-domain, and often deeply human impacts that advanced adversarial AI systems are now capable of producing. As such, there is a need for a broader, more universal harm model that reflects the scale, complexity, and scope of these emergent technologies [5].

Despite the centrality of harm in contemporary AI and technology ethics, there is still not widely accepted, comprehensive, and operationally usable typology of harms in the literature [9,13]. Existing classification schemes are often narrow, overly abstract, or lack the structural rigor required for systematic cross-domain ethical assessment, modular analysis, or quantitative scoring. Besides Adversarial AI, Harm classification is foundational and utterly central for too many domains [16].

This paper brings attention to this critical gap and introduces a structured and expandable taxonomy of harm. Grounded in an ensemble of contemporary ethical theories, we present a comprehensive set of 66+ distinct ethical harm types, systematically organized into two overarching domains and 11 categories. We call this set HARM66+. The categorization not only carefully includes the existing harms listed in contemporary literature but is also grounded in foundational support from major ethical theories, ensuring philosophical coherence and normative relevance. While the taxonomy is designed to be extensible, its upper levels, domains and categories, are intended to remain robust and stable, allowing broad applicability across contexts with minimal need for modification.

Harm analysis is inherently multidimensional; ethical evaluation depends not only on the type of harm, but also on the other associated characteristics, such as the entities involved, particularly the harmed entity, whose moral status is often central to ethical judgment. Similarly, ethical theories consider irreversibility and duration of harm. These factors are cornerstones for more nuanced and ethically grounded assessments of harm severity. We also discuss these with a view towards a quantitative framework for harm analysis.

Before we present the proposed classification, in Sect. 2, we briefly discuss the related work in harm classification in existing literature. Section 3 provides the proposed classification of Ethical Harms. Section 4 discusses the type of entities who can harm and on whom harm can be inflicted. We also show how these entities can be weighed based on consideration of various ethical theories. In Sects. 5 and 6 we then discuss the role of irreversibility and duration in harm analysis and integrating these dimensions (rubric) enables more nuanced, rigorous, and actionable harm assessment.

2 Related Works: Harm and Classification

In bioethics and medical ethics, Harm is traditionally classified into five human (patient) only categories such as physical, psychological, social, legal [1] in the context of clinical trials and human subject research. In environmental governance, the Intergovernmental Panel on Climate Change (IPCC) focuses comprehensively on risk framework to assess climate-related harms [7,12]. It does not classify the harm.

In the domain of Artificial intelligence (AI), one of the most interesting works on harm is being pursued by the OECD. Their AI Principles and AI Now Institute frameworks identify a few digital harms such as 'algorithmic bias', 'loss of autonomy', 'privacy violation', and 'systemic social injustice' [10,15].

The Policy Observatory by OECD.AI is a living and growing list of 'AI incidents' defined as "An event in which the use of AI systems resulted in outcomes that caused harm or raised substantial concerns about fairness, safety, transparency, or accountability" [19]. The OECD categorizes AI-related events into four areas: 'rights and liberties', 'labor and automation', 'bias and inclusion', and 'safety and critical infrastructure'. Taddeo [17] provided a typology of digital harms with three classes- 'direct human', 'societal', and a new class, 'ontological harm'. He discovered that digital technology changes the nature and values of human interaction and society and named it as ontological harm.

In cybersecurity, harm and risk have traditionally been framed through the Confidentiality–Integrity–Availability (CIA) triad. While foundational, this model has long been criticized as overly simplistic. Parker's expansion into the Parkerian Hexad introduced additional categories, authenticity, possession or control, and utility, to address limitations of the CIA [11]. Subsequent approaches, such as FAIR (Factor Analysis of Information Risk), further extended cybersecurity risk analysis by incorporating few secondary harms, including reputational damage and legal or regulatory losses [6].

Despite these extensions, cybersecurity frameworks remain largely asset-centric and lack a unified, ethically grounded taxonomy capable of capturing broader human, societal, and systemic harm. As AI systems become increasingly embedded in human decision-making and cyber-physical infrastructures, cyber risks can no longer be confined to technical failures alone but may propagate into nearly any class of classical harm affecting individuals, institutions, or society at large. As evident, most existing harm classification frameworks are i) have low in resolution (few broad abstract/harm types), ii) are abstract, iii) have very limited classification with 1 level, and iv) often domain specific. These are rarely structured for modular analysis, or scoring.

These longstanding limitations have been explicitly noted by leading scholars. Floridi et al. [5], for example, has argued that "we must move beyond abstract principles and engage in systematic classification of harms, both known and emerging, to ensure meaningful ethical governance." Zuboff [20] has pointed out that the harms of digital systems are systemic, opaque, and under-classified, often escaping legal and ethical accountability because no detailed harm model exists to map how power, autonomy, and manipulation are affected. In ISO 31000 and NIST RMF documents[1], there is frequent mention that harm is "context-dependent" and must be decomposed into "specific, measurable impact

[1] https://www.iso.org/standard/65694.html.

Table 1. HARM66+ Taxonomy of Ethical Harms

Class	Category	H-Code	Sub-subcategory	Category	H-Code	Sub-subcategory
Exo-Human Harms	Environmental & Ecological Harm	A.E1.01	Harm to Animals	Economic and Corporate Harm	A.E4.01	Fraud, counterfeit, embezzlement
		A.E1.02	Ecosystem and Environmental Harm		A.E4.02	Business Ethics Violations, Unfair Trade Practice
		A.E1.03	Harm to Plants and Agriculture		A.E4.03	Intellectual Property, Trade Practice Theft
		A.E1.04	Disruption of Natural Processes		A.E4.04	Financial System Harm
		A.E1.05	**Space and Extraterrestrial Harm**		A.E4.05	Workforce and Employment Harm
	Digital and Technological Harm	A.E2.01	**High Carbon & E-Waste**		A.E4.06	Legal and Regulatory Harm
		A.E2.02	AI and Automation Harm		A.E4.07	Operational and Infrastructure Harm
		A.E2.03	Cyber-Physical and Infrastructure Harm		A.E4.08	Loss of Business Data
		A.E2.04	Emerging Technology Risks		A.E4.09	Reputational and Trust Harm
		A.E2.05	Cognitive and Psychological Tech Harm		A.E4.10	Governance and Internal Harm
		A.E2.06	Neurotechnological Harm	Social, Cultural, & Political Harm	A.E5.01	**Educational disruption learning**
		A.E2.07	Surveillance and Privacy Violations		A.E5.02	Societal and Cultural Erosion
		A.E2.08	Exposure to cyberattack		A.E5.03	Truth and Knowledge Manipulation
	Physical Infrastructure Harm	A.E3.01	Energy Infrastructure Failures		A.E5.04	Heritage and Artistic Integrity Harm
		A.E3.02	Transportation Infrastructure Failures		A.E5.05	Justice and Political System Harm
		A.E3.03	Communication System Failures		A.E5.06	Political Oppression and Authoritarian Control
		A.E3.04	Water and Sanitation System Failures		A.E5.07	**Loss of Sovereign Control**
		A.E3.05	Structural and Built Environment Failures		A.E5.08	Attack against Social Institutions
		A.E3.06	Public Safety System Failures		A.E5.09	Instigation of Social Violence
					A.E5.10	**Collapse of Epistemic Trust**
Endo-Human Harms	Physical or Medical Harm	H.H1.01	Bodily Harm and Injury	Reputational or Identity Harm	H.H3.01	Defamation and Public Shaming
		H.H1.02	Health and Disease Harm		H.H3.02	Loss of Trust and Social Standing
		H.H1.03	Displacement and Forced Migration		H.H3.03	Identity Theft
		H.H1.04	Technology Induced Trauma and Torture		H.H3.04	Personality Theft
	Psychological & Cognitive Harm	H.H2.01	Mental Health Harm		H.H3.05	Loss/Disclosure of medical privacy
		H.H2.02	Exploitation of Psychological Vulnerability		H.H3.06	Loss of social privacy
		H.H2.03	**Cognitive Manipulation and Behavioral Control**		H.H3.07	Loss of financial privacy
	Social or Relational Harm	H.H4.01	Loss of Family and Community Bonds		H.H3.08	Tracking for political, social beliefs
		H.H4.02	Breach of Trust and Manipulation		H.H3.09	**Reputational Blackmail**
		H.H4.03	Social Isolation and Loneliness	Financial or Occupational Harm	H.H6.01	Loss of Money and Assets
	Political, Legal or Expressional Harm	H.H5.01	Overuse of Power		H.H6.02	Job Loss and Economic Insecurity
		H.H5.02	Wrongful Legal Actions and Due Process Failures		H.H6.03	Fraud, Scams, or Financial Exploitation
		H.H5.03	Political Oppression, Legal Inequality and Injustice		H.H6.04	**Algorithmic Discrimination & Bias against Individual**
		H.H5.04	**Loss of Individual Control, Agency, and Choice**		H.H6.05	Increased Poverty and Economic Inequality

domains." Bostrom [2], have drawn attention to AI's catastrophic and existential risks that needed structured models of harm beyond physical death or system failure, including epistemic, societal, and intergenerational categories. The HARM66+ framework addresses this unmet need by providing a structured, extensible classification that many scholars have argued is essential.

3 Classification of Ethical Harm

3.1 Definition

Ethical harm is defined as any adverse effect or damage inflicted upon individuals, groups, systems, or environments, either intentionally or not, from an action, event, or condition [1,20].

3.2 Classification

Table 1 provides the HARM66+ classification, a formal, multi-level taxonomy for analyzing harm in socio-technical systems. It organizes 66 distinct harm types into three tiers: (i) Major Harm Categories, which distinguish between Exo-Human Harms (impacting systems like environments and infrastructure) and Endo-Human Harms (affecting individuals); (ii) Intermediate Categories, which group harms into domains (e.g., environmental, technological, psychological); and (iii) Sub-subcategories, which specific harms like "Defamation," "AI Bias," or "Water System Failure." Each harm has a hierarchical identifier for granular reference and traceability.

The taxonomy is designed for mutual exclusiveness, orthogonality, normative completeness, hierarchy, stability, and parsimony. It maps all harm from 1500+ AI incidents as defined by the AI-AAIC Monitor[2]. Similarly, it maps 13 social risk categories identified from an expert review designed for mutual exclusivity, orthogonality, completeness, hierarchical traceability, stability, and parsimony of 1,781 documents [18]. HARM66+ was tested through a normative non-reducibility test: for each harm category, we evaluated whether removing that category would leave a class of ethically salient harm that could no longer be expressed without loss of moral meaning. If the removal of a category renders certain harms normatively inexpressible, i.e., irreducible to remaining categories without collapsing distinct ethical concerns, then the category is considered orthogonal. All categories in HARM66+ passed this test, indicating that each captures a distinct ethical dimension that cannot be subsumed by others without loss of normative resolution. Dotted coding allows for the extension of subclasses down to specific incident instances. The taxonomy was also stress-tested against futuristic harms portrayed in 16 major science fictions. While extensible, the taxonomy's upper levels (domains and categories) are designed for stability and broad applicability with minimal modification.

[2] https://www.aiaaic.org/aiaaic-repository.

3.3 Support from Ethical Theories

Any harm taxonomy implicitly embeds value judgments about what counts as harm and whose harm matters. How do we ensure these are not ad hoc? Aligning HARM66+ with the 11 ethical theories ensures the taxonomy is pluralistically defensible, meaning harms remain valid even when ethical priorities differ. This is essential for cross-jurisdictional and cross-cultural legitimacy. We therefore validate the normative significance of HARM66+ from a set of ethical theories to ensure that selected harms are ethically grounded, pluralistically defensible, and resistant to claims of arbitrariness, transforming the taxonomy from a technical artifact into a normative instrument.

Table 2(a) shows how major ethical theories[3] (MTEST11) view these harms, using codes for direct ("D"), indirect ("I"), conditional ("C"), or neutral ("N") consideration.

All theories considered these harms unethical, with varied reasoning. For example, Psychological and Cognitive Harm is rated directly unethical ("D") by Utilitarianism (diminishes well-being), Deontology (violates duties of respect), Virtue Ethics (reflects vice), and Rights-Based Ethics (violates mental integrity). Natural Law Theory offers only indirect ("I") support, as its primary focus is on bodily integrity.

Using a Likert scale for quantitative assessment, Table 2(b) presents statistical analysis of this support. A strong consensus (Mean ≥ 2.0, p-value < 0.05) was found for nine harm categories, including Environmental, Economic, Physical (Human), and Psychological. Moderate consensus (Mean ≥ 1.5, p-value < 0.05) was found for Digital & Technological and Financial & Occupational harms. No categories showed weak or no consensus in this analysis.

4 Entities that Can Suffer Harm

4.1 Entity Classification

Existing technical work on victim classification also remains fragmented across domains such as cybersecurity, public health, AI ethics, and legal studies, with each field offering partial taxonomies tailored to its specific harm context. In cybersecurity, frameworks such as the MITRE ATT&CK Matrix[4] and ENISA

[3] We consult 11 major ethical theories - a curated mix of normative and social ethics. The members have been widely applied in domains like healthcare, law, and governance. They include the classical triad, duty, outcome, and virtue, for individual ethics, plus social/political theories for societal (community) ethics. The set maintains internal orthogonality with purposeful overlap for robust moral triangulation, making it ideal for handling the moral uncertainty of global technology. Real-world governance, regulators, courts, institutions, and cultures draw on different moral logics simultaneously (rights, utility, care, justice, environmental value). Aligning HARM66+ with the 11 theories ensures the taxonomy remain valid even when ethical priorities differ - essential for cross-jurisdictional and cross-cultural legitimacy.

[4] https://attack.mitre.org/.

Table 2. Views of Ethical Theories on Harms and Consensus Analysis

Table 2(a). Views of Ethical Theories on Harms **Table 2(b). View Analysis**

Harm ↓ Ethical Theory →	Harm Code	Utilitarianism	Deontology	Virtue Ethics	Ethics of Care	Rights-Based Ethics	Social Contract	Rawlsian Justice	Natural Law	Environmental Ethics	Pragmatism	Existentialist Ethics	Mean	Std Dev	Z-Score	T-Statistic	P-Value	Consensus?
Environmental and Ecological Harm	A.E1.00	D	D	D	D	C	C	C	C	D	D	C	2.09	1.04	-0.43	6.64	5.78E-05	Strong
Digital and Technological Harm	A.E2.00	I	D	C	I	C	C	C	C	I	D	C	1.64	0.81	-1.68	6.71	5.31E-05	Moderate
Physical Infrastructure Harm	A.E3.00	D	D	I	C	C	D	D	D	I	D	I	2.36	0.81	0.32	9.69	2.12E-06	Strong
Economic and Corporate Harm	A.E4.00	D	D	D	C	D	D	D	C	C	D	C	2.27	1.01	0.07	7.47	2.13E-05	Strong
Social, Cultural, & Political Harm	A.E5.00	D	D	D	D	D	D	D	C	I	D	D	2.73	0.65	1.32	13.99	6.83E-08	Strong
Physical or Medical Harm	H.H1.00	D	D	D	D	D	D	D	D	I	D	D	2.91	0.30	1.82	32.00	2.09E-11	Strong
Psychological & Cognitive Harm	H.H2.00	D	D	D	D	D	C	C	I	I	D	D	2.45	0.82	0.57	9.93	1.70E-06	Strong
Reputational or Identity Harm	H.H3.00	C	C	D	D	D	C	C	I	I	D	D	2.09	0.94	-0.43	7.35	2.46E-05	Strong
Social or Relational Harm	H.H4.00	C	D	D	D	D	C	C	I	C	C	D	2.00	1.00	-0.68	6.63	5.83E-05	Strong
Political, Legal or Expressional Harm	H.H5.00	D	D	D	C	D	D	D	C	C	C	D	2.27	1.01	0.07	7.47	2.13E-05	Strong
Financial or Occupational Harm	H.H6.00	D	C	D	C	D	C	D	C	C	C	D	1.91	1.04	-0.93	6.06	1.21E-04	Moderate

threat models[5] typically categorize victims as end users, devices, or enterprise systems, emphasizing attack vectors and digital exposure rather than moral salience. In disaster risk and epidemiological ethics, classification schemes focus on human vulnerability, identifying populations by age, disability, income, or geography to prioritize resource allocation and harm mitigation [3].

AI ethics scholarship has more recently expanded moral consideration beyond individual persons to include digital systems, institutional processes, and epistemic constructs such as truth, trust, or democracy [5]. Legal frameworks, including the UN Human Rights Conventions (UN, 1948) [14] and the European Convention on Human Rights[6], classify victims primarily by group identity or protected legal status, but typically exclude non-human entities, infrastructural systems, abstract societal goods. Overall, these approaches lack either taxonomic granularity or explicit alignment with ethical theory. Understanding the victim is also fundamentally essential for evaluating harm within complex socio-technical systems. To address this gap, we also propose ENTITY7- a structured victim taxonomy spanning individuals, communities, institutions, non-human life, and abstract normative constructs to support systematic harm modeling. Table 3 presents the resulting classification of entities subject to ethical harm.

[5] https://www.enisa.europa.eu/publications/enisa-threat-landscape-2025.

[6] https://www.echr.coe.int/documents/d/echr/convention_eng.

Table 3. Entities that can be Subject to Harm

	Class	Code	Entity Subclass	Description
Living Entities	Individual Human	E1a	Living Individuals	Biological persons currently alive who can suffer the harms.
		E1b	Deceased Individuals	Deceased persons whose identity, reputation, or digital legacy may be misused or harmed.
	Groups or Communities	E2a	Identity Groups	Collectives defined by race, ethnicity, or religion, or occupation.
		E2b	Stakeholder Groups	People such as workers, employees, customers, patients, or buyers affected by product harms or misrepresentation.
		E2c	Marginalized Groups	Vulnerable or Marginalized populations structurally excluded or harmed.
		E2d	Digital Communities	Communities in common digital spaces.
		E2x	Other Groups	Other groups or herds.
	Institutions and Organizations	E3a	Public Institutions	Government bodies, agencies, or courts.
		E3b	Formal Organizations	Corporations, companies, non-profits.
		E3c	Jurisdictional Entities	Cities, counties, provinces, states, or nations - formal administrative or governing units.
		E3x	Other Organizations	Other complex organizations & ecosystems.
	Non-Human Lifeforms	E4a	Animals	Sentient or domesticated animals affected by environment or systems.
		E4b	Plants and Agriculture	Flora and crops damaged by pollution or technology.
		E4x	Other Lifeforms	Artificial / Synthetic / Alien / Extraterrestrial Lifeforms.
Inanimate & Abstract Entities	Environmental and Natural Systems	E5a	Terrestrial Ecosystems	Forests, grasslands, mountains affected by deforestation, erosion, or climate events.
		E5b	Aquatic Ecosystems	Oceans, rivers, lakes, wetlands harmed by pollution, acidification, or overfishing.
		E5c	Atmospheric and Climate Systems	Air quality, climate, weather patterns altered by carbon emissions, aerosols.
		E5d	Planetary and Extraterrestrial Systems	Outer space systems - planetary orbit, planetary atmospheres, celestial systems impacted by space debris and exploration.
	Built and Technological Entities	E6a	Physical Infrastructure	Tangible, material systems/assets — includes roads, pipelines, cables, water mains, buildings, satellites, towers, etc.
		E6b	Built Systems	Engineered systems (digital or cyber-physical) includes software, algorithms, control networks, data centers, cloud.
		E6c	Built Services	Electric utility, transportation, web services, emergency response, etc. Attack = disruption of availability, reliability, or trust in the service.
	Abstract and Social Constructs	E7a	Normative Justice Systems	Social structures for adjudicating fairness, accountability, and rights.
		E7b	Epistemic Foundations	Shared methods for determining truth (e.g., scientific method, journalism, peer review).
		E7c	Relational Trust and Legitimacy	The perceived legitimacy of actors and systems based on fair, honest, and consistent behavior.
		E7d	Democratic and Civic Norms	Practices of voting, representation, civic engagement, and constitutional freedoms.
		E7e	Faith, Belief, and Worldviews	Religious/spiritual beliefs, traditions, and ideological worldviews.
		E7f	Linguistic and Cultural Expressions	Languages, oral traditions, artistic expressions, and cultural symbols that shape identity and communication.
			(reserved)	

4.2 Victims and Ethical Theories

Victim classification plays a foundational role in ethical reasoning. Different ethical theories prioritize different classes of victims, leading to systematically divergent harm evaluations. Within the MTEST11 framework, utilitarianism (T1) assigns substantial moral weight to aggregate harms affecting groups (E2b) and ecosystems (E5a), whereas deontology (T2) and rights-based ethics (T5) emphasize harms to individuals (E1a) and protected or marginalized populations (E2a,

E2c). Ethics of care (T4), virtue ethics (T3), environmental ethics (T9), and existentialist ethics (T11) variously prioritize vulnerability, relational dependence, non-human life (E4a), and individual agency (E1a).

A robust and expressive victim taxonomy such as ENTITY7 is therefore essential for theory-aligned ethical reasoning. Table 4 summarizes how different ethical theories weight victim categories in harm evaluation.

Table 4. How Victim Classification Affects Moral Weight from the Theory

Ethical Theory	Victim Class	How Victim Classification Affects Moral Weight from the Theory
Utilitarianism	E2b (User Groups), E5a (Ecosystems)	Harms to large populations or ecosystems are morally weighty due to the cumulative suffering across many beings or systems.
Deontological Ethics	E1a (Living Individuals)	Harming individuals violates intrinsic moral duties, e.g., not lying, not killing, even if doing so may lead to beneficial outcomes.
Virtue Ethics	E2c (Marginalized Populations), E7f (Cultural Expression)	Ethical weight increases when harm reflects a lack of compassion or justice, especially toward the vulnerable or cultural heritage.
Ethics of Care	E2c (Marginalized Groups), E1a (Individuals)	Harms to those in dependent or care-based relationships are ethically severe due to the breach of relational responsibility.
Rights-Based Ethics	E1a (Individuals), E2a (Ethnic Groups)	Morally evaluates whether rights (privacy, expression, bodily integrity) are violated, prioritizing sentient or right-bearing beings.
Social Contract Theory	E3a (Public Institutions), E7d (Democratic Norms)	Harms to institutions or civic norms violate the agreed structure of society, undermining justice, law, or democracy for all.
Rawlsian Justice	E2c (Vulnerable Groups), E3c (Jurisdictions)	Justice prioritizes the most disadvantaged—harms to structurally oppressed or unrepresented populations are given moral urgency.
Natural Law Theory	E4a (Animals), E7e (Worldviews)	Harms that disrupt the natural order or purpose of life (e.g., procreation, family, nature) violate moral law, even if unintended.
Environmental Ethics	E5a–E5d (All Environmental Systems)	Assigns intrinsic moral value to non-human systems; harms to forests, oceans, or the climate are wrong even without direct human impact.
Pragmatism	E6c (Built Services), E3b (Organizations)	Ethics is judged by functional outcomes, if harm disrupts essential services or institutions, it's morally significant, regardless of theory.
Existentialist Ethics	E1a (Individuals), E7c (Relational Trust)	Suppression of autonomy, freedom, or personal authenticity are the gravest, especially when individuals are manipulated.

4.3 Harmer's Role in Ethics

While the harmer's identity is relevant for assessing blame (especially in theories like virtue ethics or ethics of care), victim classification is more central to determining a harm's moral weight. It is foundational for ethical deliberation, clarifying who is morally considerable, how harm is counted under different theories, and how competing harms are weighed. Without it, ethical evaluation risks becoming arbitrary or biased toward visible or anthropocentric harm.

5 Irreversibility of Harm

5.1 Impact

The irreversibility of harm heightens its ethical severity because it removes the possibility of repair or compensation. In contrast, reversibility allows for moral recovery and accountability, mitigating a harm's ethical weight. Ethical and legal systems often consider reversibility to determine responsibility and the permissibility of risk. Table 5 described the irreversibility characteristics of the 11 harm categories.

5.2 Irreversibility and Ethical Theories

The significance of reversibility varies across major theories, each offering distinct perspectives on its moral weight. For Utilitarianism (T1), reversibility is highly important. Reversible harms are more acceptable if they yield greater net benefits, while irreversible harms carry more ethical weight due to their lasting negative impact. Deontological ethics (T2) prioritizes adherence to moral duties, making reversibility secondary. A morally wrong action remains impermissible even if the harm is undone, as the focus is on the intrinsic morality of the act itself, not its consequences. Virtue ethics (T3) views reversibility as contextually significant. A virtuous person assesses it, and causing irreversible harm may reflect a lack of virtues like prudence, whereas reversible harms allow for moral repair.

The ethics of care (T4) finds irreversible harm particularly severe because they can irreparably damage relationships and trust. Reversible harms offer opportunities for care and repair, central to theory. Rights-based ethics (T5) considers reversibility a minor factor. The focus is on the inviolability of rights, so any violation is unethical regardless of whether the harm can be remedied. Social contract theory (T6) deems reversibility con-textually relevant. Rational agents would reject systems permitting irreversible harm, making reversible harm more tolerable under a fair social contract. Rawlsian justice (T7) assigns significant weight to irreversibility, especially for the least advantage. Irreversible harm violates fairness and cannot be offset, making reversibility crucial for long-term justice. Natural law theory (T8) disregards reversibility as a decisive factor. Actions are judged on their alignment with natural purpose, making them inherently right or wrong regardless of their consequences.

Environmental ethics (T9) considers reversibility a core concern. Harms that permanently degrade ecosystems or cause extinction are particularly egregious, making prevention of irreversible damage a paramount ethical obligation. Pragmatism (T10) focuses on practical outcomes. Reversible harms are more tolerable

Table 5. Irreversibility & Durance Characteristics of Various Harm Types

Harm Type	Irreversil Rank	H-Code	Reversibility Description	Durance	Contextual Notes
Environmental and Ecological Harm	1	A.E1.00	Often **irreversible** (e.g., extinction, ecosystem collapse). Timescales for recovery (if possible) can be centuries.	Long-Term	Ecosystem collapse, extinction, or climate shifts are typically irreversible or generational.
Physical Infrastructure Harm	2	A.E3.00	**Partially reversible**, but often expensive and time-consuming (e.g., dam failure, bridge collapse). Loss of life can make it irreversible.	Short to Medium	Data breaches or software failures may be quickly patched; privacy violations may persist.
Physical or Medical Harm	3	H.H1.00	Some permanent injuries or deaths are **irreversible**. Minor injuries may be healed, but not all.	Medium to Long	Some infrastructure can be rebuilt, but losses of life, heritage, or housing may be permanent.
Psychological & Cognitive Harm	4	H.H2.00	**Partially reversible**, but often persistent (e.g., trauma, PTSD). Treatment varies in efficacy.	Short to Medium	Economic damage is often recoverable (e.g., via restructuring or bailout), but reputational damage may linger.
Reputational or Identity Harm	5	H.H3.00	Difficult to undo fully, especially in digital age. Stigma may persist even after vindication.	Medium to Long	Cultural erosion, political disenfranchisement, or institutional distrust often span generations.
Social, Cultural, & Political Harm	6	A.E5.00	**Intergenerational** effects can be long-lasting. Societal trust, culture, or institutions may take decades to rebuild.	Short to Long	Minor injuries heal, but chronic illness or death is irreversible. Durance varies significantly.
Digital and Technological Harm	7	A.E2.00	Varies widely—loss of data or system compromise may **be reversible** via backups or patches, but impacts (e.g., surveillance) may not be.	Medium to Long	Trauma and mental illness may persist for years or lifetime, but recovery is possible for some.
Legal, Political, or Expressional Harm	8	H.H5.00	Legal redress or public apology can partially reverse the harm, but **record or damage** may persist.	Medium to Long	Digital-age reputational damage may be semi-permanent despite apologies or retraction.
Social or Relational Harm	9	H.H4.00	Personal relationships can sometimes be repaired, but full trust may not return. **Moderately reversible**.	Short to Medium	Relationships may recover, but deep betrayal or trust loss can endure. Context-dependent.
Economic and Corporate Harm	10	A.E4.00	Financial and organizational damages are often **reversible** through compensation, restructuring, or legal remedy.	Short to Long	A mistaken arrest may be resolved quickly; political suppression or exile may last decades.
Financial or Occupational Harm	11	H.H6.00	Most **reversible** by restitution, re-employment, insurance, or welfare mechanisms. Less lasting than other harms.	Short to Medium	Most financial harm is recoverable through compensation, jobs, or aid, but systemic exclusion can extend duration.

because they can be corrected, while irreversible harms limit future options, making reversibility a practical safeguard. Existentialist ethics (T11) views reversibility as significant for moral responsibility. Irreversible harms permanently define an individual's actions and moral identity, while reversible harms allow for redemption.

In summary, theories like Utilitarianism, Pragmatism, and Rawlsian Justice treat reversibility as core or con-textual. In contrast, Deontology, Rights-Based Ethics, and Natural Law Theory largely minimize it, while Existentialist Ethics links it to moral responsibility.

Here is an example of how three theories will ask the question of reversibility differently for Environmental Harm from Deforestation. Utilitarianism would calculate long-term loss vs. gain. If forest regrowth is possible, the harm may be tolerable; if irreversible (biodiversity loss), it's more ethically troubling. Deontological approaches may oppose it if they violate duties (e.g., non-maleficence, stewardship, or duties to future persons), regardless of reversibility. While Environmental Ethics will condemn it especially when species extinction or ecosystem collapse occurs, which are irreversible. Pragmatism would ask- can restoration or rewilding undo the harm? If yes, some harm may be justified conditionally.

Table 6. Importance of Irreversibility of Harm to Various Ethical Theories

Harm ↓ Ethical Theory →	Harm Code	Utilitarianism	Deontology	Virtue Ethics	Ethics of Care	Rights-Based Ethics	Social Contract	Rawlsian Justice	Natural Law	Environmental Ethics	Pragmatism	Existentialist Ethics
Environmental and Ecological Harm	A.E1.00	C	M	X	C	M	C	C	C	C	C	X
Digital and Technological Harm	A.E2.00	X	M	X	X	M	X	X	M	X	C	C
Physical Infrastructure Harm	A.E3.00	X	M	X	X	M	C	C	M	X	C	X
Economic and Corporate Harm	A.E4.00	C	M	X	X	X	C	C	M	I	C	X
Social, Cultural, & Political Harm	A.E5.00	X	M	X	C	M	X	C	M	X	X	C
Physical or Medical Harm	H.H1.00	C	M	X	C	M	X	C	M	I	C	X
Psychological & Cognitive Harm	H.H2.00	C	M	X	C	M	X	X	M	I	C	C
Reputational or Identity Harm	H.H3.00	X	M	X	C	M	X	X	M	I	X	C
Social or Relational Harm	H.H4.00	X	M	X	C	M	X	X	M	I	X	X
Political, Legal or Expressional Harm	H.H5.00	X	M	X	X	M	X	C	M	I	X	X
Financial or Occupational Harm	H.H6.00	C	M	X	X	M	C	C	M	I	C	X

Table 6 shows how the eleven ethical theories consider the reversibility of harms using four levels: Core (C), where it is central to judgment; Contextual (X), where it depends on the case; Minor (M), where it has limited influence; and Irrelevant (I), where it is not factored or dismissed an ethically unimportant. Table 5 shows the potential duration of various harm types.

6 Durance of Harm

6.1 Impact

The duration of harm, how long a negative impact persists, is also another significant factor in its ethical evaluation. Longer-lasting harms carry greater ethical weight because their effects accumulate, increasing the moral urgency for prevention. The ethical score rises sharply when harm is both long-lasting and irreversible, as this undermines future moral agency, justice, and ecological integrity. The greater the duration, the higher the moral burden. Table 5 (columns 5 & 6) and Fig. 1 show some probable context and corresponding sample durations of the eleven ethical harms, categorized as long-term (generational/permanent), medium-term (1–10 years, possibly reversible), and short-term (temporary with high recovery potential).

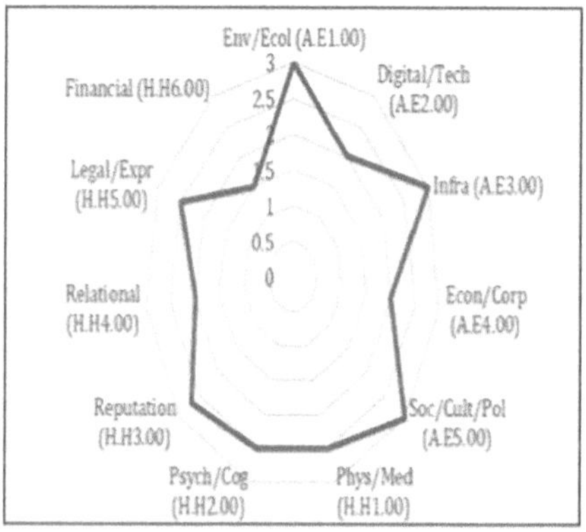

Fig. 1. Durance Scores of Various Harms (1 = short, 2 = medium, 3 = long term)

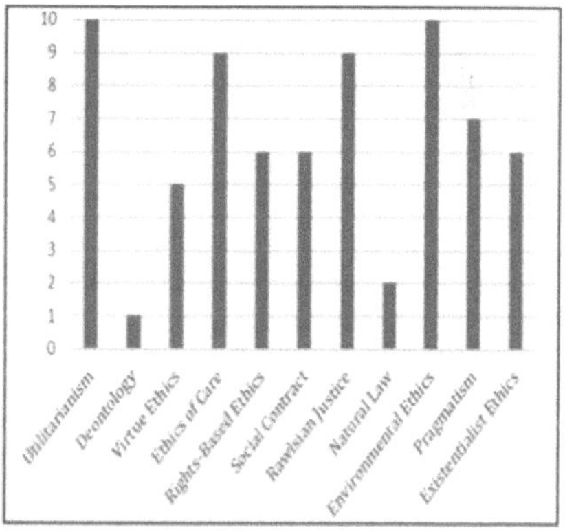

Fig. 2. Relative Importance of Durance to Various Ethical Theories

6.2 Durance and Ethical Theories

The ethical significance of harm duration varies across major ethical theories, reflecting diverse moral priorities. Several theories assign high ethical weight to harm duration. Utilitarianism (T1) is highly concerned with duration, as longer harms produce greater cumulative suffering. The ethics of care (T4) also gives high weight, as long-term harm threatens relational trust and care networks. Rawlsian justice (T7) strongly emphasizes duration, especially for persistent harms that limit opportunity for the least advantage. Environmental ethics (T9) considers duration a central concern, as harms like extinction and climate change span generations.

A second group of theories moderately considers duration. Virtue ethics (T3) views it as a reflection of the agent's character and prudence. Rights-based ethics

(T5) finds any rights violation wrong, but prolonged denial intensifies the moral breach. Social contract theory (T6) holds that persistent harms erode the legitimacy of a moral order. Pragmatism (T10) focuses on practical outcomes, making persistent and unresolved harms a greater ethical problem. Existentialist ethics (T11) sees enduring consequences as deepening the weight of an individual's moral responsibility and identity.

Notably, two theories give low importance to duration. Deontological ethics (T2) is concerned with adherence to moral duties, making an action's wrongness independent of how long its consequences last. Similarly, natural law theory (T8) evaluates acts based on their alignment with natural purpose, not the duration of their effects. Figure 2 graphically illustrates the relative weights assigned to harms' duration by these major ethical theories.

7 Use Cases for Taxonomy

A domain-level harm is a contextualized harm pattern observed within a specific socio-technical setting, while HARM66+ provides a domain-independent canonical harm vocabulary. The relationship between them is bidirectional. Forward mapping instantiates how domain harm materializes as one or more HARM66+ harm classes. Reverse mapping uses HARM66+ as a diagnostic lens to identify harms that are present, latent, or unexamined within a domain. This two-way mapping supports both harm decomposition (domain → HARM66+) and harm completeness checking (HARM66+ → domain). The mapping is non-symmetric in strength: some HARM66+ harms are directly instantiated, others conditionally or weakly activated, while additional classes may remain structurally relevant even if not empirically observed.

Example: A domain harm such as loss of student agency due to automated decision systems directly instantiates Loss of Individual Control and Agency (H.H5.04), contributes to Cognitive Manipulation and Behavioral Control (H.H2.03), and conditionally activates Algorithmic Discrimination (H.H6.04). Reverse mapping surfaces additional concerns, such as Breach of Trust (H.H4.02) or Due Process Failures (H.H5.02), that may not be explicitly named but warrant evaluation. This bidirectional framing enables harm analysis that is both context-sensitive and structurally complete, allowing comparison and mitigation across domains without collapsing domain-specific meaning.

Use Case A: When a university deploys AI for essay grading and placement, standard checks often focus on data quality, bias, and security. Mapping to HARM66+ reveals a host of additional risks such as loss of agency, psychological stress from opaque judgments, epistemic dependency, and erosion of credential trust, harms not visible through data-centric review alone.

Use Case B: For national oversight of AI in education, Education Commission applies HARM66+. It enables regulators to design much fine grained targeted, harm-specific measures, such as human-oversight requirements for agency loss or transparency obligations for epistemic harms, while providing a uniform vocabulary for standardized monitoring across institutions.

This taxonomy has also been applied to evaluate ethical risks in the data-harnessing processes of large language models (LLMs) [8].

8 Meta Harms

There is also notion of *meta-harms*, higher-order factors that shape the dynamics of harm without constituting harms themselves. Unlike first-order harms, which describe *what* is harmed, meta-harms describe *how* harms emerge, propagate, and intensify across time, scale, and interaction effects. We identify several – i) *Manifestation latency* (harms materialize only after prolonged or intergenerational delay). ii) *Feedback loops and harm compounding* (processes that initial harm reinforce themselves or interact with other harms, producing nonlinear escalation), iii) *Multiplicity* (capacity of a single system to generate multiple distinct harm types simultaneously across entities and domains), iv) *Scalability* characterizes harms that are individually minor but become severe when replicated at population or infrastructure scale. Finally, v) *invisibility and opacity* of harm. A taxonomy, such as HARM66+ can be the foundation for future analysis of such harm's dynamics.

9 Stability and Extensibility of Taxonomy

The taxonomy is designed to be both stable and extensible. Its upper levels, domains and major categories, are intentionally structured to remain robust over time, enabling broad applicability across socio-technical contexts with minimal modification. Extensibility is confined to lower-level subclasses, allowing the framework to accommodate emerging or context-specific harms without destabilizing its core structure. While the concept of harm is inevitably shaped by historical, cultural, and technological change, the taxonomy's foundational layers are intended to endure, even as terminology and subcategories evolve through continued scholarly refinement.

To address concerns of universality, most importantly, HARM66+ has been grounded in pluralistic ethical support rather than a single normative doctrine. The taxonomy is affirmed across consequentialist, rights-based, care-oriented, justice-based, and environmental ethical perspectives, reducing the risk of cultural or temporal arbitrariness. This pluralistic grounding is essential for cross-jurisdictional and cross-cultural legitimacy, long-term relevance as priorities evolve over decade.

10 Conclusions

In this paper we call for addressing a critical gap in the literature by introducing a structured and expandable taxonomy of harm. We present HARM66+ classifying over 66 distinct harm types, systematically organized into two overarching domains and 11 categories. The categorization is grounded in foundational support from major contemporary ethical theories, ensuring philosophical coherence and normative relevance. We have also discussed its normative characteristics such as irreversibility and durance, and the classification of victim entities, as these will be weighted by major ethical theories. These classifications are expected to enable more nuanced and ethically grounded assessments of harm severity, supporting rigorous applications in both analytical modeling and policy decision-making.

A specific, fine-grained, and standardized taxonomy provides three immediate benefits. Frameworks such as HARM66+ can remove ethical blind spots by surfacing latent harms, such as loss of agency, epistemic degradation, and institutional trust erosion, that compliance-oriented evaluations often miss. Second, their granularity enables harm-specific, proportionate mitigation, replacing coarse, ad hoc controls with safeguards aligned to harm severity and reversibility. Third, they establish a uniform cross-domain vocabulary, supporting consistent mitigation, comparable reporting, and interoperable data collection across institutions, sectors, and jurisdictions.

Disclosure of Interests. The authors declare that they have no competing interests that are relevant to the content of this article.

References

1. Beauchamp, T., Childress, J.: Principles of biomedical ethics: marking its fortieth anniversary, vol. 19. Taylor & Francis (2019)
2. Bostrom, N.: Existential risk prevention as global priority. Global Pol. **4**(1), 15–31 (2013)
3. Cutter, S.L., Boruff, B.J., Shirley, W.L.: Social vulnerability to environmental hazards, pp. 115–132 (2012)
4. Denning, D.E.R.: Information warfare and security, vol. 4, Addison-Wesley, New York (1999)
5. Floridi, L., Cowls, J., Beltrametti, M., Chatila, R., Chazerand, P., Dignum, V., Luetge, C., Madelin, R., Pagallo, U., Rossi, F., et al.: Ai4people–an ethical framework for a good ai society: opportunities, risks, principles, and recommendations. Mind. Mach. **28**(4), 689–707 (2018)
6. Freund, J., Jones, J.: Measuring and managing information risk: a FAIR approach, Butterworth-Heinemann (2014)
7. Keeney, R.L., Raiffa, H.: Decisions with multiple objectives: preferences and value trade-offs, Cambridge University Press (1993)
8. Khan, J.I., Prithula, S.R.: Ethical risk assessment of the data harnessing process of llm supported on consensus of well-known multi-ethical frameworks. InCanadian AI (2025)
9. Mittelstadt, B.D., Allo, P., Taddeo, M., Wachter, S., Floridi, L.: The ethics of algorithms: mapping the debate. Big Data Soc. **3**(2), 2053951716679679 (2016)
10. Now, A.: Litigating algorithms: Challenging government use of algorithmic decision systems. Tech. Rep. (2018)
11. Parker, D.B.: Fighting computer crime, Scribner (1983)
12. Pörtner, H.O., et al.: Climate change 2022: Impacts, adaptation and vulnerability, (2022). Tech. rep

13. Raji, I.D., et al.: Closing the ai accountability gap: Defining an end-to-end framework for internal algorithmic auditing. In: Proceedings of the 2020 conference on fairness, accountability, and transparency, pp. 33–44 (2020)
14. RESOLUTION, A.B.U.G.A.: 1. universal declaration of human rights adopted by un general assembly resolution 217a (iii) of 10 december 1948
15. Shelby, R., et al.: Sociotechnical harms of algorithmic systems: scoping a taxonomy for harm reduction. In: Proceedings of the 2023 AAAI/ACM Conference on AI, Ethics, and Society, pp. 723–741 (2023)
16. Slovic, P.: Perception of risk, vol. 236, American Association for the Advancement of Science (1987)
17. Taddeo, M.: The ethical governance of the digital during and after the covid-19 pandemic. Mind. Mach. **30**(2), 171 (2020)
18. Uuk, R., et al.: A taxonomy of systemic risks from general-purpose ai (2024). arXiv:2412.07780 arXiv preprint
19. Yeung, K.: Recommendation of the council on artificial intelligence (oecd). Tech. Rep., 1 (2020)
20. Zuboff, S.: The age of surveillance capitalism. In: Social Theory Re-wired, pp. 203–213. Routledge (2023)

Enhanced RAG Driven LLM-NER for Parametric Completion of Assembly Documents

Chengye Ma[1], Jiapeng Bi[2], Linsen Ma[1], and Junhao Geng[1](✉)

[1] Northwestern Polytechnical University, Xi'an, Shaanxi, China
machengye@mail.nwpu.edu.cn, gengjunhao@nwpu.edu.cn
[2] CRRC Academy Co., Ltd., Beijing 100070, China

Abstract. Large Language Models (LLMs) have demonstrated tremendous potential in the automatic generation of assembly documents. However, due to the shortcomings of traditional Retrieval-Augmented Generation (RAG) mechanisms in domain adaptability and rare entity recall rate, LLMs are still prone to entity hallucination and professional term recognition errors during the assembly instruction generation process. To address this issue, this paper proposes an enhanced RAG-driven LLM-NER parametric completion framework. By introducing domain-adaptive vocabulary enhancement and a multi-dimensional dynamic weight ranking algorithm, it achieves dual optimization of RAG retrieval efficiency and accuracy. Experimental results show that the framework significantly enhances retrieval relevance, first-round accuracy, and generation efficiency on real assembly datasets, providing a new path for the application of LLMs in the field of intelligent generation of assembly documents.

Keywords: Large Language Models · Named Entity Recognition · Assembly Documents · Document Generation · Parametric Completion

1 Introduction

As core process documents guiding assembly operations, assembly documents must be compiled in strict accordance with engineering design standards, factory process capabilities, and quality assurance systems. They are key bases for standardizing assembly processes and ensuring product quality. Modern assembly is characterized by complex structures, high system redundancy, and tight component coupling, leading to cumbersome compilation workflows, high labor costs, and susceptibility to human errors in assembly document preparation.

Currently, pre-trained large models based on the Transformer architecture have achieved remarkable results in text generation. By predicting each output token based on input tokens, they have exhibited impressive performance in text generation, semantic understanding, logical reasoning, mathematical computation, and other fields, enabling their wide application in various downstream domains [1].

H. R. Arabnia et al. (Eds.): CSCI 2025, CCIS 2968, pp. 162–170, 2027.
https://doi.org/10.1007/978-3-032-28457-0_12

However, the application of general-purpose LLMs in the highly specialized field of aerospace manufacturing still faces two core bottlenecks. Firstly, insufficient domain knowledge adaptability leads to easy deviations in the generation of professional information such as proper nouns, part numbers, and tooling parameters in the aerospace assembly field. Secondly, the limited retrieval accuracy of traditional RAG mechanisms often results in semantic mismatches or content fabrication when dealing with rare entities or sub-domain terms.

To solve this problem, a feasible approach is to use prompt engineering and fine-tuning technology. When LLMs generate assembly instructions, all parts involving instantiated entities are replaced with placeholders "*", and then instantiated entities such as component names, process parameters, and standard parts used are reinserted through other methods to form complete, correct, and rigorous assembly guidance documents. This process necessarily involves identifying attributes such as the position and type of each instantiated entity to be inserted, as well as the insertion operation of the entity itself.

Named Entity Recognition (NER), as a basic technology in natural language processing, its core task is to automatically identify named entities with specific meanings from unstructured text and classify them into predefined categories [2]. LLM-NER technology aims to leverage the reasoning and generation capabilities of LLMs to efficiently complete NER tasks. For the intelligent parametric completion of aircraft assembly instructions, LLM-NER technology can be used to identify instantiated entities to be completed and accurately locate their insertion positions in incomplete assembly documents. At present, there have been several cases of NER technology applied to solve problems in the manufacturing field, such as Liu Xuan [3], Zhao Jingxin [4], Tang Yongjian [5], etc., who have made breakthrough progress in this field respectively.

In response to the above problems, this paper proposes an enhanced RAG-driven LLM-NER parametric completion framework. In the RAG retrieval stage, the framework innovatively introduces a two-layer optimization mechanism of "domain vocabulary enhancement + dynamic weight ranking". By strengthening domain semantic perception and optimizing candidate entity ranking logic, it achieves the improvement of semantic relevance of retrieval results and context-adaptive adjustment, and ultimately improves the completion accuracy and intelligence level of parametric entities in assembly documents.

2 Framework of Parametric Completion Method for Assembly Documents

The parametric completion system for assembly documents proposed in this paper consists of three core modules: the enhanced RAG-driven instantiated entity extraction module, the LLM-NER prompt template construction module, and the human-collaborative parametric completion process module. The overall framework is shown in Fig. 1, and we will briefly describe the basic principles of the framework in accordance with the overview in the figure.

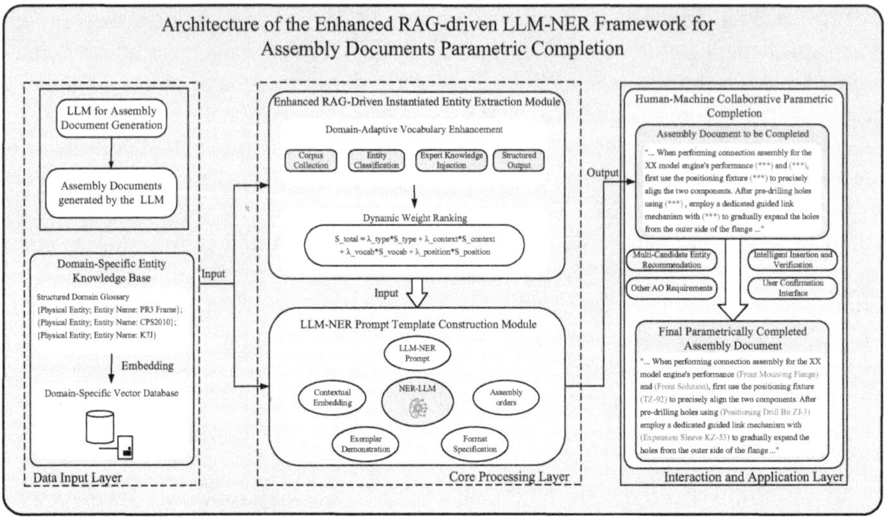

Fig. 1. Architecture of the Enhanced RAG-driven LLM-NER Framework for Assembly Documents Parametric Completion

2.1 Enhanced RAG-Driven Instantiated Entity Extraction Module

Retrieval-Augmented Generation (RAG) is a technology that integrates external knowledge bases into LLMs. It enhances prompts and generates answers by real-time retrieving relevant information from given databases or documents as the context foundation [6]. Aircraft assembly involves a large number of professional terms, concepts, and knowledge, which are usually missing in the datasets of general-purpose LLM pre-training stages. Therefore, it is necessary to additionally provide domain-specific instantiated entities as context information to large language models to achieve accurate extraction of various instantiated entities in assembly documents. This paper introduces a two-layer optimization mechanism of "domain vocabulary enhancement + dynamic weight ranking" in the RAG basic architecture, improving retrieval accuracy and domain adaptability from two dimensions: semantic feature enhancement and ranking logic optimization.

2.1.1 Domain-Adaptive Vocabulary Enhancement

The core goal of this mechanism is to solve the problem of insufficient sensitivity of general-purpose LLMs to professional terms in the assembly field. It strengthens the pertinence of semantic retrieval by constructing a structured domain vocabulary.

Its specific construction process is as follows:

- Corpus Collection: The system collects and integrates assembly documents, product design manuals, process technical specifications, and other data related to the assembly of lightweight components of unmanned aerial vehicles, establishing a comprehensive domain corpus covering assembly scenarios.

- Entity Classification: Based on assembly process logic, four core entity types are identified and divided, including Part Entity, Tooling Entity, Parameter Entity, and Standard Part Entity.
- Expert Knowledge Injection: With the participation of domain experts, a term synonym association network and hierarchical semantic relationships (such as the equivalent mapping between "carbon fiber skin" and "CFRP Skin") are constructed, and corresponding confidence weights are assigned to each term-entity type combination.
- Structured Output: Finally, a standardized domain vocabulary is formed.

This structured vocabulary is embedded into the vector database as enhanced semantic features during the RAG retrieval stage, enabling the system to accurately capture the semantic information of domain-specific vocabulary and effectively alleviate the "weak semantic perception" problem of general-purpose LLMs in professional contexts. The process is shown in Eq. (1):

$$\mathrm{v} = f_{embed}(d) \in \mathrm{R}^d \tag{1}$$

Where v represents the quantized text vector, and f _ embed denotes the vectorization method, usually implemented through an embedding LLM. After the vector database is constructed, in the actual entity extraction process, various types of instantiated entities are randomly retrieved from the vector database as references for LLM recognition and generation.

2.1.2 Multi-Dimensional Dynamic Weight Ranking Algorithm

This algorithm aims to replace the traditional single cosine similarity ranking method and construct a comprehensive scoring mechanism integrating multi-features to improve the rationality and accuracy of candidate entity ranking.

The core logic of the algorithm is to calculate the comprehensive score S_{total} of candidate entities through multi-dimensional feature fusion. The specific formula is as follows:

$$\mathrm{S}_{total} = \lambda_{\mathrm{type}} \times \mathrm{S}_{type} + \lambda_{context} \times \mathrm{S}_{context} + \lambda_{vocab} \times \mathrm{S}_{vocab} + \lambda_{position} \times \mathrm{S}_{position} \tag{2}$$

The specific definitions of each dimension feature are as follows:

S_{type} (Entity Type Matching Degree): Assigned 1.0 when the candidate entity type is completely consistent with the target entity type, otherwise 0.
$S_{context}$(Context Semantic Similarity): The Sentence-BERT model fine-tuned with domain corpus is used to calculate the semantic matching degree between the query context and the common application context of the entity.
S_{vocab} (Vocabulary Confidence Weight): Directly extract the weight value of the corresponding term in DomainVocab to highlight the priority of high-confidence domain terms.
$S_{position}$ (Position Feature Matching Degree): Based on the typical occurrence chapters of entities in assembly documents (such as introduction, process step description, bill of materials, etc.), match and score with the current query position.

The dynamic weight adjustment strategy is the core innovation of this algorithm. The weight vector $\Lambda = [\lambda_{type}, \lambda_{context}, \lambda_{vocab}, \lambda_{position}]$ is not a fixed value but adaptively adjusted according to the type of entity to be completed. For example, when completing parameter entities, the weight vector is set to $\Lambda = [0.2, 0.5, 0.2, 0.1]$, emphasizing context semantic similarity. When completing tooling entities, the weight vector is adjusted to Λ = [0.3, 0.2, 0.4, 0.1], highlighting the influence of domain vocabulary confidence weight. This dynamic adjustment mechanism makes the ranking logic more in line with the retrieval needs of different entity types, effectively improving the recall rate and accuracy of rare entities and professional terms.

2.2 LLM-NER Prompt Template Construction Module

To achieve accurate identification and annotation of entities in assembly documents, the system designs a structured prompt template to guide LLMs to complete high-quality NER tasks through standardized input. The template includes four core components:

- Role Positioning: Clarify the functional positioning of LLM as an entity recognition expert in the aerospace assembly field, strengthening domain adaptability awareness.
- Context Embedding: Supplement the entity-related information obtained by enhanced RAG retrieval as context to enrich the model's recognition basis.
- Example Demonstration: Provide annotation examples of different types of entities to help the model quickly understand the recognition rules of entities in the assembly field.
- Format Specification: Formulate standardized input and output formats to ensure the consistency and parseability of model generation results.

Through multiple rounds of test verification, the optimized prompt template significantly improves the accuracy and stability of LLMs in the entity recognition task of assembly documents, laying a solid foundation for subsequent parametric completion [7].

2.3 Human-Collaborative Parametric Completion Process

The instantiated entity retrieval-augmented generation and extraction module completes the extraction and annotation of various instantiated entities in aircraft assembly documents. This module collaborates with operators to correctly insert actual instantiated entities into specified positions, forming the final step of compiling complete aircraft assembly documents. The specific implementation involves converting the recognized assembly documents into partially editable files with human intervention. Each position requiring completion of instantiated entities provides dynamically generated input components, allowing users to select and insert correct instantiated entities at each corresponding component. As part of human intervention in intelligent parametric completion, LLM still plays a key role: based on the position, type, and context similarity of each entity to be completed identified earlier, multiple potential correct answers are selected from real instantiated entities for users to choose, avoiding manual searching by users among a large number of existing instantiated entities and significantly improving completion efficiency.

After processing by the human-intervened parametric completion module, the incomplete assembly documents are transformed into comprehensive and rigorous assembly documents, facilitating subsequent operations such as document archiving and system export.

3 Experiments and Discussion

3.1 Experiment Design

This paper takes a certain type of large passenger aircraft manufactured by an aviation manufacturing enterprise as the experimental object to verify the performance of the technical framework proposed in practical application. In the experiment, we wrote a total of 12 assembly documents for several assembly processes of the same type of aircraft,and experts in the field of aircraft assembly were invited to evaluate and guide exchanges. Throughout the experiment, we mainly focus on the following experimental parameters:

- Accuracy rate: The ratio of the correct number of instantiated entities inserted in the entire assembly document to the total number. The significance of setting this parameter is to measure the quality of the assembly document generated by the completion technology.
- Large language model used: The core of this technology lies in the LLM-NER technology applied, and different pre-trained LLMs may bring different effects, which requires additional attention.

3.2 Experimental Results Analysis and Discussion

Table 1 shows the average accuracy of each round of experiments using different LLM. From the experimental results, it can be seen that the average accuracy of the written assembly orders is maintained at more than 90% by using the completion technology of assembly orders. This shows that the use of assembly orders intelligent completion technology can have the same or better results than conventional methods while ensuring writing efficiency.

Experimental results across different LLMs showed slight variations in accuracy: DeepSeek-R1:7B performed the best overall, achieving both the highest generation speed and the highest accuracy. The reason for this is mainly due to the differences brought about by different datasets of different models in the pre-training stage: DeepSeek-R1: 7B is distilled from the DeepSeek-R1 model focusing on inference and has a MoE architecture, and the main datasets in the pre-training stage are all Chinese, so it can achieve the highest comprehensive performance in the comparison of the three models.

Table 2 shows the time consumption for completing the entity filling of each document using different methods. Method1–1 ~ 3 respectively represent the intelligent parametric completion systems using DeepSeek-R1: 7B, Qwen-8B, and Llama-7B; Method2–1 ~ 3 represent the completion results from three experts; Method3–1 ~ 3 represent the completion results based directly on NER and prompt engineering using the three LLMs. Among them, the method directly using NER and prompt engineering-based LLM actually required more time than the other two approaches. The reason lies in

Table 1. The average accuracy rate of different LLMs in each assembly documents

LLM		Qwen3: 8B	DeepSeek-R1: 7B	Llama3: 8B
assembly document's accurate rate (%)	#1	91.36	99.23	89.78
	#2	92.76	98.76	91.53
	#3	93.54	99.45	92.97
	#4	94.02	98.92	94.13
	#5	93.92	99.01	93.52
	#6	94.88	98.58	88.99
	#7	94.61	99.67	90.62
	#8	94.87	98.83	93.64
	#9	95.51	99.12	95.56
	#10	95.63	98.95	87.31
	#11	95.77	99.30	91.10
	#12	96.29	95.90	90.30
Average accurate rate (%)		94.43	98.81	91.62

the fact that although their generation speed is fast, their extremely low accuracy necessitates manual corrections for every document, thereby prolonging the overall completion time. In contrast, the time consumed by the intelligent parametric completion system was generally lower than that of the manual template filling method, demonstrating the efficiency improvement achieved by the proposed system in this study.

Table 2. Time Required for Assembly Document Completion under Different Methods

Me-thod	Assembly Document Completion Time (s)									
	1#	2#	3#	4#	5#	6#	7#	8#	9#	Average time
1–1	9.8	12.1	7.9	10.1	12.0	9.9	9.3	11.1	9.5	10.19
1–2	10.9	11.3	14.2	10.7	9.9	8.9	12.3	11.9	11.4	11.28
1–3	11.3	14.5	12.1	9.9	10.1	11.1	12.2	9.8	15.5	11.83
2–1	15.7	12.3	16.6	13.8	15.5	17.2	16.4	16.3	16.7	15.61
2–2	16.9	14.3	15.2	17.1	16.2	14.5	15.8	17.9	12.3	15.58
2–3	17.2	12.9	15.8	14.7	13.5	14.3	17.1	16.2	15.5	15.24
3–1	18.3	14.5	16.2	13.1	16.2	16.6	17.9	17.6	16.2	16.29
3–2	19.2	17.3	16.7	17.3	19.9	16.8	14.3	19.9	19.1	17.83
3–3	20.1	19.8	16.9	17.2	19.9	18.9	20.9	19.5	17.9	19.01

Based on the experimental results, it can be concluded that the intelligent completion technology of assembly orders proposed in this study can well solve the problem of illusion of instantiated entities caused by the application of large language models to the generation of aircraft assembly process documents in the field of aviation manufacturing, and can effectively promote the application of large models to intelligently generate aircraft assembly orders technical solutions.

4 Conclusion

This paper proposes a parametric intelligent completion technology applied to the writing of aircraft assembly orders, aiming to solve the problem that large language models often have hallucinations in the generation of instantiated entity names when generating aircraft assembly orders. Experimental results show that the application of this technology can solve the problem of hallucination of instantiated entities when the large model tries to generate aircraft assembly orders under the premise of ensuring the efficiency and quality of assembly order writing, so as to promote the application of large models in the field of aircraft assembly order generation.

This technology only considers the assembly stage of mechanical components of the aircraft, and in the future, it can be considered to make it more intelligent and add strong interactive and visual functions such as augmented reality (AR). For example, this technology is combined with the AR module applied to aircraft cable assembly in the future to realize the real-time identification of components and completion of orders on-site operations [8]; and the AR maintenance and recognition of complex electrical equipment of aircraft to create a visual entity recognition and order generation system [9], etc.

Acknowledgments. This research was supported by the National Center of Technology Innovation for Advanced Aviation Equipment Science Foundation under Grant No. 010601012507, the National Natural Science Foundation of China under Grant No. 52175484 and the Stable Support Project of the Ministry of Industry and Information Technology of China under Grant No. WD-2024-7-3.

References

1. Patil, R., Gudivada, V.: A review of current trends, techniques, and challenges in large language models (llms). Appl. Sci. **14**(5), 2074 (2024)
2. Liang, J., Zhang, L., Yan, S., et al.: Research progress on named entity recognition based on large language models. Comput. Sci. Explor. **10**, 1–25 (2024)
3. Liu, X., Erkoyuncu, J.A., Fuh, J.Y.H., et al.: Knowledge extraction for additive manufacturing process via named entity recognition with LLMs. Robot. Comput. Integr. Manuf. **93**, 102900 (2025)
4. Zhao, J., Ni, Z., Liu, C., et al.: Chinese named entity recognition in industry domain based on lexical enhancement. In: 2024 IEEE Cyber Science and Technology Congress (CyberSciTech), pp. 244–249. IEEE (2024)

5. Tang, Y., Hasan, R., Runkler, T.: Fsponer: Few-shot prompt optimization for named entity recognition in domain-specific scenarios. arXiv preprint https://arxiv.org/abs/2407.08035 (2024)
6. Zhao, P., Zhang, H., Yu, Q., et al.: Retrieval-augmented generation for ai-generated content: a survey. arXiv preprint https://arxiv.org/abs/2402.19473 (2024)
7. Huang, J., Lin, F., Yang, J., et al.: From prompt engineering to generative artificial intelligence for large models: the state of the art and perspective. Chin. J. Intell. Sci. Technol. **6**(2), 115–133 (2024)
8. Geng, J., et al.: A markerless AR guidance method for large-scale wire and cable laying of electromechanical products. IEEE Trans. Industr. Inform. **20**(3), 4007–4020 (2023)
9. Geng, J., et al.: Systematic AR-based assembly guidance for small-scale, high-density industrial components. J. Manuf. Syst. **79**, 86–100 (2025)

A Unified Framework for AI and Generative AI in Credit-Risk Classification and Credit Card Fraud Detection

Brooke McClinton(✉) and Shanzhen Gao

Reginald F Lewis College of Business, Virginia State University, Petersburg, VA 23806, USA
bmcc0115@students.vsu.edu

Abstract. This paper presents a unified framework for applying Artificial Intelligence (AI) and Generative AI (GenAI) to credit-risk classification and credit card fraud detection. Traditional statistical and rule-based systems struggle with nonlinear patterns, extreme class imbalance, and evolving financial behaviors. Machine learning models—including gradient boosting, ensemble methods, and neural networks—offer superior predictive performance, while GenAI techniques such as GANs, VAEs, and large language models (LLMs) enable synthetic data generation, anomaly detection, explainability, and automated documentation. Using the UCI Default of Credit Card Clients dataset as an illustrative case, the paper synthesizes validated literature, outlines a cross-domain risk-modeling architecture, and examines fairness, governance, privacy, and regulatory considerations. The study concludes with implications for responsible GenAI adoption in financial risk analytics.

Keywords: credit-risk classification · credit card fraud detection · machine learning (ML) · generative adversarial networks (GANs) · large language models (LLMs)

1 Introduction

Digital transformation has reshaped global financial systems. Consumers increasingly rely on credit cards, mobile wallets, and online banking, generating massive streams of transactional and behavioral data. These innovations enhance convenience but simultaneously amplify two major risks:

Credit card fraud: unauthorized or malicious transactions that exploit digital payment infrastructures.
Credit-risk default: borrowers failing to meet obligations due to financial instability or behavioral patterns.

Both problems are high-impact: fraud causes billions in annual global losses, while credit defaults destabilize institutions and influence regulatory capital requirements.

H. R. Arabnia et al. (Eds.): CSCI 2025, CCIS 2968, pp. 171–189, 2027.
https://doi.org/10.1007/978-3-032-28457-0_13

1.1 Limitations of Traditional Methods

Historically, financial institutions relied on:

Rule-based expert systems
Logistic regression scorecards
Manual auditing
Heuristic anomaly rules (e.g., velocity checks, merchant category flags)

Although interpretable and easy to implement, these methods are:

Static, unable to adapt to evolving fraud strategies
Linear, struggling with nonlinear relationships
Insensitive to rare events, especially when fraud prevalence is below 0.2%
Fragile under concept drift, where consumer behavior shifts over time

Bolton and Hand (2002) provide a seminal review of statistical fraud detection techniques, including peer-group analysis, Bayesian models, clustering, and distance-based outlier detection—foundational yet insufficient for modern real-time risk environments.

1.2 Rise of Machine Learning in Financial Risk Analytics

ML algorithms—particularly ensemble models such as Random Forest, XGBoost, and LightGBM—can model nonlinear interactions, handle high-dimensional data, and learn complex behavioral signatures. In fraud detection, ML systems detect anomalies that rule-based engines miss. In credit risk scoring, ML models outperform traditional scorecards across ROC AUC, precision-recall, and calibration.

A pivotal advancement is the ability to:

Automatically engineer features from sequential billing and payment histories
Capture borrower behavior across multiple financial products
Learn evolving fraud tactics through continuous retraining

1.3 Emergence of Generative AI

GenAI, including GANs (Goodfellow et al., 2014) and LLMs (GPT-4, BERT, domain-specific models), introduces transformational capabilities:

Synthetic data generation to address class imbalance
Privacy-preserving data augmentation
Unsupervised anomaly detection
Explainability through natural-language summaries
Automated credit-risk narratives and loan memos

GANs help generate realistic rare-event samples (fraud, default), enabling models to train on more representative datasets. LLMs enable understanding of text-rich data (loan descriptions, customer notes, merchant information) and produce human-readable interpretations of ML model outputs.

1.4 Contribution of This Paper

This paper delivers:

A comprehensive review of ML, GANs, LLMs, and synthetic data applications in financial risk analytics.
A unified conceptual framework bridging fraud detection and credit-risk scoring.
An applied demonstration using the UCI Default of Credit Card Clients dataset.
A discussion of governance, fairness, privacy, and regulatory concerns in AI-driven credit systems.
Implications and future research directions for GenAI in financial risk modeling.

2 Literature Review

This section provides a comprehensive synthesis of validated research across four major areas:

credit card fraud detection,
credit-risk classification,
generative AI for synthetic financial data, and
large language models in risk analysis.

2.1 Fraud Detection

Fraud detection involves identifying illegitimate financial transactions, often hidden within millions of legitimate ones. A typical fraud rate in real datasets is <0.2%, making this a severe class imbalance problem.

2.1.1 Early Statistical Approaches

Bolton and Hand (2002) remain foundational, classifying fraud detection methods into:

Statistical models (Bayesian, regression-based)
Distance-based methods (nearest neighbor, clustering)
Peer-group analysis (comparing customers to similar groups)
Anomaly detection (flagging deviations from normal behavior)

These methods offer interpretability but lack adaptability and struggle with high-dimensional, nonlinear transaction patterns.

2.1.2 Machine Learning Approaches

With the availability of large transactional datasets, ML models have become the industry standard.

Key findings from validated studies:

Dal Pozzolo et al. (2018) demonstrated that ensemble methods with specialized sampling strategies outperform logistic regression for fraud detection, particularly under extreme imbalance conditions.

Malik et al. (2022) introduced hybrid ML models combining logistic regression and tree-based ensembles, improving AUC and recall on public credit card fraud datasets.

Empirical results confirm that Random Forest, Gradient Boosting, XGBoost, CatBoost, and neural networks offer superior performance.

2.1.3 Modern Fraud Detection Trends

Modern fraud systems incorporate:

Behavioral analytics (merchant category velocity, geolocation, device fingerprints)
Streaming-based detection (real-time scoring of incoming transactions)
Adaptive models retrained on fresh data
Hybrid architecture combining rules + ML for explainability
Graph-based fraud detection, capturing relationships among accounts, devices, merchants, and IPs

Fraud detection is inherently adversarial: fraudsters continually adapt their tactics, requiring ongoing model updates, anomaly monitoring, and multi-source data integration.

2.2 Credit-Risk Classification

Credit-risk modeling predicts whether a borrower will default, enabling lenders to manage expected losses, set credit limits, and comply with Basel regulations.

2.2.1 Classical Credit Scoring

Traditional credit scoring models use:

Logistic regression
Linear discriminant analysis
FICO-style scorecards

These remain popular due to their interpretability but are limited in their ability to express nonlinear interactions among variables.

2.2.2 Machine Learning in Credit Scoring

The UCI Default of Credit Card Clients dataset (Yeh, 2009) is the most widely used academic benchmark.

Yeh & Lien (2009) compared six data-mining techniques:

Logistic Regression
K-Nearest Neighbors
Decision Trees
Support Vector Machines
Artificial Neural Networks

Neural networks achieved the highest predictive accuracy.
Subsequent systematic reviews (e.g., Noriega et al., 2023) confirm:

Gradient Boosting (XGBoost, LightGBM) excels with tabular credit risk data.
Neural networks outperform classical models when the data is sufficiently rich.
SHAP values have become standard for explainability.

2.2.3 Key Predictive Factors in Credit Risk

Well-established predictors include:

Recent payment delays (PAY_0, PAY_2)
Utilization ratios (bill amount ÷ credit limit)
Age, marital status, education
Payment history over 6 months
Bill amount trends
Credit limit stability

ML models capture interaction effects and nonlinear relationships that classical models miss.

2.3 Generative AI and Synthetic Financial Data

Generative AI, especially Generative Adversarial Networks (GANs) (Goodfellow et al., 2014), has transformed the handling of imbalanced datasets in fraud and credit-risk modeling.

2.3.1 Why Synthetic Data Matters

Fraud and defaults are rare events. GAN-based synthetic data helps:

Balance training datasets
Improve recall without harming precision
Mitigate overfitting to scarce events
Enable privacy-preserving data sharing
Support stress testing and scenario simulation

2.3.2 GAN-based Approaches

Although some early work made ambitious claims on limited data, several GAN applications have been validated in reputable venues:

GANs have shown success in structured tabular data for anomaly detection and minority-class generation.
Industry case studies show GAN-augmented classifiers outperform baseline ML models on fraud detection tasks.

Key advantages:

Learn high-dimensional nonlinear relationships
Generate realistic yet privacy-preserving samples

Challenges:

Risk of synthetic data memorizing real samples
Difficulty measuring GAN quality in tabular formats
Risk of propagating bias if trained on biased real data

2.4 Large Language Models in Financial Risk Analysis

LLMs such as GPT-4, BERT, FinBERT, and domain-specific financial transformers introduce new capabilities.

2.4.1 Applications of LLMs in Credit and Fraud

Credit-risk interpretation

LLMs generate readable explanations of borrower risk factors.
They convert SHAP values into human-readable arguments.

Fraud investigation

LLMs reconstruct fraud narratives from sequential transactions.
They classify merchant descriptions, identify anomalies, and summarize patterns.

Document intelligence

Loan applications
Customer correspondence
Regulatory filings
Merchant category descriptions

Scenario and stress testing

LLMs simulate hypothetical conditions (e.g., economic downturns).

Model governance

LLMs help generate the documentation required for compliance.

2.4.2 Limitations and Risks

LLMs also introduce concerns:

Hallucinations
Lack of numerical precision
Potential bias in text interpretation
Need for domain fine-tuning
Security concerns (prompt injection, data leakage)

Despite limitations, LLMs are among the strongest emerging tools in financial analytics.

3 Unified Conceptual Framework

Although credit card fraud detection and credit-risk classification are often treated as separate tasks, they share methodological, statistical, and operational challenges. Generative AI and ML offer cross-cutting tools that can be unified under a common framework for modern financial risk analytics.

Below is the proposed framework integrating fraud detection, credit-risk modeling, GAN-based augmentation, and LLM-driven interpretability.

3.1 Overview of the Unified Framework

The framework consists of five interconnected layers:

Data Layer — ingestion of transactional, behavioral, demographic, and text data
Processing Layer — cleaning, encoding, feature engineering, transformation
Modeling Layer — ML-based fraud & credit-risk models
Generative AI Layer — GAN/VAE synthetic data, anomaly detection
Interpretability & Governance Layer — LLMs, SHAP, fairness, regulatory compliance

3.2 Layer 1 — Data Layer

Financial risk data is inherently diverse:

(a) Structured (Tabular) Data

Transactional: amount, merchant, device, location
Repayment history: PAY_0–PAY_6
Billing patterns: BILL_AMT1–BILL_AMT6
Payment history: PAY_AMT1–PAY_AMT6
Demographics: age, gender, marital status

(b) Unstructured Text

Loan applications
Customer service transcripts
Merchant descriptions
Fraud investigation notes

(c) Behavioral/Time-Series Data

Spending velocity
Geographic inconsistency
Recency-frequency-monetary (RFM) patterns
Session-level digital behavior

(d) Graph Data

Network relationships (shared devices, IP addresses, merchants)
Entity-linking models for fraud rings

3.3 Layer 2 — Processing Layer

Financial datasets contain missing values, inconsistencies, and need transformations.

Key processing steps:

Data normalization and scaling
One-hot or target encoding
Removal of outliers using IQR or Z-score
Temporal feature extraction (sliding windows)
Segmentation of customer behavior
Fraud-specific feature engineering
Credit-risk engineered features (e.g., utilization ratio = BILL_AMT1 / LIMIT_BAL)

In the UCI Default dataset, the six-month repayment history is one of the strongest predictors of default.

3.4 Layer 3 — Modeling Layer (ML)

The modeling layer applies traditional ML algorithms to structured data.

Fraud Detection Models

Random Forest
Gradient Boosting (XGBoost, LightGBM, CatBoost)
Neural Networks (MLPs, autoencoders)
One-class SVM
Isolation Forest

Credit-Risk Models

Logistic Regression (baseline for regulatory compliance)
Gradient Boosting Decision Trees
Artificial Neural Networks
Support Vector Machines
Hybrid statistical + ML models

Performance Metrics

Fraud detection:

Precision, Recall, F1-score
ROC AUC
Precision-recall AUC
False Positive Rate (FPR)
False Negative Rate (FNR)

Credit-risk scoring:

ROC AUC
Accuracy (less important due to imbalance)
G-mean
KS Statistic
Calibration metrics

3.5 Layer 4 — Generative AI Layer

This layer integrates GANs and other generative models to improve ML modeling.

3.5.1 Synthetic Data Generation Using GANs

Fraud and defaults are minority-class events. GANs can synthesize:

Rare fraud transactions
High-risk borrower profiles
Unseen failure modes
Stress-test default scenarios

These samples balance the dataset, enabling ML models to optimize recall without excessively degrading precision.

3.5.2 Anomaly Detection with Generative Models

GANs and autoencoders detect anomalies by:

Learning normal behavior
Flagging transactions for which the reconstruction error is high
Detecting pattern drift over time

This is especially useful for zero-day fraud attacks.

3.5.3 Privacy-Preserving Synthetic Data

Synthetic financial data supports:

Model training without exposing sensitive transaction data
Data sharing across departments or institutions
Cross-border collaboration in regulated environments

GANs, coupled with differential privacy, reduce risks of memorizing real customers.

3.6 Layer 5 — Interpretability & Governance Layer

Financial ML models require explainability for internal governance and regulatory bodies.

3.6.1 SHAP-Based Interpretability

SHAP values allow forensic-level insights:

Which variables push a borrower toward default
Which features indicate fraud
Ranking of risk drivers at the customer level

E.g., in UCI credit-default modeling:

PAY_0 (most recent delinquency) is consistently top-ranked
Utilization ratio is a strong predictor
Bill-payment mismatch is a key early-warning factor

3.6.2 LLM-Assisted Explainability

LLMs convert technical outputs into readable narratives:

Risk assessment notes
Customer explanations for adverse action notices
Fraud investigation summaries
Model governance documentation

This transforms complex ML models into auditor-friendly systems.

3.6.3 Fairness and Compliance Checks

AI systems in finance must adhere to:

Equal Credit Opportunity Act (ECOA)
Fair Lending Regulation (Reg B)
GDPR data protection requirements
Basel model-risk management guidelines

LLMs help automate compliance reporting and fairness audits.

4 Case Study: UCI Default of Credit Card Clients

This section presents a detailed application of the unified framework using the UCI dataset (Yeh, 2009), one of the most widely used public credit-risk datasets.

4.1 Data Description

The dataset contains 30,000 credit card clients from Taiwan with:

Demographics

SEX
EDUCATION
MARRIAGE
AGE

Credit Attributes

LIMIT_BAL (credit limit)
PAY_0 to PAY_6 (delinquency history)
BILL_AMT1 to BILL_AMT6
PAY_AMT1 to PAY_AMT6

Target Variable

default.payment.next.month (1 = default, 0 = non-default)

Observations

Default cases ≈ 22.1% (less extreme than fraud datasets)
Strong temporal dependency across 6 months of repayment data

Some variables require **recording** (EDUCATION and MARRIAGE categories are inconsistent).

4.2 Baseline ML Models

Baseline models include:

Logistic Regression

Still widely used due to interpretability but limited for nonlinear interactions.

Decision Trees

Provides interpretability but is prone to overfitting.

Random Forest

Handles nonlinear relationships and reduces overfitting.

XGBoost / LightGBM

State-of-the-art for structured tabular data.

Artificial Neural Networks

Capture complex interactions; best performance reported in Yeh & Lien (2009).

4.3 GAN-Based Synthetic Data Augmentation

Though synthetic data must be carefully validated, GAN-based augmentation offers benefits:

Enhance class balance
Protects privacy

Enables scenario simulation
Improve minority-class recall

Evaluation must check:

Distribution similarity
Coverage of minority-class patterns
Non-leakage of real individuals

4.4 LLM-Assisted Interpretability and Documentation

LLMs can generate:

Credit-risk narratives explaining why a borrower is high-risk
Adverse action notices required by regulation
Risk summaries for internal auditors
Natural-language justifications grounded in SHAP explanations

This allows ML output to be integrated directly into human workflows.

5 Discussion

The deployment of AI and Generative AI in fraud detection and credit-risk classification brings substantial benefits—but also introduces risks, regulatory challenges, and governance requirements. This section analyzes five strategic dimensions:

Model Governance
Fairness and Ethical Considerations
Privacy and Synthetic Data Risks
Model Risk Management (MRM)
Practical Deployment Considerations

Each dimension is essential for using ML and GenAI responsibly in high-stakes financial environments.

5.1 Model Governance

Financial institutions operate in a highly regulated environment where model risk must be monitored and controlled.

5.1.1 Regulatory Expectations

Across major jurisdictions:

United States:

Federal Reserve SR 11–7 (Model Risk Management)
OCC and FDIC guidance
Fair Credit Reporting Act (FCRA)
Equal Credit Opportunity Act (ECOA, Regulation B)

Europe:

European Banking Authority (EBA) Guidelines
GDPR (data protection)
EU AI Act requirements for "high-risk" AI systems

Global (Basel):

Basel Committee principles for sound credit-risk modeling
Stress testing and model validation expectations

AI models—especially neural networks and GANs—must comply with:
Documentation standards
Validation and independent review
Monitoring for drift
Change management protocols
Auditability

5.1.2 Governance Challenges

AI introduces governance challenges that are not present in traditional statistical models:

Lack of intuitive interpretability
Complex training pipelines
Sensitivity to hyperparameters
Ensemble and deep-learning opacity
Data leakage risks
Need for continuous retraining

As a result, the governance layer must include:

Explainability tools (e.g., SHAP)
LLM-generated documentation
Reproducible pipelines
Metadata tracking (data versions, model versions, parameters)

5.2 Fairness and Ethical Considerations

Credit decisions influence access to financial resources, employment, housing, and quality of life. Ensuring fairness is not only a regulatory requirement; it is an ethical imperative.

5.2.1 Sources of Bias

Bias can be introduced through:

Historical discrimination embedded in data
Proxies for protected attributes (e.g., ZIP code, income)
Imbalanced representation in the training set
Feature engineering choices
Model architecture preferences
Feedback loops (riskier groups receiving lower credit limits)

5.2.2 Fairness Techniques

Approaches to mitigating bias include:

Pre-processing Methods

Rebalancing data
Removing sensitive variables
Adversarial debiasing through GANs
Synthetic minority oversampling

In-processing Methods

Fairness-constrained optimization
Regularization penalties
Adversarial training to remove sensitive information

Post-processing Methods

Adjusting decision thresholds per demographic group
Reject inference for rejected applicants
Calibration techniques

5.2.3 Transparency and Explainability

Regulators require explanations for:

Adverse credit decisions
Disparate impacts
Model reasoning during audit events

LLMs can help generate consistent explanations but must be grounded in actual model outputs to avoid hallucinations.

5.3 Privacy and Synthetic Data Risks

Synthetic data—especially GAN-generated—can support privacy, but only if properly validated.

5.3.1 Benefits of Synthetic Data

Enables data sharing across teams
Allows training ML models without exposing personal data
Supports external collaboration (regulators, consultants, researchers)
Facilitates simulation of rare or extreme events

5.3.2 Risks of Synthetic Data

However, poorly controlled synthetic data can leak sensitive information.

Potential failures:

Memorization: GAN replicates real individuals
Inference attacks: adversaries reconstruct training data
Distribution drift: synthetic data differs significantly from real data
Amplified bias: synthetic samples reflect biased patterns in originals

5.3.3 Privacy-Safe Synthetic Data Techniques

To address risks:

Differentially private GANs (DP-GANs)
Membership inference testing
Distance metrics (Wasserstein, KS tests)
Synthetic-to-real performance validation
Cross-correlation checks

Synthetic data cannot be deployed without rigorous validation of privacy and utility.

5.4 Model Risk Management (MRM)

AI models introduce new failure modes requiring careful risk management.

5.4.1 Types of Model Risk

Prediction risk: wrong decisions, especially false negatives in fraud
Data risk: missing or corrupted data, mislabeling, leakage
Operational risk: delays, deployment failures, misconfigured thresholds
Behavioral drift: fraud patterns evolve; credit behavior changes
Adversarial risk: attackers target ML weaknesses

5.4.2 Drift Monitoring

Both fraud and credit-risk models require:

Population Stability Index (PSI)
Characteristic Stability Index (CSI)
Input distribution monitoring
Output distribution monitoring
Concept drift detection algorithms

5.4.3 Human-in-the-Loop Systems

Due to regulatory and ethical constraints:

High-risk decisions must be reviewed by analysts
ML/GenAI outputs support, but do not replace, decision-makers
LLM-generated summaries help humans understand complex model behavior

Human oversight ensures accountability.

5.5 Operational Deployment Considerations

Deploying ML and GenAI models at scale requires robust infrastructure.

5.5.1 Real-Time Fraud Detection

Key requirements:

Low-latency scoring (≤100 milliseconds)
High throughput
Online learning or incremental updates
Model fallback logic
Multi-model ensembles

5.5.2 Credit-Risk Scoring Deployment

Requirements differ:

Batch scoring at monthly or daily intervals
Clear documentation for underwriting teams
Integration with loan origination systems
Stress testing for economic scenarios

5.5.3 LLM Integration in Enterprise Systems

LLMs play supportive roles:

Automated risk memos
Customer communication
Documentation generation
Model explainability reports
Fraud investigation summaries

LLMs must be deployed with:

Prompt safeguards
Retrieval-augmented grounding
Access controls
Logging for audit purposes

Summary of Discussion Section
AI and GenAI offer unprecedented capabilities, but come with real challenges:

Governance is essential for regulatory compliance
Fairness must be actively monitored and corrected
Synthetic data must be validated for privacy and utility
Model risk must be managed across the model lifecycle
Deployment at scale requires engineered infrastructure
Human oversight remains a critical component

6 Conclusion and Future Work

Artificial Intelligence (AI) and Generative AI (GenAI) have become integral to modern financial risk analytics. Their applications in credit card fraud detection and credit-risk classification demonstrate superior predictive capabilities, flexibility, and adaptability compared to traditional rule-based and statistical models. Machine learning (ML) models—particularly gradient boosting methods, ensemble learning, and neural networks—consistently outperform logistic regression on both fraud and credit risk tasks, especially when the data is high-dimensional or nonlinear.

Generative AI introduces new opportunities to address class imbalance, enhance privacy-preserving data sharing, improve anomaly detection, and support model interpretability. GAN-generated synthetic data can enrich training sets, while large language models (LLMs) can generate readable explanations, automate documentation, and support analysts through natural-language risk summaries and fraud narratives.

However, the adoption of AI and GenAI requires careful governance. Financial risk modeling operates under strict regulatory constraints, including ECOA, GDPR, Basel guidelines, and model risk management frameworks. Issues of fairness, transparency, privacy, and robustness must be addressed through both technical and organizational solutions. Synthetic data must be evaluated carefully to avoid leakage; advanced ML models must be accompanied by explainability mechanisms; and LLMs must be grounded in validated model outputs to prevent fabrication or hallucination.

Looking ahead, several promising directions emerge:

6.1 Multimodal Credit and Fraud Models

Future models will combine numerical, behavioral, textual, and graph-based data into unified architectures. This will improve risk detection accuracy and reflect how real-world fraud networks operate.

6.2 Fairness-Aware Training for Financial Models

Integrating fairness constraints directly into training—via adversarial debiasing, constrained optimization, and synthetic data balancing—will support equitable lending and reduce disparate impact.

6.3 Privacy-Preserving Synthetic Data

Differentially private GANs, secure multiparty computation, and federated learning will allow institutions to train models collaboratively without sharing raw sensitive data.

6.4 Integrated Fraud-and-Credit Risk Engines

Most current systems separate fraud and credit-risk modeling. Future platforms will unify them, enabling joint detection of suspicious transactions and early indicators of default behavior, informed by shared behavioral patterns.

6.5 Human-Centered AI for Decision Support

LLMs will increasingly play a role in presenting ML model outputs to underwriters, auditors, and fraud investigators. Human-in-the-loop systems will remain essential to avoid automation bias.

6.6 Regulatory-Ready AI Pipelines

As global AI regulations mature (EU AI Act, U.S. AI Executive Orders), risk models will increasingly require built-in documentation, explainability, transparency, and auditability—with LLMs supporting real-time model governance.

In conclusion, AI and GenAI offer transformative opportunities for financial risk modeling. When combined with responsible governance frameworks and fairness-aware design, these technologies can significantly improve fraud detection, enhance credit-risk scoring, promote financial inclusion, and support safer digital financial ecosystems.

Acknowledgments. This paper was inspired and developed through research supported by the National Science Foundation under Grant No. 2217346, conducted during the second author's residency at the National Data Science Alliance in Atlanta, GA, as part of the Research Affinity Cohorts (RACs) program held from June 17–28, 2024. The second author received full funding to participate in this program, as well as in the 2024 NCSSM Data Science Summer Institute (July 8–12, 2024) and the 2025 NCSSM Data Science Summer Institute (June 23–26, 2025), where he gained valuable insights that significantly contributed to the development of this work.

Portions of this paper were presented and refined during discussions at Brown University in May 2025. The second author gratefully acknowledges Dr. Fan Wu, the technical lead of the RACs program, and expresses sincere appreciation to his teammates—Dr. Nytelsha Bookert, Dr. Xiqiang Zheng, and Dr. Blessing Ojeme—for their collaboration and support. He also benefited greatly from insights gained at the fifth National HBCU Blockchain and FinTech Conference, hosted by Morgan State University in Atlanta, GA (October 3–4, 2024), and the sixth National HBCU Blockchain and FinTech Conference, hosted by North Carolina A&T State University in Greensboro, NC (September 19–20, 2025).

Further refinements to this paper were made following presentations and constructive feedback received at workshops hosted by Brown University (May–August 2025), the International Business Pedagogy Workshops at Georgia State University (May 2025), and the NCSSM Data Science Summer Institutes (July 8–12, 2024, and June 23–26, 2025).

References

Bolton, R.J., Hand, D.J.: Statistical fraud detection: a review. Stat. Sci. **17**(3), 235–255 (2002). https://doi.org/10.1214/ss/1042727940

Dal Pozzolo, A., Boracchi, G., Caelen, O., Alippi, C., Bontempi, G.: Credit card fraud detection: a realistic modeling and a novel learning strategy. IEEE Trans. Neural Netw. Learn. Syst. **29**(8), 3784–3797 (2018). https://doi.org/10.1109/TNNLS.2017.2736643

Goodfellow, I. et al.: Generative adversarial nets. In: Advances in Neural Information Processing Systems (NeurIPS) (2014)

Malik, H., Farooq, W., Haider, S.I., Rehman, M.A., Alqahtani, T., Zaman, M.: Credit card fraud detection using a new hybrid machine learning algorithm. Mathematics. **10**(9), 1480 (2022). https://doi.org/10.3390/math10091480

Noriega, J.P., Rivera, L.A., Herrera, J.A.: Machine learning for credit risk prediction: a systematic literature review. Data. **8**(11), 169 (2023). https://doi.org/10.3390/data8110169

Yeh, I.C.: Default of Credit Card Clients [Dataset]. UCI Machine Learning Repository (2009). https://doi.org/10.24432/C55S3H

Yeh, I.C., Lien, C.H.: The comparisons of data mining techniques for the predictive accuracy of probability of default of credit card clients. Expert Syst. Appl. **36**(2), 2473–2480 (2009). https://doi.org/10.1016/j.eswa.2007.12.020

Artificial Intelligence - Algorithms and Application

The Rationale and Implementation of a Collection-Based Probabilistic Inference Engine

Feng-Jen Yang(✉)

Department of Computer Science, Florida Polytechnic University, Lakeland, FL 33805, USA
fyang@floridapoly.edu
https://floridapoly.edu

Abstract. Probabilistic inference is a cornerstone of artificial intelligence, particularly in decision-making under uncertainty. Among the most widely used methods is the Naïve Bayes method, which has proven effective in many intelligent systems due to its simplicity and computational efficiency. However, despite its widespread adoption, Naïve Bayes relies on assumptions that may not hold in all real-world scenarios. One key limitation is its treatment of evidence as a set of unique elements, ignoring the frequency or intensity of repeated observations. This simplification can lead to unrealistic or inaccurate results in domains where repetition carries significant meaning. This paper explores and implements a mathematical extension of the Naïve Bayes method that can be developed into a more nuanced approach to evidence representation that better aligns with real-life applications.

Keywords: Probabilistic Inference · Collection-Based Inference · Multinomial Distribution

1 Introduction

Classical probabilistic inference and classification methods in artificial intelligence are largely rooted in the Naïve Bayes framework. This method has been extensively developed and applied across a wide range of intelligent systems, particularly for tasks involving decision support under uncertainty. Its appeal lies in its simplicity, scalability, and ease of implementation. However, despite its strengths, Naïve Bayes is not without flaws. One notable issue is its mathematical unsoundness in certain real-world problem domains, where its assumptions fail to capture the complexity of actual data. This can result in computational outcomes that are unrealistic or misleading.

As many AI developers and researchers are aware, the Naïve Bayes method is based on Bayesian probability theory and set operations [1,2]. In this framework, observable events are treated as elements within a set, and the boundaries of the

H. R. Arabnia et al. (Eds.): CSCI 2025, CCIS 2968, pp. 193–203, 2027.
https://doi.org/10.1007/978-3-032-28457-0_14

set define the scope of the problem domain. During the inference, some events can serve as evidence supporting the occurrence of other events, which are treated as hypotheses. Probabilistic inference is then performed by calculating posterior probabilities by using prior probabilities and class likelihoods, depending on what evidence has been considered to support what hypothesis [3–6].

In practical applications, the prior probability represents the likelihood that a hypothesis is true before any evidence is introduced. It is an unconditional probability, reflecting our initial beliefs or assumptions. Once evidence is observed, we update this belief to obtain the posterior probability, which is a conditional probability indicating how likely the hypothesis is true given the available evidence.

While this Bayesian framework is conceptually sound and widely used, its implementation often involves a simplification that can be problematic. Specifically, evidence is typically treated as a set, meaning each piece of evidence is counted only once, regardless of how many times it might appear in a real-life instance. Mathematically, this implies that repeated occurrences of the same evidence within a sample do not increase its influence on the posterior probability. Although this assumption simplifies computation and avoids overfitting, it may not accurately reflect real-world dynamics where repetition and intensity matter.

Take, for example, the task of spam email detection. If an email contains sensitive words such as reward, gift, or inheritance multiple times, it is more likely to be spam than an email where these words appear only once. Ignoring the frequency of such terms could lead to underestimating the true likelihood of spam. Similarly, in perfume manufacturing, two perfumes made from the same ingredients but with different proportions will smell distinctly different. The quantity and repetition of ingredients directly affect the final product, and treating them as mere presence/absence elements would overlook these important nuances.

These examples illustrate that the traditional set-based representation of evidence may not be sufficient in domains where frequency, intensity, and context play a critical role. To address this limitation, I propose a shift in paradigm from representing evidence and hypotheses as a set of elements to representing them as a collection of items, where repetition is explicitly acknowledged and quantified. This approach allows for a more accurate and nuanced interpretation of data, leading to better performance in applications where the strength or frequency of evidence is essential.

Therefore, while the Naïve Bayes model provides a solid foundation for probabilistic reasoning, its practical implementation must evolve to accommodate the complexities of real-world data. By refining how evidence is represented and incorporated, AI systems can achieve more reliable and context-aware decision-making, ultimately improving their effectiveness across a broader range of applications.

2 Literature Review

Probabilistic inference has long been a foundational technique in artificial intelligence, particularly for reasoning under uncertainty. The classical Naïve Bayes method, grounded in Bayesian probability theory, has been widely adopted due to its simplicity and computational efficiency. However, its assumption of feature independence and its treatment of evidence as a set, where repetition is ignored, can limit its applicability in real-world scenarios.

Recent research has explored various extensions and alternatives to traditional probabilistic inference frameworks. Boreale and Collodi [7] proposed a parallelisable, small-step operational approach for guaranteed inference in probabilistic programs, emphasizing the need for formal semantics and scalable sampling methods. Their work highlights the importance of computational tractability in probabilistic reasoning, aligning with the goals of collection-based inference.

In the context of multinomial distributions, recent advances have focused on scalable and expressive modeling. The SBMGT framework introduced by researchers at PPoPP 2025 [8] presents a scalable Bayesian multinomial group testing method, which is particularly relevant to the collection-based approach that relies on multinomial probability computations. Similarly, the work by Aitchison and colleagues [9] on Bayesian multinomial logistic-normal models offers a robust framework for handling multivariate count data, reinforcing the value of modeling evidence frequency in probabilistic classification.

These studies collectively point to a growing recognition of the limitations inherent in traditional set-based inference and the need for models that can accommodate the complexity of real-world data. The collection-based probabilistic inference framework proposed in this study builds upon these insights by integrating evidence frequency into the inference process, thereby enhancing the expressiveness and accuracy of probabilistic reasoning in domains where repetition and intensity are critical.

3 The Collection-Based Probabilistic Inference

In many real-world problem domains, inference cases are not simply based on the presence or absence of evidence, but rather on collections of evidence where individual pieces may occur multiple times. Consider a scenario in which each inference case consists of a collection of k distinct types of evidence, and each type of evidence may appear multiple times within the same case. Furthermore, each collection of evidence may support one or more of m possible hypotheses.

The central question in such a framework becomes:

Given that evidence E_1 occurs e_1 times, E_2 occurs e_2 times, ..., and E_k occurs e_k times in a particular inference case, what is the probability that a specific hypothesis H_i is true, where $1 \leq i \leq m$?

Mathematically, this setup corresponds to a multinomial experiment, where each trial result is one of k possible outcomes E_1, E_2, ..., and E_k, and the posterior probability to be computed by using the supportive evidence that $E_1 = e_1$, $E_2 = e_2$, ..., and $E_k = e_k$ to support a hypothesis H_i, where the total number of trials are $(e_1 + e_2 + ... + e_k)$.

To compute the posterior probability that hypothesis Hi is true given the observed evidence counts, the following mathematical steps are derived:

- **Step 1. Apply the Bayesian Theorem:**

$$P(H_i|(E_1 \cap E_2 \cap ... \cap E_n)) = \frac{P(H_i \cap (E_1 \cap E_2 \cap ... \cap E_n))}{P(E_1 \cap E_2 \cap ... \cap E_n)}$$

- **Step 2. Expand the Numerator and Denominator:** The numerator is the joint probability of the hypothesis and the evidence, and the denominator is the total probability of the evidence across all hypotheses:

$$\frac{P(H_i \cap (E_1 \cap E_2 \cap ... \cap E_n))}{P(E_1 \cap E_2 \cap ... \cap E_n)} = \frac{P((E_1 \cap E_2 \cap ... \cap E_n) \cap H_i)}{\sum_{j=1}^{m} P((E_1 \cap E_2 \cap ... \cap E_n) \cap H_j)}$$

- **Step 3. Apply the Product Rule:** Each term in the numerator and denominator can be rewritten using the product rule:

$$\frac{P\Big(((E_1 = e_1) \cap (E_2 = e_2) \cap \cdots \cap (E_k = e_k)) \cap H_i\Big)}{\sum_{j=1}^{m} P\Big(((E_1 = e_1) \cap (E_2 = e_2) \cap \cdots \cap (E_k = e_k)) \cap H_j\Big)} = \frac{P\Big(((E_1 = e_1) \cap (E_2 = e_2) \cap \cdots \cap (E_k = e_k)) \mid H_i\Big) \times P(H_i)}{\sum_{j=1}^{m} P\Big(((E_1 = e_1) \cap (E_2 = e_2) \cap \cdots \cap (E_k = e_k)) \mid H_j\Big) \times P(H_j)}$$

- **Step 4. Model the Likelihood Using the Multinomial Distribution:** Assuming the evidence counts e_1, e_2, ..., e_k follow a multinomial distribution conditioned on each hypothesis H_i, the likelihood term becomes:

$$\frac{P\Big(((E_1 = e_1) \cap (E_2 = e_2) \cap \cdots \cap (E_k = e_k)) \mid H_i\Big) \times P(H_i)}{\sum_{j=1}^{m} P\Big(((E_1 = e_1) \cap (E_2 = e_2) \cap \cdots \cap (E_k = e_k)) \mid H_j\Big) \times P(H_j)} = \frac{\frac{n!}{e_1! \times e_2! \times \cdots \times e_k!} \times p_1^{e_1} \times p_2^{e_2} \times \cdots \times p_k^{e_k} \times P(H_i)}{\sum_{j=1}^{m} \left(\frac{n!}{e_1! \times e_2! \times \cdots \times e_k!} \times p_1^{e_1} \times p_2^{e_2} \times \cdots \times p_k^{e_k} \times P(H_j)\right)}$$

where:

- e_i is the number of times E_i occurs in the inference case.
- n is the total number of items in this inference case, i.e., $n = e_1 + e_2 + \cdots + e_k$.
- p_i is the probability that E_i occurs in the samples where hypothesis H_i is true.
- p_j is the probability that E_j occurs in the samples where hypothesis H_j is true.

The factorials in these computations account for the fact that the order of evidence occurrences is irrelevant in this collection-based approach. This ensures that all permutations of the same evidence observed are treated equally.

The denominator in the posterior probability performs a normalization across all possible hypotheses, ensuring that the sum of posterior probabilities for all hypotheses given the same evidence collection equals 1. This is essential for maintaining the probabilistic integrity of the model [7].

This collection-based inference model provides a more realistic and flexible framework for probabilistic reasoning in domains where the frequency of evidence matters. Unlike traditional set-based approaches that ignore repetition, this method captures the intensity and distribution of evidence, leading to more accurate and context-sensitive classification outcomes.

4 An Implementation

The collection-based inference method presents a practical and scalable alternative to traditional probabilistic reasoning frameworks, particularly in contexts where computational efficiency and adaptability are paramount. One of the method's most compelling advantages lies in its ability to manage complexity effectively, enabling integration into real-world systems that demand rapid decision-making and robust performance. This includes applications in areas such as decision support systems, intelligent agents, and data-driven diagnostics, where both accuracy and responsiveness are critical.

Despite its conceptual richness and expressive power, the method remains computationally tractable. It avoids the overhead typically associated with exhaustive probabilistic computations by leveraging structured collections of evidence and hypotheses. This design choice allows for streamlined inference processes that can be executed efficiently, even in environments with limited computational resources.

To demonstrate the feasibility and practical utility of this approach, consider its application to a hypothetical dataset, as shown in Table 1. The dataset involves three distinct pieces of evidence, denoted as E_1, E_2, and E_3, and two competing hypotheses, H_1 and H_2. Each evidence item may occur multiple times, and the frequency of these occurrences plays a crucial role in shaping the inference outcome.

To illustrate the feasibility of this approach, I apply it to a hypothetical dataset is given in Table 1, which involves three evidence and two hypotheses denoted as E_1, E_2, E_3, H_1, and H_2, respectively.

Let us now examine a specific inference scenario: suppose that E_1 appears 2 times, E_2 appears once, and E_3 appears 3 times. Given this distribution of evidence, the question arises: *how likely is each hypothesis,* H_1 *and* H_2*, to be true?*

This case provides a concrete illustration of how the collection-based inference method can be applied to assess the relative plausibility of competing hypotheses based on observed evidence patterns.

Table 1. A Hypothetical Dataset

No.	Evidence	Hypothesis Hold
1	E_3, E_3	H_1
2	E_1, E_2, E_3, E_1	H_1
3	E_3, E_2, E_3	H_1
4	E_1, E_2, E_2	H_1
5	E_2, E_2, E_2	H_1
6	E_3, E_1	H_1
7	E_1, E_3	H_2
8	E_2, E_3, E_3, E_2	H_2
9	E_1, E_3, E_3, E_3	H_2
10	E_1, E_1, E_2, E_1	H_2

4.1 Performing Inference

From the provided dataset, we can systematically determine the following prior probabilities, which serve as foundational elements for further analysis:

- The given dataset has a total of 10 samples, and 6 of them indicate that H_1 is true. So, the prior probability $P(H_1) = 6/10$.
- The given dataset has a total of 10 samples, and 4 of them indicate that H_2 is true. So, the prior probability $P(H_2) = 4/10$.

In this inference case, E_1 appears 2 time, E_2 appears 1 time, and E_3 appears 3 times. So,

- $e_1 = 2$.
- $e_2 = 1$.
- $e_3 = 3$.
- $n = 2 + 1 + 3 = 6$.

Among the 6 samples where H_1 is true, E_1 totally appears 4 times, E_2 totally appears 7 times, and E_3 totally appears 6 times. So,

- $p_1 = 4/(4+7+6) = 4/17$
- $p_2 = 7/(4+7+6) = 7/17$
- $p_3 = 6/(4+7+6) = 6/17$

Among the 4 samples where H_2 is true, E_1 totally appears 5 times, E_2 totally appears 3 times, and E_3 totally appears 6 times. So,

- $p_1 = 5/(5+3+6) = 3/14$
- $p_2 = 3/(5+3+6) = 6/14$
- $p_3 = 6/(5+3+6) = 6/14$

Plugging these numbers and probabilities into the posterior probability computation of this inference case, the inference result is computed as follows:

$$P(H_1|((E_1 = 2) \cap (E_2 = 1) \cap (E_3 = 3)))$$
$$= \frac{\frac{6!}{2! \times 1! \times 3!} \times (\frac{4}{17})^2 \times (\frac{7}{17}) \times (\frac{6}{17})^3 \times \frac{6}{10}}{\frac{6!}{2! \times 1! \times 3!} \times (\frac{4}{17})^2 \times (\frac{7}{17}) \times (\frac{6}{17})^3 \times \frac{6}{10} + \frac{6!}{2! \times 1! \times 3!} \times (\frac{5}{14})^2 \times (\frac{3}{14}) \times (\frac{6}{14})^3 \times \frac{4}{10}}$$
$$= 0.4113$$

In a similar manner, the posterior probability of H_2 is computed as follows:

$$P(H_2|((E_1 = 2) \cap (E_2 = 1) \cap (E_3 = 3)))$$
$$= \frac{\frac{6!}{2! \times 1! \times 3!} \times (\frac{5}{14})^2 \times (\frac{3}{14}) \times (\frac{6}{14})^3 \times \frac{4}{10}}{\frac{6!}{2! \times 1! \times 3!} \times (\frac{4}{17})^2 \times (\frac{7}{17}) \times (\frac{6}{17})^3 \times \frac{6}{10} + \frac{6!}{2! \times 1! \times 3!} \times (\frac{5}{14})^2 \times (\frac{3}{14}) \times (\frac{6}{14})^3 \times \frac{4}{10}}$$
$$= 0.5887$$

4.2 The Python Code

The Python code, shown below, demonstrates how the collection-based inference engine can be operationalized to perform the necessary computations. This implementation serves as a proof of concept, showing that the theoretical framework can be translated into practical algorithms capable of supporting inference tasks in dynamic and data-intensive environments.

```
"""
This code is self-contained and ready to run. The computations
are based on the dataset given in Table 1.
"""

import math
```

```
def compute_posterior(evidence_counts, evidence_probs, prior_probs):
    """
    Computes the posterior probability for each hypothesis using
    the multinomial formula.

    Parameters:
    - evidence_counts: List of counts [e1, e2, ..., ek]
    - evidence_probs: List of dictionaries
                      [{H1: p1, H2: p1}, ..., {H1: pk, H2: pk}]
    - prior_probs: Dictionary {H1: P(H1), H2: P(H2)}

    Returns:
    - Dictionary of posterior probabilities for each hypothesis
    """

    hypotheses = list(prior_probs.keys())
    n = sum(evidence_counts)
    factorial_n = math.factorial(n)
    denominator = 0
    numerators = {}

    for h in hypotheses:
        product = 1
        for i, e_i in enumerate(evidence_counts):
            p_i = evidence_probs[i][h]
            product *= (p_i ** e_i) / math.factorial(e_i)
        numerator = factorial_n * product * prior_probs[h]
        numerators[h] = numerator
        denominator += numerator

    posterior_probs = {h: numerators[h] /
                          denominator for h in hypotheses}
    return posterior_probs

# Example input
evidence_counts = [2, 1, 3]  # e1=2, e2=1, e3=3
evidence_probs = [
    {'H1': 4/17, 'H2': 5/14},  # Probabilities of E1 under H1 and H2
    {'H1': 7/17, 'H2': 3/14},  # Probabilities of E2 under H1 and H2
    {'H1': 6/17, 'H2': 6/14}   # Probabilities of E3 under H1 and H2
    ]
prior_probs = {'H1': 0.6, 'H2': 0.4}  # Prior probabilities of
                                      # H1 and H2

# Compute and print posterior probabilities
posterior = compute_posterior(evidence_counts, evidence_probs,
            prior_probs)
for h, prob in posterior.items():
    print(f"Posterior probability of {h}: {prob:.4f}")
```

The output generated by the program is presented below and serves as a direct reflection of the computations performed using the collection-based inference framework:

```
Posterior probability of H1: 0.4113
Posterior probability of H2: 0.5887
```

This output not only validates the correctness of the implementation but also illustrates how the theoretical concepts introduced earlier can be translated into practical results. By examining the output, we can observe how the framework processes the input data, applies probabilistic reasoning, and arrives at meaningful conclusions that support decision-making in real-world scenarios.

5 Possible Strategic Applications

The collection-based probabilistic inference engine demonstrates broad potential in strategic applications. First, in threat detection and classification, it can analyze communications, sensor data, or cyber activities to identify recurring suspicious patterns, improving detection accuracy. Second, for intelligence analysis, the approach integrates multi-source reports and evaluates the likelihood of hostile actions based on repeated keywords or events. In battlefield decision support systems, it processes real-time sensor inputs such as drone imagery and radar signals, using repeated detections to infer enemy positions and intentions. For cybersecurity and intrusion detection, the engine identifies abnormal traffic and repeated attack behaviors, reducing false positives and prioritizing high-risk alerts. In logistics and resource allocation, it predicts equipment failures and supply needs by analyzing maintenance records and usage frequency, optimizing resource deployment. In surveillance and reconnaissance, the model distinguishes random noise from meaningful repeated observations, enhancing situational awareness. Finally, in psychological operations and information warfare, it monitors social media and communication channels to detect repeated narratives or propaganda, inferring influence campaigns and shifts in public sentiment to provide actionable insights for decision-makers.

6 Conclusion

Classical probabilistic inference in artificial intelligence has long been rooted in set-based approaches, which offer a structured and mathematically rigorous foundation for reasoning under uncertainty. These methods have proven effective across a range of applications, particularly in domains where data and relationships can be neatly categorized and modeled. However, as the complexity and diversity of real-world problem spaces continue to grow, especially in data-rich environments, traditional set-based frameworks often fall short in capturing the nuanced interdependencies and variability inherent in such contexts.

This study presents a significant advancement by introducing a collection-based inference framework that extends the capabilities of classical probabilistic reasoning. Unlike set-based models, which are typically constrained by rigid boundaries and assumptions, the collection-based approach allows for a more flexible and expressive representation of uncertainty. By developing a robust mathematical rationale for this extension, the framework not only preserves the foundational strengths of probabilistic inference but also enhances its adaptability to a wider array of problem domains.

The primary contribution of this work lies in its ability to extract deeper, more context-sensitive insights from complex datasets. This enriched form of reasoning supports more informed and reliable decision-making, particularly in systems where uncertainty is pervasive and data is heterogeneous. As a result, the proposed methodology holds promise for improving the performance and trustworthiness of decision support systems across fields such as healthcare, finance, autonomous systems, and beyond.

Furthermore, the collection-based framework opens new avenues for theoretical exploration and practical implementation. It invites a reexamination of existing probabilistic models and encourages the development of novel algorithms that can leverage the expressive power of collections. Ultimately, this approach represents a meaningful step toward more intelligent, adaptive, and context-aware artificial intelligence systems capable of operating effectively in today's increasingly complex and data-driven world.

References

1. Bayes, T.: An essay towards solving a problem in the doctrine of chances. Philos. Trans. Roy. Soc. **53**(1), 370–418 (1763)
2. Yang, F.-J.: Eliciting an overlooked aspect of Bayesian reasoning. ACM SIGCSE Bull. **39**(4), 45–48 (2007). https://doi.org/10.1145/1345375.1345412
3. Yang, F.-J.: Crafting a lightweight Bayesian inference engine. In: The World Congress on Engineering 2016 (WCE 2016), vol. II, pp. 612–617, London, UK (2016)
4. Yang, F.-J.: A general purpose probabilistic inference engine. In: IAENG Transactions on Engineering Sciences - Special Issue for the International Association of Engineers Conferences 2016, pp. 33–44. World Scientific, Singapore (2017). https://doi.org/10.1142/9789813226203_0004
5. Yang, F.-J.: An implementation of Naive Bayes classifier. In: International Conference on Computational Science and Computational Intelligence (CSCI), pp. 301–306, Las Vegas, NV, USA (2018). https://doi.org/10.1109/CSCI46756.2018.00065
6. Yang, F.-J.: A SQL-based probabilistic inference. In: 2023 Congress in Computer Science, Computer Engineering, and Applied Computing (CSCE 2023), pp. 241–245, Las Vegas, NV (2023). https://doi.org/10.1109/CSCE60160.2023.00043
7. Boreale, L., Collodi, L.: Guaranteed inference for probabilistic programs: a parallelisable, small-step operational approach. ACM Trans. Probabil. Mach. Learn. (2024, accepted). https://doi.org/10.1145/3769870
8. SBMGT Research Team: SBMGT: scaling bayesian multinomial group testing. In: PPoPP '25: Proceedings of the 30th ACM SIGPLAN Annual Symposium on Principles and Practice of Parallel Programming, pp. 512–523, Las Vegas, NV, USA (2025). https://doi.org/10.1145/3710848.3710861

9. Silverman, T.D., Roche, K., Holmes, Z.C., David, L.A., Mukherjee, S.: Bayesian multinomial logistic normal models through marginally latent matrix-T processes. J. Mach. Learn. Res. **23**(1), 255–296 (2022). https://doi.org/10.5555/3586589.3586596

An Employee-Centric Holistic Framework for AI Adoption in Digital Transformation of SMEs

Kimmo Kiiveri(✉), Jyrki Liesivuori, Seppo Virtanen, and Jouni Isoaho

University of Turku, FI-20014 Turku, Finland
kimmo.kiiveri@utu.fi

Abstract. Small and medium-sized enterprises (SMEs) struggle to translate artificial intelligence (AI) investments into sustained value because prevailing maturity models are technology-centric and weakly connected to employee behavior. We propose the Employee-Centric AI Maturity Model (EAIMM), a parsimonious six-dimension framework—Strategy, Organization, People, Data, Technology, Environment—anchored in absorptive capacity (ACAP). EAIMM maps each dimension to ACAP's acquire–assimilate–transform–exploit stages and foregrounds absorptive micro foundations as the main drivers of adoption. The framework yields testable propositions and offers a research agenda to operationalize and validate the EAIMM, providing actionable guidance for SMEs

Keywords: Artificial Intelligence · Digital Transformation · AI Maturity Model

1 Introduction

Digital transformation (DT) represents the strategic integration of digital technologies into all areas of an organization, fundamentally altering how it operates and delivers value [1]. While DT sets the stage for Artificial Intelligence (AI) adoption by providing essential digital infrastructure, data, and cultural readiness, its implementation in Small and Medium-sized Enterprises (SMEs) remains complex due to their unique resource constraints, informal structures, and interaction-focused environments [2, 3].

In Europe, micro-enterprises constitute 93.5% of all companies and account for 29.4% of the workforce, whereas all SMEs represent 99.8% of companies and employ 64.4% of the workforce [4]. Considering the significant role of SMEs in the economy, it is crucial that they be empowered to succeed in AI adoption, allowing them to drive innovation and support economic growth.

Corvello et al. [5] define antifragile SMEs as having core properties of entrepreneurial orientation, context insightfulness, and operational dexterity. Antifragile SMEs perceive crises as opportunities for transformation with the leverage of digital tools. Doz and Kosonen [6] highlight the importance of strategic sensitivity, collective commitment, and resource agility in acting swiftly with active leadership.

SMEs have different sizes and often face a complex set of interrelated barriers, including resource constraints, skill gaps, employee resistance, and ethical or security

H. R. Arabnia et al. (Eds.): CSCI 2025, CCIS 2968, pp. 204–227, 2027.
https://doi.org/10.1007/978-3-032-28457-0_15

concerns [7, 8]. Contrary to the common belief that small and medium-sized enterprises (SMEs) are inherently innovative due to their size, Macri et al. [9] suggest that a more complex interplay of economic and internal organizational factors shapes this behavior. Addressing these challenges requires a balanced approach and ability to adapt. The most successful models require both strong leadership and management, as well as an awareness of socioeconomic and political environments [10].

Technology adoption in both DT and AI is not a singular event but a continuous iterative process [1]. Traditional models, while useful starting points, often prove too general by concentrating heavily on technology implementation while neglecting the complex interplay between human behavior and organizational dynamics [11, 12]. These models may also often assume a pro-change bias, overlooking the significant role of resistance to change, particularly resistance driven by fear of job insecurity, in hindering successful technology adoption [13–15]. Recent AI adoption research further reinforces this view, demonstrating that technological implementation alone is insufficient and that a holistic approach is needed [16, 17].

This study explores the multifaceted environments of DT and AI adoption, particularly in SMEs. It integrates established theoretical models addressing individual acceptance and resistance to provide a comprehensive understanding of the sociotechnical dynamics that underpin technology adoption in SMEs.

The proposed employee-centric holistic model is presented with the intention of bridging the gap between theoretical models and the unique realities of AI adoption in SMEs, such as limited resources, informal structure, and high reliance on individual employees. Ultimately, for SMEs to succeed in the digital era, they must embrace a holistic strategy that integrates technology with a focus on their people, their culture, their strategic goals, and the specific context in which they operate, to guarantee the effectiveness and long-term impact of all initiatives.

Research questions:

RQ1: What is the interplay of different dimensions of technology adoption, such as strategy, culture, people, data, and technology, in the context of SME AI adoption?
RQ2: What are the key factors that determine employee acceptance or resistance to AI adoption?
RQ3: How can AI maturity models be designed to achieve a better balance between theoretical comprehensiveness and practicality for SMEs?

2 Digital Transformation

It is essential to clarify the key constructs related to DT. Digitization is the technical process of transforming analog information into digital format. Digitalization, as the sociotechnical transformation impacting social and institutional contexts within organizations [18] facilitated by digitization, serves as a foundational process that leads to DT. DT is the comprehensive integration of digital technologies into all areas of an organization, fundamentally altering how it operates and delivers value to customers [1]. Each construct has a distinct perspective: digitization has a technological perspective, digitalization has an organizational perspective, and DT has a strategic perspective.

DT extends beyond the mere implementation of technologies; it represents a strategic shift that leverages technological advancements to create new revenue streams, enhance efficiency, and foster innovation [19]. Crucially, DT also requires structural adjustments, often including changes in leadership, organizational structures, and corporate culture [20]. Employees are also affected by the change, requiring enhanced capabilities, including digital literacy and new skill sets, to adapt to the changing digital landscape [21, 22].

DT is an ongoing process of adaptation to a constantly evolving digital landscape to meet the expectations of customers, employees, and partners [23]. In this context, DT forms a crucial foundation for AI adoption. DT provides the necessary digital infrastructure, data assets, and culture of readiness, all of which are essential preconditions for leveraging AI effectively. AI can also act as a driver of DT, creating new opportunities to improve data assets, enhance automation, and drive innovation. In this study, we adopted a holistic perspective of AI utilization, recognizing that while AI can be applied to address isolated tasks, the true power of AI lies in driving fundamental changes that enhance the overall organization.

Digital technologies, often encapsulated by the acronym Social, Mobile, Analytics, Cloud, and Internet of Things (SMACIT) [22], are inherently disruptive, as they possess the potential to reshape entire industries and markets [24]. This disruptive nature aligns with Christensen's [25] Disruptive Innovation Theory, which describes how innovations initially target niche markets, but eventually displace established competitors. The adoption of SMACIT technologies can be particularly transformative for SMEs, potentially enabling them to compete with larger enterprises by leveraging their agility and innovation capacity. However, realizing this potential requires deliberate strategic choice and careful implementation.

Vial [1] further identified three types of disruptions caused by digital technologies: changes in consumer behavior and expectations, disruption of the competitive landscape, and increased data availability. While digital technologies offer the technical feasibility of creating new value propositions, realizing this potential requires significant structural changes within organizations [20]. Changes may involve adapting leadership approaches from top-down to more collaborative and empowering leadership, restructuring roles and responsibilities to suit digital processes, embedding a culture of innovation and adaptability, and enhancing employees' skills to work with new technologies. These structural changes are essential, as new technological infrastructure alone may not guarantee DT success.

A systematic literature review of DT through the TOE model [26] found that over 50% of authors describe DT using terms like "technology," "business model," "change," and "process," which are fundamental dimensions of any DT effort. Although AI is noted for its impact on the environment, regulatory and other environmental factors have received significantly less attention. This gap underscores that DT is not solely about integrating technology, but is a multifaceted process involving changes in technology, business models, processes, and environmental contexts.

Different perspectives further highlight the complexity of DT. For instance, Schallmo and Williams [27] emphasize a strategic shift in how organizations formulate strategies and create value. In contrast, Hanelt et al. [28] view DT as an organizational change driven by the widespread diffusion of digital technologies, and Vial [1] describes it as an operational process aimed at improving entities through combinations of digital tools. Recognizing and integrating these diverse perspectives is essential, particularly for SMEs, to effectively navigate the complexities of digital transformation.

2.1 AI Adoption as a Continuation of Digital Transformation

While AI represents a powerful technological advancement, maximizing its benefits necessitates a broader organizational transformation that goes beyond simply adopting new technology. Effective data management is critical for successful AI implementation. In larger organizations that handle substantial quantities of data, extensive data-handling processes may be involved. However, smaller organizations and startups, often with less data, possess the unique advantage of leveraging AI more rapidly and with greater agility [29], owing to a reduced need for elaborate processes. This highlights the varying needs and benefits of AI, based on the scale and resources of different organizations.

AI adoption exceeds previous digital transformations in both its potential impact and challenges. Unlike earlier digital technologies that primarily digitize existing processes or grow from simple applications, such as disruptive innovation theory, AI introduces autonomous learning and decision-making capabilities that can fundamentally reshape business models and industry landscapes [30]. The AI's ability to process unstructured data, recognize patterns, and predict outcomes enables levels of personalization and automation previously unattainable, necessitating a rethinking of organizational strategies and value propositions.

The transformative potential of AI adoption demands higher levels of data maturity, ethical considerations regarding AI decision-making, and significant shifts in workforce skills and roles [31]. Thus, the full adoption of AI requires organizations to undertake more profound structural and cultural changes [32, 33] than those associated with prior digitalization initiatives. Moreover, AI adoption has profound implications for society at large, influencing employment patterns, privacy concerns, ethical standards, and legal frameworks, thereby extending its impact beyond organizational boundaries [34].

AI adoption can be viewed as a complement to and a continuation of the DT process [1, 35]. As businesses digitize their operations, they accumulate vast amounts of data, creating an ecosystem that supports AI integration [36]. During the earlier phases of digital transformation, the digital infrastructure established, such as cloud computing platforms, data analytics capabilities, and interconnected devices, provide the foundational elements for AI. AI allows organizations to extract deeper insights from their data, automate complex tasks, and improve the accuracy and efficiency of decision-making. Therefore, AI adoption can be viewed as the next logical step in DT, building upon existing technological advancements to drive further innovation, cost reduction, and competitive advantage [37].

However, AI itself can be a driver for DT, pushing organizations towards necessary technological and cultural shifts when full digital transformation has not yet been achieved. For example, AI's inherent demand for robust data frameworks, scalable infrastructure, and data-literate workforces often mandates that organizations address these aspects to successfully adopt AI [36].

While smaller SMEs can offer advantages such as faster decision-making, organizational size alone does not guarantee success in adopting AI or any other digital transformation, and a complex set of organizational and external factors need to be taken into account [9]. Furthermore, while transformation is often presented as inherently beneficial, the implementation of AI, as with all change efforts, may not lead to positive outcomes for all stakeholders, and the potentially negative impacts must be considered, especially with employees [33].

3 Methodology

This study systematically explores DT and AI adoption by employing a structured methodology to synthesize relevant literature. First, we established foundational concepts related to DT and AI, setting a clear context for subsequent analyses. This involves exploring established models, providing clear definitions of digitization, digitalization, and digital transformation, and introducing key AI models. Following this, we present a review of acceptance and resistance models for DT, examining how these theories describe individual acceptance of and resistance to new technologies.

To explore the landscape of maturity models in this area, we conduct a review of Digital Maturity Models (DMMs) specifically tailored to small and medium-sized enterprises (SMEs). The search for DMM literature was performed using Google Scholar and ResearchGate, with a focus on publications between 2019 and 2024. A preliminary search in Google Scholar using the query Allin title: ("DMM" OR "digital maturity model") AND (SME OR "small and") yielded limited results, leading to a broader search on ResearchGate using ("DMM" OR "digital maturity model") AND (SME OR "small and"). The first 100 results from ResearchGate were screened by title, abstract, and article availability, resulting in seven relevant sources.

Subsequently, our investigation focused on the AI Maturity Models (AIMMs). Since very few papers had SME in the title, they were not included in the search query. A Google Scholar search using the query Allin title: AI OR "artificial intelligence" maturity model returned 51 results, of which 35 were selected for in-depth review. Further searches in ResearchGate using the query AI OR "artificial intelligence" AND "maturity model," limited to publications between 2019 and 2024, resulted in a further 200 results. After removing duplicates identified on Google Scholar, 18 relevant sources were retained. Unfortunately, only a few studies address SMEs.

The combined pool of 35 AIMM references (Google Scholar) and 18 AIMM references (ResearchGate) was assessed for scope, with narrowly focused, industry-specific papers being excluded. Eight studies were retained based on their methodological approach (Systematic Literature Reviews), a focus on SMEs, or applicability to a broader scope (public sector or manufacturing), to ensure insights that were applicable to SMEs in general, not taking into account the company size, organization, resources, industry, or location. The methodology is presented in Fig. 1.

This process, combined with an overview of the individual acceptance and resistance models and foundational definitions, provides a comprehensive background for the formulation of the model. This allows us to understand the varied dimensions that are considered in the current maturity models, resistance, and how this information can be brought together to contribute to the field with a focus on SMEs.

Limitations of this study are that the selected articles for DM and AIMM were only in English and global, neglecting country-specific issues. By contrast, in resistance model studies, many were local and in a specific industry. The effect of organizational size was not identifiable.

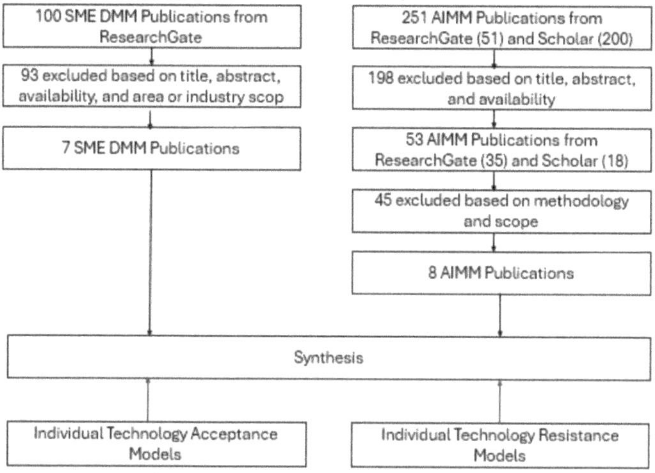

Fig. 1. The methodology used in the study

4 Key Models for DT in SMEs

To fully capture all the dimensions and theories affecting DT in SMEs, it is crucial to investigate diverse maturity models that specifically address their unique characteristics. Traditional digital maturity models often overlook SME-specific attributes, highlighting the need for a more tailored approach. By thoroughly examining these varied models for digital maturity, AI adoption, acceptance, and resistance, it is possible to create a holistic model tailored to SMEs.

4.1 Digital Maturity Models for SMEs

Before delving into AI maturity models, it is essential to examine digital maturity models (DMMs), particularly how they apply to SMEs. Understanding the nuances of digital maturity in SMEs provides a foundational context for exploring AI adoption and integration within these organizations.

Mick et al. [38] created a DT roadmap from the sustainability perspective. The proposed SDE-SMEs roadmap includes a maturity model with the following dimensions: Digital Technologies, Customer Focus, Organizational Culture, Organizational Governance, People, and Sustainability. The study's literature review had dimensions from previous DT roadmaps, and those that focused on SMEs had the following dimensions: Digital Technologies (7), Customer Focus (4), Organizational Culture (4), Organizational Governance (4), People (5), and Sustainability (1). Based on this study, while SMEs are often described by their small size and informal environments, they do not automatically benefit from digital transformation. Standard digital maturity models often fall short for SMEs, and a more tailored approach that considers unique contexts, stakeholder engagement, and management awareness is required.

Omol et al. [39] looked at 31 different studies for maturity models in developing economies from a resource strain point of view, in which technology was a predominant dimension, but there is also a need for digital culture and mindset within the organization. The dimensions they recognized from literature with more than two references were Technology (14), Strategy (14), People (11), Product (9), Operations (9), Organization (8), Customer (7), Culture (4), Transformation Management (4), and Leadership (3). A product as a dimension reflects that higher organizational maturity also impacts product maturity. They found that, for resource-dependent SMEs, digital maturity is significantly influenced by both limited internal capabilities and external dependencies, requiring models that go beyond internal factors.

Re et al. [2] conducted a case study of 18 Italian SMEs and identified eight distinct dimensions to be considered in assessing digital maturity: digital strategy and roadmap, employee skills and culture, organizational flexibility and adaptability, information technology, integration, customers, external environment, and performance and benefits. This model identifies the internal factors of skill, culture, adaptability, and flexibility. However, it also recognizes that external factors, including the SME's size relative to its customers and suppliers, influence the effectiveness of technology adoption, as smaller SMEs may find it simpler to adopt changes driven by larger entities, whereas larger SMEs have the potential to lead to adoption in their supply chains. Broader socioeconomic factors, legal frameworks, and exogenous events, such as COVID-19, also shape digital maturity.

Hyperautomation has emerged as an important AI use case in the context of DT. Defined as a business-driven approach to automate as many business and IT processes as possible, hyperautomation holds the potential to transform organizations by increasing efficiency, improving decision-making, and enabling employees to focus on strategic initiatives [40]. Barriers to hyperautomating SMEs include organizational and cultural challenges, technological limitations, individual resistance, resource constraints, and external influences [41]. These barriers are not isolated but interconnected, highlighting the need for a holistic approach rather than simply a technological solution.

Klotz and Eisenbart [10] developed a new digital maturity model by reviewing 60 existing ones. They extracted 34 dimensions from the 27 models that they found most relevant and condensed them into a seven-dimensional model. The model places Digital Strategy at its center, supported by six other dimensions. Culture and leadership are highlighted as fundamental dimensions and need to be defined top-down, whereas digital technologies are seen more as enablers than drivers. The organization and processes dimension requires streamlining and a focus on efficiency. Employees and digital skills are essential to support the transformation but can be accessed by training or outsourcing, where necessary. Customer focus on personalized engagement and communication through digital channels. Finally, products and innovation are linked to the competitive advantage of a customized, self-developed digital platform.

Van Tonder et al. [42] identify 29 studies with 23 different DMMs for SMEs. The nine key dimensions included in DMMs to measure the level of digitalization in SMEs' business models are Strategy (15), People (13), Organization (11), Technology (9), Culture (8), Customers (8), Leadership (7), Processes (6), and Products (6). Many models are focused on the manufacturing industry; therefore, generalization to other sectors might not apply. To accurately assess digital maturity in SMEs, this study underscores the need for a multifaceted model that moves beyond a purely technological focus and accounts for strategic, cultural, organizational, and human elements, acknowledging that the nuances and complexities in each dimension affect operationalization.

Haryanti et al. [43] used elements of the SPICE framework to build a digital maturity model for SMEs, namely, DX-SAMM. SPICE (ISO/IEC 15504) is a framework for assessing the capability levels of an organization's processes. While not intended for strategic alignment or the combination of other management standards, SPICE's core approach to assessing process maturity and elements from associated standards, such as Capability Maturity Model Integration (CMMI) and ISO 9001 quality management system, informed the development of DX-SAMM. The DX-SAMM self-assessment model features six maturity levels: level 0, Incomplete, Level 1, Performed, Level 2, Managed, Level 3, Established, Level 4, Predictable, and Level 5, optimized. In their study of 44 different DMMs, seven core dimensions were identified: Organization and Structure, Technology, Strategy, Culture, Customer, Process Transformation/Business Process, and Employee. SMEs should utilize a maturity model that emphasizes a defined digital strategy, leadership, and supportive culture, which is implemented with a focus on technology, processes, skills, and capabilities, driven by a focus on customers and product innovation.

Despite the ability to readily identify the distinctive characteristics, integrating these traits into DMMs remains challenging. SMEs operate in unpredictable, informal, and interaction-focused environments that standard DMMs do not always adequately address. The dimensions within DMMs often become SME-specific only after successful implementation, as it is through practical applications that the models adapt to the unique contexts of SMEs. This adaptation is largely facilitated by interactions with various stakeholders, which can be a major differentiator between SMEs and larger companies, and DT relies heavily on management ability and awareness [3]. Consequently, stakeholder engagement plays a crucial role in shaping DMMs to meet the specific needs, highlighting the importance of considering these unique environmental factors from the outset. The dimensions are listed in Table 1.

Table 1. The dimensions from DMMs

Dimension	Occurrences	Count
Strategy	Digital Strategy and Roadmap, Digital Strategy, Strategy (31)	33
People	Employee Skills and Culture (Employee aspect), Individual Resistance, Employees and Digital Skills, Employee, People (24)	28
Technology	Information Technology, Technological Limitations, Technology (24), Technology & Infrastructure	27

(*continued*)

Table 1. (*continued*)

Dimension	Occurrences	Count
Organization	Organizational Flexibility and Adaptability, Organizational and Cultural Challenges (Organizational aspect), Organization and Processes, Organization and Structure, Organization (19), Transformation Management (4)	27
Customer	Customers, Customers (9), Customer (8)	18
Process	Organization and Processes (Process aspect), Process Transformation/Business Process, Processes (6), Operations (9)	17
Culture	Employee Skills and Culture (Cultural aspect), Organizational and Cultural Challenges (Cultural aspect), Culture and Digital Leadership, Culture (13)	16
Product	Products (15)	15
Leadership	Leadership (10)	10
Environment	External Environment, External Influences	2

4.2 AI Maturity Models

Maturity models in the context of AI capabilities exhibit considerable variations in scope, typology, and design. Sadiq et al. [44] categorized AI maturity models (AIMMs) based on their purpose: descriptive models that aim to provide an understanding of the current state of AI adoption; prescriptive models that offer a roadmap for progression; and comparative models that enable benchmarking. They highlighted the prevalent focus on descriptive and comparative models within the field, leaving prescriptive approaches with less emphasis. Beyond this, the models vary in their maturity level assignments and representation, utilizing either discrete or continuous scales [44]. These variations highlight the lack of a unified standard when considering the maturity of AI capabilities.

The AIMM by Alsheiabaini et al. [45], drawing upon earlier models such as CMMI, proposes a model with five maturity levels across four dimensions: AI functions, representing the tools and technologies; Data Structure, reflecting the nature of data; People, encompassing the individuals creating AI technologies; and Organization, referring to the organizational characteristics. However, while this model provides a useful starting point, the broad nature of the organizational dimension suggests that other models that delineate different facets of this area may be useful.

In the AIMM for the public sector [46], a more granular model with five maturity levels and eight dimensions is defined: Technology and Tools, Data, People and Competencies, Organization and Processes, Leadership and Strategy, Financial, Law, Ethics and Trust, and Security and Privacy. This model highlights a more comprehensive approach to defining AI ecosystems. However, while these models may offer a high level of detail, expert feedback suggests that they are too complex and difficult to practically apply within organizations.

Gentsch's [47] algorithmic maturity model shifts the focus towards the phases of enterprise transformation, with four levels: non-algorithmic, semi-automated, automated, and super-intelligent enterprise, where AI takes decisions autonomously. The model is supported by five dimensions: Strategy, People/Organization, Decisions, Data, and Analytics. While Gentsch highlights the need for a Chief Data Officer (CDO) and the potential for a Chief Artificial Intelligence Officer (CAIO) to gain competitive advantage, it is important to note that the execution of AI initiatives is more important than simply implementing new executive titles. The model emphasizes that achieving AI maturity is a journey with gradual progression through different stages.

Based on interviews, Wallraff et al. [48] identified four levels of AI scaling: gathering the first AI experience, implementation of initial use cases, coordinated realization of use cases, and integrated company-wide implementation. Five dimensions were identified: The Strategy & Organization; Culture & Change Management plays an essential role in scaling AI and in the way knowledge is captured and shared; The Resources & Processes dimension places emphasis on how to transition to an AI-supported company while making efficient use of resources and procedures; Data also includes consideration of legal aspects; and Technology & Infrastructure refers to the technological infrastructure that a company needs to scale AI technologies. The solid scaling of AI applications requires a holistic and strategic approach, and the success of technological change largely depends on whether employees are actively willing to use AI. The model also acknowledges the significance of regulatory compliance and ethical guidelines. A surprising outcome of the research was the strong emphasis on the 'human' dimensions of culture and strategy as being fundamental for AI success, often overshadowing the perceived importance of technology, which is seen as an enabler and not a driver.

Hansen et al. [16] proposed an AI Capability Maturity Model (AICMM) drawing on a literature review and empirical design. The model focuses on practicality through five maturity levels and a zero level, which are aligned with the TOE model's dimensions of technology (Data and Infrastructure), organization (Strategy, People, and Culture), and external factors (Ethics and regulations, and Pressures and motivation). The study highlights that AI adoption requires an iterative journey, where success depends on a balanced approach where a strong strategic direction is supported by cultural change and effective data and resource management, while also empowering people to embrace and support new practices.

Sonntag et al. [17] created a Smart Manufacturing Maturity Model (SMMT) for AI deployment, specifically for manufacturing companies. In this case, manufacturing was considered sufficiently wide to be suitable for general use. The model builds on the analysis of 11 maturity models. The SMMT model comprises five maturity levels and five dimensions: Culture and Competencies, Strategy, Data, Organization and Processes, and Technology. The study identifies the data dimension as central to AI deployment capability, while Culture & Competencies and Strategy dimensions often lag because of limited AI literacy and unclear strategic direction. The findings highlight that successful implementation requires a balanced focus on all of these interrelated dimensions and that technology alone is not sufficient for effective AI scaling.

Reichl and Gruenbichler [12] analyzed 16 distinct AI maturity models applied across various industries and identified key patterns in their structure and terminology. The models typically consisted of four to six stages, with five stages being the most common, featured in nine out of 16 models. Notably, the Lichtenthaler model [49] introduces two additional stages: Isolated Ignorance as a precursor and Intuitive Ingenuity as the final stage. Frequently recurring terms for the stages included Initial, Assessing, Determined, Managed, Optimized, Operational, and Integrated, though some models employed numerical labels or scales ranging from "Low to High" or "Novice to Master." The dimensions identified with more than one reference were Data (9), Strategy (8), People and Competences (7), Organization (6), Technology (5), Process (3), Culture (2), Budget (2), Analysis (2), Products and Services (2), Regulation (2), Resources (2), and Staff/Workforce (2).

However, the findings reveal that while these models provide valuable frameworks for evaluation, they often lack practical tools, actionable recommendations, and comprehensive validation. Furthermore, inconsistencies in terminology and geographic- or industry-specific adaptations limit their universal applicability. This study emphasizes the need for standardized models, user-friendly assessment tools, and greater transparency in design to improve the effectiveness of these models. The dimensions identified across these frameworks is provided in Table 2.

Table 2. The dimensions from AIMMs

Dimension	Occurrences	Count
People	People, People and Competencies (7), People/Organization (people aspect), Staff/Workforce (2)	13
Strategy	Strategy (9), Leadership and Strategy, Strategy & Organization, Pressures and Motivation	12
Data	Data (9), Data Structure, Data and Infrastructure	11
Organization	Organization (7), Organization and Processes, People/Organization (organization aspect), Strategy & Organization	10
Technology	Technology (6), Technology and Tools, Technology & Infrastructure	8
Resources	Resources (2), The Resources & Processes, Budget (2), Financial	6
Process	Process (3), Organization and Processes, The Resources & Processes	5
Culture	Culture (3), Culture & Change Management, Culture and Competencies	5
Analysis	Analysis (2), Analytics, Decisions, AI Functions	5
Regulation	Regulation (2), Law, Ethics and Trust, Ethics and Regulations	4

Analyzing the consolidated dimensions from Tables 1 and 2 combined in Fig. 2 reveals a consistent emphasis on core factors influencing both digital and AI maturity models. Both model types consistently emphasize the importance of strategy, people, technology, and organizations as central dimensions. While Technology remains a key dimension in DMMs, Data has become increasingly prominent in AIMMs, reflecting the data-centric nature of AI. This distinction reflects the unique challenges of AI adoption, in which the ability to utilize data effectively is essential for achieving a higher level of adoption.

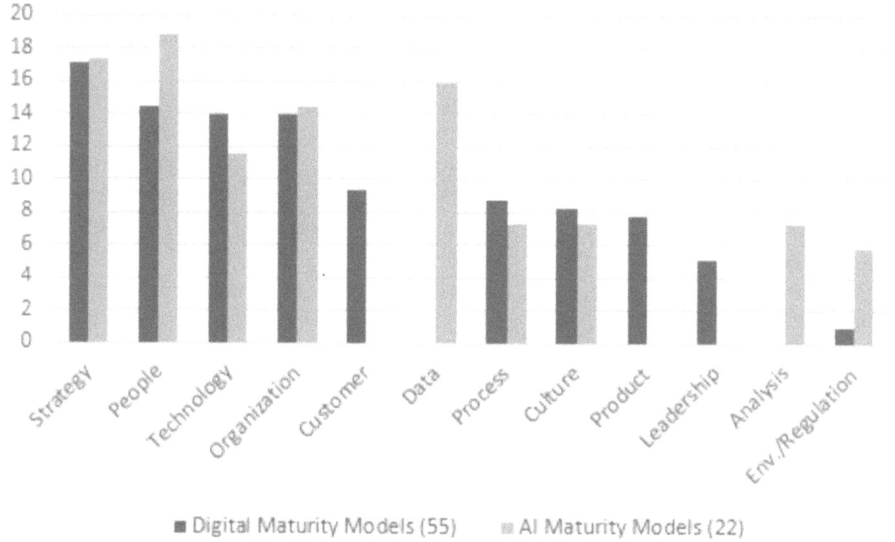

Fig. 2. Dimensions in percentage from digital and AI maturity models

5 Individual Acceptance and Resistance Models for DT

Multiple models attempt to explain the individual acceptance of DT in organizations, differing in their focus, theoretical basis, and methodological approaches. While sharing common ground, these models vary in how they approach the interplay of psychological, organizational, and technological factors impacting individuals' willingness and ability to embrace change. In SMEs, the impact of each individual is magnified, making their perspectives and actions critically influential on the overall success of the organization.

5.1 Acceptance Models

The diffusion of innovation theory [50] suggests that adoption is driven by the relative advantage, compatibility, complexity, trialability, and observability of a technology. The Technology Acceptance Model (TAM) [51], based on the Theory of Reasoned Action [52], posits that Perceived Usefulness (PU) and Perceived Ease of Use (PEOU) are the core drivers of adoption. Technologies with user-friendly designs are more readily adopted because they reduce cognitive and technical barriers, making the transition to new tools seamless and efficient [53].

Similarly, the Personal Resource Adaptation Model (PRAM) [54, 55] and Employee Digitalization Acceptance Model (EDAM) [56] highlight the importance of internal psychological elements such as proactivity, growth mindset, and trust in the technology itself. However, while PRAM and EDAM highlight psychological factors, these models are limited in scope, as they may not fully address ethical implications, workforce anxieties, and the complexity of the implementation process.

The Unified Theory of Acceptance and Use of Technology (UTAUT) [57] and the Job Demands-Resources (JD-R) models [58, 59] emphasize the role of the organization in facilitating adoption through supportive leadership, resource allocation, and a culture of innovation. In particular, the UTAUT2 model [60] extends the original UTAUT model by acknowledging that the experiential, economic, and habit-forming motivations of individuals impact the adoption of technology, demonstrating that technology use extends beyond pure functionality and also includes emotional and habitual dimensions.

These models highlight various dimensions, yet a synthesis of both employee- and customer-focused perspectives was created from the UTAUT and UTAUT2 by Blut et al. [61], which shows how context, such as the employee role and type of transaction, influences user behavior. This highlights the importance of considering many factors to understand individual responses to DT. The interplay between individual psychological factors, organizational support, and technological capabilities is a consistent theme across all models.

5.2 Resistance Models

While TAM, UTAUT, and its extensions offer valuable frameworks for understanding technology adoption, they often fail to address the complexities of resistance to change. This gap is evident in limited and fragmented literature on resistance and user rejection.

Laumer and Eckhardt [11] showcase the existing research on user resistance to information technology, identifying a lack of a unified theoretical model and noting that resistance is not simply the reverse side of acceptance. As Taherdoost [62] highlights, other perspectives, such as the Theory of Reasoned Action (TRA), the Theory of Planned Behavior (TPB), and Diffusion of Innovations theory (DOI), also contribute to understanding technology acceptance by incorporating different elements such as perceived benefit, social norms, and ease of use; however, these models primarily focus on technology adoption and have a cognitive bias, overlooking other human factors and the role of the organization and ethical considerations.

To better understand the multilayered nature of resistance to new technologies, a variety of perspectives are needed, each highlighting the complex interplay of cognitive, social, and environmental factors. Status Quo Bias Theory [63] states that employees may resist new technologies because they prefer established routines, even if those new tools offer potential benefits, a fact that can be attributed to job security fears that the user may have [14, 15]. Similarly, the equity implementation model [64] relates resistance to a perceived inequitable distribution of costs and benefits.

The Theory of Planned Behavior (TPB) [65] emphasizes how attitudes, social norms, and perceived behavioral control influence an individual's intention to adopt, or not to adopt. However, a well-known limitation is the intention-behavior gap. Measuring this gap can be challenging because intentions are usually captured through self-report surveys, which can have complicated interpretations [66].

However, these theories do not acknowledge the influence of power and political perspectives [67], demonstrating how resistance can stem from shifts in organizational power structures and challenges to existing hierarchies [11]. Importantly, Laumer and Eckhardt, building upon Lapointe and Rivard's [68] multilevel model, highlight that resistance is not a singular event but builds over time, potentially leading to active or passive rejection, depending on the unique conditions of the employee and the organization.

Although resistance may result in different actions up to a sabotage [69], various viewpoints show that resistance is not always inherently negative and can sometimes act as a filter to stop poor implementations. To alleviate resistance, it is necessary to understand its interconnected and multifaceted nature, and Kotter's [70] 8-step organizational change model provides some direction for this, as it is built around fostering employee engagement and reducing resistance to change through strong leadership and with a clear, consistent vision, but it has limitations in fast-paced environments [71]. To address resistance fully, further research should integrate the various drivers of resistance and explore their interconnectedness.

6 An Employee-Centric AI Maturity Model for SMEs

The Employee-Centric AI Maturity Model (EAIMM) comprises six dimensions selected as a parsimonious, non-overlapping set repeatedly observed in digital/AI maturity research: Strategy, Organization, People, Data, Technology, and Environment. These dimensions align with the absorptive-capacity (ACAP) cycle of acquire, assimilate, transform, and exploit [72]. This is presented in Fig. 3.

The role of each dimension within the ACAP cycle can be defined as follows:

Strategy is foundational to the acquire and exploit phases. It defines the scope and scale of knowledge acquisition by establishing target use-cases, articulating the value logic, defining risk, and specifying success metrics.

Organization provides the necessary orchestration for the acquire and transform phases, while also supporting assimilation and exploitation. This is achieved through the definition of roles, the establishment of boundary spanners and governance structures, and the implementation of routines for scaling pilots and redesigning processes.

People represent the micro foundations of adoption and are critical during the assimilate and exploit stages. Factors such as skills, self-efficacy, role clarity, and a supportive learning climate are the drivers of perceived usefulness and ease of use, which in turn determine the day-to-day uptake of new technologies.

Data, as the primary asset for AI, is central to the assimilate and transform phases. The quality, accessibility, interoperability, and compliance of data are prerequisites for assimilating and transforming AI into functional workflows.

Technology serves as the tooling and integration layer, primarily impacting the assimilate and subsequent exploit phases. The architectural design and usability of technology are what make AI practically applicable within existing work systems.

Environment encompasses the external constraints and enablers, such as policies, regulations, partner density, and market turbulence, that exert influence across all stages of the ACAP cycle.

To maintain a parsimonious, causal, and testable model, additional labels often found in similar frameworks have been subsumed within these six dimensions. For instance, "leadership" is embedded within Strategy, "process" within Organization, and "culture", aspects like learning climate and psychological safety, within the People dimension. The consideration of "customers" is positioned as an outcome of capability deployment rather than a distinct capability itself.

Boundary spanners are people or teams who bridge otherwise separate groups, for example between production and IT department. In SME AI adoption, boundary spanners bridge critical organizational and knowledge gaps by performing functions well-established in the literature: they translate external signals and technical jargon into actionable use-cases, coordinate across internal silos to align data requirements with real workflows, and guide proof-of-concepts into routine use [73, 74]. Reflecting the informal structures common in SMEs, boundary spanning is often an emergent rather than a formal role. These individuals typically occupy hybrid positions, such as a line manager with data fluency. A primary advantage of this model is the potential to accelerate the ACAP cycle; however, this reliance on informal expertise also creates a significant risk of key-person dependency.

The term "orchestration capabilities" is employed in place of "management" to more accurately reflect the nature of AI adoption in small and medium-sized enterprises (SMEs). This process is not merely about supervising resources but involves a dynamic, cross-boundary sequencing of activities. Given that the value of AI is contingent upon complementary resources such as data, technology, and process usability, "orchestration" better captures the complexities of sequencing and making choices with limited resources. This aligns with theoretical perspectives on dynamic capabilities and resource management [75–77]. Crucially, this sequencing extends to the human dimension. As [78] demonstrates, overcoming resistance requires orchestration of the affective environment, where managers engage in a temporally coordinated process of understanding, tuning, and reappraising employee emotions rather than relying on isolated persuasion tactics.

The concept of "absorptive micro foundations" refers to the employee-level enablers that facilitate the ACAP cycle in practice. Building on the foundational work on absorptive capacity [79–81], this perspective emphasizes that individual-level factors are crucial for organizational learning and adaptation. These micro foundations, which include skills, self-efficacy, role clarity, psychological safety, a conducive learning climate, and positive perceptions of usefulness and ease of use similar to PU and PEOU, are the primary drivers of the assimilation and transformation phases and are essential for sustaining exploitation in daily work routines. This focus on individual human capabilities is supported by frameworks such as Hamel's pyramid of human capabilities [82].

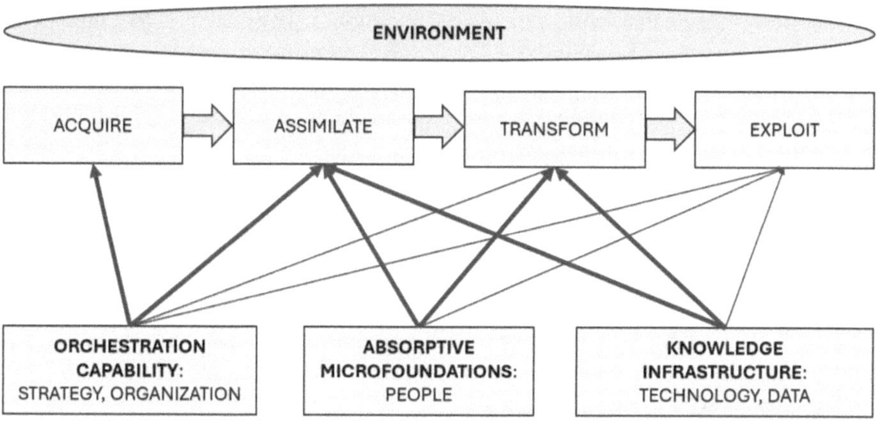

Fig. 3. The ACAP model and dimensions. Thicker arrow presents stronger association

6.1 Empowerment through Data and People

AI enhances employees' ability to turn data into insights by providing more intuitive interfaces such as natural language querying or automated visualizations, simplifying data interaction, and empowering employees. This democratization of data enables individuals, without traditional data analytics or coding skills, to independently explore and use data for insights which strengthens the assimilation and transformation.

Although data play a fundamental role in AI adoption, they also present challenges. These include ensuring robust data security, protecting user privacy, meeting regulatory compliance obligations, mitigating systemic biases within datasets, and addressing persistent issues regarding overall data quality. However, the data deficiencies identified during this process provide valuable insights that drive continuous improvements in data quality and governance increasing the knowledge.

A central element of the EAIMM is the pivotal role of employees, captured in the people's dimension. As emphasized in the upper layer of Hamel's pyramid, employee capabilities cannot simply be acquired but must be embraced and empowered through a clear strategy supported by usable data-based technology. Zahra and George [72] argue that moving from potential capacity to realized capacity requires social interaction mechanisms. Practically, this means investing in micro foundations: role clarity for boundary spanners, communities of practice, democratized data, and time for learning, so employees can absorb, combine, and apply AI-enabled insights.

7 Results and Discussion

To answer RQ1, this study examined how DT and AI adoption intersect within SMEs, building upon existing DMMs and AIMMs. While previous research provides a solid foundation, EAIMM specifically highlights the existing research gap in understanding the interplay of various dimensions in technology adoption and how these dimensions work together to drive user acceptance, which is essential to technology adoption.

This study reveals that while data and technology are key enablers, the interplay of strategic vision, organizational culture, employee competencies, and data maturity truly drives success in SME AI adoption. These dimensions must be considered in tandem, and should not be viewed as individual steps in the implementation process. The dimensions are insufficient in isolation, and the synergy between them creates a more powerful foundation [83].

In terms of RQ2, this study highlighted that a bottom-up, informal, persuasive, and empathic approach with employees characterizes SMEs, as noted in [84]. Culture and employee skills and attitudes play a large role in SMEs, where strong employee engagement and alignment with organizational values can accelerate AI adoption. Hamel and Prahalad [82] highlight how employees are important for success: the strategic vision should be shared, the employees are a strategic resource, and they should be empowered.

Crucially, however, the potential for user resistance, which can stem from perceived inequity, power shifts, or disruption of established processes, needs to be actively addressed to avoid hindering the adoption of new technologies. A supportive culture can ease concerns regarding job displacement and promote a forward-looking mindset. Skilled employees who perceive AI tools as both useful and user-friendly can become champions of new technologies, enhancing the organization's overall maturity.

Considering RQ3, the existing literature on DMMs and AIMMs reveals a common emphasis on certain core dimensions: strategy, people, data, technology, organization, and environment. While these aspects are consistently identified across various frameworks and models [85], the unique conditions of SMEs, such as limited financial and human resources, informality, and high interaction intensity, necessitate a more nuanced and context-sensitive approach, particularly regarding the practicality of AI maturity models in these contexts.

Despite the increasing popularity and practical utility of DMMs, there remains a significant gap in the empirical research linking DMM-based assessments to company performance, raising questions about the practical validity of these models. The lack of standardized definitions hinders both reliable measurement and academic investigation of the relationship between digital maturity and business outcomes [86].

Although this absence makes it scientifically challenging to determine the direct impact of DMMs on company performance, practitioners rely on in-field observations that consistently suggest positive effects on an organization's competitive advantage, highlighting the practical value perceived by users, despite theoretical gaps. Consequently, managers often perceive a beneficial connection between digital maturity and performance, which explains the widespread adoption of DMMs in practice and their influence on shaping digital strategies [87].

To address these gaps, we anchor the EIAMM in ACAP as the causal spine that explains how SMEs can convert external AI knowledge into outcomes. This anchoring turns the dimensions from a descriptive checklist into a process theory with testable propositions, thereby improving parsimony and pathways for empirical validation.

7.1 Propositions

To explain how orchestration and micro foundations affect the effectiveness in SMEs, the following propositions are made:

P1 (Acquire → Assimilate): Orchestration capability positively moderates the relationship between external knowledge acquisition and assimilation quality.
P2 (Assimilate → Transform): Data quality mediates the effect of assimilation on process/workflow transformation.
P3 (Transform → Exploit): PU and PEOU mediate the relationship between transformation and routinized exploitation.
P4 (Micro foundations): Employee micro foundations positively predict assimilation and transformation, controlling for firm size and industry.
P5 (Environment): Environmental turbulence and regulatory pressure strengthen the value of orchestration for the acquire and transform stages (contextual moderation).
P6 (Size): The marginal returns to orchestration on acquire/transform are greater in smaller SMEs due to lower coordination costs.

Collectively, these propositions outline a multi-level and contingent model of AI adoption within the ACAP framework. They posit that success is not driven by a single factor but by the dynamic interplay between top-down orchestration capabilities and bottom-up, employee-level micro foundations. This set of propositions thereby provides a clear, testable roadmap for future empirical research to validate the model and offer nuanced, actionable guidance for SMEs.

8 Conclusion

Although AI maturity models are critical for guiding AI adoption and scaling, their effectiveness is hindered by the lack of standardization and practical tools [12]. Studies [16, 17] underscore the importance of culture, strategy, and data, yet demonstrate how many models still overemphasize technological aspects. Responding to this gap, the EAIMM was developed as a model tailored specifically for SMEs and rooted in Hamel's pyramid of human capabilities [82].

The employee-centric holistic model introduced here is a valuable tool for understanding and navigating the adoption of AI in SMEs. It recognizes the interconnected nature of DT and AI adoption, where one improvement can drive progress in other areas, offering a holistic perspective that is often missing in existing AIMMs. It is grounded in existing theories and studies, yet adapted to SME-specific characteristics, such as resource constraints, informal structures, and employees.

In the ACAP-anchored EAIMM, the six interconnected dimensions separate into orchestration layers (Strategy, Organization) and employee–/capability-centric enablers (People, Data, Technology), with Environment moderating all stages; this yields actionable guidance by tying each dimension to acquire, assimilate, transform, and exploit. The model also recognizes SME heterogeneity and supports flexible, context-sensitive implementation: orchestration supplies structure and pacing, while employee enablement with technology and data readiness create the conditions, microfoundations, for adoption while keeping strategies adaptive and human-centric from initial knowledge intake through scaled usage.

As a theory-building contribution, our primary aim has been to achieve parsimony and causal clarity rather than to develop immediate measurements. This work is, therefore, conceptual and does not yet offer empirical measures or validation. We encourage and challenge future research to operationalize these constructs and test the propositions we have advanced. Researchers should translate this framework into testable models, for example, by developing reliable indicators for orchestration capability, data readiness, the effect of boundary spanners, and absorptive microfoundations. These indicators could then be evaluated through methods such as multi-informant SME surveys combined with usage logs and governance artifacts.

To validate the model, we recommend assessing its convergent, discriminant, and predictive validity using structural equation modeling (SEM/PLS) or multi-level designs, linking operationalized constructs to key outcomes such as transfer effectiveness and post-transfer performance in SMEs. These quantitative efforts should be expanded through comparative and longitudinal studies across diverse sectors and regulatory contexts to establish the EAIMM's generalizability and boundary conditions. Furthermore, to complement this structural validation, future qualitative research could provide a more granular understanding of how the microfoundations of absorptive capacity, such as individual cognitive frameworks and communication routines, manifest specifically within the unique context of SME AI adoption.

Disclosure of Interests.. The authors have no competing interests to declare that are relevant to the content of this article.

References

1. Vial, G.: Understanding digital transformation: A review and a research agenda. J. Strateg. Inf. Syst. **28**(2), 118–144 (2019). https://doi.org/10.1016/j.jsis.2019.01.003
2. Re, N.U., Ghezzi, A., Balocco, R., Rangone, A.: Supporting the digitalization of SMEs through maturity models. In: Proceedings of ECIE 2023: 18th European Conference on Innovation and Entrepreneurship (ECIE 2023), pp. 1–9 (2023)
3. Li, L., Su, F., Zhang, W., Mao, J.-Y.: Digital transformation by SME entrepreneurs: A capability perspective. Inf. Syst. J. **28**(6), 1129–1157 (2018). https://doi.org/10.1111/isj.12153
4. Di Bella, L., Katsinis, A., Lagüera-González, J., Odenthal, L., Hell, M., Lozar, B.: Annual report on European SMEs 2022/2023. Publications Office of the European Union (2023). https://doi.org/10.2826/69827
5. Corvello, V., Verteramo, S.: Antifragility in small and medium service enterprises. In: Excellence in Services University of Salerno: 24th International Conference Proceedings. Salerno, Italy (2021)
6. Doz, Y., Kosonen, M.: Fast Strategy: how Strategic Agility Will Help you Stay Ahead of the Game. Harvard Business School Press, Boston (2008)
7. Maratis, J., Ramadan, A., Rahmania Az Zahra, A., Ahsanitaqwim, R., Bennet, D.: Navigating the challenges of digital transformation in traditional organization. APTISI Trans. Manag. **8**(3), 186–194 (2024). https://doi.org/10.33050/atm.v8i3.2349
8. Restrepo-Morales, J.A., Ararat-Herrera, J.A., López-Cadavid, D.A., et al.: Breaking the digitalization barrier for SMEs: A fuzzy logic approach to overcoming challenges in business transformation. J. Innov. Entrep. **13**, 84 (2024). https://doi.org/10.1186/s13731-024-00429-w
9. Macrì, D.M., Tagliaventi, M.R., Bertolotti, F.: A grounded theory for resistance to change in a small organization. J. Organ. Chang. Manag. **15**(3), 292–310 (2002). https://doi.org/10.1108/09534810210429327

10. Klotz, P., Eisenbart, B.: MEs talk—Startups walk: how startups can benefit from the use of a digital maturity model for SMEs as part of digital transformation? In: Verkuil, A.H. (ed.) Start-up Cultures in Times of Global Crises, pp. 59–74. Springer, Cham (2024). https://doi.org/10.1007/978-3-031-53942-8_5
11. Laumer, S., Eckhardt, A.: Why do people reject technologies? A review of user resistance theories. In: Dwivedi, Y.K., Wade, M.R., Schneberger, S.L. (eds.) Information Systems Theory. Integrated Series in Information Systems, vol. 28, pp. 63–86. Springer, New York (2012). https://doi.org/10.1007/978-1-4419-6108-2_4
12. Reichl, G., Gruenbichler, R.: Maturity models for the use of artificial intelligence in enterprises: a literature review. In: Proceedings of the 19th International Scientific Conference on Industrial Systems, pp. 486–502 (2023)
13. Brynjolfsson, E., McAfee, A.: The Second Machine Age: Work, Progress, and Prosperity in a Time of Brilliant Technologies. W. W. Norton & Company, New York (2014)
14. Granulo, A., Fuchs, C., Puntoni, S.: Psychological reactions to human versus robotic job replacement. Nat. Hum. Behav. **3**(10), 1062–1069 (2019). https://doi.org/10.1038/s41562-019-0670-y
15. Hötte, K., Somers, M., Theodorakopoulos, A.: Technology and jobs: A systematic literature review. Technol. Forecast. Soc. Chang. **194**, 122750 (2023). https://doi.org/10.1016/j.techfore.2023.122750
16. Hansen, H.F., Lillesund, E., Mikalef, P., Altwajry, N.: Understanding artificial intelligence diffusion through an AI capability maturity model. Inf. Syst. Front. **26**, 2147–2163 (2024). https://doi.org/10.1007/s10796-024-10528-4
17. Sonntag, M., Mehmann, S., Mehmann, J., Teuteberg, F.: Development and evaluation of a maturity model for AI deployment capability of manufacturing companies. Inf. Syst. Manag. **42**(1), 37–67 (2024). https://doi.org/10.1080/10580530.2024.2319041
18. Tilson, D., Lyytinen, K., Sørensen, C.: Digital infrastructures: the missing IS research agenda. Inf. Syst. Res. **21**(4), 748–759 (2010). https://doi.org/10.1287/isre.1100.0318
19. Fitzgerald, M., Kruschwitz, N., Bonnet, D., Welch, M.: Embracing digital technology: A new strategic imperative. MIT Sloan Manag. Rev. **55**(2), 1–12 (2014)
20. Hess, T., Matt, C., Benlian, A., Wiesboeck, F.: Options for formulating a digital transformation strategy. MIS Q. Exec. **15**(2), 123–139 (2016)
21. Kane, G.C., Palmer, D., Phillips, A.N., Kiron, D., Buckley, N.: Strategy, not technology, drives digital transformation. MIT Sloan Manag. Rev. **57**(1), 1–25 (2015)
22. Bharadwaj, A., El Sawy, O.A., Pavlou, P.A., Venkatraman, N.: Digital business strategy: toward a next generation of insights. MIS Q. **37**(2), 471–482 (2013) https://doi.org/10.25300/MISQ/2013/37:2.3
23. Teichert, R.: Digital transformation maturity: A systematic review of literature. Acta Univ. Agric. Silvic. Mendel. Brun. **67**(6), 1673–1687 (2019). https://doi.org/10.11118/actaun201967061673
24. Karimi, J., Walter, Z.: The role of dynamic capabilities in responding to digital disruption: A factor-based study of the newspaper industry. J. Manag. Inf. Syst. **32**(1), 39–81 (2015). https://doi.org/10.1080/07421222.2015.1029380
25. Christensen, C.M.: The innovator's Dilemma: when New Technologies Cause Great Firms to Fail. Harvard Business School Press, Boston (1997)
26. Toscano Jara, J., Loza Aguirre, E., Segura, M., Franco Crespo, A.: Digital transformation: a systematic literature review from the theoretical lenses of the TOE framework. Innovar. **34**(93), e98245 (2023). https://doi.org/10.15446/innovar.v34n93.98245
27. Schallmo, D.R.A., Williams, C.A.: Digital Transformation Now!: Guiding the Successful Digitalization of Your Business Model. Springer, Cham (2018). https://doi.org/10.1007/978-3-319-72844-5

28. Hanelt, A., Bohnsack, R., Marz, D., Antunes Marante, C.: A systematic review of the literature on digital transformation: insights and implications for strategy and organizational change. J. Manag. Stud. **58**, 1159–1197 (2021). https://doi.org/10.1111/joms.12639
29. Holmström, J.: From AI to digital transformation: the AI readiness framework. Bus. Horiz. **64**(6), 683–692 (2021). https://doi.org/10.1016/j.bushor.2021.03.006
30. Brynjolfsson, E., McAfee, A.: The business of artificial intelligence: what it can—and cannot—do for your organization. Harv. Bus. Rev. **95**(4), 3–11 (2017)
31. Raisch, S., Krakowski, S.: Artificial intelligence and management: the automation–augmentation paradox. Acad. Manag. Rev. **46**(1), 192–210 (2021). https://doi.org/10.5465/amr.2018.0072
32. Constantinides, P., Henfridsson, O., Parker, G.G.: Platforms and infrastructures in the digital age. Inf. Syst. Res. **29**(2), 381–400 (2018). https://doi.org/10.1287/isre.2018.0794
33. Davenport, T.H., Ronanki, R.: Artificial intelligence for the real world. Harv. Bus. Rev. **96**(1), 108–116 (2018)
34. Floridi, L., Cowls, J., King, T.C., Taddeo, M.: How to design AI for social good: seven essential factors. Sci. Eng. Ethics. **26**, 1771–1796 (2020). https://doi.org/10.1007/s11948-020-00213-5
35. Brock, J.K.-U., von Wangenheim, F.: Demystifying AI: what digital transformation leaders can teach you about realistic artificial intelligence. Calif. Manag. Rev. **61**(4), 110–134 (2019). https://doi.org/10.1177/1536504219865226
36. Bughin, J. et al.: Artificial Intelligence: the Next Digital Frontier? McKinsey Global Institute (2017)
37. Ransbotham, S., Kiron, D., Gerbert, P., Reeves, M.: Reshaping business with artificial intelligence: closing the gap between ambition and action. MIT Sloan Manag. Rev. **59**(1), 1–22 (2017)
38. Mick, M.M.A.P., Kovaleski, J.L., Mick, R.L., Chiroli, D.M.D.G.: Developing a sustainable digital transformation roadmap for SMEs: integrating digital maturity and strategic alignment. Sustainability. **16**(20), 8745 (2024). https://doi.org/10.3390/su16208745
39. Omol, E., Mburu, L., Abuonji, P.: Digital maturity action fields for SMEs in developing economies. JECET. **12**(3), 101–114 (2023). https://doi.org/10.24214/jecet.B.12.3.10114
40. Gartner: Top 10 strategic technology trends for 2020. Gartner Inc. (2020). https://www.gartner.com/smarterwithgartner/gartner-top-10-strategic-technology-trends-for-2020. Accessed 9 Dec 2025
41. Tomitza, C. et al.: How to achieve hyperautomation? Towards a maturity model for SMEs. In: Wirtschaftsinformatik 2024 Proceedings, paper, vol. 109, (2024)
42. van Tonder, C., Bossink, B., Schachtebeck, C., Nieuwenhuizen, C.: Key dimensions that measure the digital maturity levels of small and medium-sized enterprises (SMEs). J. Technol. Manag. Innov. **19**(1), 110–130 (2024). https://doi.org/10.4067/S0718-27242024000100110
43. Haryanti, T., Rakhmawati, N.A.R., Subriadi, A.P.: The extended digital maturity model. Big Data Cogn. Comput. **7**(1), 17 (2023). https://doi.org/10.3390/bdcc7010017
44. Sadiq, R.B., Safie, N., Abd Rahman, A.H., Goudarzi, S.: Artificial intelligence maturity model: A systematic literature review. PeerJ Comput. Sci. **7**, e661 (2021). https://doi.org/10.7717/peerj-cs.661
45. Alsheiabaini, S., Cheung, Y., Messom, C.: Towards an artificial intelligence maturity model: From science fiction to business facts. In: PACIS 2019 Proceedings, paper, vol. 46, (2019) https://aisel.aisnet.org/pacis2019/46
46. Dreyling, R., Lemmik, J., Tammet, T., Pappel, I.: An artificial intelligence maturity model for the public sector: A design science approach. TalTech J. Eur. Stud. **14**(2), 49–72 (2024). https://doi.org/10.2478/bjes-2024-0016
47. Gentsch, P.: AI business: framework and maturity model. In: AI in Marketing, Sales and Service. Palgrave Macmillan, Cham (2019). https://doi.org/10.1007/978-3-319-89957-2_3

48. Wallraff, B. et al.: Scaling AI: Empirical Maturity Model and Practical Guidance for Companies. CBS International Business School (2024) https://www.researchgate.net/publication/385303428
49. Lichtenthaler, U.: Five maturity levels of managing AI: from isolated ignorance to integrated intelligence. J. Innov. Manag. **8**(1), 39–50 (2020). https://doi.org/10.24840/2183-0606_008.001_0005
50. Rogers, E.M.: Diffusion of Innovations. Free Press of Glencoe (1962)
51. Davis, F.D.: Perceived usefulness, perceived ease of use, and user acceptance of information technology. MIS Q. **13**(3), 319–340 (1989). https://doi.org/10.2307/249008
52. Fishbein, M., Ajzen, I.: Belief, attitude, intentions, and behavior: an introduction to theory and research. Contemp. Sociol. **6**(2), 244–245 (1975)
53. Venkatesh, V., Davis, F.D.: A theoretical extension of the technology acceptance model: four longitudinal field studies. Manag. Sci. **46**(2), 186–204 (2000). https://doi.org/10.1287/mnsc.46.2.186.11926
54. van den Heuvel, M., Demerouti, E., Schaufeli, W.B., Bakker, A.B.: Personal resources and work engagement in the face of change. In: Houdmont, J., Leka, S. (eds.) Contemporary Occupational Health Psychology: Global Perspectives on Research and Practice, vol. 1, pp. 124–150. Wiley (2010). https://doi.org/10.1002/9780470661550.ch7
55. van den Heuvel, M., Demerouti, E., Bakker, A.B.: How psychological resources facilitate adaptation to organizational change. Eur. J. Work Organ. Psy. **23**(6), 847–858 (2014). https://doi.org/10.1080/1359432X.2013.817057
56. Höyng, M., Lau, A.: Being ready for digital transformation: how to enhance employees' intentional digital readiness. Comput. Hum. Behav. Rep. **11**, 100314 (2023). https://doi.org/10.1016/j.chbr.2023.100314
57. Venkatesh, V., Morris, M.G., Davis, G.B., Davis, F.D.: User acceptance of information technology: toward a unified view. MIS Q. **27**(3), 425–478 (2003). https://doi.org/10.2307/30036540
58. Bakker, A.B., Demerouti, E.: The job demands–resources model: state of the art. J. Manag. Psychol. **22**(3), 309–328 (2007). https://doi.org/10.1108/02683940710733115
59. Crawford, E.R., LePine, J.A., Rich, B.L.: Linking job demands and resources to employee engagement and burnout. J. Appl. Psychol. **95**(5), 834–848 (2010). https://doi.org/10.1037/a0019364
60. Venkatesh, V., Thong, J.Y., Xu, X.: Consumer acceptance and use of information technology: extending the unified theory of acceptance and use of technology (UTAUT2). MIS Q. **36**(1), 157–178 (2012). https://doi.org/10.2307/41410412
61. Blut, M., Chong, A.Y.L., Tsigna, Z., Venkatesh, V.: Meta-analysis of the unified theory of acceptance and use of technology (UTAUT): challenging its validity and charting a research agenda in the red ocean. J. Assoc. Inf. Syst. **23**(1), 13–95 (2022) https://doi.org/10.17705/1jais.00719
62. Taherdoost, H.: A review of technology acceptance and adoption models and theories. Procedia Manuf. **22**, 960–967 (2018). https://doi.org/10.1016/j.promfg.2018.03.137
63. Kim, H.W., Kankanhalli, A.: Investigating user resistance to information systems implementation: A status quo bias perspective. MIS Q. **33**(1), 567–582 (2009). https://doi.org/10.2307/20650309
64. Joshi, K.: A model of users' perspective on change: the case of information systems technology implementation. MIS Q. **15**(2), 229–242 (1991). https://doi.org/10.2307/249384
65. Ajzen, I.: The theory of planned behavior. Organ. Behav. Hum. Decis. Process. **50**(2), 179–211 (1991). https://doi.org/10.1016/0749-5978(91)90020-T
66. Sheeran, P.: Intention–behavior relations: A conceptual review. Eur. Rev. Soc. Psychol. **12**(1), 1–36 (2002). https://doi.org/10.1080/14792772143000003

67. Markus, M.L.: Power, politics, and MIS implementation. Commun. ACM. **26**(6), 430–444 (1983). https://doi.org/10.1145/358141.358148
68. Lapointe, L., Rivard, S.: A multilevel model of resistance to information technology implementation. MIS Q. **29**(3), 461–491 (2005). https://doi.org/10.2307/25148692
69. Cieslak, V., Valor, C.: Moving beyond conventional resistance and resistors: an integrative review of employee resistance to digital transformation. Cogent Bus. Manag. **12**(1), 2442550 (2024). https://doi.org/10.1080/23311975.2024.2442550
70. Kotter, J.P.: Leading change: why transformation efforts fail. Harv. Bus. Rev. **73**(2), 59–67 (1995)
71. Carreno, A.: An analytical review of John Kotter's change leadership framework: A modern approach to sustainable organizational transformation. SSRN (2024). https://doi.org/10.2139/ssrn.5044428
72. Zahra, S.A., George, G.: Absorptive capacity: A review, reconceptualization, and extension. Acad. Manag. Rev. **27**(2), 185–203 (2002). https://doi.org/10.5465/amr.2002.6587995
73. Levina, N., Vaast, E.: The emergence of boundary spanning competence in practice: implications for implementation and use of information systems. MIS Q. **29**(2), 335–363 (2005). https://doi.org/10.2307/25148684
74. Marrone, J.A.: Team boundary spanning: A multilevel review of past research and proposals for the future. J. Manag. **36**(4), 911–940 (2010). https://doi.org/10.1177/0149206309353945
75. Teece, D.J.: Explicating dynamic capabilities: the nature and microfoundations of (sustainable) enterprise performance. Strateg. Manag. J. **28**(13), 1319–1350 (2007). https://doi.org/10.1002/smj.640
76. Barney, J.B. et al.: Resource orchestration to create competitive advantage: breadth, depth, and life cycle effects. J. Manag. **37**(5), 1390–1412 (2011). https://doi.org/10.1177/0149206310385695
77. Dhanaraj, C., Parkhe, A.: Orchestrating innovation networks. Acad. Manag. Rev. **31**(3), 659–669 (2006). https://doi.org/10.5465/amr.2006.21318923
78. Franck, H., Gylfe, P., Vuori, T., Vaara, E.: Middle managers' regulation of the emotions of others in strategy implementation: A process perspective. J. Manag. Stud. (2025). https://doi.org/10.1111/joms.70027
79. Cohen, W.M., Levinthal, D.A.: Absorptive capacity: A new perspective on learning and innovation. Adm. Sci. Q. **35**(1), 128–152 (1990). https://doi.org/10.2307/2393553
80. Volberda, H.W., Foss, N.J., Lyles, M.A.: Absorbing the concept of absorptive capacity: how to realize its potential in the organization field. Organ. Sci. **21**(4), 931–951 (2010). https://doi.org/10.1287/orsc.1090.0503
81. Felin, T., Foss, N.J., Heimeriks, K.H., Madsen, T.L.: Microfoundations of routines and capabilities: individuals, processes, and structure. J. Manag. Stud. **49**(8), 1351–1374 (2012). https://doi.org/10.1111/j.1467-6486.2012.01052.x
82. Hamel, G., Prahalad, C.K.: Competing for the Future. Harvard Business School Press, Boston (1994)
83. Arroyabe, M.F., Arranz, C.F.A., Fernandez De Arroyabe, I., de Arroyabe, F.: J.C.: Analyzing AI adoption in European SMEs: A study of digital capabilities, innovation, and external environment. Technol. Soc. **79**, 102733 (2024). https://doi.org/10.1016/j.techsoc.2024.102733
84. Zoppelletto, A., Bullini Orlandi, L., Zardini, A., Rossignoli, C., Kraus, S.: Organizational roles in the context of digital transformation: A micro-level perspective. J. Bus. Res. **157**, 113563 (2023). https://doi.org/10.1016/j.jbusres.2022.113563
85. Motjolopane, I., Chanza, M.: Digital transformation dimensions for evaluating SMEs' readiness for big data analytics and artificial intelligence: A review. Int. J. Res. Bus. Soc. Sci. **12**(7), 583–595 (2023). https://doi.org/10.20525/ijrbs.v12i7.2837

86. Thordsen, T., Murawski, M., Bick, M.: How to measure digitalization? A critical evaluation of digital maturity models. In: Hattingh, M., Matthee, M., Smuts, H., Pappas, I., Dwivedi, Y., Mäntymäki, M. (eds.) Responsible Design, Implementation and Use of Information and Communication Technology, LNCS, vol. 12066. Springer, Cham (2020). https://doi.org/10.1007/978-3-030-44999-5_30
87. Thordsen, T., Bick, M.: A decade of digital maturity models: much ado about nothing? IseB. **21**, 947–976 (2023). https://doi.org/10.1007/s10257-023-00656-w

Designing Fair and Transparent AI Systems for Automated Debate Judging Across Multiple Formats

Jizhou Tong, Adwaith Kalugotla, Bhuvanesh Bagam, Qing Zhang, and Yunkai Liu(✉)

Gannon University, Erie, PA 16541, USA
{tong001,kalugotl001,bagam002,zhang043,liu006}@gannon.edu

Abstract. This paper presents a fair-by-design framework for automated debate adjudication emphasizing identity neutrality, role-aware attribution, and reproducible scoring. The pipeline anonymizes speakers, restricts evaluation to whitelisted turns, and anchors evidence to turn IDs. A large language model generates rubric-based, multi-dimensional assessments aligned with World Schools Debate criteria, with robustness improved via multi-sample aggregation ($k = 3$) and uncertainty reporting. Evaluated on the NSDA Nationals 2023 Final ("This House supports a people's democratic right to secede") using transcripts only, results show clustered speaker totals and a narrow team margin matching the official decision. Transparency, format-agnosticism, and limitations motivating multi-event, multimodal evaluations are discussed.

Keywords: Automated Debate Judging · Large Language Models · Fair and Transparent Evaluation · Attributional Scoring · Cross-Format Evaluation

1 Introduction

Debating has long been recognized as a valuable practice for developing critical thinking, logical reasoning, and persuasive communication. Competitive debate is practiced through diverse formats, each characterized by distinctive structural conventions and pedagogical objectives. Lincoln–Douglas debates emphasize value-based propositions in a one-on-one format. Public Forum debates feature two-person teams addressing current issues for general audiences. Policy debates involve two teams debating policy resolutions with an emphasis on depth of evidence and cross-examination. Parliamentary and World Schools debates are fast-paced, team-based formats that dominate many international tournaments [1,2]. Although this diversity enhances the educational and competitive dimensions of debate, it simultaneously complicates adjudication since judges must adjust to varying rubrics, time allocations, and performance standards.

Research Track Acronym: CSCI-RTAI.

H. R. Arabnia et al. (Eds.): CSCI 2025, CCIS 2968, pp. 228–240, 2027.
https://doi.org/10.1007/978-3-032-28457-0_16

Human adjudication has traditionally served as the backbone of debate evaluation, yet it suffers from recurring challenges. Judges must process arguments in real time, track rebuttals, and provide structured feedback under time pressure. This often leads to bias, inconsistency, and cognitive overload, which reduces fairness and transparency in outcomes. Even expert adjudicators frequently diverge in their scoring of logical coherence, persuasiveness, and rhetorical delivery, leading to unpredictable results [3,4]. These challenges highlight the need for more objective and reproducible adjudication methods.

Advances in large language models (LLMs) have enabled new approaches to automated debate evaluation. Recent systems illustrate this potential. Debatrix introduces an iterative chronological analysis method with multi-dimensional scoring and demonstrates improved accuracy in debate outcome prediction [5]. Agent4Debate employs a multi-agent framework in which different agents perform search, analysis, and rebuttal, with judgments synthesized into final evaluations [6]. The Blitz Debate system combines content-based and emotional analysis to achieve high predictive accuracy in winner determination [7]. Collectively, these approaches demonstrate the feasibility of AI-driven adjudication.

Nevertheless, important challenges remain. Prior studies have demonstrated that large language models (LLMs) exhibit positional and lexical biases such as a tendency to prefer arguments occurring later in a debate or expressed with particular lexical choices [4]. Analyses of LLM-as-a-Judge frameworks have also identified limitations including limited explainability and vulnerability to over-persuasion, which reduce the robustness of automated adjudication [8]. To address these issues, existing work has proposed approaches such as multi-agent evaluation panels, ensemble scoring mechanisms exemplified by ChatEval [9], and calibration strategies designed to mitigate systematic bias [4].

While these studies demonstrate promising progress, most existing systems are constrained by narrow evaluation scopes or limited adaptability to diverse debate settings. Some approaches concentrate primarily on outcome prediction, while others emphasize argument-level analysis without addressing broader adjudication needs. Moreover, concerns regarding fairness, explainability, and bias mitigation remain insufficiently resolved, raising doubts about the reliability of fully automated judging in competitive contexts [4,8,9]. These gaps suggest that debate evaluation requires systems that not only assess arguments with high accuracy but also adapt seamlessly to the structural requirements of different debate traditions.

To address these challenges, this work introduces a unified adjudication framework that emphasizes fairness, interpretability, and cross-format flexibility. Instead of focusing on a single debate style, the framework is designed to accommodate multiple formats which include Lincoln–Douglas, Public Forum, Policy, Parliamentary, and World Schools debates by means of modular evaluation components. By combining speech transcription, argument structure detection, and multi-criteria scoring, the framework aims to produce judgments that are both transparent and reproducible. Beyond technical improvements, the proposed system is envisioned as a foundation for more scalable and trustworthy AI-assisted adjudication, supporting educational training, competitive fairness, and hybrid human–AI judging environments.

This paper makes four primary contributions:

1. It proposes a fair-by-design, format-agnostic adjudication framework that prioritizes identity neutrality, role-aware attribution, and reproducible scoring.
2. It introduces a modular architecture that supports multiple debate traditions (Lincoln–Douglas, Public Forum, Policy, Parliamentary, and World Schools) through adaptable evaluation components.
3. It incorporates transparency-enhancing mechanisms such as turn-level anonymization, evidence anchoring, and multi-sample uncertainty reporting, that address persistent concerns of bias and explainability in automated judging.
4. It presents an empirical evaluation on a high-stakes World Schools Debate final, demonstrating preliminary convergent validity with official panel decisions and highlighting rebuttal-heavy segments as the main source of evaluative variance.

2 Background

Competitive debate manifests in a variety of structured formats, each reflecting distinctive educational objectives, speech structures, and adjudicative standards. This diversity enriches the pedagogical value of debate by cultivating different dimensions of reasoning and communication, yet it simultaneously creates significant challenges for the establishment of fair and consistent adjudication systems.

Lincoln–Douglas debate (LD) is a one-on-one format focused on ethical and philosophical issues. It originated in the United States with the explicit goal of training students to analyze normative frameworks and value-based propositions. A standard LD round consists of alternating constructive speeches, cross-examinations, and rebuttals, with speaking times ranging from three to seven minutes. The evaluative focus lies on the clarity and coherence of value frameworks, logical consistency, and the ability to weigh competing moral claims. Unlike evidence-heavy formats, LD emphasizes philosophical rigor, making judges' interpretation of normative criteria especially influential in determining outcomes [1,10].

Policy debate is recognized as the most research-intensive format. Conducted in a two-on-two structure, it revolves around an annual resolution, typically advocating for or against a policy change. Each side delivers constructive speeches, engages in cross-examination, and presents rebuttals, with the entire round lasting more than ninety minutes. Debaters are expected to marshal extensive evidence, design counterplans, and deploy disadvantages and topical critiques. Evaluation in Policy debate focuses on evidentiary depth, analytical rigor, and strategic refutation. Judges must weigh complex flows of information and competing technical arguments, often under considerable cognitive load [1,11].

Public Forum debate (PF) was developed to be accessible to non-specialist audiences. Teams of two debate current events on resolutions that change monthly or bi-monthly. The speech sequence includes constructive cases, rebuttals, three "crossfire" exchanges, summaries, and closing speeches. PF emphasizes clarity, persuasion, and responsiveness to general audiences rather than technical jargon or excessive evidence density. Judges are instructed to adopt the perspective of an informed layperson, prioritizing logical clarity, rhetorical effectiveness, and the comparative weighing of arguments. As such, PF is widely used in secondary education to cultivate public speaking skills and civic engagement [1,12].

Parliamentary debate adopts a markedly different orientation by prioritizing spontaneity and rhetorical agility. With preparation times often limited to fifteen to twenty minutes, Parliamentary formats reward quick thinking, humor, and adaptability. The British Parliamentary (BP) style, which is widely practiced in international university debating circuits, consists of four teams with two on the Government side and two on the Opposition side, and each team is composed of two speakers. Teams not only advance their own cases but also compete against allied and rival teams within the same round. Judging emphasizes the comparative contribution of each team to the round, including originality, engagement with prior speeches, and persuasiveness relative to the other three teams [13,14].

World Schools Style debate blends features of Parliamentary and prepared formats, and it is the official format of the World Schools Debating Championships. Each side fields three speakers, delivering both substantive and reply to speeches. A distinctive element is the use of Points of Information (POIs) that is short interjections during opponents' speeches designed to test responsiveness and demonstrate active engagement. WSS balances accessibility and rigor by combining prepared motions with impromptu topics. Judging criteria typically include matter (argument quality), method (organization and structure), and manner (style and delivery), which makes it a holistic assessment of debate performance [15].

Taken together, these formats demonstrate the richness of competitive debate as an educational practice that cultivates critical thinking, evidence-based research, and persuasive communication. Yet this very diversity also creates substantial challenges for adjudication. The structural parameters of the formats vary widely: in Public Forum debate, speeches are concise and oriented toward accessible persuasion, whereas in Policy debate, speeches are significantly longer and evidence-intensive, requiring judges to process highly technical arguments across extended rounds. The evaluative emphases also diverge. Policy debate prioritizes evidentiary depth, technical refutation, and the strategic comparison of policy impacts. Lincoln–Douglas debate emphasizes the construction and defense of coherent philosophical frameworks, demanding that judges assess value hierarchies and ethical reasoning. Public Forum debate rewards clarity and audience adaptation, encouraging judges to evaluate persuasiveness from the perspective of an informed layperson. Parliamentary formats emphasize rhetorical agility and spontaneity, with minimal preparation time, making delivery style and adaptability decisive factors in evaluation. World Schools Style debate integrates a unique layer of interactivity through Points of Information (POIs), which are brief interjections offered by opposing speakers during substantive speeches and are considered by judges as indicators of responsiveness and engagement. Beyond these structural and evaluative differences, the educational goals of each format diverge: Policy debate fosters technical research skills and mastery of complex policy literature; Lincoln–Douglas debate trains students in moral philosophy and abstract reasoning; Public Forum debate develops civic literacy and accessible public speaking; Parliamentary and World Schools formats encourage teamwork, spontaneity, and intercultural communication. This diversity enriches the pedagogical landscape of debate but simultaneously renders consistent adjudication difficult, as judges must adapt their evaluative standards to each format. For this reason, the prospect of designing a unified or automated judging system requires not only accuracy in assessing arguments but also adaptability to the heterogeneous norms and expectations embedded within different debate traditions.

3 Related Work

Research on automated debate evaluation has drawn from multiple traditions including computational argumentation, natural language processing, and multi-agent systems. Early approaches combine argumentation theory with graph-based representations, which enable automatic mapping of claims, premises, and rebuttals for scoring consistency. Recent neural architecture extends these methods by integrating transformer-based language models with argument graph networks, demonstrating improvements in both argument detection and relational analysis [3].

Several systems illustrate the potential of large language models (LLMs) as debate judges. Debatrix employs an iterative chronological analysis that evaluates debates across multiple dimensions which include logical validity, evidentiary support, and rhetorical clarity. By combining structured scoring rubrics with stepwise LLM analysis, Debatrix has reported higher alignment with expert human judgments compared to baseline classifiers [5]. Similarly, Agent4Debate adopts a multi-agent framework which assigns specialized roles such as information retrieval, argument construction, and rebuttal generation to different agents. The final judgment is synthesized from their outputs, achieving competitive performance in debate outcome prediction [6].

Beyond structured evaluation systems, studies of AI persuasiveness have shown that LLMs can outperform humans in certain rhetorical contexts. Experimental evidence indicates that models like GPT-4 are often rated as more persuasive than human debaters, which raise both opportunities and ethical concerns regarding the deployment of AI judges in competitive settings [8]. Complementary research on bias in LLM evaluators has identified systematic vulnerabilities including positional bias (favoring later arguments) and lexical bias (favoring particular phrasing), which reduce the reliability of automated adjudication [4].

More recently, researchers have explored mechanisms to mitigate the shortcomings of LLM-based evaluators. Proposals include multi-agent evaluation panels and ensemble scoring approaches, as exemplified by ChatEval, which integrates multiple LLM judges to reduce individual model bias [9]. Other studies have suggested calibration strategies and transparency-enhancing mechanisms to improve fairness and explainability in AI adjudication [4]. These developments underscore that while AI-driven judging is feasible, ensuring fairness, adaptability, and accountability remains an open research challenge.

4 Proposed System Design

The proposed framework is designed as a multi-layered pipeline that integrates automatic transcription, structured parsing, modular segmentation, and large language model (LLM)-based evaluation. The architecture is motivated by two central challenges in automated debate adjudication: (i) the diversity of debate formats, which demand adaptable evaluation protocols, and (ii) the limitations of existing AI judges, which often lack robustness, fairness, and explainability. To address these issues, the system introduces design features that ensure modularity, extensibility, and bias mitigation. Figure 1 illustrates the system architecture.

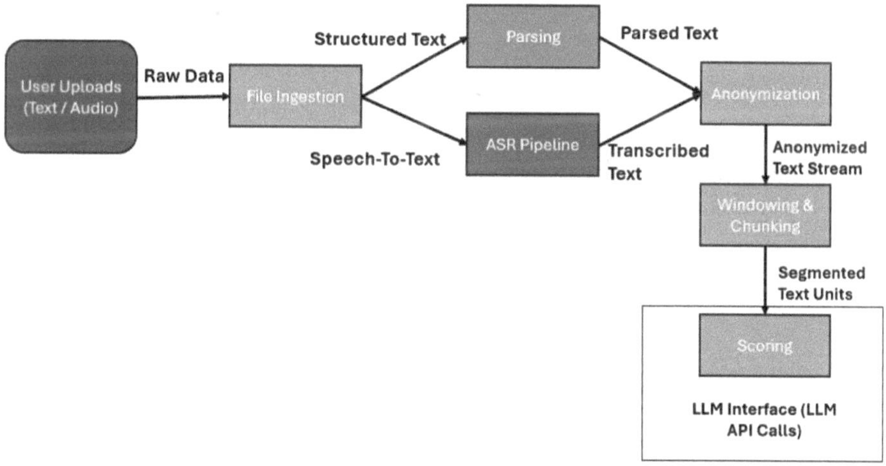

Fig. 1. System Architecture of the Proposed Debate Evaluation Framework.

4.1 Input and Ingestion Layer

The architecture accommodates dual-modality input by accepting either archival transcripts or live audio recordings, thereby reflecting the practical constraints of competitive debate settings. Textual files in formats such as TXT, DOCX, and PDF are processed by the File Ingestion module, which performs structured text extraction to produce standardized plain-text representations while incorporating error-handling for formatting inconsistencies and extraction noise. Audio recordings in MP3, WAV, or M4A formats are directed to the ASR Pipeline, which executes speech-to-text conversion using Whisper. An optional diarization subcomponent further attributes the resulting transcribed text to individual speakers, yielding structured transcripts under noisy, multi-speaker conditions.

4.2 Preprocessing and Normalization Layer

All ingested content is transformed into parsed text, subsequently normalized into turn-level units that associate each utterance with its speaker, argumentative role (e.g., constructive, rebuttal, closing), and temporal markers. To reduce bias associated with speaker identity, the Anonymization module replaces personal identifiers with neutral placeholders such as speaker A and speaker B, thereby producing an anonymized text stream. This design ensures that subsequent computational evaluation is directed toward argumentative content rather than stylistic or identity-based features.

4.3 Segmentation and Chunking Layer

Given that debate transcripts typically exceed the context window limitations of contemporary large language models, the system applies a segmentation strategy implemented by the Windowing & Chunking module. The anonymized text stream is divided into segmented text units through overlapping windows aligned with speaker turns. This dual alignment mechanism preserves discourse continuity, prevents truncation of arguments, and ensures that units of analysis correspond to coherent argumentative contributions. The resulting sub-documents are thus optimized for both semantic fidelity and model compatibility.

4.4 Evaluation and Scoring Layer

The segmented text units are transmitted to the Scoring component via the LLM Interface (API calls). Each unit undergoes intermediate summarization to distill claims, rebuttals, and evidentiary support, thereby reducing verbosity while generating structured artifacts for comparative analysis. The scoring module subsequently produces predictions across multiple evaluative dimensions, including logical coherence, evidential quality, rhetorical clarity, and temporal management. To enhance robustness and transparency, the system employs repeated sampling with pairwise comparisons between speakers, yielding mean scores, variance estimates, and probabilistic win–loss outcomes. The explicit reporting of uncertainty intervals further strengthens interpretability, enabling adjudicators to assess outcomes in terms of confidence levels rather than deterministic point values.

4.5 Distinction from Existing Systems

Unlike existing debate-assistance tools that focus on transcription accuracy or surface-level scoring, the proposed design integrates real-time transcription, format-agnostic segmentation, bias-aware scoring, and uncertainty reporting. This combination allows the system to generalize across debate formats (Lincoln–Douglas, Public Forum, Policy, Parliamentary, World Schools) while maintaining robustness and transparency.

5 Experimental Setting and Automated Judging System

We evaluate a fully automated, AI-based judging pipeline on a real, high-stakes World Schools Debate (WSD) round: the NSDA Nationals 2023 – World Schools Debate Final on the motion "This House supports a people's democratic right to secede." WSD is a three-on-three parliamentary format with six constructive speeches and two replies delivered in the order P1 → O1 → P2 → O2 → P3 → O3 → O-Reply → P-Reply, where "P" denotes Proposition speakers and "O" denotes Opposition speakers. The system consumes the verbatim transcript only (no audio or video) and anonymizes speakers (A–F) to mitigate identity-based bias. To enforce attributional fairness, scoring is restricted to a turn-whitelisting procedure as shown in Fig. 2, whereby each speaker is evaluated solely on their assigned turns. Under this design, all quoted evidence must be explicitly anchored by `turn_id` and must originate from the speaker's own whitelisted turns, ensuring traceability, preventing cross-role score contamination, and enabling systematic auditability of the adjudication process.

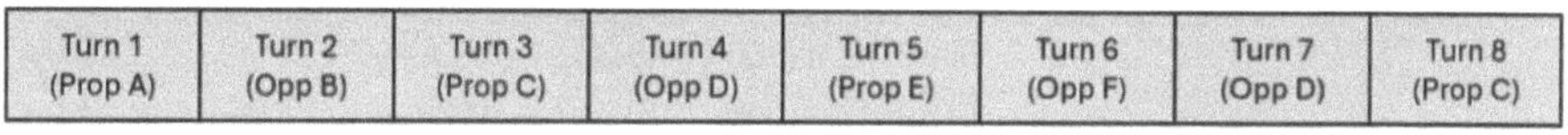

Fig. 2. Illustration of turn-whitelisting in the NSDA Nationals 2023 World Schools Debate Final. Each speaker is mapped to specific turns (A→T1; B→T2; C→T3, T8; D→T4, T7; E→T5; F→T6), and evaluation is confined to these turns to preserve attributional fairness.

Scoring follows WSD-aligned eight-dimensional rubric as shown in Table 1, with total equal to the sum of all dimensions (maximum 100):

Scores are produced by a large language model (LLM) accessed via API under a fixed WSD-aligned instruction set. To mitigate stochastic variation, the pipeline performs multi-sample aggregation with $k = 3$ independent evaluations per speaker at temperature = 0.2. For each rubric dimension, it reports the mean (score) and the standard deviation (uncertainty). We leave top-p at the API default (1.0 at run time), implying no nucleus truncation; decoding parameters and prompts are documented to support reproducibility.

Table 1. World Schools Debate Rubric Dimensions and Scales.

Dimension	Scale	Definition
Thesis Quality	0–20	Strength and clarity of stance and burdens; principled framing consistent with the motion.
Reasoning & Logic	0–20	Coherence of causal links, weighing, and comparative analysis across lines of clash.
Rebuttal (Clash)	0–20	Responsiveness to opponents' main claims; prioritization and resolution of decisive issues.
Evidence Use	0–15	Integration of transcript-anchored quotations (with `turn_id`) and appropriate factual references.
Clarity & Persuasion	0–10	Structural clarity, signposting, rhetorical effectiveness, and crystallization.
Time Management	0–5	Coverage and pacing within the allotted time.
Teamwork & Alignment	0–5	Cohesion with teammates and fulfillment of role expectations (1st/2nd/3rd, reply).
Courtesy & Rules	0–5	Adherence to WSD etiquette and procedural norms.

During scoring, the pipeline does not have access to human panel decisions, speaker awards, or tournament metadata. Team outcomes are computed post-aggregation by summing the three speakers' total scores on each side (Proposition vs. Opposition). This yields a system decision independent of human results, which is suitable for subsequent validity analyses.

5.1 Per-Speaker Scores

To mitigate stochastic variation, results are aggregated over k = 3 independent evaluations per speaker. Table 2 presents the per-speaker scores under this setting.

Table 2. Automated speaker scores on the NSDA Nationals 2023 WSD Final (transcript-only, WSD rubric, $k = 3$).

Speaker	Turns	Thesis	Reason	Rebuttal	Evidence	Clarity	Time	Team	Courtesy	Total
A	T1	18.00	17.00	12.00	13.00	8.33	4.33	4.00	5.00	81.67
B	T2	16.00	17.00	18.00	12.00	8.00	4.00	4.00	5.00	84.00
C	T3, T8	17.67	16.67	15.00	13.00	8.00	4.00	4.00	5.00	83.33
D	T4, T7	17.00	17.33	15.67	12.67	8.00	4.00	4.00	5.00	83.67
E	T5	17.67	16.67	14.67	13.00	8.67	4.00	4.00	5.00	83.67
F	T6	14.00	15.00	17.00	10.00	8.00	4.00	4.00	5.00	77.00

As shown in Table 2, the distribution of totals (77.00–84.00) indicates a competitive final typical of World Schools Debate. Speaker B attains the highest overall score (84.00), driven by a notably strong Rebuttal (18.00), while Speakers A and E lead in Thesis (≈18.00 and 17.67). Evidence Use concentrates around 12–13, which is consistent with late-round dynamics where principled framing, weighing, and clash resolution often dominate over extensive external citation in a transcript-only setting. Courtesy is uniformly 5.00, exhibiting a ceiling effect at this competitive level; the dimension is retained for transparency and rule compliance but has limited discriminative power here.

5.2 Score Uncertainty

Standard deviations are calculated over k = 3 samples per speaker to quantify uncertainty. Table 3 reports variance across dimensions.

Table 3. Standard deviation of automated scores ($k = 3$ samples per speaker).

Speaker	Thesis	Reason	Rebuttal	Evidence	Clarity	Time	Team	Courtesy	Total
A	0.00	0.00	0.00	0.00	0.47	0.47	0.00	0.00	0.94
B	0.00	0.00	0.00	0.00	0.00	0.00	0.00	0.00	0.00
C	0.47	0.47	0.82	0.00	0.00	0.00	0.00	0.00	1.70
D	0.00	0.94	0.47	0.47	0.00	0.00	0.00	0.00	1.89
E	0.47	0.47	0.47	0.00	0.47	0.00	0.00	0.00	1.89
F	0.00	0.00	0.82	0.00	0.00	0.00	0.00	0.00	0.82

As shown in Table 3, uncertainty is generally low across dimensions. Variance concentrates in Rebuttal (for instance, Speakers C and F at 0.82), aligning with expectations for clash-heavy exchanges in which small changes in phrasing or emphasis can affect evaluative judgments. The observed dispersion remains small relative to the scale ranges, supporting the choice of k-sample aggregation (k = 3) for stable central estimates. In settings with greater stylistic variability or looser prompts, larger k could further reduce residual variance.

5.3 Team-Level Aggregation

Table 4. Team-level aggregates and automated decision.

Side	Members	Total	Mean
Proposition	A + C + E	248.67	82.89
Opposition	B + D + F	244.67	81.56
Margin (Prop – Opp)	—	+4.00	+1.33

As shown in Table 4, the system yields a narrow Proposition win (+4.00). Although Speaker B is the top individual scorer, the Opposition aggregate is reduced by Speaker F's lower total. By contrast, the Proposition side exhibits a more uniform distribution of speaker performance (≈82–84 each). This single-round observation is consistent with the WSD emphasis on role fulfillment and cross-speaker coherence: team victory correlates more strongly with balanced contributions than with reliance on a single standout.

5.4 Convergent Validity with Panel Decision

The automated team-level decision favors Proposition by +4.00 points and matches the official panel's verdict on the NSDA Nationals 2023 World Schools Debate Final. We treat this as preliminary convergent validity, not an accurate claim. Given the small margin, rubric differences, and our transcript-only, anonymized, turn-whitelisted setup, stronger claims require multi-round, multi-event evaluation with pre-registered protocols and statistical tests of concordance. When human speaker's points are accessible, we will examine per-dimension agreement and bias to identify systematic deviations.

6 Ethics and Transparency

This study evaluates an automated debate judging pipeline on a public, high-stakes World Schools Debate (WSD) round (NSDA Nationals 2023 Final). Because participants are typically minors, we process transcript text only, remove personal identifiers via speaker anonymization (A–F), and enforce turn whitelisting so each speaker is scored exclusively on their own turns. The system has no access to panel decisions or tournament metadata at scoring time, reducing information leakage. Model usage is disclosed: an API-served LLM with fixed prompts, temperature 0.2, default nucleus setting, and $k = 3$ independent samples per speaker with mean and standard-deviation reporting. We do not release raw personal data and make no attempt to re-identify individuals. Findings are reported at aggregate levels; the system is presented as a research prototype, not a drop-in replacement for human adjudication. Potential risks include automation bias and distribution shift; mitigations include transcript-only inputs, attributional constraints, uncertainty estimates, and explicit guidance against single-round over-interpretation. Future work will include preregistered, multi-event evaluations and bias audits across formats.

7 Limitation

This study is limited by its single-event scope (one NSDA Nationals 2023 WSD final), transcript-only inputs that omit delivery and non-verbal cues, and validation restricted to team-level agreement without per-speaker or per-dimension human scores. The WSD-aligned rubric and weights embed normative choices and exhibit ceiling effects on some dimensions, making close margins sensitive to alternative weightings. Scores are contingent on a particular LLM, prompts, and decoding with $k = 3$, which raise reproducibility and cost considerations. Despite anonymization and turn whitelisting, residual bias may persist, the system is tailored to WSD, and public knowledge of the rubric could invite strategic gaming in operational use.

8 Conclusion

This work presents a fair-by-design, format-agnostic framework for automated debate judging that emphasizes transparency, attributional fairness, and reproducibility. The pipeline standardizes inputs at the turn level, anonymizes speakers, and constrains attribution through turn whitelisting; quoted evidence is explicitly anchored to turn identifiers, and multi-dimensional scoring is paired with uncertainty estimates derived from multi-sample aggregation. Modular adapters and role-aware parsing support multiple debate traditions including Lincoln–Douglas, Public Forum, Policy, Parliamentary, and World Schools, while preserving the ability to audit individual decisions.

A transcript-only evaluation on a high-stakes World Schools Debate round (NSDA Nationals 2023 Final) shows that the system yields tightly clustered per-speaker scores and a narrow team-level margin favoring Proposition, coinciding with the official adjudication. This outcome functions as preliminary convergent validity under identity-neutral, transcript-only constraints rather than as an accuracy claim. The analysis further indicates that balanced cross-speaker performance is captured by the rubric and that variance concentrates in rebuttal-heavy segments, supporting the use of repeated sampling and explicit uncertainty reporting.

The design choices intentionally trade multimodal richness for auditability: excluding audio and video removes delivery cues but clarifies what is being scored and why. Governance features include speaker anonymization, turn-anchored evidence, fixed prompts, and versioned outputs. Together, these provide a reproducible basis for review and error analysis across events and formats.

Looking forward, broader validation across tournaments and debate styles is necessary to strengthen robustness and generalizability. The construction of multimodal evaluation datasets that align transcripts with audio and delivery features represents a key next step, enabling finer-grained analysis of persuasion and style. Complementary studies of inter-rater agreement between automated judgments and panels of human adjudicators, as well as the design of hybrid human–AI adjudication models, provide promising pathways toward accountable, scalable, and pedagogically valuable applications of automated debate judging.

References

1. DebateUS: High school debate events and formats. https://debateus.org/high-school-debate-events-and-formats. Accessed 14 Sept 2025
2. Cathy, A.: Types of debate. https://myperfectwords.com/blog/debate-writing/types-of-debate. Accessed 14 Sept 2025
3. Ruiz-Dolz, R., Heras, S., García-Fornes, A.: Automatic debate evaluation with argumentation semantics and natural language argument graph networks. *arXiv preprint* arXiv:2203.14647 (2024). https://doi.org/10.48550/arXiv.2203.14647
4. Wang, P., et al.: Large language models are not fair evaluators. *arXiv preprint* arXiv:2305.17926 (2023). https://doi.org/10.48550/arXiv.2305.17926

5. Liang, J., et al.: Debatrix: multi-dimensional debate judge with iterative chronological analysis based on LLM. *arXiv preprint* arXiv:2403.08010 (2024). https://doi.org/10.48550/arXiv.2403.08010
6. Zhang, Y., Yang, X., Feng, S., Wang, D., Zhang, Y., Song, K.: Can LLMs beat humans in debating? A dynamic multi-agent framework for competitive debate. *arXiv preprint* arXiv:2408.04472 (2024). https://doi.org/10.48550/arXiv.2408.04472
7. Zhang, L., Gonzalez, C.: An AI-driven debate judging system using emotional and content analysis based on artificial intelligence and machine learning. In: Artificial Intelligence, Soft Computing and Application Trends 2025, pp. 217–230. Academy & Industry Research Collaboration (2025). https://doi.org/10.5121/csit.2025.150419
8. Kenton, Z., et al.: On scalable oversight with weak LLMs judging strong LLMs. https://doi.org/10.48550/arXiv.2407.04622
9. Chan, C.-M., et al.: ChatEval: towards better LLM-based evaluators through multi-agent debate. *arXiv preprint* arXiv:2308.07201 (2023). https://doi.org/10.48550/arXiv.2308.07201
10. DebateDrills: Lincoln-Douglas debate format. https://www.debatedrills.com/lincoln-douglas/debate-format. Accessed 14 Sept 2025
11. Stanford National Forensic Institute: Choosing the right debate program. https://snfi.stanford.edu/select-the-right-program. Accessed 14 Sept 2025
12. Wikipedia: Public forum debate. https://en.wikipedia.org/wiki/Public_forum_debate. Accessed 14 Sept 2025
13. Wikipedia: British parliamentary style. https://en.wikipedia.org/wiki/British_Parliamentary_Style. Accessed 14 Sept 2025
14. Wikipedia: Parliamentary style debate. https://en.wikipedia.org/wiki/Parliamentary_style_debate. Accessed 14 Sept 2025
15. Wikipedia: World schools style debate. https://en.wikipedia.org/wiki/World_Schools_Style_debate. Accessed 14 Sept 2025

Automatic Workout Tracking Using Support Vector Machine with Frequency Feature Engineering

John Brose, Cheol-Hong Min, and Hassan Salamy(✉)

University of Saint Thomas, Saint Paul, MN 55105, USA
{brose4585,cmin,hsalamy}@stthomas.edu

Abstract. This study presents a machine learning-based approach for automating workout tracking through wearable sensors. Using accelerometer data collected from a MetaMotion 10-axis sensor, various weightlifting exercises, such as bench presses, pull-ups, and squats, were identified and classified using MATLAB Classification Learner. Feature extraction focused on both time and frequency domain features, with an emphasis on optimizing classification performance by carefully selecting and adjusting the features used. Multiple machine learning algorithms were tested to determine the most effective classifier, with Support Vector Machines (SVM) proving to be the best-performing model, achieving an accuracy of 93.6%. The trained classifier was then implemented in a MATLAB application to predict exercise activities in real time, providing accurate time and set tracking for each workout. Future improvements could include integrating additional sensors, expanding the training dataset, and enhancing system robustness across multiple users.

Keywords: Human Activity · Activity Recognition · Machine Learning

1 Introduction

The widespread use of Inertial Measurement Units (IMUs) and wireless sensor networks has transformed how we interact with technology, providing a notable example of how technology impacts various aspects of our lives. Specifically, advancements in technology have enabled continuous and real-time monitoring, whether in human activity recognition structural monitoring [1], or surveillance applications [2]. The integration of hardware technologies with software advancements in machine learning and artificial intelligence enabled the automatic detection, recognition, and tracking of various data generated from wireless sensor networks [10]. Specifically, the development of personal wireless sensor network applications has expanded into daily activity recognition and various physical activity monitoring, particularly in healthcare, for athletes in various sports, and personal fitness.

H. R. Arabnia et al. (Eds.): CSCI 2025, CCIS 2968, pp. 241–255, 2027.
https://doi.org/10.1007/978-3-032-28457-0_17

In healthcare, such sensors and integration with machine learning have proved to be an extremely useful tool, even for understanding physical movement and movement-related disorders [4]. Sensors could be placed in different locations throughout the body to understand other types of behavioral characteristics, to observe the progression of the behavioral changes, etc. [5]. In most IMU-based sensor system applications, the target is to provide a response or feedback to the intended user [4].

In sports and personal fitness applications, IMU-based sensors are very popular due to their size and ability to recognize various human movements. In Human Activity Recognition (HAR), there is a plethora of applications developed using wearable sensors that are developed in various forms such as watches, cuffs, rings, etc., for personal fitness and wellness [6].

A typical IMU sensor is equipped with a 3-axis accelerometer sensor, a 3-axis magnetometer sensor, and a 3-axis gyroscope sensor. Depending on the application, a single sensor module or a combination of the three sensors is used to sense, and a machine learning algorithm is coupled to classify or analyze the collected data. Its efficacy in using a subset of the sensors was also analyzed and discussed in several papers [7,8]. Also the number of sensor units can be reduced when an appropriate sensor location is used to monitor the activities [8,9]. Weight training monitoring is also another application where reduced sensors and sensing units can be applied to track different weightlifting exercises. Traditionally, manual tracking methods were used to monitor the activity, therefore requiring an additional person to keep track of the routines and the number of repetitions, or relying on self tracking, which is cumbersome for the trainee and could lead to sub-optimal training results. Several researchers have explored methods to track weight lifting exercises by utilizing wearable sensors and machine learning algorithms [10]. Some studies have shown that the wrist is a great location for the wearable sensor as it provides the least interference with the exercise, which can also sense various upper body exercises [10].

Signal processing plays an essential role in understanding the data. Time-domain characteristics and frequency-domain characteristics create features that can be used to train specific upper body exercises and feature engineering to optimize feature selection and to improve classification accuracy of the machine learning models [11]In this study, subsets of upper body weightlifting exercises are explored and classified for automatic workout monitoring.

Weightlifting is a phenomenal activity with numerous benefits for the human body, including increased energy [12,14], stamina, and strength [16]. However, being consistent with working out can be one of the most challenging yet essential things to reap these benefits. Activity tracking allows users to track the number of sets, exercises performed, and rest time between sets during a workout. These benefits would motivate users to closely monitor their workouts and attend the gym consistently, as a competition-like mentality can make tasks easier to pursue and maintain [12]. This study investigates the implementation and application of machine learning algorithms for classifying data from a sensor module. It also demonstrates the implementation of the classifier with new datasets to track activities. Workout tracking was the activity of interest, which was monitored using a sensor attached to the body. With the raw sensor data, various features were extracted and stored in an array to train a new classifier. The classifier was trained using algorithms from the ClassificationLearner in MATLAB. Multiple algorithms were tested and analyzed until the most effective classifier could be identified. Not only were the algorithms iterated upon, but the feature selection was explored. Various features were studied in depth to determine their impact on the classifier's accuracy, and some were modified to better suit the dataset. Once the classifier was chosen, it was exported from classificationLearner into a MATLAB script. Finally, new training data, previously unseen by the classifier, was introduced and classified. After additional post-processing of the classifier output, the new workout training data was correctly classified into the predetermined workouts, and the time spent on each activity was also tracked.

2 System Architecture

The system consists of an Apple Watch-like wrist wearable device that collects data and a corresponding classification model. Data was collected for bench press, cable rows, squats, pullups, shoulder press, tri pushdowns, and rest. It consisted of two main parts: the sensor and the actual activities monitored. This section provides details about the wearable system.

2.1 Wearable Sensor

The wearable sensor used was the MetaMotion 9-Axis IMU and Environment Monitoring Sensor. It has various monitoring tools, including a barometer, ambient light sensor, magnetometer, and gyroscope. However, the investigation into workout activity tracking found that the only sensor needed was the 3-axis accelerometer [17], giving access to x, y, and z accelerometer data. Neither the gyroscope nor the magnetic sensor proved to be useful in workout monitoring. One sensor system was placed on the right wrist of the user for optimal tracking during exercises. Since the exercises involve simultaneous and synchronous movement of the arms, it was decided that only one sensor was needed and was placed on the right wrist. Due to the activities monitored, placement was not restricted to the dominant arm. Most strength training activities, such as the bench press or shoulder press, involve a lot of arm movement but very little lower-body movement. Therefore, the identification may have been impossible if the sensor had been placed anywhere other than the arm. The sampling rate chosen was 50 Hz. The reason for this was twofold. Firstly, weightlifting is generally a slow movement (in terms of acceleration); for instance, one repetition in a workout takes at least one second. Thus, having an oversampled rate of 50 Hz should provide sufficient information per repetition without including excess data. Secondly, supporting research from "Accuracy of human motion capture system for sport applications" [13] states that a sample rate of 50 Hz to 250 Hz is sufficient for effective movement tracking in sport applications. Since weightlifting falls within the slower category of sports, a sample rate of 50 Hz was chosen.

2.2 Activities

The workout activities were obtained from Microsoft Exercise Recognition from Wearable Sensors database [14], Kaggle Gym Workout IMU Dataset [15], and additional exercises performed while wearing the MetaMotion sensors were bench presses, pull-ups, squats, shoulder presses, cable rows, tricep pushdowns, and rest periods. All these exercises typically involve a unique body movement, meaning that the classifier should be able to distinguish between them. Rest was a very important activity to track, as it helped distinguish sets and indicate how quickly the user was progressing through the workout. The activities were tracked by having a user start the accelerometer tracking, then perform one of the exercises for 30 s, rest for 30 s, perform the next exercise for 30 s, and repeat this process until all exercises were completed.

2.3 Classification Extraction and Events

Once the workout activities were tracked and all classifier training was completed, the system's goal was to utilize the trained classifier to process a new set of data and track a series of events associated with the classified activities. Specifically, focusing on the total time of each exercise is mainly helpful for rest, as the rest time between sets is a vital strength training parameter, as well as the number of sets performed per exercise. A block diagram illustrating the system architecture flow is displayed in Fig. 1.

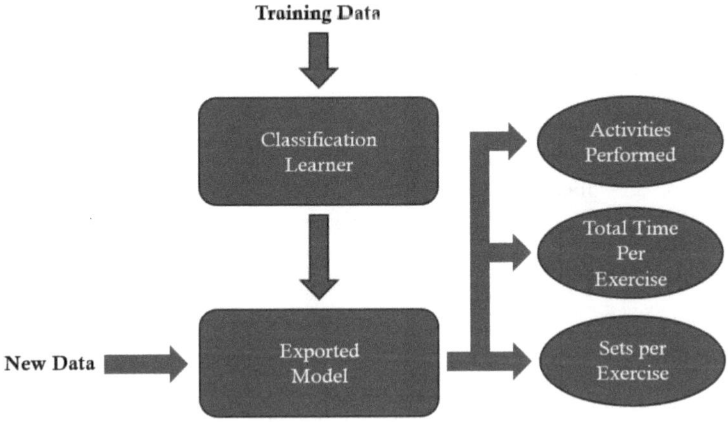

Fig. 1. Block diagram of the proposed system

3 Classification Algorithm

Once activities were performed and data were collected, the next step was to select which features to extract. The time domain features used were mean, variance, zero crossings, and cross-correlation. The mean was taken to obtain the relative position of the sensor as it changed throughout the activities. Variance and zero crossings similarly assisted in identifying any cyclical, repeating movements. Both were taken in case one outperformed the other when classification was done. Initially, the frequency domain features extracted were five equal power bands. The DC portion of the data was extracted from the frequency domain by feeding the data through a high-pass filter and then omitting the 0 Hz DC frequency when classifying the data. The initial results showed that the power was concentrated in the lower frequency bands, as most of the energy for the exercises was located in the lower ranges. However, more in-depth explanations will be discussed in the latter sections. All identified features were extracted from windowed data, using a window size of 100 data points and a 50% overlap. This both reduced the size of the data and allowed for the implementation of the algorithm to be applied to blocks of data, which meant that each feature array element equated to one second.

Since the classification of the data set was performed on various features, a formal approach was used to optimize the algorithm selection and overall classifier for any new feature array input. When features were extracted and input into MATLAB for training, an iterative process was conducted to determine the optimal algorithm for the dataset. First, one classifier from each of the popular supervised learning algorithms, such as Decision Tree, k-Nearest Neighbors, Naive Bayes, and Random Forest, was explored to obtain a baseline result. The ML algorithm that performed the best was then further evaluated and explored to obtain improved results for each well-known classifier. Since the study focuses on optimizing SVM for better performance, results from other classifiers are omitted in the discussions. For most combinations of features, Support Vector Machines (SVMs) usually performed the best. Further analysis was conducted to identify the best-performing algorithm. This process was then conducted a myriad of times for various feature combinations, either excluding some or including others. Another useful approach to increase overall performance was to understand what kind of performance was ideal for tracking exercises. Specifically, the rest activity can easily be identified in the application since it does not occur randomly in the middle of a set. Thus, when examining the confusion matrix, as shown in Fig. 2, certain misclassifications are allowable and preferred over one exercise being misidentified as another exercise.

4 Feature Optimization

Feature optimization was a critical aspect of improving the performance of the system. However, the time-domain feature parameters could not be easily modified, as they involve changing the window sizes. What was explored was the combinations of extracted time-domain features. Therefore, frequency features, i.e., the power bands, were optimized the most. This allowed the accuracy of the classifier to increase by 1%.

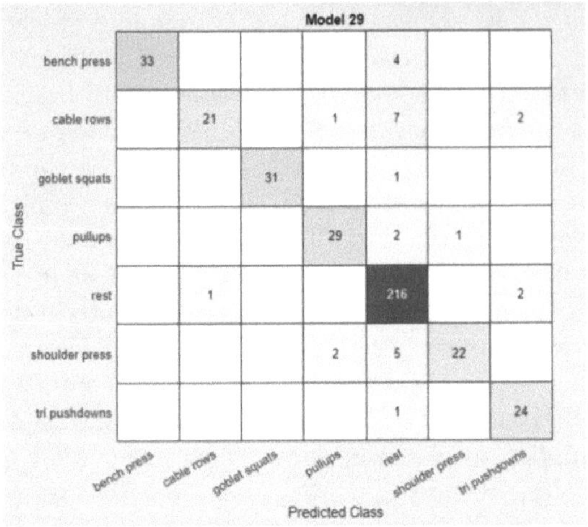

Fig. 2. Confusion matrix shows the classification results using SVM

4.1 Equal Power Bands

The first feature selection method used five equal power bands. With this feature array and using the iterative process described earlier, the overall accuracy obtained was 92.8% with the SVM model. To achieve this result, Variance, Power Band 1 (1–5 Hz), Power Band 4 (15–20 Hz), and Power Band 5 (21–25 Hz) were excluded during training. The reason for excluding power bands 1, 4, and 5 was due to the distribution of total power across those bands, as they showed no significant variation for the observed activities.

An issue arises with the power band arrangement when using equal power bands of 5 Hz each. The total amount of energy is much too high in Power Band 1 due to the characteristics of the exercises in the study. It causes the power band 1 to always have high energy for every activity, rather than only having high power for some activities, making it an ineffective metric for distinguishing between activities. This is due to the fact that weight training actions are comprised of slow, controlled motions. A potential adjustment would be to spread out the lower frequencies more and remove the transitions. The following section presents and discusses a method for spreading out the lower frequencies into multiple bands.

4.2 No High-Frequency Power Bands

The above section demonstrated that overall performance could be improved by expanding on the low ends of the frequency, and it also revealed that transitions were primarily an issue when they overshadowed the magnitude of energy during an activity. With this in mind, the low frequency was expanded even further for power bands. Only specific discrete frequency values were captured for power bands 1, 2, and 3. Power Band 1 observed only the 0.5 Hz frequency,

Power Band 2 with a 1 Hz frequency, and Power Band 3 with a 1.5 Hz frequency. Secondly, to address the inaccuracies presented by the transitions, frequencies from 15.5 Hz to 25 Hz were completely omitted from the feature list, as frequency activities primarily occurred at lower frequencies. Therefore, Power Band 4 contained information between 2–4 Hz, and Power Band 5 contained information between 4.5–15 Hz. Such considerations would reduce inaccuracies due to transitions and enhance overall performance. With these adjustments, the overall accuracy obtained was 93.6%, which is an appreciable 1% increase over the last couple of trials. The algorithm used was the quadratic SVM, and variance and Power Band 4 were excluded due to the dominant movements from pre- and post-individual exercise routines.

4.3 "Best Model" No Transitions, No High Freq Power Bands

The last iteration performed on the feature list involved completely omitting the pre- and post-actions, i.e., noise, that were causing many issues in both the power bands and variance features. The power band setup was the same as in the previous section, excluding the high frequencies and expanding the low frequencies to the greatest extent possible, as this had originally yielded the best results.

The impact of excluding the transitions from the power bands was noticeable. When investigating Power Band 1, it became clear that the previously observed sharp transition, which would skew the data, is no longer present. By performing this operation across the entire dataset, the linear SVM achieved an overall accuracy of 96.3%, which was the best observed performance for the exercises. Furthermore, no features were excluded from the final model, indicating that the variance, as well as the issues with the power bands, were due to the transitions.

5 Results

With the feature selection explored to achieve optimal performance and the best classifier chosen for the highest overall performance, the next step was to export the trained model and test it with new sets of data. The broad overview of what the application set out to accomplish was to take in new data, extract features from the new data, use the exported model from classificationLearner to predict exercise based on the new data, and from there extract two events, the amount of time per exercise and the number of sets per exercise performed.

The in-depth response of the used model is shown in Fig. 3. Label 1 corresponds to *rest*, 2 for *bench press*, 3 for *pullups*, 4 for *squats*, 5 for *shoulder press*, 6 for *cable rows*, and 7 for *tri pushdowns*. In addition, other activities that are not part of the exercise routine, i.e., pre-, post- and in-between exercises and rest periods, were labeled as rest. A few things to note from the above figures are that the misclassifications usually occur as the rest. Such misclassification is advantageous because the rests are not performed during the middle of the set, and post-processing of the predicted output can be done to ensure that short

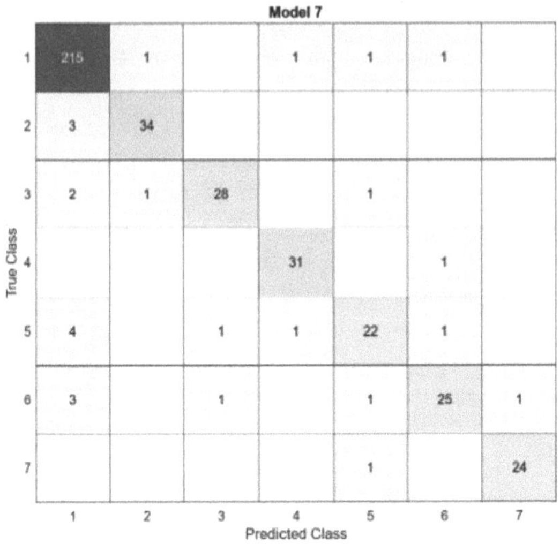

Fig. 3. Confusion matrix for the model used in testing

blips of rest are easily omitted. Every other activity has a low false negative rate, which yields excellent results, while the lowest accuracy is observed for the shoulder press. It is because the shoulder press is similar to the bench press, as observed from wrist-worn sensors, which means it is misclassified more often.

Figure 4 showcases the model's practical application, accurately identifying each activity from the continuously executed exercises. The model handles transitions between the exercises well and closely follows the activity sequence, demonstrating robustness. Notably, there are no misclassifications during the activities, further reinforcing the model's accuracy in real-world scenarios.

Table 1. Used Model actual time vs predicted time

Activity	Actual Time (min)	Predicted Time (min)
Rest	3.65	3.70
Shoulder Press	0.48	0.47
Tri Pushdowns	0.42	0.42
Pullups	0.53	0.52
Squats	0.53	0.53
Cable Rows	0.52	0.52
Bench Press	0.62	0.60

After manually verifying and calculating the actual and predicted times of each exercise, it is clear how closely the model matches the actual output, as

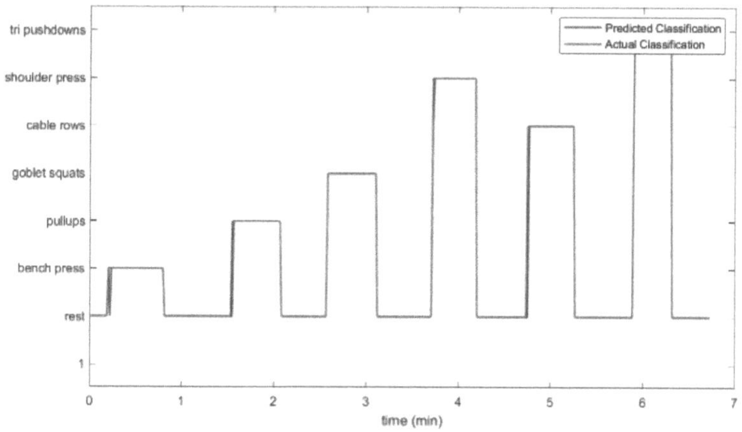

Fig. 4. Used Model Actual vs. Predicted Application

shown in Table 1. Each activity is highly accurate in terms of time, demonstrating that this method of identification is reliable and feasible.

5.1 Training Model Adjustments

While the model discussed above performed well, the model presented in Sect. 4.3 performed better, with an overall accuracy of 96.3%. The model was tested by exporting the trained model from ClassificationLearner to a MATLAB script, and the same data used to train the model was processed through the MATLAB script to evaluate the model's performance. However, this performance did not carry over to the testing, as discussed below.

As shown in Fig. 5, the model did not perform as well as expected. The biggest downside is clearly illustrated in the figure, where removing the transitions results in the model's inability to identify transitions as a single type of exercise. By including the transitions, the classifier can perform better classifications, rather than often misclassifying. This model still presents a good overall performance. However, the transitions are frequently misidentified, and the *cable rows* exhibit a higher rate of misclassification. It shows that the model is not overfitting to the training data. Otherwise, it would have correctly identified the exercises with minimal error.

5.2 Analyzing New Test Data

Analyzing new test data started with designing a different workout procedure to evaluate and test the model's performance and capabilities. The test procedure covered all the exercises in a regular workout routine: one set, rest for 30 s, and repeat. Three sets per exercise were performed.

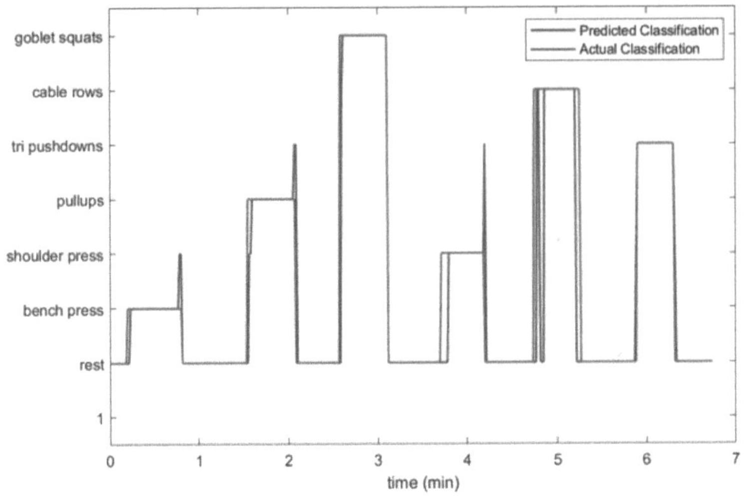

Fig. 5. "Best" Model Actual vs. Predicted Application

When the model was tested with a new set of test data, it did not perform as well as it did with the validation set described in the above section. Many misclassifications occurred, with several instances of one exercise being misclassified as another. This was likely due to the difference in daily workout activity, which can vary from user to user. Specifically, when using an accelerometer, if the rate or speed of the exercise differs from what was previously trained, the model may struggle to identify it. However, one positive thing to note is that the activities overlap closely visually in terms of the macro view. The rest periods are, for the most part, clearly rest, and exercises, for the most part, match the expected exercise, which demonstrates the model's adaptability to different workout routines and user behaviors while also highlighting areas for improvement.

With these observations in mind, the solution to obtaining a clean output from the garbled output of Fig. 6 was done by post-prediction data processing. This was done by making a few assumptions about how the data was taken. For one, an assumption was made that there was rest time between every working set. Secondly, it was assumed that the exercise could not be switched in the middle of a set. From these two assumptions, the data could be separated into workout sets by examining when the data went from rest to non-rest at the start and then from non-rest back to rest at the end. In a series of exercise sets, each segment had its own predicted workouts, and the maximum summed predicted exercise was labeled as the exercise for that entire set; thus, majority voting was applied. For example, if a set consisted of 50 classifications of shoulder presses and 35 classifications of bench presses, the entire set would be classified as shoulder presses.

Figure 7 illustrates the outcome of the post-prediction data processing, which is highly effective. The predicted and actual classifications match almost per-

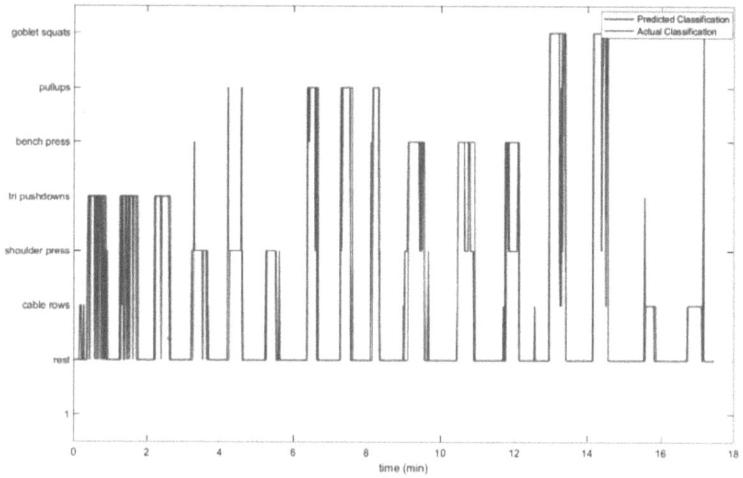

Fig. 6. New Test Data Predicted vs. Actual

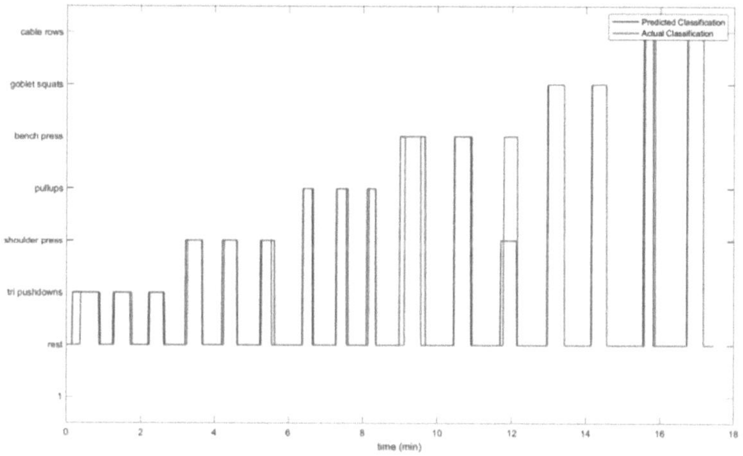

Fig. 7. New Test Data Post Processing Predicted vs. Actual

fectly, except for one major misclassification, where bench press is incorrectly classified as shoulder press. However, this is due to the two exercises being very similar in terms of sensor location and movement during the exercise. The difference fell to the speed at which the exercise was performed, and that bench press must have been performed at a speed similar to that of the shoulder press. Another issue that accumulates over time is the slight errors in transitioning from rest to exercise. The transitions are always slightly different and will eventually add up to a more significant difference, but overall, the performance is acceptable.

5.3 Extracting Data After Predictions

The new data could be input and identified consistently. The next step in the application was to extract a few key parameters about the workout session described earlier. Specifically, time tracking of each exercise and set tracking of each exercise. Time tracking was accomplished by understanding that the sample time was 50 Hz, with a window size of 100 samples and 50% overlap. This meant that each feature vector corresponded to 100 samples of data, and incremented by one second of data. Moreover, since each feature vector leads to one prediction from the model, each output from the model equals 1 s of data, or, in other words, one second of that exercise. Therefore, the application sums the predicted exercises from the model (after post-processing) to obtain the total time of each exercise for the entire session. Therefore, the average time per set for any given exercise can be identified, since it equals total time divided by the number of sets per exercise. Both of which had already been calculated.

Table 2. New Dataset: Set Tracking

Activity	True-Sets	Predicted-Sets	Predicted-Sets Smoothed
Rest (Average Time /Set (min))	0.67	0.29	0.63
Shoulder Press	3	28	4
Tri Pushdowns	3	16	3
Pullups	3	12	3
Squats	2	7	2
Cable Rows	2	15	2
Bench Press	3	11	2

Table 3. New Dataset: Time Tracking

Activity	True-Time (min)	Predicted-Time (min)	Predicted-Time Smoothed (min)
Rest	11.4	11.4	10.7
Shoulder Press	1.03	1.87	1.68
Tri Pushdowns	1.43	1.03	1.6
Pullups	0.8	0.75	0.78
Squats	0.88	0.68	0.85
Cable Rows	0.72	0.92	0.72
Bench Press	1.27	0.82	1.13

Tables 2 and 3 show the results of the model in terms of events that were tracked. The data is separated into three columns. True data, based on how the data was taken; predicted data, which was found from feeding the predicted outputs from the model directly into the event detection code; and predicted data smoothed, which was the predicted outputs from the model smoothed into nice blocks of working out and then sent through the event detection code.

The set tracking performance was promising, provided that a few stipulations were met. If the data was not smoothed (Predicted Sets column), then the set tracking was unpredictable. This was because any one misclassification led to an entire two-set misclassification. There would be a transition from one activity to another and back to the same activity, leading to many more sets than occurred. However, the sets were highly accurate once the predicted activities were smoothed using post-data processing. The only exercise with an issue was the bench press, which was incorrectly classified as a shoulder press. This was discussed multiple times because the shoulder and bench press exercises share similar wrist motions when performed. Therefore, the recorded accelerometer data is very similar.

The time tracking table yields similar results to the set tracking, with overall accuracy closely matching the true time, indicating that the model performed well. The smoothing predicted time was not significantly better than the raw predicted time. It makes sense because one misclassification only leads to one second of time inaccuracy due to the window size. The predicted time matched the true time, although the sum of all exercises had roughly a half-minute error. On the other hand, the smoothed predicted time performed better, giving an overall closer match, with a few worst-case exercises that gave similar results to the raw predicted time. The shoulder press, bench press, and rest were the times that showed the most significant error. It was because there was a case where an entire set of shoulder presses and bench presses was misclassified. On the other hand, rest time was off by quite a bit due to the cumulative error effect across the entire dataset. Every transition from rest to exercise led to a few seconds of error, which accumulated across all sets, resulting in nearly one minute of error. Although this error is significant, for weight lifting monitoring, it is close enough since the exact rest time is not critical.

6 Conclusion and Future Work

The workout exercise tracking yielded numerous positive outcomes. The feature selection process required an extensive investigation into the dispersion of power bands to accurately determine which frequencies to include and prioritize. It was discovered that lower frequencies were the most significant, and the 0.5 Hz power bands proved to be particularly useful. The last two frequency bands were combined with some of the higher frequencies. However, the highest frequencies were eliminated because they contributed additional noise to the system, which was not consistently present during the exercises. Transitions were also closely analyzed to see if omitting them would yield better results. Classifying the dataset using a feature list was another critical aspect of the project that required extensive optimization. Several machine-learning algorithms were explored; however, the key factor was the selection of features. The classifiers were evaluated against various feature combinations, and the best-performing classifier was identified by examining overall accuracy, as well as the confusion matrix, particularly in relation to allowing misclassifications during rest periods. The classifier that delivered the best results, regardless of feature combination, was the quadratic Support Vector Machine. This model achieved an overall accuracy of 93.6% for application use. The final stage of the project involved implementing the model in MATLAB to analyze a new dataset and predict untrained exercise sets. The application extracted features from the new dataset and processed them through the model, generating predicted exercises. Although the initial predictions were somewhat noisy and included many misclassifications, post-data processing considerably cleaned up the results by making certain assumptions about how the data was collected. Once the data was organized into exercise sets, it was analyzed for events, specifically the number of sets per exercise and the total time spent on each exercise. Both events were tracked with high accuracy; all sets were classified correctly except for one misclassification of a bench press as a shoulder press. The total time recorded for each exercise was very accurate, with the exception of the misidentified exercise, which was off by approximately one set's worth of time. While the work presented in this paper demonstrates that the workout tracking was successful and repeatable even with new input data, there are several areas for improvement to enhance the system's robustness. One recommendation is to place multiple sensors in different locations on the body for better performance. Specifically, adding another accelerometer on the chest would significantly improve exercise identification and reduce misclassifications, such as the bench press being confused with the shoulder press. Another improvement would be to use a larger training dataset for the classification learner and the resulting model. A larger dataset would account for more variations, such as differences in exercise execution speed and minor positional changes that may occur. Finally, collecting a much larger dataset from additional gym users would enhance the classifier's robustness, making it more effective for new users who weren't included in the original training dataset.

References

1. Huynh, H.H., Min, C.-H.: Rotating machinery fault detection using support vector machine via feature ranking. Algorithms **17** (2024). https://doi.org/10.3390/a17100441
2. Yang, F., Gou, L., Cai, X.: Pedestrian positioning scheme based on the fusion of smartphone IMU sensors and commercially surveillance video. IEEE Sens. J. **22**(5), 4697–4708 (2022). https://doi.org/10.1109/JSEN.2022.3140791
3. Nguyen, H., et al.: Using inertial sensors to automatically detect and segment activities of daily living in people with Parkinson's disease. IEEE Trans. Neural Syst. Rehabil. Eng. **26**(1), 197–204 (2017)
4. Cao, P., Min, C.-H.: Fall prediction in people with Parkinson's disease. In: 44th Annual International Conference of the IEEE Engineering in Medicine and Biology Society (EMBC), Glasgow, Scotland, United Kingdom pp. 1502–1505 (2022). https://doi.org/10.1109/EMBC48229.2022.9872013

5. Min, C.-H., Tewfik, A.H.: Semi-supervised event detection using higher order statistics for multidimensional time series accelerometer data. In: 2011 Annual International Conference of the IEEE Engineering in Medicine and Biology Society, Boston, MA, USA, pp. 365–368 (2011). https://doi.org/10.1109/IEMBS.2011.6090119
6. Brickwood, K., et al.: Consumer-based wearable activity trackers increase physical activity participation: systematic review and meta-analysis. JMIR Mhealth Uhealth **7**(4), e11819 (2019)
7. Ascioglu, G., Senol, Y.: Activity recognition using different sensor modalities and deep learning. Appl. Sci. **19**(13), 10931 (2023). https://doi.org/10.3390/app131910931
8. Attal, F., et al.: Physical human activity recognition using wearable sensors. Sensors **15**(12), 31314–31338 (2015). https://doi.org/10.3390/s151229858
9. Shi, L., et al.: Real-time human activity recognition system based on capsule and LoRa. IEEE Sens. J. **21**(1), 667–677 (2021)
10. de Beukelaar, T.T., Mantini, D.: Monitoring resistance training in real time with wearable technology: current applications and future directions. Bioengineering **10**(9), 1085 (2023). https://doi.org/10.3390/bioengineering10091085
11. Preatoni, E., Nodari, S., Lopomo, N.F.: Supervised machine learning applied to wearable sensor data can accurately classify functional fitness exercises within a continuous workout. Front. Bioeng. Biotechnol. (2020). https://doi.org/10.3389/fbioe.2020.00664
12. Warburton, D.E.R., Nicol, C.W., Bredin, S.S.D.: Health benefits of physical activity: the evidence. CMAJ : Can. Med. Assoc. J. = J. de l'Association medicale canadienne **174**(6), 801–809 (2006). https://doi.org/10.1503/cmaj.051351
13. van der Kruk, E., Reijne, M.M.: Accuracy of human motion capture systems for sport applications; state-of-the-art review. Eur. J. Sport Sci. **18**(6), 806–819 (2018). https://doi.org/10.1080/17461391.2018.1463397
14. Morris, D., Saponas, T.S., Guillory, A., Kelner, I.: RecoFit: using a wearable sensor to find, recognize, and count repetitive exercises. In: ACM Proceedings of the SIGCHI Conference on Human Factors in Computing Systems, pp. 3225–3234 (2014)
15. Sairam, S.: Gym workout IMU dataset. Kaggle. https://www.kaggle.com/datasets/shakthisairam123/gym-workout-imu-dataset
16. Helms, E.R., Eric, R., et al.: Methods for regulating and monitoring resistance training. J. Hum. Kinet. **74**, 23–42 (2020). https://doi.org/10.2478/hukin-2020-0011
17. Milanko, S., Jain, S.: LiftRight: quantifying strength training performance using a wearable sensor. Smart Health **16** (2020)

Twinleaf Chain: A Variable Precision Rough Set Based Approach for Classification

Xuguang Chen(✉)

Saint Martins University, Lacey, WA 98503, USA
xchen@stmartin.edu

Abstract. The Twinleaf Chain, a novel classification model, is introduced in this paper. It integrates Variable Precision Rough Set (VPRS) theory with a decision tree–like structure and can potentially handle uncertainty and imprecision in supervised learning tasks. The Twinleaf Chain employs a symmetrical, chain-like architecture resembling a centipede, where each non-leaf node selects one or more features using λ-dependency to maximize the assignment of samples into positive, negative, or boundary regions of the VPRS model. This design allows the model to iteratively classify clear samples while deferring ambiguous instances to lower layers, ensuring more accurate handling of uncertain data. The construction process involves discretizing data and selecting relevant features via λ-dependency for each non-leaf node, thereby maintaining the model's consistency and robustness. During evaluation, test samples traverse the chain, with unclassifiable instances flagged in the boundary region for further analysis. Compared to standard decision trees, the Twinleaf Chain exhibits a more regular structure, which can enhance interpretability, stability, and overall performance, particularly in noisy or complex datasets.

Keywords: Variable Precision Rough Sets · Decision Trees · Twinleaf Chain · λ-Dependency · Classification · The type of the submission Short Paper · **Acronym of the most relevant symposium** Research Track on Artificial Intelligence (CSCI-RTAI)

1 Introduction

In the realm of supervised machine learning, classification tasks often involve dealing with uncertain or noisy data, where traditional models like decision trees may produce complex, irregular structures that hinder interpretability and efficiency. Decision trees, popularized by algorithms such as ID3 and C4.5 [1, 2], build hierarchical rules based on feature splits to partition data into classes. However, their reliance on metrics such as information gain leads to asymmetries: branches vary in depth due to differing data purity, and node branching depends on feature cardinality, resulting in unpredictable tree shapes.

Rough set theory, introduced by Pawlak [3, 4], offers a robust framework for handling vagueness without requiring probabilistic assumptions, focusing on approximations of

H. R. Arabnia et al. (Eds.): CSCI 2025, CCIS 2968, pp. 256–266, 2027.
https://doi.org/10.1007/978-3-032-28457-0_18

sets through lower and upper bounds. The Variable Precision Rough Set (VPRS) model [5] extends this by incorporating precision parameters (lower limit l and upper limit u) to define positive (POS), negative (NEG), and boundary (BND) regions, allowing for controlled tolerance of misclassification. This is particularly useful in real-world datasets where exact partitions are infeasible.

Motivated by these strengths, this paper proposes the Twinleaf Chain, a specialized decision tree variant, grounded in VPRS model. Unlike conventional trees, it enforces a symmetrical, linear chain structure with twin leaves per layer, using λ-dependency [6] for feature selection to optimize region assignments. This design addresses key limitations: it promotes regularity for better visualization, handles multi-feature splits for enhanced discrimination, and defers boundary samples systematically.

In Sect. 2, the VPRS model, λ-dependency, and the architecture of a Twinleaf Chain are introduced. The steps of constructing a Twinleaf Chain are covered in Sect. 3. Section 4 discussed several practical issues like discretization and stopping criteria, and guidelines for classifying new samples. A summary and future work is Sect. 5.

2 Background

2.1 Variable Precision Rough Set Model and λ-Dependency

In machine learning, a dataset is typically represented as an information table as shown in Table 1, where rows correspond to individual samples (instances or observations) and columns correspond to features (attributes or predictors). In supervised learning tasks, the table also includes a column for the target variable (label). Based on the values of one or more features, samples in the information table can be grouped into classes, also referred to as elementary sets.

Rough set theory, originally introduced by Pawlak, provides a mathematical framework for analyzing such data, and the Variable Precision Rough Set (VPRS) model [5] is one of its widely used extensions. The VPRS model is defined over a universe of objects U. In practice, however, it is often assumed that only a finite subset $U' \subseteq U$ is observable. For example, the samples in Table 1 may be viewed as U', with binary-valued features and labels.

The collection of elementary sets induced by some or all of the condition attributes C forms a partition of U, denoted U/C. For instance, the attributes $C1$ and $C3$ can form a partition having the following four elementary sets: $E_{C1=0,\,C3=0} = \{s1, s3, s4, s5, s7\}$. $E_{C1=0,\,C3=1} = \{s8, s9\}$, $E_{C1=1,\,C3=0} = \{s6, s11, s12\}$, and $E_{C1=1,\,C3=1} = \{s2, s10, s13, s14\}$. Similarly, the decision attribute(s) D induce a partition U/D, which defines the decision categories. For a single binary decision attribute, this partition consists of two disjoint sets: the target set X and its complement $\neg X$, i.e., $U/D = \{X, \neg X\}$. In Table 1, if the samples of interest are those with $D = 1$, then $X = \{s1, s3, s5, s7, s9, s11, s12, s14\}$. In the VPRS model, the prior probability of X is given by $P(X) = |X|/|U|$, where $|\cdot|$ denotes cardinality. Each equivalence class E is associated with two probabilities: $P(E) = |E|/|U|$ and $P(X|E) = |X \cap E|/|E|$. Here, the probability $P(E)$ reflects the relative size of E, while the probability $P(X|E)$ measures its overlap with the target set X. Since the full universe U is typically unavailable, these probabilities are estimated from the finite

Table 1. An Information Table

Sample	C1	C2	C3	C4	D
s1	0	1	0	0	1
s2	1	0	1	1	0
s3	0	1	0	0	1
s4	0	0	0	0	0
s5	0	1	0	0	1
s6	1	0	0	0	0
s7	0	0	0	1	1
s8	0	0	1	0	0
s9	0	0	1	1	1
s10	1	0	1	0	0
s11	1	0	0	0	1
s12	1	0	0	1	1
s13	1	1	1	1	0
s14	1	1	1	1	1

subset U'. For example, based on the data in Table 1, $P(E_{C1=0,C3=0}) \approx 0.3571$ and $P(X|E_{C1=0,C3=0}) = 0.8$.

Table 2 λ-dependency based on C1 and C3

	C1	C3	P(E)	P(X|E)	Samples
E0	0	0	0.3571	POS	s1, s3, s4, s5, s7
E1	0	1	0.1429	BND	s8, s9
E2	1	0	0.2144	BND	s6, s11, s12
E3	1	1	0.2857	NEG	s2, s10, s13, s14
$l = 0.3$, $u = 0.8$, $\lambda_{C1,C3} \approx 0.3542$					

The VPRS model defines rough approximations of the target set X using two precision parameters: the lower limit l and the upper limit u. These parameters determine whether an elementary set belongs to the positive region denoted as POS_u, the negative region denoted as NEG_l, or the boundary region denoted as $BND_{l,u}$. Formally, they can be defined below [4, 5].

- $E \in NEG_l(X)$, if $P(X|E) \leq l$,
- $E \in POS_u(X)$, if $P(X|E) \geq u$,
- $E \in BND_{l,u}(X)$, if $1 < P(X|E) < u$.

Correspondingly, the POS_u, NEG_l, and $BND_{l,u}$ can be defined below [4, 5].

- $NEG_l(X) = \bigcup\{E: P(X|E) \leq l\}$,
- $POS_u(X) = \bigcup\{E: P(X|E) \geq u\}$,
- $BND_{l,u}(X) = \bigcup\{E: l < P(X|E) < u\}$.

For example, with $l = 0.3$ and $u = 0.8$ in Table 2, we obtain:

- $E_{C1=0,C3=0} \in POS_u$,
- $E_{C1=0,C3=1}$ and $E_{C1=1,C3=0} \in BND_{l,u}$,
- $E_{C1=1,C3=1} \in NEG_l$.

Finally, following [6, 8], the λ-dependency of the target set X on the condition attributes C is defined as

$$\lambda\ (X|C) = \frac{\sum P(E) \mid P(X|E)--P(X) \mid}{2P(X)(1-P(X))} \quad (1)$$

where $2P(X)(1\text{-}\ P(X))$ is used for the purpose of normalization, and the elementary sets E are belonging to the partition U/C, i.e. $E \in POS_u(X|C) \cup NEG_l(X|C)$. This quantity measures the normalized deviation of the conditional probabilities $P(X|E)$ from the prior probability $P(X)$ over the positive and negative regions of the target set X. In constructing a node of the Twinleaf chain, the λ-dependency serves as a criterion for selecting subsets of condition attributes that maximize classification of samples into positive or negative regions. For instance, in Table 1, according to (1), λ-dependency of the partition formed by $C1$ and $C3$ is $\lambda_{C1,C3} \approx 0.3542$. Correspondingly, $\lambda_{C2,C4} \approx 0.4583$, which indicates that the attribute subset $\{C2,C4\}$ achieves a higher discriminatory power.

2.2 Decision Trees and Twinleaf Chains

Decision trees are a supervised learning algorithm for classification and regression tasks [1, 2]. They work by learning a set of decision rules from training data, forming a tree-like model. Starting from the root node, the tree will branch out based on decisions, leading to leaf nodes that deliver predictions. The shape of a decision depends on the feature distribution in the training data and the splitting criteria, typically showing these characteristics:

Asymmetry The splitting process in a decision tree hinges on metrics like information gain, which depend on the data's specific distribution. This leads to trees where branches can vary a lot in depth and width, making the structure lopsided. For example, some branches might stop splitting early if the data is already "pure," while others keep splitting deeper due to complex data.

Uneven Depth Stopping conditions - like max depth, minimum sample size, or purity thresholds - mean leaf nodes can sit at different depths. For instance, if a subset of samples can be fully classified early, that branch might only be one or two levels deep, while others need more splits to handle trickier samples.

Varying Branch Counts The number of branches at a node depends on the feature's range of values. For discrete features, the number of branches often matches the number

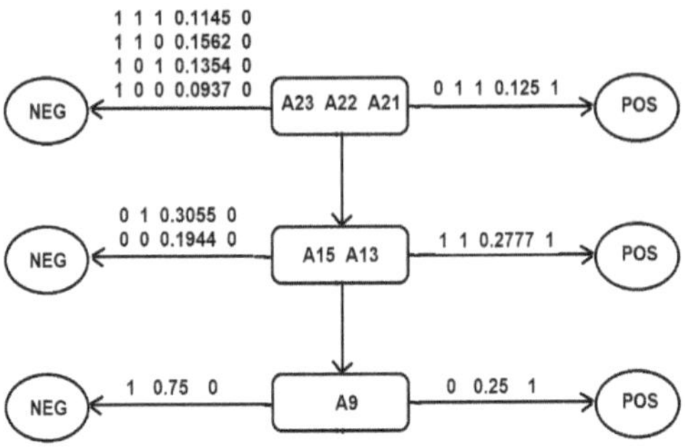

(a) A Sample Twinleaf Chain without a BND

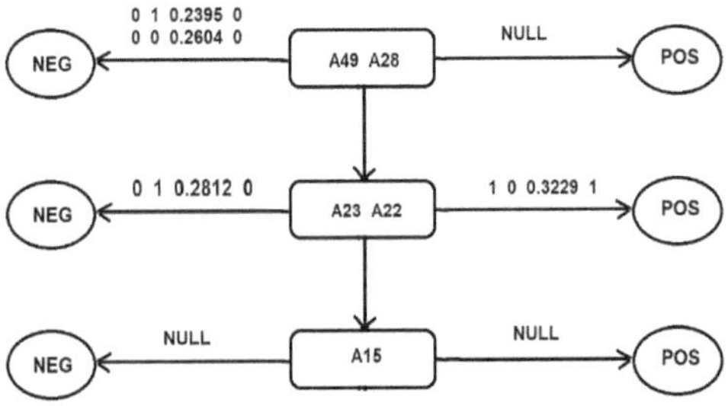

(b) A Sample Twinleaf Chains with a BND

Fig. 1. Two Sample Twinleaf Chains

of possible values, so nodes can have different branch counts, adding to the tree's irregular shape.

Compared to a standard decision tree, a Twinleaf Chain can be considered a special decision tree with some unique features below.

Regular Shape As shown in Fig. 1, a Twinleaf Chain has a neat, symmetrical structure that looks like a centipede. It starts at the root node and splits into layers. Each layer has a central non-leaf node flanked by two leaf nodes - one on each side. Except for the top and bottom nodes, each non-leaf node connects to the non-leaf nodes above and below it, forming a chain-like structure.

Feature Selection In a regular decision tree, each node picks a single feature that best splits the subset of training samples at that node. In a Twinleaf Chain, each node can pick either a single feature or a group of features to make the best split. The way to choose the most relevant features for each node can be different. There are different approaches to choose these features, for instance using the λ-dependency method described in [6].

Leaf Node Meaning In a standard decision tree, leaf nodes represent the final classification label for a subset of samples. In a Twinleaf Chain, the leaves represent regions from the VPRS model: POS (positive) or NEG (negative). The POS region includes samples with the classification label we care about, while the NEG region has samples that don't fit that label. There's also a BND (boundary) region for samples that can't be sorted into POS or NEG based on the current node's features. The samples in BND get passed to the next layer's nodes for further classification using different features.

3 Building a Twinleaf Chain

This section outlines the process of constructing a Twinleaf chain, with a particular focus on its steps. The issues that need further consideration during the tree construction will be discussed in the next section. The algorithm describing these steps is summarized in Table 3.

Step 0: Preprocess sample dataset Prior to constructing the Twinleaf Chain, the sample dataset must be cleaned to remove incomplete instances (e.g., samples with missing values in any condition or decision attribute). If a condition attribute contains continuous values, it should be discretized into binary values (0 or 1) using an appropriate discretization method, such as the approach described in [8]. Similarly, if a condition attribute is non-numeric, it must be converted into binary form (0 or 1) using a suitable encoding technique (e.g., one-hot encoding or binary mapping).

After preprocessing, we assume that the dataset contains N samples, among which a subset X represents the target samples of interest, i.e., those with decision attribute value equal to 1. Each sample is described by m condition features, denoted as C, and one decision attribute D. Both the condition features C and the decision attribute D take values in $\{0, 1\}$.

Step 1: Constructing the root node First, the prior probability $P(X)$ is calculated according to the formula provided in Sect. 2, which serves as the baseline measure of the likelihood that a randomly chosen sample belongs to the target class. This step provides an initial reference point for subsequent conditional probability computations.

Next, from the full set of m condition features, the most informative p features are selected. The goal of this selection is to obtain a subset of features that can best partition the dataset into a collection of **elementary sets** - that is, subsets of samples that share identical feature-value combinations. These elementary sets then are examined to determine whether they belong to the positive region (POS), the negative region (NEG), or the boundary region (BND) according to the definitions of POS, NEG, and BND in the VPRS model. Feature selection is therefore a crucial step, as it directly affects both the granularity of the partitioning and the accuracy of subsequent classification. In practice, various feature-selection methods can be applied; such as, the λ-dependency based on approach described in [6] can be used for this purpose.

Once the p features are chosen, the dataset is partitioned into elementary sets accordingly. Each elementary set, denoted as E_i, corresponds to a unique feature-value combination, and two probabilities are computed for it: $P(E_i)$ and $P(X|E_i)$. Based on these probabilities, the set is classified into one of three regions following the VPRS model. Finally, the feature–value combinations that define the POS and NEG elementary sets, together with their associated values of $P(E_i)$ and $P(X|E_i)$, are annotated on the edges connecting the root node to its two children in the Twinleaf chain structure. This annotation provides explicit interpretability: the edges not only represent a branching decision but also carry probabilistic information about the reliability of that decision.

Table 3. Algorithm of Building a Twinleaf Chain

Input:
- Sample set S = {s1, s2, ..., sn}
- Each sample has condition attributes C = {c1, c2,..., cm} and decision attribute $D = \{0, 1\}$.**Output:**
- A tree structure in which each non-leaf node has three children:
- Left → POS region
- Right → NEG region
- Middle → BND region (recursively expanded)

Preprocess(S, C, D):
- remove incomplete samples
- discretize continuous attributes
- convert non-numeric attributes to binary {0, 1}

BuildChain(S, C, D):
- if S is empty then return
- delect most relevant unused features by, e.g., λ-dependency
- partition S into elementary sets E using selected features
- for each E in partition do
• Compute P(E) and P(X|E)
• Classify E into POS, NEG, or BND (VPRS model)
- remove samples in POS and NEG sets from S
- mark the condition attributes having been used to form the partition
- create non-leaf node with three children:
• Left → POS region
• Right → NEG region
• Middle → BuildChain(BND samples)

return node

For instance, consider the example in Fig. 1(a), where the selected features are *A*23, *A*22, and *A*21. Since each feature can take values 0 or 1, the combination of these three features generates 2^3 = 8 elementary sets. Among these, the set with *A*23 = 0, *A*22 = 1, and *A*21 = 1 is classified into the POS region, as its conditional probability $P(X|E)$ exceeds the threshold. Accordingly, the feature values (0, 1, 1) along with its computed $P(E)$ and $P(X|E)$ are annotated on the edge leading from the root to the POS child. Similarly, the elementary set with *A*23 = 1, *A*22 = 1, and *A*21 = 1 is classified into the NEG region, and its corresponding information is added to the NEG edge. By contrast, the elementary sets defined by (0, 1, 0), (0, 0, 1), and (0, 0, 0) fall into the BND region, as their conditional probabilities are inconclusive. These BND sets are not annotated in the figure, because their samples are retained for constructing the next non-leaf node in the Twinleaf chain, where the classification process continues recursively with a different subset of features.

Through this procedure, the Twinleaf chain grows node by node. At each stage, the POS and NEG samples are progressively removed from the dataset, while only the BND samples remain for further refinement. This hierarchical construction allows the model to focus its discriminative power precisely on the uncertain regions, thereby improving both the accuracy and interpretability of the final classification.

After these steps, the dataset is further refined in preparation for subsequent node construction. Specifically, the features used at the root are marked so that they will not be reconsidered. At the same time, samples that have already been assigned to POS or NEG elementary sets are also excluded from further use in constructing lower-level nodes.

Step 2: Constructing the remaining nodes The subsequent nodes are built in the same manner as the root node: select the most suitable features from the updated dataset and partition the samples accordingly. For example, in the second node, the chosen features are $A15$ and $A13$. The samples are then divided into elementary sets, which are subsequently assigned to different regions. If no elementary set can be placed into POS or NEG, then, as illustrated Fig. 1b, the corresponding edge is labeled with NULL.

4 Issues in Twinleaf Chain Construction

This section discusses several issues to be considered when constructing a Twinleaf chain.

4.1 Selecting the Most Effective Features for Each Non-leaf Node

As with conventional decision trees, the features of each sample used for building a Twinleaf chain must be discretized; otherwise, a discretization process is required. Various discretization methods can be applied, such as entropy-based discretization, the Minimum Description Length (MDL) principle, ChiMerge (chi-square test), or soft-cut [8]. Depending on the dataset, a continuous feature may be discretized into binary values (e.g., 0 and 1) or into multiple categories. A key concern arises when the two classes overlap along a feature dimension, leading to samples falling near the cut points. In such cases, if the discretized value of a feature lies very close to the cut point, the assigned interval label may not reflect the true class of the sample. To address this, methods such as soft-cut can be employed, where only samples sufficiently distant from the cut points can be discretized, thereby reducing the likelihood of misclassification.

The successfully discretized data forms the dataset used to build the Twinleaf chain. At each node, the goal is to select the most suitable subset of features from all available ones. A straightforward but impractical approach would be to exhaustively evaluate all possible feature combinations, which is computationally prohibitive. Instead, more efficient alternatives may be adopted, such as feature engineering techniques (constructing composite features), multivariate or oblique decision trees (linear combination splits), or rule-based partitions involving multiple features.

In addition, the λ-dependency measure [8] can also be used as a criterion for feature selection. Specifically, when applying the λ-dependency, it assumes that each sample in the sample dataset is described by p features. First, each feature is used individually to classify the dataset under the VPRS model. That is, each time, samples are grouped into elementary sets according to the selected feature, and these sets are then assigned to the POS, NEG, or BND regions as required by the VPRS model. The λ-dependency of each feature is then computed, and the feature with the highest λ-dependency is identified.

Next, this feature is paired with each of the remaining features to form two-feature subsets, which are used again to partition the dataset into elementary sets, assign them into POS, NEG, and BND, and compute λ-dependency values. The subset yielding the highest λ-dependency is retained. The process is then repeated iteratively: the selected subset is expanded by adding one more feature at a time, and the λ-dependency values are compared.

Stopping conditions for this process can be set manually. For example, the search may terminate once the selected subset contains three features. Alternatively, a comparative criterion can be applied: for instance, when the λ-dependency value of the best subset with q features is sufficiently close to that of the best subset with $q + 1$ features, the process can be stopped.

4.2 Stopping Conditions for Constructing a Twinleaf Chain

In theory, if each non-leaf node uses exactly one feature, then the length of a Twinleaf chain would be up to the total number of features in the dataset. In practice, however, each non-leaf node corresponds to a subset of all features, and thus the stopping conditions for constructing a Twinleaf chain can be defined as follows.

Condition 1: All samples have been classified (i.e., BND is empty) If the current feature subset at a non-leaf node assigns all samples into either POS or NEG, with BND being empty as shown in Fig. 1a, it can be considered that all samples are correctly classified. In this case, construction stops.
Condition 2: All samples are in BND (i.e., both POS and NEG are empty) If the current feature subset assigns all samples into BND, while both POS and NEG are empty as in Fig. 1(b), it indicates that the remaining unused features cannot further distinguish the samples. Construction stops at this point.
Condition 3: POS, NEG, and BND are all non-empty but features are exhausted If the current feature subset partitions samples into POS, NEG, and BND simultaneously, but no further features remain (i.e., all condition attributes have been used), then construction also stops.

4.3 Classifying Test Samples with a Constructed Twinleaf Chain

Once a Twinleaf chain has been constructed, it must be evaluated on test data and subsequently used for sample classification. The procedure is similar to that of a decision tree and proceeds as follows:

Step 1: Preprocessing For each new sample, preprocess the features as needed. For example, continuous features should be discretized prior to classification.
Step 2: Traversing the chain Starting from the root node, traverse down the chain according to the sample's feature values. At each non-leaf node, compare the feature values of the new sample with the possible values defined by the node's feature subset, and assign the sample accordingly. For instance, in the Twinleaf chain illustrated in Fig. 1(a), if a new sample has feature values $A23 = 0$, $A22 = 1$, $A21 = 1$, it is assigned to POS at the root. Similarly, if $A23 = 0$, $A22 = 1$, $A21 = 0$, the sample is assigned to BND. In the latter case, the sample does not receive a class label immediately, but

instead proceeds to the next non-leaf node where additional features (e.g., $A15, A13$) are examined. If $A15 = 1$ and $A13 = 1$, the sample is then assigned to POS.

If a sample reaches the last non-leaf node and is still assigned to BND, it is considered unclassifiable. For example, in the Twinleaf chain shown in Fig. 1(b), the final non-leaf node connects to two leaves with edges labeled NULL. Regardless of whether $A15$ is 0 or 1, the sample remains in the BND region, indicating that it cannot be classified.

5 Summary and Future Work

This paper has presented the Twinleaf Chain, an innovative classification framework that merges Variable Precision Rough Set (VPRS) theory with a structured decision tree model to address challenges in handling uncertain data. The Twinleaf Chain distinguishes itself from traditional decision trees through its regular, centipede-like architecture: symmetrical layers with central non-leaf nodes flanked by twin leaves, enabling multi-feature selection for optimal splits. The construction process proceeds iteratively from the root to the terminal nodes and involves operations such as sample discretization, feature selection, and sample classification based on the VPRS model. Its classification process can follow the standard steps of a regular decision tree while incorporating additional techniques when traversing the entire chain.

While the Twinleaf Chain shows promise for binary classification, several avenues for future work remain. First, empirical evaluations on large-scale datasets, such as those from the UCI repository, are needed so as to comprehensively assess its performance using metrics like accuracy, F1-score, and computational time. Second, extending the framework to multi-class problems by generalizing POS/NEG regions to multiple labels could broaden its applicability, potentially through multi-valued λ-dependency variants. Third, performance improvements could be explored, for example, by integrating ensemble methods such as bagging multiple chains or boosting for BND-focused refinement to enhance robustness. Additionally, handling continuous features more dynamically - perhaps via fuzzy rough sets-would support real-time scenarios. Finally, parallel implementations for big data applications, as suggested in related work, merit further investigation.

References

1. Quinlan, J.R.: Induction of decision trees. Mach. Learn. **1**(1), 81–106 (1986)
2. Quinlan, J.R.: C4.5: Programs for Machine Learning. Morgan Kaufmann Publishers Inc., San Francisco (1993)
3. Pawlak, Z.: Rough sets. Int. J. Comput. Inf. Sci. **11**(5), 341–356 (1982)
4. Pawlak, Z.: Rough Sets: Theoretical Aspects of Reasoning about Data. Kluwer Academic Publishers, Dordrecht (1991)
5. Ziarko, W.: Variable precision rough sets model. J. Comput. Syst. Sci. **46**(1), 39–59 (1993)
6. Ziarko, W.: Partition dependencies in hierarchies of probabilistic decision tables. In: RSKT 2006, LNAI, vol. 4062, pp. 42–49 (2006)
7. Chen, X., Ziarko, W.: Experiments with rough set approach to face recognition, the special issue of the. Int. J. Intell. Syst. (2010)

8. Chen, X.: A rough set approach to recognize images of handwritten digits. In: Proceedings of the 2022 ACM Southeast Conference (ACMSE 2022), pp. 42–49 (2022)
9. Wei, W., Hui, M., Zhang, B., Scherer, R., Damaševičius, R.: Research on decision tree based on rough set. J. Internet Technol. **22**(6), 1385–1394 (2021)
10. Wei, J.M., Wang, M.Y., You, J.P.: VPRSM based decision tree classifier. Comput. Inform. **26**(6), 663–677 (2007)
11. Wu, S., Wang, M.Y., You, J.P.: A classification prediction method using rough set and decision tree. In: Proceedings of the 8th International Conference on Computing and Artificial Intelligence (ICCAI 2022). ACM (2022)
12. Sharma, R., Jain, P., Shrivastava, S.K.: An optimize decision tree algorithm based on variable precision rough set theory using degree of β-dependency and significance of attributes. Int. J. Comput. Sci. Inform. Technol. **3**(3) (2012)
13. Minz, S., Jain, R.: Rough set based decision tree model for classification. In: Skowron, A., Polkowski, L. (eds.) Data Warehousing and Knowledge Discovery (DaWaK 2003) Lecture Notes in Computer Science, vol. 2737, pp. 172–181 (2003)
14. Minz, S., Jain, R.: Construction of decision trees based on entropy and rough sets under tolerance relation. Int. J. Comput. Intell. Syst. (2007)

Authorship Verification for Academic Dishonesty in the Era of AI

Jun Jang[1(✉)] and Thai Le[2]

[1] Oxford High School, Oxford, MS 38655, USA
jun.green.jang@gmail.com
[2] Indiana University, Bloomington, IN 47405, USA

Abstract. With the rise of Generative AI (Gen AI), academic dishonesty in the classroom has skyrocketed, with over 90% of students now admitting to using it for schoolwork. While Gen AI offers various learning benefits, empirical studies show that an overreliance on these tools erodes creativity and critical thinking, which are crucial for the future of innovation. Unfortunately, the existing solutions often fall short. AI detectors merely analyze one text at a time, failing to account for students' previous writings. Meanwhile, many authorship verification (AV) models fail to analyze the nuances in writing styles that truly distinguish authorship. To fill this existing gap, this paper proposes a novel authorship verification system designed as a feature vector model. The model incorporates a comprehensive feature set, which is a combination of standard transformer-based token-level features (e.g., POS tag patterns) and handcrafted stylistic features (e.g., vocabulary richness & sentence structure variation). Extracted from a variety of datasets and real-world high school student essays, these combined vectors were used to train and evaluate multiple machine learning (ML) binary classification models for detecting academic dishonesty. The proposed comprehensive AV feature vector model filled the critical gaps in the literature, outperforming the standard token-based approaches by over 25%, while also enhancing transparency through interpretable decisions. This research highlights the importance of writing style features within a holistic, feature-based AV system. It assists educators across the globe in detecting academic dishonesty more reliably and responsibly, preserving critical thinking and innovation in the age of AI.

Keywords: Authorship Verification (AV) · Machine Learning (ML) · Artificial Intelligence (AI)

1 Introduction

Studies show that a majority of high school students admit to cheating on test, homework, and writing [3,15]. More than a third also admit that they used the

Research Track on Artificial Intelligence (CSCI-RTAI).

H. R. Arabnia et al. (Eds.): CSCI 2025, CCIS 2968, pp. 267–282, 2027.
https://doi.org/10.1007/978-3-032-28457-0_19

Internet to plagiarize assignments [7]. Authorship verification (AV) is the process of deciding whether or not two input texts are written by the same author. The importance of authorship verification lies in its wide applications. It can be used for plagiarism detection, writing style analysis, forensic investigations, academic integrity, intellectual property disputes, content moderation, literary studies, and more.

In recent years, the emergence and popularity of generative artificial intelligence (Gen AI) is changing the way we work and live, whether it is finding answers, searching for information, automating tasks, or generating content. In writing, large language models (LLMs, a type of Gen AI) such as OpenAI's GhatGPT [13], Google's Gemini [5], and Meta's Llama [12] have gone through a breakthrough. They can generate entire articles on a professional level based on prompts, not to mention school homework, reports, or essays. While AI can help improve productivity significantly, researchers find that it has adverse effects on critical thinking skills and creativity which students learn from writing [4]. In this era of Gen AI, therefore, the importance and demand for authorship verification is ever increasing, especially in academic settings.

This paper presents an authorship verification software we developed for teachers (or anybody) who are in need of detecting academic (or any) dishonesty or impersonation. The software takes two input texts, one with author verified (ground truth) and the other with author unknown or suspicious. First, the software pre-processes input texts by eliminating unnecessary information for accurate extraction of writing style. It then extracts their writing styles into a set of attributes (*features* in technical terminology) we developed. Finally, the software runs the trained machine learning classification model. The result is "yes" or "no" with probability if two inputs are written by the same author. Figure 1 visualizes the overall flow of the authorship verification process. Note that the training of the machine learning model is described separately in Sect. 2.3.

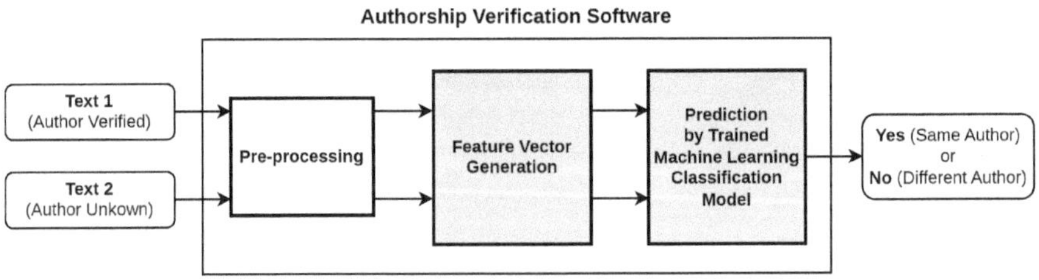

Fig. 1. Authorship verification workflow proposed in this project.

2 Methods

We view "writing style" as a collection of features. *Feature* is a widely used terminology in machine learning that refers to an individual measurable property or characteristic of data. For the purpose of authorship verification, features impact the accuracy and quality of predictions. We developed a set of novel features that we believe captures writing style effectively.

This research consists of three main parts. The first is data collection and preprocessing in which we utilized two separate datasets and prepared the data for feature extraction and classification (Sect. 2.1). The second is feature development and extraction (Sect. 2.2). Stylistic choices vary from author to author—some share certain styles, while others do not—making it impossible for any single feature to distinguish all authors perfectly. Thus, we came up with a set of features, each of which helps identify a group of authors from others. The final step is to select the machine learning technique that yields the highest accuracy in authorship verification (Sect. 2.3).

2.1 Data and Preprocessing

For training and testing, we have decided to utilize the Reuters_50_50 dataset [10]. The dataset is a subset of the Reuters Corpus Volume I (RCV1) [9], previously used in author identification studies. We selected the 50 authors with the largest number of articles, specifically choosing texts that were labeled with at least one subtopic under the CCAT (corporate/industrial) category. This selection helps reduce the influence of topic differences when distinguishing between texts. The training set comprises 2,500 texts (50 per author), and the test set includes an additional 2,500 texts (50 per author) that do not overlap with the training set. As [10] mentions, the topic variety reduce the influence of topicality for our authorship verification task, which is particularly useful to ensure no bias towards a particular topic at hand. Furthermore, the dataset is suitable for the intended academic environment as the news articles from Reuters resemble a high formality.

To better replicate an academic setting, we collected real essays from high school students. Our dataset comprises eight distinct authors, each contributing between two and six essays. While the dataset is relatively small and may lead to weaker model performance, we believe its use is essential, as it closely aligns with the academic context our model is designed for. This dataset consists of essays ranging from a variety of topics including history, english, and technology to minimize topic bias for AV.

Preprocessing is a crucial step in authorship verification and other machine-learning tasks. For our AV model, we implemented several preprocessing steps to effectively prepare the data for feature extraction and classification. For every piece of writing, we constructed an "entry" consisting of the original preprocessed text, a list of POS tags, POS tag chunked groups, detailed subtree expansions of the POS tag chunks, and the extracted tokens. To create the initial preprocessed text, we ensure that all of the text is lowercase for correspondence with the transformer-based feature extractions. We used NLTK's Treebank Brill tagger, which was trained on the Penn Treebank tagset, to generate the list of POS tags. Similarly, we used NLTK's RegexpParser and Treebank tokenizer for creating the POS tag chunks & chunk subtrees and tokenizing respectively [1]. Each entry was passed into feature extraction to prevent computational redundancy and fit the transformers for feature extraction.

For many of the newly developed features, we used spaCy's NLP pipeline [16] to perform seperate tokenization, part-of-speech (POS) tagging, and dependency parsing, among other tasks. This allows us to extract specific linguistic features from the tokenized version of the raw texts as detailed in Sect. 2.2. Furthermore, we also developed a custom sentencizer to accurately split sentences while handling specific punctuation rules and quoted text. For instance, it prevents splitting at punctuation within common abbreviations like "Mr." or "Dr.," avoids treating semicolons and colons as sentence boundaries, and correctly handles punctuation appearing within or adjacent to quotation marks. Essentially, it ensures accurate sentence splitting to preserve features that rely on a sentence-by-sentence structure.

For features such as sentence length & variety and sentence structure variation, we ensure that no "empty" sentences are counted as sentences by filtering them out. This is critical as an empty sentence would skew sentence length and variety features as it has a length of 0, and an empty sentence would be incorrectly classified as "Unclassified." For our vocabulary levels feature, we preprocessed an existing Kaggle CEFR dictionary dataset [8] to create our own CEFR dictionary classifier.

Another crucial step of preprocessing is managing the datasets particularly for creating unique and balanced pairs. In a binary classification task, models output either 0 or 1 and must be trained on corresponding labels. In our AV model, a label of 1 signifies that both texts in a pair are authored by the same person (positive instance), while a label of 0 indicates they are written by different authors (negative instance). To train our model on both 0 and 1 labels, we create unique and balanced text pairs as part of our preprocessing pipeline.

The unique and balanced text pairs generating process goes as follows. First, we shuffle all of the texts for each individual author, and then a specified portion (0.5) of texts are "reserved" and set aside to create different-author pairs. The remaining texts are then iteratively paired to form unique same-author pairs, each labeled as positive (1). Then, we collect the reserved texts from all authors and randomly sample pairs across different authors to form all the different-author pairs, each labeled as negative (0). Since the total number of same-author pairs might not match the different-author pairs, we balance the dataset by selecting an equal number of same-author and different-author pairs. This is critical as an imbalance of positive and negative instances may develop biases while training classification models. Thus, we have generated unique and balanced text pairs to construct a well-rounded, unbiased dataset that represents both positive and negative instances. The code for generating unique and balanced pairs was obtained and modified from this public repository[1].

[1] https://github.com/swan-07/authorship-verification.

2.2 Features

We developed a set of novel features in this project. They range from the statistics of sentence and vocabulary to advanced predictability and readability used in natural language processing. The list of features represented in numeric values form a *feature vector* which becomes input to ML models to make predictions. The following subsections describe the features we developed and used in this project. Following the development of our novel set of writing style features, we incorporated an additional feature set from a state-of-the-art feature vector model by [17], which implemented a feature vector model proposed by [18]. Many of their features are Transformer-based and rely on Scikit Learn's TFIDFVectorizer to compute the TF-IDF vectors from the documents.

Vocabulary Levels. Writers commonly differ in their vocabulary usage, particularly in the range of words they select. It is even more case in academic settings. For example, student A does not use high-level vocabulary as much as student B does.

In order to determine what constitutes "higher" and "lower" level vocabulary, we utilize the Common European Framework of Reference for Languages (CEFR) [2]. CEFR categorizes vocabulary level in terms of language proficiency in six levels, from A1 to C2, which can be regrouped into three broad levels: basic, independent, and proficient users.

To compare the distributions in CEFR levels, we developed our own CEFR dictionary classifier using a dataset on Kaggle[2]. To ensure proper correspondence between the dataset's dictionary and the input text's words, we preprocessed the dataset's word entries using spaCy's lemmatization pipeline[3]. This lemmatization is useful for cases when the input text has words such as "abolished," which are not in their base form. Lemmatization converts "abolished" into its base form, "abolish," and appends it to the dictionary accordingly. Thus, all variations of the base word "abolish" are properly categorized through the dictionary correspondence.

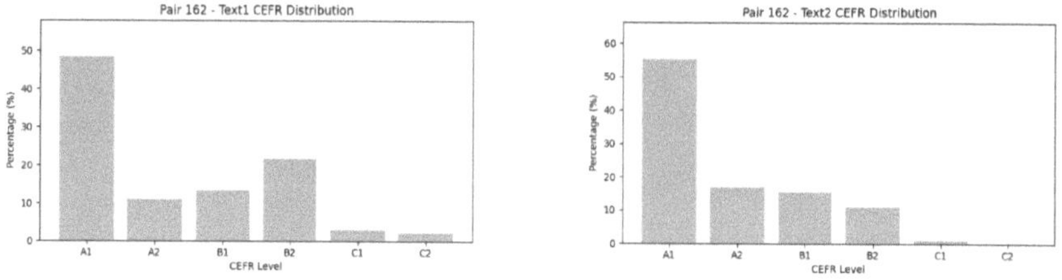

Fig. 2. Distribution of vocabulary levels of two authors.

[2] https://www.kaggle.com/datasets/nezahatkk/10-000-english-words-cerf-labelled.
[3] https://spacy.io/api/lemmatizer.

Figure 2 shows an example of two different distributions of two texts of different authors. The author on the left uses vocabulary at higher levels more than the author on the right. However, we observed that vocabulary level distributions may be considerably influenced by the topic at hand. For instance, an author's selection of vocabulary would heavily differ from a formal, technical report to a fictional kids' story. By default, our classifier accounts for unique topical vocabulary by classifying them as unknown words. Furthermore, our AV model is specified to an academic setting where we presume a consistent level of formality. The formality context in combination with the unknown word classification limits the extent to which topical vocabulary would skew the distribution.

To numerically capture the differences in vocabulary level distributions between two texts, we have chosen to use Jensen–Shannon divergence (JS divergence)[4]. JS divergence is a symmetric version of the Kullback Leibler divergence (KL divergence) and was selected to numerically compute a balanced measure of the differences between two distributions without introducing directional bias. This symmetry ensures that the JS divergence measures how much the CEFR vocabulary distributions of two texts deviate from each other regardless of the order. The JS divergence value is then inserted into a feature vector for classification (See Fig. 5).

Sentence Length and Variety. Writers can also be differentiated based on the characteristics of their sentences, namely the sentences' lengths and variety. Figure 3 shows an example of different average sentence length across different authors. To determine the average sentence length, we simply calculated the mean of the sentence lengths within a given text. To numerically capture the variation in sentence lengths, we utilized additional statistical methods, standard deviation (SD) and entropy.

The combination of standard deviation and entropy provides a comprehensive analysis of sentence length variation, as each metric captures distinct aspects of the distribution. Standard deviation measures the dispersion of sentence lengths around the mean, indicating how consistently an author maintains sentence length. Entropy, on the other hand, assesses the diversity and unpredictability of sentence lengths by evaluating how evenly sentence lengths are distributed across different values. High entropy indicates a wide variety of sentence lengths with minimal repetition, while low entropy implies that sentence lengths were concentrated around specific values. The absolute difference between two texts' SD and entropy values are inputted into the feature vector.

Sentence Structure Variation. Another way in which writers differ is through the sentence structures they tend to form. We define the distinct structures of sentences into six categories: *simple, compound, complex, compound-complex, fragment, and unclassified.* Table 1 explains and exemplifies all of the different sentence structures that we classify each input text under.

[4] https://en.wikipedia.org/wiki/Jensen-Shannon_divergence.

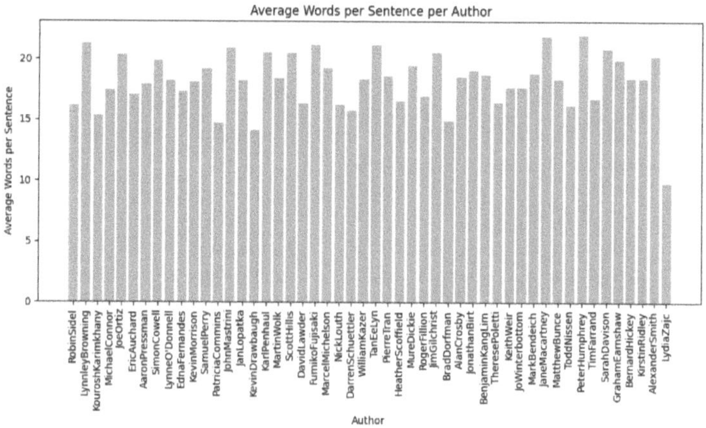

Fig. 3. Average sentence length across different authors.

Following the categorization of each sentence from an input-text, we analyze the variability of their distribution. This variability serves as another useful feature for distinguishing writing styles. To numerically capture the differences in sentence structure distributions between two texts, we have once again chosen to use JS Divergence.

This feature is particularly useful for authorship verification, as an author's inherent sentence structure preferences should remain consistent regardless of the topic. Assuming a certain level of formality in an academic setting, the feature effectively captures writing style while minimizing the influence of subject matter. The JS divergence value is then inserted into a feature vector for classification.

Perplexity. Perplexity is a widely used, important metric in natural language processing (NLP) that measures a probabilistic model's ability to predict a given text. It is used to measure a language model's performance by analyzing how well it predicts the next word or token based on the previous words. In the context of writing style analysis, we have chosen perplexity to measure the predictability and flow of an author's text. A low perplexity indicates that the given text follows a more predictive and fluent sequence of words.

To calculate perplexity, we first tokenized the text which was split into overlapping chunks to handle long inputs. This was required as the model that we chose was GPT2 from HuggingFace Transformers[5], which had a maximum input of 1024 tokens. The loss (how well the model predicts the next token) was extracted using GPT2, and the perplexity was calculated as e^{loss}. The final value for each text is the average perplexity across all chunk(s), representing the fluidity of the writing. The absolute difference between two texts values are inputted into the feature vector.

[5] https://huggingface.co/docs/transformers/en/model_doc/gpt2.

Table 1. Categories of sentence structures with examples and attributes.

Sentence Structure	Example	Attributes
Simple	Authorship verification analyzes unique writing styles to identify authors.	Contains a single independent clause.
Compound	Authorship verification utilizes statistical methods, and it incorporates machine learning algorithms.	Combines two independent clauses joined by a coordinating conjunction (no dependent clauses).
Complex	Although authorship verification is highly accurate, it sometimes faces challenges with short texts.	Contains one independent clause and at least one dependent (subordinate) clause.
Compound-Complex	Authorship verification examines various linguistic features, and it employs both standard deviation and entropy to enhance precision, ensuring reliable results.	Contains at least two independent clauses and at least one dependent clause.
Fragment	A critical tool in authorship verification.	An incomplete sentence that lacks either a subject, a verb, or a complete thought.
Unclassified	Authorship verification methods are essential for literary analysis they help determine the true author of anonymous works which can resolve disputes.	Does not neatly fit into standard categories due to structural issues such as run-ons or improper punctuation.

Readability. Relating to perplexity, readability further measures the flow and understandability of text. We have chosen the Flesch reading ease test[6] to quantify the readability of writing. Text readability with the Flesch reading ease score is evaluated based on sentence length and word complexity. Thus, this feature differentiates authors by quantifying their preferences for either simpler language or a more complex composition of text. For example, author A who has a clear and straightforward writing style may consistently have high Flesch reading ease scores while author B who has a tendency of challenging language might have lower scores. Furthermore, the Flesch reading ease score further helps in differentiating certain authors who have a tendency to overwrite, form overly complex sentences, or excessively use advanced vocabulary. The absolute difference between two texts' readability scores are inputted into the feature vector.

Adverbial Placement. Another stylistic choice that sets authors apart is their adverbial placement. Analyzed in [11], adverbial placement proved to be a valuable feature for stylometric analysis, "attributing authorship with an average

[6] https://en.wikipedia.org/wiki/Flesch-Kincaid_readability_tests.

accuracy of 9.7% greater." We developed our version of the feature by classifying adverbial placement into one of five categories as shown in Table 2.

Table 2. Categories, examples, and description of adverbial placement feature.

Category	Example	Definition
Sentence-initial	*Quickly, he packed his bags.*	The adverb appears at very the beginning of the sentence.
Preverbal	*He quickly packed his bags.*	The adverb is placed immediately before the main verb.
Postverbal	*He packed quickly to catch the train.*	The adverb is placed immediately after the main verb.
Sentence-final	*He packed his bags quickly and efficiently.*	The adverb appears at the end of the sentence.
Other	*Quickly? He always does things at the last minute!*	An unconventional placement that doesn't fall into the other categories.

The premise behind adverbial placement analysis is that there are numerous grammatically "acceptable" placements for the same adverb. Therefore, authors could unconsciously develop a preference for certain placements, providing insight for authorship verification. We quantify these preferences by using JS Divergence once again to distinguish two different probability distributions.

However, it is crucial to note that we do not assume that every adverb has the semantic or grammatical freedom to be placed wherever. [11] also confirms that not all adverbs can be grammatically positioned anywhere in a sentence. Certain adverbs, particularly manner adverbs like *beautifully, quietly, and carefully*, can appear in various positions. However, some adverbs, like *nearly*, are forced to positions near the verb phrase for pronounceability, while adverbs, like *daily*, are typically placed at the end of the sentence. Although these constraints exist, we have proceeded with this feature as it's been empirically useful for distinguishing writing style.

Additional Features. In addition to our novel handcrafted features, we incorporated a set of high-dimensional stylistic features derived from the work of [18], as implemented by [17]. These features were primarily transformer-based and computed using scikit-Learn's `TFIDFVectorizer`. Their features included POS-tag n-grams, special characters, frequency of function words, average number of characters per word, distribution of word-lengths (1–10), vocabulary richness, POS-tag chunks, POS chunk construction, stop-word and POStaghybrid tri-grams, and part-of-speech tag ratios. The explanations for these features can be found in [18].

However, due to the high dimensionality of these vectors (within the thousands), we applied Principal Component Analysis (PCA) to compress the feature space while retaining the most informative components. The PCA was only applied to these pipeline-based features, and the resulting principal components were then concatenated with our handcrafted features to form the final input vector for classification. We incorporated these features to enhance the robustness of our overall feature set.

2.3 Training

Authorship verification can be best modeled as a binary classification problem deciding whether two texts are written by the same author or not. For each given text pair picked from dataset, the differences between the feature vectors of the two texts become inputs to the ML binary classifier training [18].

We trained 5 classification models (Logistic Regression, Random Forest, SVM, KNN, Gradient Boosting) with dataset described in Sect. 2.1. It is a supervised learning. The more data/training it has, it makes better predictions in general.

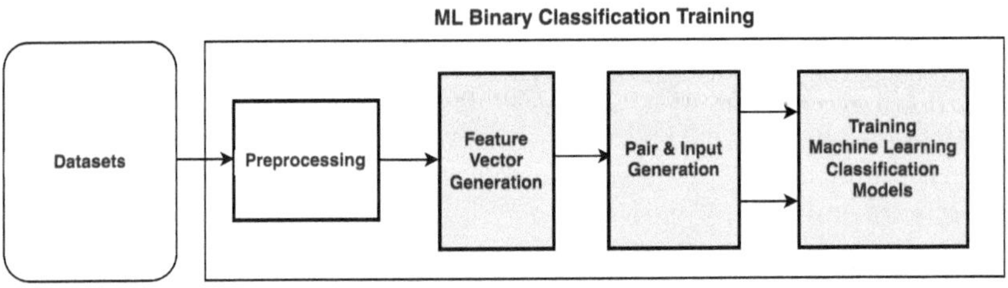

Fig. 4. Machine learning classifier training process.

Figure 4 shows the flow of training process. First, input texts are preprocessed as described in Sect. 2.1. Each feature may need different preprocessing. We implemented all preprocessing from scratch if software library function is not available. Second, feature vectors for each text in dataset are computed and generated. Third, we generate a ML classifier input file that contains a list of the differences between the feature vectors of a pair of two texts. Figure 5 shows the first few lines of an input file which machine classification models are trained with.

3 Evaluation and Results

We implemented the proposed software tool on a local machine using Python programming language. Several Python software packages were used for some

```
perplexity_diff,avg_sentence_length_diff,sentence_length_sd_diff,sentence_length_entropy_diff,sentence_type_js,cefr_js,readability_diff,adverbial_js,label
8.323531150817871,3.888888888888889,0.92776603811755,0.1125315118139091,0.084809524231682,0.0074153688780625,2.5399999999999999,0.0144978432054849,0
1.3407669067382812,1.203947368421051,1.824145664811004,0.7867073412581114,0.191013186548526 2,0.0244711436560091,18.14,0.0379771648948254,0
7.340461730957031,4.563829787234043,0.0876307319554516,0.3751855437369533,0.0551137481853086,0.0088530248729579,5.180000000000007,0.1759403346700623,0
10.50737762451172,0.219875776397517,3.1480347176898045,0.1605656222141176,0.1262313029926429,0.00803574310375,8.860000000000007,0.3809300909097882,0
18.212194442749023,8.458333333333332,0.160557221593935,0.3647869792568117,0.0490351756535127,0.0051939615030569,9.82,0.0001777289232848,0
3.018453598022461,6.899999999999999,3.077532766836665,1.256564762130953,0.1809336367244812,0.0210873682140548,14.65,0.162691183110389,1
4.118534088134766,3.088275862068965,0.9801296933251392,0.2463615912615413,0.0110466460291047,0.0039685263343672,7.450000000000003,0.121468690637158,0
8.078222274780273,6.833333333333332,1.1528579688777203,0.2505031564347577,0.012469401172858,0.0216971852036662,19.500000000000007,0.310181433694384 3,0
2.252796173095703,0.0200668896321083,0.8881418906387761,0.115927839994967,0.0537609605401499,0.0098542724263497,8.869999999999997,0.0709982159901287,1
12.381624221801758,1.209134615384 6168,0.7846317453249103,0.1804524092107451,0.1109796065736351,0.0091168610432772,1.93,0.0294374749652919,0
7.716712951660156,2.001344086021508,1.2381946544868845,0.0265585575303672,0.0671210446365455,0.0073669779077371,10.800000000000004,0.4137811483306705,0
6.842006683349609,4.304545454545455,0.5889786394675394,0.7443146580682467,0.1444237607243756,0.0053436338892249,5.789999999999999,0.0579142617625023,1
```

Fig. 5. Contents of input file to ML classifier.

parts of the project. For ML classification models, we used a well known scikit-learn Python package [14]. We also used the spaCy Python package [6] for text processing such as classification, tokenization, stemming, tagging, parsing, etc.

Instead of a fixed train-test split, we decided to apply k-fold cross-validation to evaluate the model's performance [14]. The dataset is divided into k equal parts called "folds." From this partitioning, the model is trained on $k-1$ folds and tested on the remaining fold. The process is repeated k times where each fold is used for testing once. The final evaluation metrics are averaged across all folds to give a more reliable evaluation of the model's performance. This method was chosen due to the smaller datasets that we had as it reduces the risk of overfitting or positive/negative bias from a single train-test split. Furthermore, when evaluating different machine learning models, we used grid search for tuning hyperparameters to optimize model performance [14].

3.1 Detection Accuracy

To evaluate the performance of the model, we have decided to utilize 5 common machine learning performance metrics for classifiers: AUC, c@1, $F_{0.5}$, F1, and Brier Score. These metrics were particularly relevant for AV tasks for PAN at CLEF, which focused on a balanced assessment of model performance[7]. The "overall" metric in the last column is simply the average of the five metrics across each model. The performance evaluation of only the handcrafted feature set is summarized in Table 3 and 4. On the contrary, the performance evaluation of using the full concatenated feature set is summarized in Table 5 and 6.

3.2 Analysis

Top Performing Model. Examining Table 5, the tuned Gradient Boosting model outperformed most of the other models across every single evaluation metric, closely followed by the SVM classifier. It's likely due to Gradient Boosting's ability to capture non-linear relationships between features. Unlike simpler models such as Logistic Regression, which assumes a linear relationship between the features and classification, Gradient Boosting follows an approach where it uses an ensemble of decision trees. This allows the model to learn from subtle variations in text-based features which is important for AV tasks, especially on

[7] https://pan.webis.de/shared-tasks.html#authorship-verification.

Table 3. Evaluation on Reuters-50-50 (Handcrafted Features Only).

Classifier	Version	AUC	c@1	$F_{0.5}$	F1	Brier Score	Overall
Logistic Regression	Default	0.589	0.558	0.565	0.589	0.755	0.611
Logistic Regression	Tuned	0.593	0.563	0.569	0.590	0.755	0.614
Random Forest	Default	0.614	0.585	0.578	0.582	0.758	0.623
Random Forest	Tuned	0.613	0.579	0.580	0.582	0.759	0.623
SVM	Default	0.606	0.582	0.576	0.588	0.758	0.622
SVM	Tuned	0.607	0.567	0.564	0.585	0.757	0.616
KNN	Default	0.556	0.546	0.547	0.550	0.717	0.583
KNN	Tuned	0.570	0.548	0.551	0.559	0.738	0.593
Gradient Boosting	Default	0.617	0.586	0.588	0.596	0.758	0.629
Gradient Boosting	Tuned	0.627	0.591	0.594	0.612	0.761	0.637

Table 4. Evaluation on Student Essays (Handcrafted Features Only).

Classifier	Version	AUC	c@1	$F_{0.5}$	F1	Brier Score	Overall
Logistic Regression	Default	0.400	0.460	0.346	0.347	0.709	0.453
Logistic Regression	Tuned	0.446	0.520	0.371	0.373	0.724	0.487
Random Forest	Default	0.466	0.525	0.445	0.443	0.726	0.521
Random Forest	Tuned	0.497	0.516	0.449	0.422	0.726	0.522
SVM	Default	0.513	0.450	0.354	0.356	0.739	0.482
SVM	Tuned	0.502	0.490	0.350	0.344	0.728	0.483
KNN	Default	0.554	0.576	0.532	0.503	0.711	0.575
KNN	Tuned	0.582	0.560	0.499	0.463	0.736	0.568
Gradient Boosting	Default	0.411	0.498	0.360	0.336	0.556	0.432
Gradient Boosting	Tuned	0.478	0.478	0.292	0.284	0.610	0.429

Table 5. Evaluation on Reuters-50-50 (Full Feature Set).

Classifier	Version	AUC	c@1	$F_{0.5}$	F1	Brier Score	Overall
Logistic Regression	default	0.712	0.654	0.654	0.655	0.784	0.692
Logistic Regression	tuned	0.711	0.651	0.650	0.653	0.782	0.689
Random Forest	default	0.733	0.670	0.667	0.650	0.790	0.702
Random Forest	tuned	0.739	0.670	0.675	0.646	0.791	0.704
SVM	default	0.742	0.679	0.678	0.661	0.795	0.711
SVM	tuned	0.753	0.681	0.682	0.654	0.798	0.714
KNN	default	0.700	0.648	0.646	0.661	0.772	0.685
KNN	tuned	0.716	0.660	0.658	0.673	0.783	0.698
Gradient Boosting	default	0.748	0.677	0.683	0.651	0.796	0.711
Gradient Boosting	tuned	0.745	0.674	0.680	0.647	0.790	0.707

Table 6. Evaluation on Student Essays (Full Feature Set).

Classifier	Version	AUC	c@1	$F_{0.5}$	F1	Brier Score	Overall
Logistic Regression	Default	0.743	0.680	0.659	0.674	0.753	0.702
Logistic Regression	Tuned	0.799	0.715	0.704	0.709	0.759	0.737
Random Forest	Default	0.784	0.695	0.710	0.717	0.780	0.738
Random Forest	Tuned	0.788	0.702	0.724	0.717	0.779	0.742
SVM	Default	0.771	0.710	0.707	0.749	0.776	0.743
SVM	Tuned	0.733	0.539	0.368	0.426	0.740	0.561
KNN	Default	0.666	0.641	0.649	0.703	0.702	0.672
KNN	Tuned	0.641	0.589	0.616	0.671	0.726	0.649
Gradient Boosting	Default	0.701	0.629	0.625	0.587	0.678	0.644
Gradient Boosting	Tuned	0.745	0.647	0.636	0.609	0.725	0.672

a larger dataset like Reuters_50_50. Gradient Boosting's ability to dynamically adjust feature importance also likely played a significant role. Since all of the features extracted are likely not all equally informative, Gradient Boosting's ability to select and give higher importance to the most relevant ones and weigh down the less-important ones likely improved the performance.

Comparison to State-of-the-Art Models. There have been previous attempts at authorship verification in the current literature which have similar approaches. One of the state-of-the-art attempts was [17]. It implemented a feature vector model derived from [18]. Since they used the same Reuters_50_50 dataset, we can directly compare the performance.

Table 7. Comparison of Performance Evaluation on Reuters_50_50 Dataset.

Model	AUC	c@1	$F_{0.5}$	F1	Brier Score	Overall
Logistic Regression (Handcrafted Only)	0.593	0.563	0.569	0.590	0.755	0.614
Gradient Boosting (Handcrafted Only)	0.627	0.591	0.594	0.612	0.761	0.637
Logistic Regression (Full Feature Set)	0.711	0.651	0.650	0.653	0.782	0.689
Gradient Boosting (Full Feature Set)	0.745	0.674	0.680	0.647	0.790	0.707
Wan (2024) Logistic Regression	0.565	0.525	0.548	0.613	0.557	0.562

As shown in Table 7, all four of our models - Logistic Regression and Gradient Boosting using both handcrafted and full features sets - outperformed [17] across nearly all evaluations. Most notably, the two models utilizing the full features set achieved significantly better overall performance across all of the performance evaluation metrics compared to Wan's model. While their model showed a slightly higher F1 score compared to our handcrafted-only feature sets, we still outperformed their model across every other evaluation metric with the handcrafted set. These results are very promising as our smaller set of handcrafted features already outperforms the state-of-the-art model on the same dataset. It demonstrates that our handcrafted writing style features are stronger indicators of authorship. Even more promisingly, when combined key features from the state-of-the-art approach with our full feature set, we observed a substantial boost in classification performance across every single evaluation metric.

Performance Comparisons Between Default and Tuned Hyperparameters. Although Grid Search was implemented for an initial attempt to fine-tune the hyperparameters, the Random Forest model's performance remained roughly the same. The SVM models' performance surprisingly decreases from the default to tuned parameters. This could be explained by the fact that for more complex models like Random Forest and SVM, increased complexity from tuning may not translate into better generalization. Furthermore, in some cases, the default parameters were likely already near-optimal which is why the features did not strongly respond to further tuning.

High School Student Essay Dataset Performance. While our initial evaluation on the high school student essay dataset (Table 4) weren't the greatest using only the handcrafted feature set, we observe a significant improvement after incorporating the full feature set. The previously lacking performance can largely be attributed to the small dataset size along with the diverse essay topics.

Since this dataset only consisted of 8 distinct authors, each having anywhere between 2-6 essays, we were only able to generate a small number of unique and balanced pairs to train/test our model. This constraint led to extremely high variance in performance across K-folds during cross-validation evaluations. For example, in one of our K folds for our Gradient Boosting (default) model, we achieved an overall performance evaluation of 0.716 on the student essay dataset, which significantly outperforms the averaged result in our Gradient Boosting (tuned) model for the Reuters dataset, which was 0.637. However, the average performance across the K folds was weighed down by particular folds where we achieved a mere overall performance of 0.235.

However, after incorporating the full feature set for our models, we observed a substantial and consistent improvement in performance across all classifiers as shown in Table 6. This suggests that the additional features enabled the models to capture deeper stylistic patterns which helped mitigate the limitations posed by a small dataset and varied topics.

4 Conclusion

We presented an authorship verification tool designed to help detect potential writing dishonesty. A novel set of handcrafted stylistic features is developed and implemented in Python. Multiple well-established machine learning classifiers were then trained and evaluated on both a standard benchmark dataset and a real-world high school student essays.

Our experiments show very encouraging results. Not only did our handcrafted feature set outperform the state-of-the-art feature vector model on the same dataset, but after integrating key features from that model into our full feature set, we achieved a substantial boost in classification performance across all evaluation metrics. This indicates that our combined feature set and novel writing style features capture richer linguistic and stylistic patterns.

Even on the smaller and more variable high school essay dataset, our models had strong performances when using the full feature set, suggesting that these features are effective at identifying authorship. We expect the performance to further improve as we collect additional high school essays in the future.

References

1. Bird, S., Klein, E., Loper, E.: Natural Language Processing with Python. O'Reilly Media Inc. (2009). https://www.nltk.org/book/
2. Council of Europe (online): Common European framework of reference for languages (2005). https://www.coe.int/en/web/common-european-framework-reference-languages
3. Edutopia (Online): Fostering students' academic integrity (2024). https://www.edutopia.org/article/preventing-academic-dishonesty-high-schoo/
4. Gerlich, M.: AI tools in society: impacts on cognitive offloading and the future of critical thinking. Societies **15**(1) (2025). https://doi.org/10.3390/soc15010006, https://www.mdpi.com/2075-4698/15/1/6
5. Google (online): Gemini (2025). https://gemini.google.com/
6. Honnibal, M., Montani, I., Van Landeghem, S., Boyd, A.: spacy: Industrial-strength natural language processing in Python (2020). https://doi.org/10.5281/zenodo.1212303
7. International Center for Academic Integrity: Cheating in High School (2025). https://academicintegrity.org/resources/facts-and-statistics. Accessed Jan 2025
8. Kaggle (online): Cefr levelled English texts (2025). https://www.kaggle.com/datasets/amontgomerie/cefr-levelled-english-texts
9. Lewis, D.D., Yang, Y., Rose, T.G., Li, F.: Rcv1: a new benchmark collection for text categorization research. J. Mach. Learn. Res. **5**, 361–397 (2004)
10. Liu, Z.: Reuter_50_50 [Dataset] (2006). https://doi.org/10.24432/C5DS42
11. Lukin, E., Roberts, J.C., Berdik, D., Mugar, E., Juola, P.: Adjectives and adverbs as stylometric analysis parameters. Int. J. Digit. Human. **5**(2), 233–245 (2023). https://doi.org/10.1007/s42803-023-00065-y
12. Meta (online): Llama (2025). https://www.llama.com/
13. OpenAI (online): ChatGPT (2025). https://openai.com/chatgpt/overview/
14. Pedregosa, F., et al.: Scikit-learn: machine learning in Python. J. Mach. Learn. Res. **12**, 2825–2830 (2011)
15. Renuka Sharma, K.M., Vyas, V.: Investigating academic dishonesty among business school students using fraud triangle theory and role of technology. J. Educ. Bus. **99**(2), 69–78 (2024). https://doi.org/10.1080/08832323.2023.2260925
16. spaCy (online): spacy library architecture (2025). https://spacy.io/api

17. Wan, S.: Transparent authorship verification with machine learning models (2024). https://math.mit.edu/research/highschool/primes/materials/2024/Wan.pdf
18. Weerasinghe, J., Singh, R., Greenstadt, R.: Feature vector difference based authorship verification for open-world settings. In: CLEF (Working Notes), pp. 2201–2207 (2021)

A Contract-Driven, On-Device AI Pipeline for Accessible Campus Wayfinding

Sebastian Moyano Lopez, Johan Mantilla Pancho, Erika Anrrango Pastillo, and Myriam Hernandez-Alvarez(✉)

Departamento de Informática y Ciencias de la Computación, Escuela Politécnica, Nacional, Ladrón de Guevara E11-25 y Andalucía, Quito 170525, Ecuador
myriam.hernandez@epn.edu.ec

Abstract. This paper presents a contract-driven, fully on-device pipeline and instrumentation framework for real-time perception→planning→interaction on commodity Android phones for accessible campus wayfinding for people with visual impairments (VI). The paper studies systems-level integration of perception, planning, and human-AI interaction on Android smartphones via a contract-driven architecture, which is machine-readable and allows for turning recognition into action. A perception JSON stream includes objects plus kinematics like direction, speed, ordinal distance, and performance counters. The GeoJSON schema is the output of the planning module. This data comprises a WGS84 LineString with labeled waypoint Points providing accessibility semantics and route info.

Keywords: On-device inference · accessible augmented reality · mobile computer vision · NavMesh/A* path planning · GeoJSON data contracts

1 Introduction

Navigating large university campuses remains difficult for people with low vision because available map apps are not designed as accessibility-first walking assistants. Their interfaces require that the user make a high visual effort due to dense maps, small labels, and low-contrast layers that are hard to perceive for a person with low visual acuity. Also, voice prompts are brief and infrequent, lacking continuous, context-aware guidance. Navigation in a campus also demands micro-navigation, indicating which entrance, which side of a path, ramp vs. stairs, etc. Popular apps rarely encode accessibility semantics to distinguish those kinds of details. Finally, many mobile UIs fall short of WCAG basics [1] like contrast, touch targets, and descriptive labels. Those UIs leave critical gaps for low-vision individuals trying to find their way on a university campus.

The demands of such an application cut across computer vision, planning, and human-computer interaction. Other work optimizes a single component in isolation, for example, object detector accuracy or AR rendering, rather than the full loop, including perception, planning, and interaction running on a smartphone. As a result, there is limited evidence about how the trade-offs among accuracy, latency, and usability play out in real field conditions.

H. R. Arabnia et al. (Eds.): CSCI 2025, CCIS 2968, pp. 283–296, 2027.
https://doi.org/10.1007/978-3-032-28457-0_20

This study addresses that gap by developing an accessible campus-scale wayfinding as a real-time sense→decide→act problem executed entirely on the phone. We present PoliUbícate [2], a mobile AI system that integrates: (A) on-device object detection, using YOLOv8/TFLite/OpenCV that produces lightweight kinematics like direction, speed, and distance, which deliver JSON data; (B) an accessible AR interface, using Unity/AR Foundation/ARCore, which is aligned with WCAG 2.1 and enriched with text-to-speech (TTS); and (C) a planning layer that computes best walkable routes on a 3D campus model, applying Unity NavMesh with the A* algorithm for route finding [3]. To optimize routes, deliver a GeoJSON, route.json that includes LineString plus labeled waypoint Points. The JSON and GeoJSON data are consumable by the AR component. The system can be deployed in mid-range Android devices and was engineered and evaluated under the constraints of a working university campus, specifically the Escuela Politécnica Nacional campus.

Our interest is the study of interfaces between modules. The accessible AR prompts are human-facing, and the contract-driven JSON obtained from object perception and the GeoJSON obtained from the planner are machine-facing. The modules turn raw recognition into timely, comprehensible actions. This orientation concerns the integration of AI with other technologies, intelligent user interfaces, and the evaluation of AI tools using reproducible protocols on real hardware and in real environments.

To guide the research, we pose questions related to each component of the system. The first questions are related to perception (RQ-A1/A2/A3): Is the on-device detector robust under unit/integration tests, do its kinematic signals (direction, speed, distance) provide actionable semantics, and is the socket layer resilient without app crashes?

The second group of questions is related to accessible AR (RQ-B1/B2/B3): Does the AR UI reach acceptable usability (SUS $\geq$ 70), and do iterative design changes (higher contrast marker-based guidance, improving TTS prompts) address user-observed confusion while supporting real-time JSON consumption without slowdown or breakage?

The third group of questions is associated with planning and geospatial considerations (RQ-C1/C2): How can the planner deliver AR-ready GeoJSON with labeled waypoints, and what is the measurable discrepancy between Unity/NavMesh routes and an external ground truth like Google Earth Pro, on a specific path?

Answering these questions matters because it yields baselines produced in the field (EPN campus), using accuracy and mAP to report model performance for object detection, with measurements across the whole pipeline from the camera through the user action, on mid-range Android devices, with users who rely on accessible interfaces. Even though we used simulated low-vision participants, which is a limitation, the metrics still capture the system behaviors. The app used was "Tengo Baja Visión" [4] configured to the 'loss of visual acuity' symptom at 50%.

What is important is that our strategy delivers reusable artifacts, such as detector and navigation data contracts, with accessibility semantics that the AR UI can use. The remainder of the paper reviews related work, details the system architecture and methods, and presents results that answer the research questions above, followed by a section with conclusions and future work.

2 Related Work

We have study related work across three strands: perception on phones, AR-based guidance, and campus-scale path planning. Our objective is to use what is recommended and to detect possible gaps to be researched. In AR for this type of application, the studies suggest that multimodal feedback with TTS improves comprehension and task completion. The application papers use these recommendations and report acceptable usability under controlled conditions. Also, we can detect gaps like efficiency concerns with cloud-dependent pipelines that could be problematic in real-world use. Additionally, most mobile computer vision work only reports model accuracy or demo overlays, without exposing the structure of the interfaces that turn detections into guidance. One way to obtain this information is to use machine-readable semantics that other components can consume.

In path planning and geospatial modeling, campus and indoor navigation systems commonly use Unity/ARFoundation with NavMesh/A* or graph-based methods. Yet planners seldom export routes in AR-ready, semantically rich formats, and even less often validate route geometry against an external geodetic baseline. However, for accessible wayfinding, semantics at the waypoint level, like ramps, stairs, and obstacles, are important. They are a way to deliver to the AR clients the route information in the form of polylines and labels without ad-hoc glue code. Moreover, a real-time loop from perception to action on smartphones, tied together by reproducible contracts and measured with deployment-relevant metrics, is underexplored. Our work is designed to fill precisely those gaps.

In Table 1, we show the related work with key findings. We adopt the lessons from the literature and close the missing link by wiring a WCAG-aligned AR client to the mentioned live contracts: perception JSON (objects + kinematics + perf) and planner GeoJSON (LineString + labeled waypoints) that the phone can use in real time on mid-range hardware.

Our computer vision module adds actionable kinematics and outputs them as a machine-readable stream. Additionally, we also document robustness, with unit/integration tests, non-blocking sockets, and short-history cleanup. This way, the other components of AR and planning can rely on stable, timely signals.

Also, we implement a NavMesh/A* planner that exports a GeoJSON WGS84 LineString with labeled waypoint Points, which is directly consumable by AR, and we quantify route discrepancy on a controlled campus path. This way, we aim to turn planning from a black box into a verifiable, contract-driven component.

3 Materials and Methods

This section details the system overview with the design sense, decide, and act. Each of the three component is described, as well as the methodology used for the interfaces, environments, and measurements needed to replicate or extend the study, which include module contracts (JSON and GeoJSON), logging fields and timing points, route definitions and ground-truth preparation, and the exact metrics used to evaluate perception, planning, AR interaction, and the end-to-end system.

Table 1. Related Work

Reference / Year	Modality / Platform	Task / Participants	Key findings
[5] / 2024	AR accessibility guidelines; alternative sensory cues; assistive-tech compatibility	Meta-synthesis (no participants)	Distills practical guidelines (clean visuals, consistency, alternative sensory cues, compatibility with screen readers). Supports a multimodal, WCAG-aligned AR UI.
[6] / 2021	Methodological mapping of navigation/assistive systems	Literature survey	Shows many studies rely on simulated/blindfolded participants rather than VI users; motivates transparent reporting of simulation-based usability (our limitation statement).
[7] / 2021	Smartphone navigation with audio + haptics (indoor/outdoor)	App-centric evaluations	Confirms that multimodal cues improve navigation; aligns with our marker-based + TTS AR approach on phones.
[1] / 2020	Large-scale Android accessibility audit	Automated/empirical inspection	Finds persistent gaps (contrast, touch targets, labels). Justifies WCAG 2.1–aligned overlays and proactive prompts (e.g., "Enable GPS").
[8] / 2023	Review of scene-understanding for blind/VI users	Synthesis (no participants)	Highlights privacy/latency issues with cloud vision; underpins our decision to run on-device CV and stream compact JSON semantics.
[9] /2023	Mobile AR (Unity/Vuforia; Android; TTS)	36 VI students; survey outcomes; energy reported	Validates audio prompts and real deployment constraints (e.g., energy). Stops short of contract-driven, end-to-end integration (perception→planner→AR).
[10] /2021	Smartphone AR + Maps	9 VI users; outdoor wayfinding	Demonstrates phone AR in the wild; lacks AR-ready route contracts and measured end-to-end integration.

(*continued*)

Table 1. (*continued*)

Reference / Year	Modality / Platform	Task / Participants	Key findings
[11] / 2023	Comprehensive detector survey	Meta-analysis	Traces evolution toward lightweight/anchor-free heads (e.g., YOLOv8), framing suitability for on-device inference.
[12] / 2020	Optimized YOLOv3	Algorithmic/efficiency paper	Early evidence that engineering tweaks (backbone/heads/post-proc) yield practical gains without heavy compute—supports our optimize-for-mobile stance.
[13] / 2024	Real-time CV pipelines	Meta-review	Emphasizes latency/reliability in production pipelines, reinforcing our unit/integration testing and socket resilience focus.
[14] / 2021	Embedded CV under resource limits	Field PoC	Shows the feasibility of embedded CV under tight power/compute—analogous to phone-side constraints and need for robust pipelines.
[15] / 2008	Geodetic validation baseline	No participants	Establishes external baseline utility (e.g., Google Earth Pro) for checking route length/shape, supporting our geospatial-fidelity evaluation.

3.1 System Overview

In the first component, perception is executed on-device using CV techniques. An Android application runs YOLOv8n exported to TFLite at an input size of 640 × 640. The data is preprocessed using OpenCV. Postprocessing gives detection records, with information about class, confidence, and bounding box (bbox). Also, this component includes a lightweight ObjectAnalyzer that adds direction (mapped to cardinal headings), speed (px/s), and an approximate distance proxy. Additionally, a short history is maintained and cleared after inactivity to bound memory.

The second component is for planning. It is deployed as a 3D model of the campus, which is built in Unity. NavMesh surfaces define walkable areas and constraints; paths are computed with the A* algorithm over the navigation graph to obtain the optimal routes. Planned routes are exported to a portable format and consumed by AR.

The AR UI is implemented using the Unity app, including AR Foundation and ARCore. The design of the component aims to deliver a UI with high contrast for easier reading, and it is complemented with TTS prompts that are the main guide for navigation. The UI follows WCAG 2.1 principles regarding contrast, typography, and touch targets.

Finally, the inter-module communication is developed using an exchange between modules via sockets with compact JSON messages for the link between perception and AR components. Also, GeoJSON files and streams are used for communication between the planner and the AR components. This contract-driven design decouples modules and ensures the same AR client can ingest live detections and planned routes without code changes. This is the major contribution of the present work.

The system can run on mid-range Android phones. The application was developed using Android Studio, TFLite runtime, OpenCV, Unity 3D, AR Foundation & ARCore, and standard JSON and GeoJSON parsers. The setting was the outdoor university campus of Escuela Politécnica Nacional in Quito, with varied lighting during morning, afternoon, and evening, its open plazas and narrower corridors, and typical pedestrian traffic. The connectivity conditions were with mixed Wi-Fi and cellular connections. The socket layer was tested under packet loss and reconnection scenarios.

3.2 Perception (Component A): Model and Signals

For perception, we used Yolov8n in its TFLite version for Android, with 640×640 input and standard non-maximum suppression (NMS) for duplicate boxes removal, with $\mathrm{IoU} = 0.45$, with a confidence threshold of 0.25. The object analyzer computes the kinematics for the tracked detection the kinematics, including direction (1), speed (2), and distance (3).

$$\mathrm{Direction}: \theta = \mathrm{atan2}(\Delta y, \Delta x) \text{ map to } \{N, NE, E, SE, S, SW, W, NW\} \tag{1}$$

Where Δx, Δy: frame-to-frame displacement in image pixels.

$$\mathrm{Speed}: v = \|\Delta p\| / \Delta t \tag{2}$$

Where: $\Delta p = (\Delta x, \Delta y)$; $\|\Delta p\|$ is its Euclidean norm, and Δt: time between frames in seconds. Speed is in px/s.

$$\text{Distance hint}: \text{d\_hint} = f(\text{bbox size, depth cues}) \tag{3}$$

Where bbox is the bounding box, and the distance hint is an ordinal cue, mapped like {far, medium, and near}.

The consistency of kinematics (stability) is enforced with a history cleanup that removes stale tracks, with a TTL $\approx$ 5 5 s to avoid ghost objects. It means that if an object is not seen in the last 5 s, its track and kinematics are removed.

Detection JSON (per frame). The per-frame message of the perception module sends a contract that turns raw detection into machine-readable guidance signals. AR parses this JSON the same way on any phone and sends it to the AR logic. An example is shown next.


```
{
"ts_ms": 1690000000000,
```

```
"perf": {"fps": 29.8, "t_infer_ms": 18.4, "cpu_pct": 42},
"objects": [
{"id": 12, "cls": "pedestrian", "conf": 0.86,
"bbox": [x, y, w, h],
"dir": "E", "speed_px_s": 95.3, "distance_hint": "near"}
]
}
```


3.3 Accessible AR UI (Component B)

The visual design of the AR UI present high-contrast, marker-based overlays, such as "stairs", "ramp", "entrance", rather than persistent route polylines, which were found to be confusing in earlier iterations.

The audio signals consist of TTS prompts with concise phrasing that give announcements of proximity near waypoints and obstacles. TTS complements platform screen readers, and the AR UI avoids interfering with OS-level services for accessibility.

For the interaction, it uses large touch targets with readable typography. The system provides proactive notices where needed.

3.4 Planning and Export (Component C)

For planning, the system uses NavMesh and the A* algorithm for finding the best route. Navmesh is used to generate the campus 3D model that encodes walkable surfaces. A* computes paths under accessibility constraints like ramps preferred over stairs.

Routes are exported as a FeatureCollection, using the GeoJSON route contract:

route_info: {start_point, end_point, total_distance_meters}
features[0]: LineString with WGS84 coordinates of the route polyline
features[i > 0]: Point waypoints labeled with accessibility semantics:

§ properties: {waypoint_index, type, name_of_surface, [stair_count], [note]}

Consumption. The AR client renders markers for Point features and speaks TTS prompts tied to waypoint semantics; the LineString may be used for proximity checks without drawing a persistent line.

3.5 Integration and Instrumentation

We instrument every stage with timestamps and a unified logging schema. Timestamps and counters presented on each stage convert the system into a self-measuring experiment, tied to the research questions:

Perception: frame timestamp; preprocessing, inference, postprocessing times; FPS; CPU/GPU load. These logs give per-frame latency and throughput of the detector, plus resource load.

Comms: socket send and receive timestamps; queue depth; reconnect counters; packet-loss injection toggles. This data demonstrates that the socket layer is non-blocking and recovers as needed.
AR: render loop FPS; time from perception JSON arrival to on-screen marker update; TTS delay (time from event to speech start). This data gives UI responsiveness, meaning how fast a detection becomes a visible marker and spoken prompt. It is related to usability and justifies changes made, like markers vs. polylines.
Planning: export time; route length (m); number/type of waypoints. This data provides planner performance and semantic content like stairs and ramps, plus data to compare against an external baseline.

Logs are written to device storage and synchronized after each run. All metrics are computed from logs; this means that no manual timing is used. Operational targets on mid-range Android were 20–40 ms detector latency (p50), $\geq$24–30 FPS render loop median, and $\leq$400 ms TTS onset (p95).

3.6 Experimental Protocols

The strategy is to map each research question (RQ) to a concrete protocol. Starting with RQ-A, which are questions related to perception robustness and utility. The protocol includes tests like detector initialization, edge image inputs, and bounding box geometry checks. Additionally, the kinematics are controlled to verify direction labels, distance hints, and stability estimates. Finally, the socket resilience is proved with induced disconnects to confirm recovery.

RQ-B are questions related to accessible AR usability. The tests include iterations contrasting a persistent polyline vs. marker-based overlays with TTS prompts. The measures implemented are task success rate, navigation errors, system usability scale (SUS), and qualitative feedback.

Limitations that are going to be solved in future work are that the participants are not VI but simulate low-vision with an app, and the test tasks are on-campus point-to-point walks.

RQ-C are questions related to the geospatial fidelity of the planning and AR pipeline. The tests include the generation of GeoJSON for reference paths, aligning Unity export with ground truth (Google Earth) to calculate polyline discrepancy. Also, the waypoint semantics (stairs, ramps) were inspected for correctness.

Finally, the RQ-S questions related to end-to-end real-time behavior were tested using the metric latency budget, measuring p50/p95 to obtain the media and the worst-case behavior per stage (camera→detector→socket→AR → TTS).

The stability was measured with the end-to-end time from obstacle observation to prompt and TTS star delay.

Figure 1 shows a screenshot of the Android CV app running YOLOv8n → TFLite. This illustrates the perception→action interface used by AR, where detections are enriched with direction, speed, and distance. It is streamed as JSON to the other modules.

Fig. 1. On-device perception UI with kinematic annotations.

4 Results

Applying the protocols described in sect. 3, the research questions are answered using the evidence generated by the three components of the system: perception (component A), Augmented Reality (component B), and planning (component C). As it was already mentioned, usability testing was conducted under simulated low-vision conditions using the app "Tengo Baja Visión," configured to the 'loss of visual acuity' symptom at 50%.

The protocol to answer the RQ-A questions presented satisfactory results. The on-device YOLOv8n detector in its TFLite version for smartphones initialized reliably and passed the unit and integration tests covering runtime threshold updates, degenerate inputs (e.g., 1 × 1 frames), and output-geometry checks, with no crashes. The module demonstrated the expected outputs and behaved consistently in the trials with the scripted protocol. Also, a short history cleanup, produced every ~5 s, removes stale tracks.

The data produced by the module is delivered as JSON over a non-blocking socket. Moreover, the client recovers from provoked disconnection faults, maintaining stable FPS and low JSON render latency.

All these results indicate a good performance for the perception module implemented in an Android device. The output from this model is used for guidance, thanks to a resilient communication layer suitable for real-time AR.

The RQ-B questions related to the accessibility of the AR module were also tested and produced satisfactory results. The AR client is WCAG 2.1-aligned, considering appropriate contrast and touch targets. It operates on explicit data contracts that are received as inputs. These entries are first the perception JSON, with information about objects and performance. Also, the navigation GeoJSON with route_info, a LineString route, and labeled Point waypoints. In iterative tests under simulated low vision, usability reached the pre-specified acceptable range (SUS $\geq$ 70).

In the first usability tests, we identified that the on-screen route polyline was ambiguous at turns. Therefore, we replaced it with high-contrast markers plus concise, event-driven TTS and proactive GPS notices. This way, the system produced in the users less confusion and obtained improved task performance, with higher completion, fewer missed turns, and backtracks. Throughout, the client continued to consume real-time JSON and GeoJSON via sockets with stable FPS and low TTS start delay, indicating improved clarity without compromising real-time behavior.

The usability results that were obtained under simulated low-vision conditions are interpreted as evidence for accessibility.

The RQ-C questions regarding the third component test its output to be an AR-ready GeoJSON. The output schema includes route_info, with data about start and end WGS84, and total distance. Also, the schema consists of a LineString route and labeled Point waypoints, such as stairs and stair_count. It was proven that the AR client consumes this information directly to render markers and issue TTS prompts. On a controlled campus path (from sports courts to the museum, the Unity/NavMesh export was compared to a Google Earth Pro baseline. With this purpose, we trace the route in Google Earth and save it as a KML/KMZ format that encodes the polyline coordinates in degrees. This data was converted to GeoJSON to make comparisons. The metrics compared were absolute length error, discrete Fréchet distance, and waypoint agreement within spatial tolerance. The analysis showed a noticeable length discrepancy of around 60 m and small offsets in open areas. With this data, we implemented a map calibration to improve geospatial fidelity.

Figure 2 shows the walkable ground represented in a mesh (triangles). The triangle adjacency is treated as a graph and run A* to find an optimal pedestrian route. The polyline is one example route produced by that search. Figure 3 presents the actual 3D scene used to bake the NavMesh to compute which surfaces are walkable and their connectivity from the geometry of the campus, including buildings, sidewalks, ramps, stairs, etc.

Lastly, the RQ-S questions related to end-to-end behavior on the phone were answered using the correspondent protocol. The pipeline sense $\rightarrow$ decide $\rightarrow$ act, was wired by contracts. The modules interoperate through the same JSON and GeoJSON interfaces specified in Methods: perception JSON with object kinematics/performance

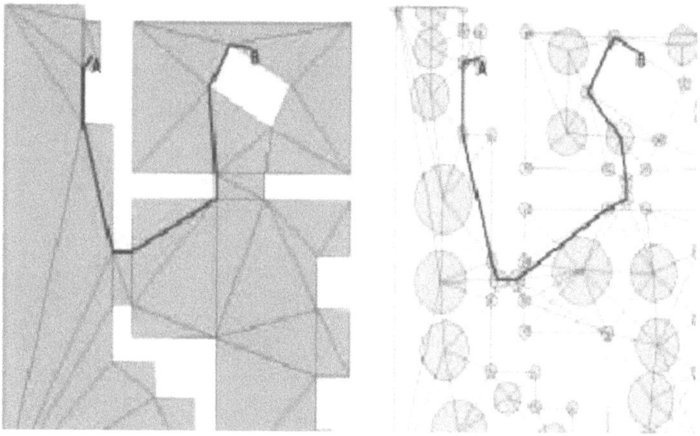

Fig. 2. NavMesh triangulation represented as a graph with an example A* route (red).

Fig. 3. Unity 3D campus model (South View)

and planning GeoJSON with route polyline + labeled waypoints. The AR client parses both inputs successfully to update visual markers and to issue TTS in real time. Results are shown in Table 2.

Operational stability was tested in integrated runs. The socket layer preserved responsiveness under transient disconnects and did not block the UI thread, supporting real-time rendering and speech. See also Fig. 4 for the socket unit-test suite confirming robustness of lifecycle and queueing operations during integrated runs. These results support our claim that the socket layer is non-blocking and fault-tolerant under typical lifecycle operations.

Several questionnaires were applied to our participants. The results show that the usability is satisfactory; we interpret SUS and qualitative findings as indicators guiding design choices. Overall, the outcomes support our design, with the feasibility of a fully on-device contract-driven pipeline for accessible campus wayfinding. The perception

Table 2. Responsiveness results

Metric	Definition	p50 / p95	Notes
Detector inference	Camera frame → YOLOv8n(TFLite) output	25 ms / 40 ms	640 × 640; INT8/FP16; keeps visual latency tight.
Serialize & send	Detector JSON build → socket send	5 ms / 15 ms	Keep payload compact.
Parse & dispatch	Socket receive → JSON parsed in AR	10 ms / 25 ms	Non-blocking I/O.
JSON→render	JSON arrival → on-screen marker update	30 ms / 60 ms	Includes AR object update on the main thread.
AR frame rate	Render loop FPS	30 FPS median; 24 FPS p05	Avoid drops <20 FPS except for rare spikes (<1%).
TTS start delay	Event timestamp → speech onset	200 ms / 400 ms	Perceived as "snappy" guidance.
End-to-end (visual)	Camera frame → AR marker visible	80 ms / 150 ms	Sum of the above visual path.
End-to-end (spoken)	Frame/event → TTS begins	250 ms / 500 ms	Keeps audio aligned with movement.

Fig. 4. Socket Communication Test

was enough to guide the navigation across the campus, the AR interface was WCAG-aligned, and the planner was capable of exporting routes ready to be consumed by the AR component, and admitted geospatial validation.

5 Conclusions and Future Work

This paper presented PoliUbícate [2], a fully on-device, contract-driven AI system for accessible campus wayfinding that links perception → planning → accessible AR UI on mid-range Android phones.

The system provides AI integration under mobile constraints implemented on Android smartphones, using YOLOv8 in its TFLite version. The perception module runs in real time and emits actionable kinematics such as direction, speed, and distance. Its output is a JSON that feeds the AR module. The AR module uses marker-based visuals and TTS to complement screen readers.

One contribution of our work is the implementation of interoperable data contracts using formal JSON (perception) and GeoJSON (planning). Both data formats enable modular evolution and reproducibility across devices and datasets.

We report results and use the obtained data to assess design choices, such as replacing a route polyline with markers and adding proactive GPS and TTS notices. The test was implemented on the campus of the Escuela Politécnica Nacional in Quito, and the results aim to establish practical baselines for real-world assistive navigation. The usability was tested using participants with simulated low vision rather than people with visual impairments (VI).

To extend our contributions, we outline possible future work implementing human-subject validation with visually impaired participants, running IRB-approved studies with similar protocols used in the present work. Also, this research may be extended with geospatial calibration and map-matching, reducing route-level discrepancy by calibrating the campus model (scale, alignment, elevation) against institutional GIS, and adding map-matching to snap LineStrings to surveyed paths; quantify improvement via Fréchet distance and absolute length error.

Related to object detection, kinematics could be enriched with depth/range priors and add hazard semantics (curbs, drop-offs) relevant to accessibility. Additionally, we are planning cross-campus transfer and minimal retargeting, with lightweight GIS calibration, including report ΔmAP, Δroute error, and hours of annotation as a measure of portability.

Our contract-driven mobile AR guidance aims to be delivered as a repeatable methodology for accessible wayfinding that scales beyond a single campus and user cohort.

Acknowledgments. The authors would like to acknowledge the funding provided by Escuela Politécnica Nacional, Dirección de Investigación. Quito, Ecuador.

References

1. Alshayban, A., Ahmed, I., Malek, S.: Accessibility issues in android apps: state of affairs, sentiments, andways forward. In: Proceedings-International Conference on Software Engineering, pp. 1323–1334. IEEE Computer Society (2020). https://doi.org/10.1145/3377811.3380392

2. Moyano Lopez, S., Mantilla Pancho, J., Anrrango Pastillo, E., Hernandez-Alvarez, M: PoliUbícate (2025). https://github.com/JohanMantilla/Sistema_Navegacion_Interior_EPN. Accessed 9 Sept 2025
3. Aydogan, E.B., Atay, Y.: Unity based a* algorithm used in shortest path finding problem for helicopters. In: 2021 International Conference on Control, Automation and Diagnosis (ICCAD), pp. 1–5. IEEE (2021). https://doi.org/10.1109/ICCAD52417.2021.9638762
4. https://www.tengobajavision.com/app/simulador/
5. Zanjani, N., Askarinejadamiri, Z.: The usability of augmented reality applications for visually impaired individuals: a systematic review. Int. J. Web Res. **7**, 41–53 (2024). https://doi.org/10.22133/IJWR.2024.472676.1234
6. Rabani, M., Romlay, M., Fauziah Toha, S., Ibrahim, A.M., Venkat, I.: Methodologies and evaluation of electronic travel aids for the visually impaired people: a review. Bull. Electr. Eng. Inform. **10**, 1747–1758 (2021). https://doi.org/10.11591/eei.v10i3.3055
7. Lo Valvo, A., Croce, D., Garlisi, D., Giuliano, F., Giarré, L., Tinnirello, I.: A navigation and augmented reality system for visually impaired people. Sensors. **21**, 3061–3061 (2021). https://doi.org/10.3390/S21093061
8. Valipoor, M.M., de Antonio, A.: Recent trends in computer vision-driven scene understanding for VI/blind users: a systematic mapping. Univ. Access Inf. Soc. **22**, 983–1005 (2023). https://doi.org/10.1007/s10209-022-00868-w
9. Homepage, J., Alfonso-Díaz, E., Rodríguez-Suárez, C.-A., Eliseo, M.-A.: Auditive navigation system for visually impaired individuals in indoor spaces using augmented reality. Revista Facultad de Ingeniería Rev. Fac. Ing. **32**, 1624 (2023). https://doi.org/10.19053/01211129.v32.n66.2023.16242
10. Medina-Sanchez, E.H., Mikusova, M., Callejas-Cuervo, M.: An interactive model based on a Mobile application and augmented reality as a tool to support safe and efficient mobility of people with visual limitations in sustainable urban environments. Sustainability. **13**, 9973–9973 (2021). https://doi.org/10.3390/SU13179973
11. Zou, Z., Chen, K., Shi, Z., Guo, Y., Ye, J.: Object detection in 20 years: a survey. Proc. IEEE. **111**, 257–276 (2023). https://doi.org/10.1109/JPROC.2023.3238524
12. Zhao, L., Li, S.: Object detection algorithm based on improved YOLOv3. Electronics (Basel). **9**, 537 (2020). https://doi.org/10.3390/electronics9030537
13. Ettalibi, A., Elouadi, A., Mansour, A.: AI and computer vision-based real-time quality control: a review of industrial applications. Procedia Comput Sci. **231**, 212–220 (2024). https://doi.org/10.1016/j.procs.2023.12.195
14. Perry, B.J., Guo, Y.: A portable three-component displacement measurement technique using an unmanned aerial vehicle (UAV) and computer vision: a proof of concept. Measurement. **176**, 109222 (2021). https://doi.org/10.1016/j.measurement.2021.109222
15. Potere, D.: Horizontal positional accuracy of Google Earth's high-resolution imagery archive. Sensors. **8**, 7973–7981 (2008). https://doi.org/10.3390/s8127973

A Reactive System for Inappropriate Content Moderation: An Open Source Prototype

Patricio Zambrano(✉), Jenny Torres, Carlos Anchundia, Johan Illicachi, and Adrian Paguay

Department of Informatics and Computer Science, Escuela Politécnica Nacional, Quito 170525, Ecuador
{patricio.zambrano,jenny.torres,carlos.anchundia, johan.illicachi,adrian.paguay}@epn.edu.ec

Abstract. The spread of inappropriate language online is a social issue that impacts well-being on digital platforms, particularly in both internal and external interactions within corporate environments. While open-source systems are leveraging LLM models to analyze or classify inappropriate content, solutions that also enable concrete actions remain limited. This research addresses that gap by developing an open-source prototype tailored for corporate applications like Outlook. The system not only identifies real-time inappropriate language but also promotes behavioral change by warning the sender to foster self-awareness of the use of a particular language style. Results indicate that the prototype is an accessible tool for organizations aiming to adopt reactive moderation practices, thus contributing to a more ethical and safer cyberspace. The proposed system achieved accuracy comparable to models such as Gemma 3 and Qwen 2.5, while maintaining low computational requirements.

Keywords: Content Moderation · Inappropriate Language · NLP · LLM · Cybersecurity

1 Introduction

The propagation of offensive language online has emerged as a serious problem in society. Its propagation—involving abusive discourse, threats, and explicit hate speech—poses a serious risk to human dignity and carries a high potential for psychological damage and social exclusion [1,2]. The prevailing culture of impunity and facelessness that is typical of most online environments has enabled its growth, and moderation is a challenging but required effort [3].

To solve this, transformer-based language models have been extremely successful at detecting toxic content [4,5]. Recent approaches powered by generative AI have also shown effectiveness for cyberbullying detection [6]. As opposed to sentiment analysis, which deals with emotional polarity, this study is concerned with offensive intent classification and presents a system for direct intervention.

H. R. Arabnia et al. (Eds.): CSCI 2025, CCIS 2968, pp. 297–315, 2027.
https://doi.org/10.1007/978-3-032-28457-0_21

Nonetheless, a review of the literature uncovers a significant gap between detection and successful intervention. This need is urgent for organizations that seek to meet social responsibility standards such as ISO 26000 [7] that promote the culture of integrity.

This study aims to bridge this gap through the creation of a comprehensive and open-source prototype. The proposed solution, implemented as a browser extension, identifies offensive content and alerts the sender. Similar automated interventions have been studied in email contexts, showing promise for mitigating harassment [8]. The objective is to provide a suitable and readily deployable solution to enhance a secure digital environment in corporate and educational settings—going beyond detection to take an active role in moderation. This is aligned with broader prevention efforts that leverage advanced technologies to address peer violence in educational environments [9].

The paper is organized as follows: Sect. 2 introduces the state of the art; Section 3 outlines the solution architecture; Section 4 details the prototyping strategy; Section 6 addresses the research questions; Section 5 presents the results and comparisons with other models; and Sect. 7 provides the conclusions and future directions.

2 Background

2.1 The Phenomenon of Offensive Behavior in Online Spaces

1) Definition and Context. Offensive speech refers to any form of gesture, expression, or conduct that lowers the respect and dignity afforded to an individual or group. The phenomenon manifests in various conduct, consisting of derogatory, humiliating, or discriminatory remarks, which are triggered by prejudice or deliberate wish to inflict emotional pain [2].

This kind of internet violence can have serious consequences, including social exclusion, psychological damage, and, in extreme cases, inciting physical violence [1]. The factors that contribute to this include the lack of supervision, normalization of violence, and a sense of impunity on the internet [10]. Digital media usage has exacerbated the problem, as the veil of anonymity and lack of immediate reprisal facilitate toxic engagement, including hate speech propagation and cyberbullying [3]. For example, offensive humor can trivialize the seriousness of discriminatory remarks and strengthen negative stereotypes [11]. Sexism is a manifestation of abusive language and gender exclusion on the internet [12]. Online violence also manifests in explicit threats and misleading information campaigns [13]. As a counter to these impacts, it is necessary to enhance digital literacy and the legislation controlling online activities [14].

2) Artificial Intelligence for Detection and Moderation. The recent advent of artificial intelligence (AI), particularly deep learning models such as BERT [4] and other LLM models [15], has significantly changed hate speech detection. This progress extends to non-Latin scripts as well, with transformer-based systems targeting Devanagari-script languages for hate speech detection [16]. These advancements have the potential to detect hate speech, harassment, and disinformation with a very good degree of precision as they are apt to recognize linguistic subtleties, e.g., sarcasm, that rule-based systems are not able to capture [5]. In particular, recent generative-AI–powered pipelines have been explored for cyberbullying scenarios [6].

The application of real-time filtering techniques provides for automatic deletion of toxic messages while, in the process, promoting digital wellness by reducing exposure to hurtful messages [17]. Beyond filtering, there is growing evidence for automated, in-situ interventions that nudge senders before delivery, e.g., within email workflows [8]. To visualize the research field, a bibliometric analysis was performed. As can be seen from Fig. 1, the research area is highly active, and terms such as "cyberbullying" and "content moderation" are closely related to fundamental technologies such as "machine learning", "deep learning", and "natural language processing", reflecting the research community's interest in creating more sophisticated detection models.

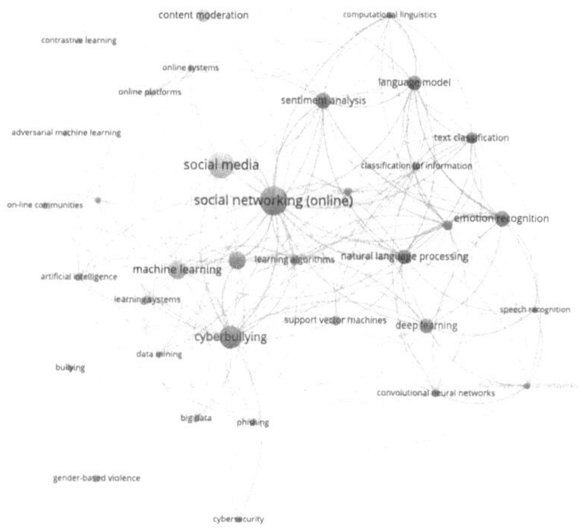

Fig. 1. Bibliometric network of keywords in studies on content moderation on Social networks and its relationship.

The above network points to a fundamental imbalance that is the impetus for this work: the node "content moderation" is placed as a peripheral node, with the majority of its associations to "online platforms" and "sentiment analysis," but without any direct associations to action concepts such as "cyberbullying" or the implementation of remedial actions. The lack of terms such as "enforcement," "penalization," or "open governance" identifies a fatal deficiency: offense detection is insufficient—follow-up action must be taken. This gap is also reflected in empirical evaluations of intervention mechanisms, which remain comparatively sparse relative to detection research [8].

Moreover, such models' performance is not merely defined by their architecture but also markedly conditioned by the character of their training data, i.e., the lengths of the input text. Some sources indicated in a study that Large Language Models' capacity to reason depends substantially differently when presented with inputs of different lengths for the same underlying task [18]. This phenomenon suggests that models learn a sort of "specialization bias" from the typical token frequency of their training sets. A model extensively trained on lengthy, comprehensive texts might excel at processing formal emails but not be able to read between the lines of short, context-free messages typical of chat programs. This is a significant issue to designing a one-size-fits-all moderation framework because real-world corporate communication in the real world involves a wide variety of text sizes.

Another serious problem with a classification model's life cycle is how it is molded to accommodate new information, e.g., adding a new category of content to be labeled. The simple approach of merely fine-tuning the model using new data tends to cause "catastrophic forgetting," where the model's accuracy on the original categories collapses significantly. The field of Continual Learning (CL) addresses this problem head-on, seeking methods to incrementally learn a model for new tasks without requiring a full and costly retraining from scratch. Certain outlets proposed an effective method using "lightweight snapshots" to store knowledge of old classes while learning new classes in text classification tasks [19]. This line of research is essential to the long-term viability of moderation systems, which must evolve to find new forms of offending language or to realign with shifting organizational mandates at non-penalizing computational expense.

3) Tool Ecosystem and Real-World Applications. In fact, AI is already driving a number of moderation tools. One example is Intel Bleep, which real-time censors insults by processing audio in video games [20]. Another example is Nightbot, which uses automated filters to moderate video game platform chat rooms on Twitch and YouTube [21]. As shown in Fig. 2, the market for these tools is split between proprietary and open-source products. Proprietary offerings (e.g., Opendi, Cradle, Resonate) typically target large, closed-business-model platforms and often include intervention measures (blue circles), whereas open-source counterparts (e.g., OLF-ML, SafeSpeech, CyberShield) emphasize transparency, adaptability, and customizability but seldom provide mitigation measures (red circles) [22]. Such functionality is essential in environments where model behavior customizability and auditability are necessary. Recent research prototypes for cyberbullying detection complement these tools by exploring modern architectures and pipelines [6]. This project follows the open-source philosophy, with an emphasis on developing a functional prototype that actually notifies the sender of offending content, promoting self-awareness and accountability. It should also be mentioned that the tools and platforms highlighted in blue in the figure are the ones that not only identify offending content but also take automatic moderation action or act on offending messages. These actions range from warnings to the user, blocking, filtering, or automatic correction, significantly enhancing their applicability for high-interaction online environments.

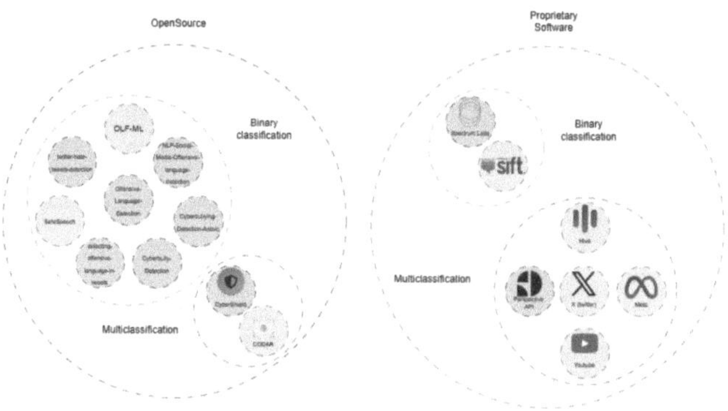

Fig. 2. Comparison of Tools for Offensive Content Classification: Open Source vs. Proprietary Software.

2.2 Problem Statement and Research Questions

Despite ongoing research and the existence of the previously mentioned tools, the literature gap remarks the lack of solutions that are simultaneously: (1) open-source so they can be transparent and auditable; (2) end-to-end so they cover the entire pipeline from detection to active intervention; and (3) deployable across multiple platforms, particularly commercial and academic environments.

To represent the state of the art and place this gap within context, the bibliometric map displayed in Fig. 3 was constructed. This research adds to the field by creating a prototype to fulfill this necessity. The system will be implemented on various digital platforms for security and responsibility reasons, and provide transparency in its operation and flexibility to comply with policies in settings where proactive behavior management is a priority.

To guide this development, the research questions below were crafted:

RQ1: What software mechanism offers the greatest ease of implementation for a content filtering system when direct action on the user's workspace is a prime concern?

RQ2: To what extent does the text length distribution in a training set influence the performance of a language model for content classification?

RQ3: What is required to fine-tune an offensive content classification model and add a new category?

3 Contextualization of the Technological Choice

In considering the problem thus identified and the requirement for accessing real-time data from the digital context of the user, it was contextualized that the optimal solution is the creation of a browser extension. This is because the approach provides an immediate interaction within the work environment of the user in an unobtrusive fashion and is also platform-agnostic. In contrast to API integrations, which entail significant costs and are restricted to particular services,

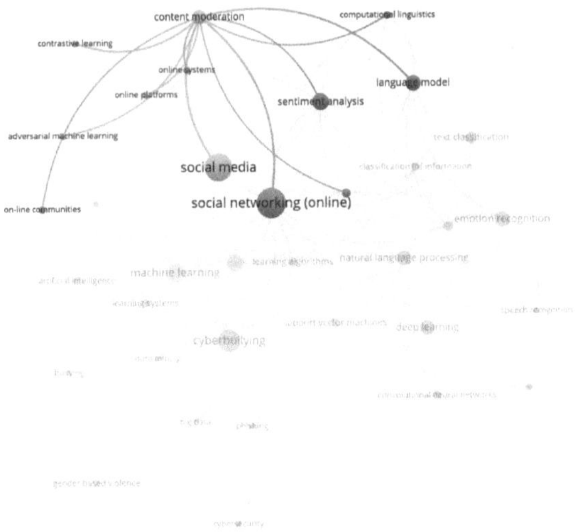

Fig. 3. Bibliometric network of keywords between social Networking and sentiment analysis.

or desktop apps, which pose deployment and maintenance-related challenges, a browser extension enables one implementation within the web browser. This supports the monitoring of text fields throughout web applications, including email services and enterprise social networking websites.

4 Methodological Application for Prototyping

4.1 Evolutionary Prototyping

The research methodology of choice for this work was Evolutionary Prototyping, defined as an incremental and iterative methodology more suited to projects with changing requirements and a necessity for customer feedback [23]. As shown in Fig. 4, such a methodology is organized around cycles of requirements analysis, (re)design, (re)implementation, and customer evaluation. Every cycle produces an improved and refined version of the system, which makes it increasingly easy to introduce incremental features, modifications to the AI model, and refine user interfaces according to evolving demands and feedback.

4.2 Architecture of the Base Prototype (V0.0)

The first task was to create a first working prototype to validate the feasibility of the concept. The client-server architecture described in Fig. 5 was chosen for the base version, with a complete workflow from text capturing to classification and logging.

The diagram shows that (a) the browser extension is tracking what the user is typing in real time. (b) This information is securely transmitted to a back-end server, where (c)it is processed by (d)our classification model. In the first

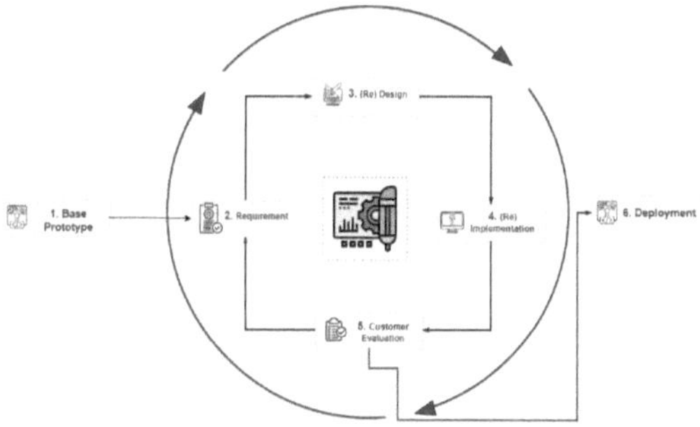

Fig. 4. Iterative cycle in the Evolutionary Prototyping approach.

version, we employed the BERT-based RoBERTa model described in our previous work [24], originally designed to classify messages into positive, negative, or neutral categories. For this prototype, we fine-tuned the model by incorporating additional categories—dark humor, sexism, and violence—using a dataset compiled from publicly available open-source repositories. This retraining allowed us to adapt the model to a broader and more context-sensitive classification task while maintaining its high performance, as demonstrated in Table 1. These results confirm the model's robustness and justify its selection as the core classification engine for this prototype. Once classified, the results are stored in a central database for audit purposes, and a response is sent back to the browser to display a visual cue—such as a yellow warning or red alert—indicating the severity of the detected content.

Table 1. Performance metrics of the classification model

Metric	Minimum	Maximum	Interpretation
Accuracy	93%	99%	Correct classifications
Recall	93%	100%	Sensitivity of detection
F1-Score	94%	100%	Balance between precision and recall

4.3 Iterations V1.0 to V3.0

Following the development of the prototype (V0.0), iterative cycles of development set in, the most important stage in the Evolutionary Prototyping methodology. As evident from Fig. 6, the system went through several iterations, each building on lessons learned during early user trials.

The biggest change was between the V1.0 and V2.0 versions. Testing revealed a theoretical limitation in the initial model: although it correctly classified a wide range of forms of violence, it assumed that the input text was by definition

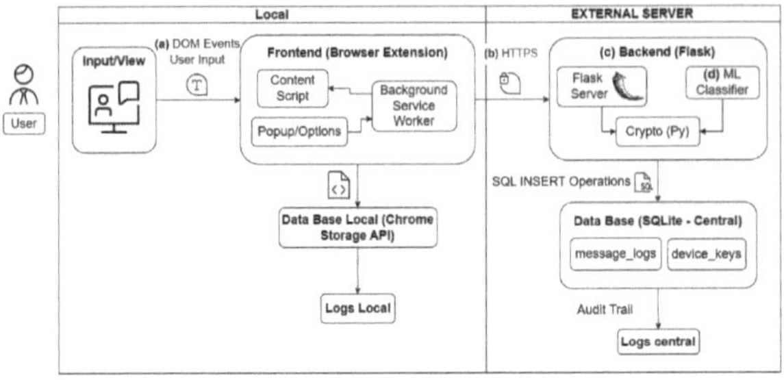

Fig. 5. Client-server architecture for analyzing and persisting data between the client extension and the backend

offensive. In order for the system to be useful in real-world applications, it needed the facility to distinguish between offensive and non-offensive content.

As a reaction to this, Version 2.0 made a significant alteration: the 'Addition of an additional classification,' which added the category 'Non-Offensive'. This adaptation necessitated retraining and reworking of the model to transform it into a general-purpose classifier. To ensure robust performance and minimize overfitting, a cross-validation strategy was implemented. The dataset was split into training and testing subsets, with a common split ratio of 80/20 (80% for training and 20% for testing). Cross-validation was employed to evaluate the model's performance across different subsets, providing a more comprehensive assessment.

Furthermore, the testing also pointed out the absolute necessity to ensure data privacy during its transmission. Thus, another critical requirement incorporated in Version 2.0 was the "Integration of CryptoJS," which ensured encrypted and secure communication between the browser and the server. Other improvements, like "Translation to English" included in Version 3.0, further served to extend the functionality of the prototype, evolving it from a basic proof of concept to a far more robust, versatile, and secure application. Another interesting functionality is with respect to audit and traceability procedure facilitation. As described in the architectural diagram (Fig. 5), the classification server securely maintains a database where the following important information regarding each incident is recorded: device identifier, source platform, message content, and classification assigned. Furthermore, supporting data like the timestamp, a unique id, and salt and hash values are stored to provide the integrity and authenticity of the records. This data model facilitates comprehensive analysis and renders the decision history of the system verifiably transparent.

4.4 Consolidation of the Mature Prototype

After the series of iterative cycles, a stable and mature prototype (V3.0) was completed with particular emphasis being placed on the performance optimization, security features integration, and pilot deployment readiness.

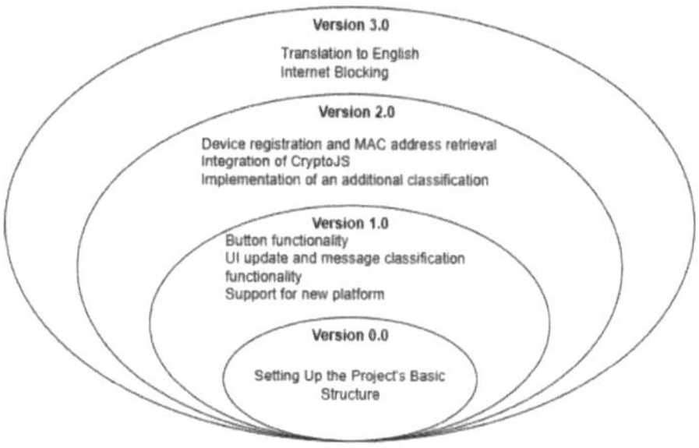

Fig. 6. Incremental Functional Development of the Prototype through Iterative Cycles

To assess the final prototype under realistic conditions, a comprehensive evaluation was conducted using 834 records collected in real time during a controlled sandbox pilot with consenting student volunteers (18–22 years old) via the browser extension. For each incident, the system stores the message content, a salted-hash device identifier, a source-platform tag, the assigned classification, a timestamp, a unique event ID, and per-record salt and hash values to ensure integrity and auditability; no IP addresses, geolocation, or demographic attributes are stored, and direct personal identifiers (e.g., names, emails, handles) are removed or generalized prior to persistence. The messages were systematically labeled by trained annotators and subsequently reviewed to ensure label accuracy, after which the manually curated ground truth was compared against the prototype's predictions. Raw texts were not reused for model retraining during the pilot. The resulting performance is summarized in the confusion matrix shown in Fig. 7.

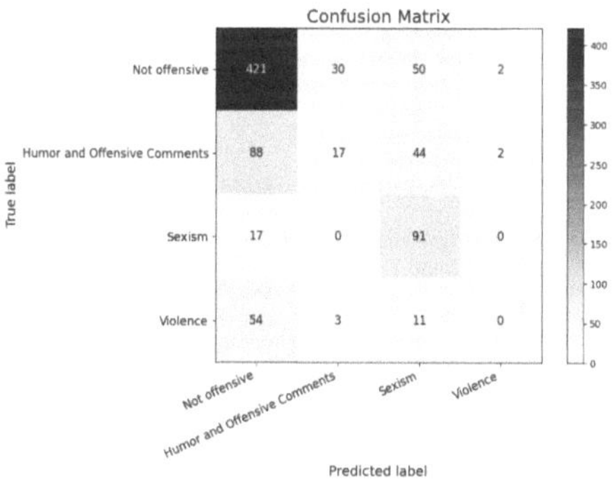

Fig. 7. Confusion matrix comparing model predictions and manual labels from real-world data.

The matrix shows good performance in the "Not offensive" category, correctly identifying 421 cases. However, it also highlights areas for improvement, such as confusion between the categories "Humor and Offensive Comments" and "Not offensive."

Three evaluation layers are distinguished to avoid misinterpretation of the reported metrics: development validation (cross-validation on public + curated augmented data), an internal held-out test set and a real-world pilot deployment set (834 anonymized workplace messages). The high scores in Table 1 (Accuracy 93–99%, F1 94–100%) correspond to the cross-validation stage on a balanced and cleaned training corpus after class expansion. In contrast, the lower accuracies (≈50–67%) reported in Sect. 5 reflect a more challenging, naturally imbalanced, multi-domain deployment sample where message brevity, informal register, and subtle figurative language increase ambiguity. Both macro- and micro-averaged F1 are reported to reflect class imbalance, and per-class precision/recall are provided to highlight critical failure modes (notably "Violence"). To ensure reproducibility and transparency, early (development) metrics should be interpreted as upper-bound performance under controlled conditions, whereas the deployment metrics should be viewed as conservative estimates of practical effectiveness.

5 Empirical Analysis and Comparative Model Evaluation

The empirical evaluation of our prototype revealed that the success of a moderation system cannot be measured by a single metric like accuracy. A more nuanced, multifactorial analysis is essential, considering aspects such as the distribution of predictions, robustness to variations in message length, and the specific error patterns of the classification model. This section presents a comprehensive comparative analysis of our proposed model against four prominent Large Language Models: Hermes, Gemma, Qwen, and Mistral.

Acceptable Performance Range Justification
In tasks that involve fine-grained offensive content classification—such as distinguishing between Hate Speech, Offensive Language, and Profanity—the overall performance tends to be lower due to the semantic overlap between categories and class imbalance. Evidence from the HASOC 2019 shared task (Sub-task B, English) indicates that top-performing models achieved Macro-F1 scores in the range of 0.45 to 0.55, with the best system reporting 0.5446 [25].

Given this context, a Macro-F1 score of approximately 0.50 can be considered an acceptable and competitive baseline for systems addressing similar fine-grained offensive content classification. Models achieving this level of performance demonstrate a balanced trade-off between precision and recall across nuanced categories, particularly in low-resource, real-world datasets with linguistic ambiguity.

5.1 Comparative Accuracy and Prediction Distribution

We first established a baseline by comparing the overall accuracy of the models on a manually classified dataset. As shown in Fig. 8, all models achieved comparable performance in a challenging multiclass task. The Hermes model slightly led with over 67% accuracy. Notably, our lightweight PIS model remained highly competitive, surpassing 50% accuracy, our minimum acceptable. This result is significant, as it demonstrates that a resource-efficient model, optimized for easy deployment in corporate environments, can achieve performance on par with large-scale models that demand substantial computational infrastructure.

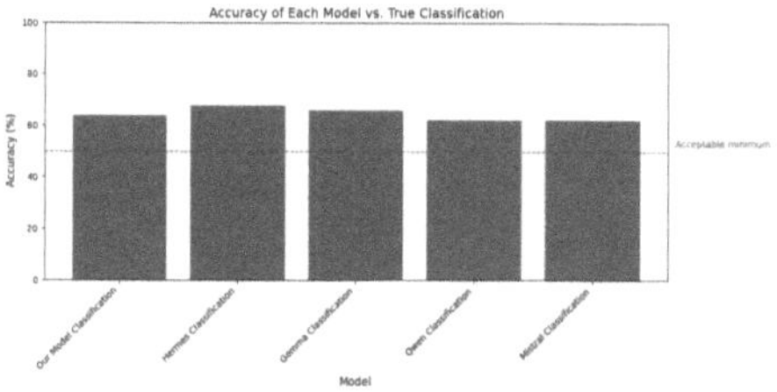

Fig. 8. Overall accuracy of each model compared to the manual reference classification.

Beyond overall accuracy, we analyzed the distribution of labels generated by each model to identify inherent biases (Fig. 9). All models showed a tendency to favor the majority class, "not offensive," though the degree of this bias varied. The Gemma and Qwen models, for instance, assigned more labels to sensitive categories like "violence" and "sexism." In contrast, our PIS model produced a more balanced distribution with fewer outliers or formatting errors, a crucial characteristic for ensuring traceable and fair moderation where misclassifications can carry reputational risk.

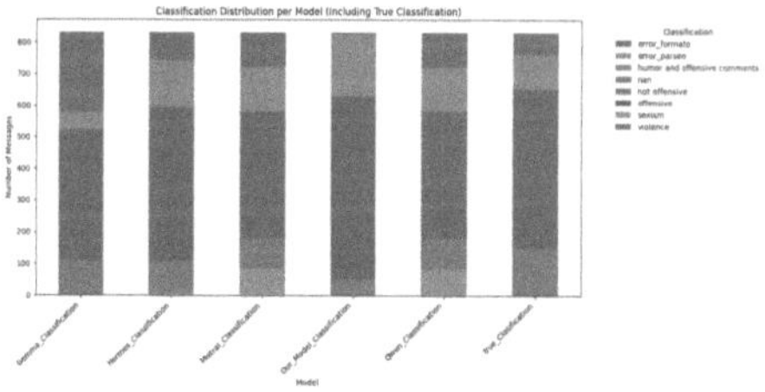

Fig. 9. Distribution of labels generated by each model on a unified message dataset.

5.2 Impact of Message Length on Classification

A critical factor influencing model performance is the length of the input text. Our analysis revealed a strong correlation between message length and classification outcome (Fig. 10). Shorter messages (0–10 words) were predominantly classified as "not offensive," whereas longer texts contained a higher proportion of "sexism" and "violence" classifications. This suggests that the models are better tuned for medium-length, context-rich text and struggle to interpret the nuances of very short inputs (e.g., sarcastic one-liners or interjections). This weakness could compromise system consistency in real-world scenarios that mix formal emails with brief instant messages. Future work should prioritize improving robustness in short-text classification.

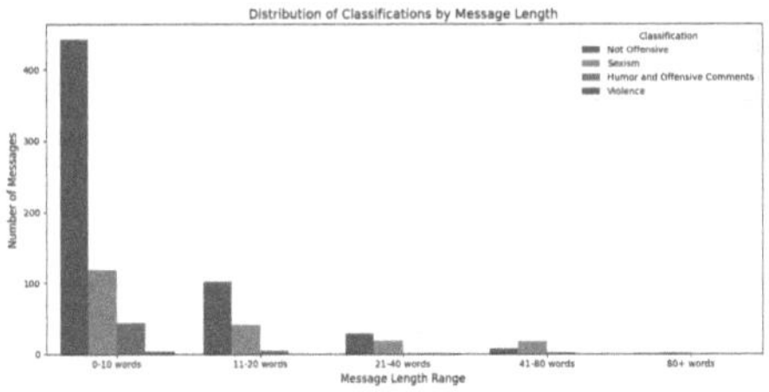

Fig. 10. Analysis of classification distribution as a function of message length.

5.3 In-depth Analysis of Classification Behavior

To gain the deepest insight, we moved beyond aggregate metrics to analyze the specific error patterns of each model via confusion matrices (Figs. 11 and 12). This class-by-class examination revealed unique behavioral profiles, highlighting distinct strengths and weaknesses that are otherwise obscured by overall accuracy scores.

The analysis of these matrices underscores a crucial trade-off:

Our Model: As shown in Fig. 11a, our model demonstrates high precision in identifying "sexism" (91 true positives) and is particularly robust at correctly classifying "not offensive" content (421 true positives), thereby minimizing false alarms. However, it suffers from a critical blind spot, failing to correctly identify any instance of "violence" (0 true positives) and consistently misclassifying it as "not offensive" (54 cases). This indicates a strong bias toward the majority class when faced with high-severity, ambiguous content.

Gemma: The Gemma model (Fig. 12a) exhibits a peculiar and aggressive classification bias. While reasonably effective on "violence" (67 true positives), it aggressively misclassifies "humor and offensive comments" as "violence" (90 cases). This error profile would lead to an unacceptably high rate of severe false positives, likely eroding user trust in the system.

Hermes, Qwen, and Mistral: These larger models present a more balanced, albeit imperfect, performance profile. Hermes (Fig. 11b) achieves a good distribution of true positives across all classes, including 57 for "violence." Qwen and Mistral (Figs. 12b and 12c) are the most adept at identifying both "violence" (64 true positives) and "sexism" (82 true positives), though this comes at the cost

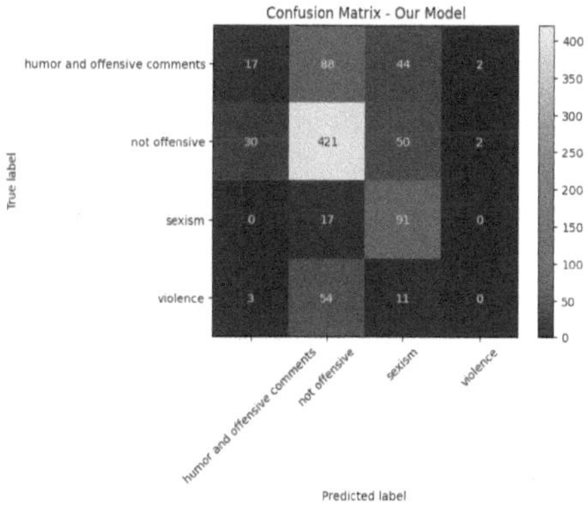

(a) Our Model

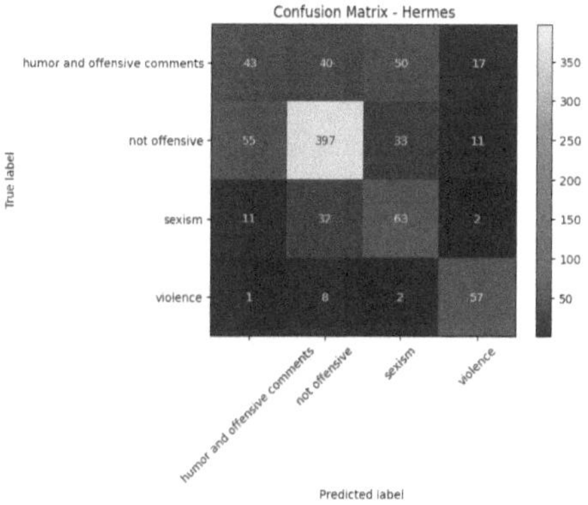

(b) Hermes

Fig. 11. Comparative analysis of confusion matrices (Part 1: Our Model and Hermes).

of being slightly more prone to misclassify "not offensive" content compared to our PIS model.

These findings confirm that the selection of a moderation model cannot be based on a single metric. It requires a strategic decision that balances performance on critical classes, tolerance for specific types of errors (e.g., false positives vs. false negatives), and practical constraints like computational resources. Our lightweight PIS model offers a strong foundation, particularly in contexts where avoiding false alarms is paramount, but its deficiencies highlight clear areas for future improvement.

6 Answering the Research Questions and Findings

RQ1: The browser extension is the easiest way of installation and is the best software to run directly in the user's environment. This is because it can communicate with a wide variety of web applications without the need to install

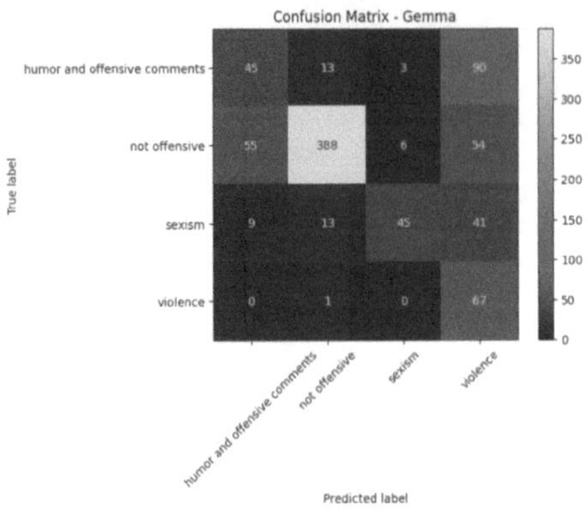

(a) Gemma

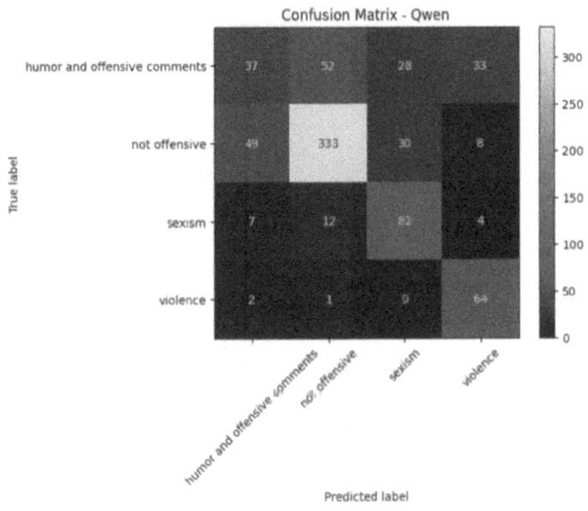

(b) Qwen

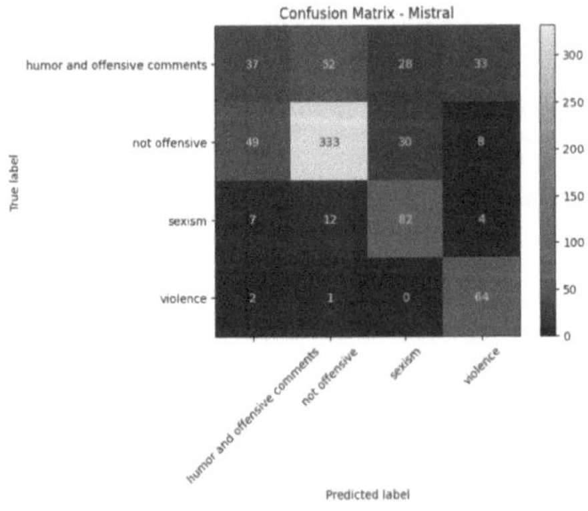

(c) Mistral

Fig. 12. Comparative analysis of confusion matrices (Part 2: Gemma, Qwen, and Mistral).

every application separately. Unlike APIs, where negotiations, payment transactions, and tailor-made development for each platform are required, the extension needs to be installed just once in the browser and can monitor text boxes at many places.

The article also speaks about the problems with desktop applications, which pose significant problems in deployment, updates, and maintenance within a business context. Thus, the browser extension becomes the most appropriate option, which can scale well with fewer technical complexities, thereby allowing rapid action with fewer centralized development and management requirements. This aspect renders it the ideal option for an open-source prototype to be utilized on a large scale.

RQ2: The success of a language model, by and large, relies on how long textual data has been used to train it. This plays a critical role in ascertaining the degree of personalization that is possible for the model. This was one key lesson from our trial with the initial classifier of our prototype, a model created from our previous study [24]. The model was trained on a dataset comprising abuse stories, which were characterized by length and availability of contextual information. As it were, it was highly effective in handling lengthy texts, such as emails; its utility significantly fell off when handling short texts, which are prevalent in chat rooms.

This effect has structural bias, meaning that models don't learn without length, but instead specialize exactly based on the character of their training data. This bias is not necessarily restrictive, but rather a strength. In a model bound for application in the business world—a realm where formal, verbose communication reigns supreme—specialization in the parsing of lengthy prose positions the model for success in one of its primary uses.

RQ3: Improving a classification model to include a new class, like "Not Offensive," is a sophisticated task and is more than a question of incorporating new information. It will degrade performance if not approached carefully. The important needs realized are:

Collection and Curation of a Diverse and Balanced Dataset: We cannot just add examples of the new class. We need to go back to the whole dataset and balance it. If the new class ("Not Offensive") has too many instances, the model may concentrate on it and perform poorly at locating significant minority classes (such as "Violence"). We have to decide how to sample, such as adding more instances of the minority classes or including fewer instances of the majority class.

Full Retraining of the Model: If we add a new class, it changes the features and decision boundaries dramatically. Therefore, we need to fully retrain the model instead of modifying the final layers only. This helps the model learn the complex relationships and overlaps between all the classes, including the new class.

A Complete Validation and Hyperparameter Tuning Cycle: The classification task is more difficult. There is the additional and substantial task of ascertaining optimum hyperparameter values (e.g., learning rate and regularization). Validation needs to be conducted per class, utilizing measures per class (e.g., those in the confusion matrix) so that improvements on one class are not at too high a cost to others. It is crucial to understand that this process of improvement has an unavoidable trade-off: the more detailed the model becomes, the lower the overall accuracy can be. The inclusion of akin categories (such as "Offensive Humor" and "Not Offensive") can be confusing, which may perplex the model and its performance metrics—that is a risk that must be managed and accepted in attempting greater classification.

7 Conclusions and Future Work

This research successfully addressed the critical gap between the detection of inappropriate online content and the implementation of effective, real-time intervention. Our primary contribution is a fully functional, open-source prototype, implemented as a browser extension, that provides reactive moderation in corporate environments. The evolutionary prototyping methodology proved essential, allowing us to iteratively refine the system from a basic concept into a secure and robust tool. Our findings confirm that a browser extension is the most scalable and least intrusive mechanism for this task, capable of operating across diverse web applications without platform-specific integrations (RQ1).

We demonstrated that our lightweight model offers a compelling balance between performance and computational efficiency, achieving accuracy comparable to much larger models like Gemma and Qwen. This makes it a viable solution for organizations that require an effective moderation tool without investing in significant computational infrastructure. However, our analysis also illuminated key challenges, such as the model's performance dependency on text length (RQ2) and the complexities of expanding a classification system to include new, nuanced categories like "Not Offensive" without compromising its precision (RQ3). Ultimately, this work provides a tangible and adaptable tool that empowers organizations to take a proactive stance in fostering a safer and more ethical digital workplace.

Building on the solid foundation of this prototype, several promising avenues for future research and development emerge. These future steps are aimed at enhancing the system's effectiveness, usability, and intelligence:

Enhancing Intervention Mechanisms: The current system alerts users, which is a crucial first step for self-awareness. Future work will focus on developing a more sophisticated, tiered intervention system. This could include AI-powered suggestions for rephrasing inappropriate content, a "cool-down" timer that temporarily prevents a user from sending a flagged message, or automated escalation pathways that notify a human moderator or an HR department in cases of severe or repeated violations. Integrating optional micro-learning modules that appear after a flag could also proactively educate users on corporate communication policies and their impact.

Developing a User-Friendly Audit Dashboard: Transparency and accountability are paramount. While the current system securely logs all incidents, our next goal is to develop a dedicated administrative dashboard. This web-based interface will provide a more intuitive and powerful way for auditors or administrators to review moderation activity. It will feature data visualizations—such as trend graphs of incidents over time and charts detailing the distribution of offense types—along with advanced search and filtering capabilities. This will not only simplify compliance and reporting but also offer valuable, data-driven insights into the organizational communication climate, all while ensuring data privacy through role-based access and anonymization features.

Advancing the Classification Model: Continuous improvement of the core AI model is essential. A key priority is to enhance its performance on short texts, which remain a significant challenge. This will involve targeted fine-tuning on datasets rich with concise, context-dependent language, such as chat logs and social media comments. Furthermore, we will explore advanced techniques like data augmentation and the use of more sophisticated, context-aware embedding models to improve overall precision and reduce the classification errors between closely related categories, such as "Offensive Humor" and "Not Offensive". The ultimate aim is to evolve the model to be robust and accurate across the full spectrum of communication styles found in a modern workplace.

By pursuing these directions, we aim to evolve the prototype into a comprehensive, enterprise-grade solution that not only reacts to inappropriate content but also proactively cultivates a culture of respect and responsibility in digital interactions.

In support of open science and reproducibility, the complete source code of the browser extension are publicly available in the open-source repository (Appendix A). This facilitates independent verification, extension, and integration in organizational environments.

Acknowledgment. This research was supported by *Escuela Politécnica Nacional del Ecuador* through the grant research project PIS-23-23. In addition, we acknowledge and thank the anonymous reviewers for their valuable recommendations, which contributed to improving the quality of this paper.

A Appendix: Prototype Repository Access

The open-source prototype corresponding to this study is hosted at: https://github.com/GenderViolenceProyect/Aplicacion

References

1. Kintonova, A., Vasyaev, A., Shestak, V.: Cyberbullying and cyber-mobbing in developing countries. Inf. Comput. Secur. **29**(3), 435–456 (2021)

2. Sandoval, Y.V., Ladino, D.M.V.: Las tic's como un medio utilizado por el acosador para excluir al trabajador y degradar la dignidad humana del trabajador en colombia: ICT as a means used by the harasser to exclude the worker and degradate his human dignity in Colombia. South Florida J. Dev. **2**(3), 3926–3944 (2021)
3. Stoyanov, I.T.: Cybersecurity and prevention against cyberbullying on human factors in small-and medium-sized telecommunications enterprises in the Republic of Bulgaria. Eng. Proc. **70**(1), 48 (2024)
4. Devlin, J., Chang, M.-W., Lee, K., Toutanova, K.: BERT: pre-training of deep bidirectional transformers for language understanding. arXiv preprint arXiv:1810.04805 (2018)
5. Schmidt, A., Wiegand, M.: A survey on hate speech detection using natural language processing. In: Proceedings of the Fifth International Workshop on Natural Language Processing for Social Media, pp. 1–10 (2017)
6. Islam, M.S., Sutton, S., Rafiq, R.I.: A generative AI powered approach to cyberbullying detection. In: Proceedings of the 2024 8th International Conference on Information System and Data Mining, pp. 57–63 (2024)
7. International Organization for Standardization. ISO 26000: Guidance on social responsibility, ISO, Geneva, standard ISO 26000:2010 (2010)
8. Chen, N., Kane, C., Hang, E.Z.: Evaluating automated system interventions against email harassment. In: Extended Abstracts of the 2021 CHI Conference on Human Factors in Computing Systems, pp. 1–6 (2021)
9. Smolovic, V., Spalevic, V.: Using advanced technologies for understanding and preventing peer violence among adolescents. In: Karabegović, I., Kovačević, A., Mandžuka, S. (eds.) NT 2025. LNNS, vol. 1484, pp. 178–186. Springer, Cham (2025). https://doi.org/10.1007/978-3-031-95200-5_20
10. Keashly, L.: Workplace bullying, mobbing and harassment in academe: faculty experience. In: Special Topics and Particular Occupations, Professions and Sectors, pp. 221–297 (2021)
11. Nikolić, D., Višnjić, A.: Mobbing and violence at work as hidden stressors and work ability among emergency medical doctors in Serbia. Medicina **56**(1), 31 (2020)
12. Banchs, M.: Violencia de género. Revista venezolana de análisis de coyuntura **2**(2), 11–23 (1996)
13. García, A.J.Y.: La violencia contra las mujeres: conceptos y causas. Barataria. Revista Castellano-Manchega de Ciencias Sociales **18**, 147–159 (2014)
14. Expósito, F., Moya, M., et al.: Violencia de género. Mente y cerebro **48**(1), 20–25 (2011)
15. Vaswani, A., et al.: Attention is all you need. In: Advances in Neural Information Processing Systems, pp. 5998–6008 (2017)
16. Aodhora, S.R., Ahsan, S., Hoque, M.M.: Cuet_hateshield@ nlu of Devanagari script languages 2025: transformer-based hate speech detection in Devanagari script languages. In: Proceedings of the First Workshop on Challenges in Processing South Asian Languages (CHiPSAL 2025), pp. 260–266 (2025)
17. Zhang, Z., Robinson, D., Tepper, M.: Cyberbullying detection with multiple data sources. In: Proceedings of the 10th ACM Conference on Web Science, pp. 271–280 (2018)
18. Levy, M., Jacoby, A., Goldberg, Y.: Same task, more tokens: the impact of input length on the reasoning performance of large language models. In: Proceedings of the 62nd Annual Meeting of the Association for Computational Linguistics (Volume 1: Long Papers), pp. 15:339–15:353. Association for Computational Linguistics, Bangkok, Thailand (2024). https://aclanthology.org/2024.acl-long.818

19. Wang, J., Dong, D., Shou, L., Chen, K., Chen, G.: Effective continual learning for text classification with lightweight snapshots. In: Proceedings of the Thirty-Seventh AAAI Conference on Artificial Intelligence (AAAI-23). Association for the Advancement of Artificial Intelligence (2023). https://ojs.aaai.org/index.php/AAAI/article/view/26206
20. Duffield, W.: Intel bleep: customizable content moderation in gaming. Cato Institute, United States of America (2021). https://coilink.org/20.500.12592/0pk614. Accessed 09 Jun 2025. COI: 20.500.12592/0pk614
21. NightDev: Nightbot. NightDev. Consultado el 17 de junio de 2025 (2025). https://nightbot.tv/
22. Hoffmann, M., Nagle, F., Zhou, Y.: The value of open source software. In: Harvard Business School Strategy Unit, Working Paper 24-038 (2024). SSRN https://ssrn.com/abstract=4693148 or http://dx.doi.org/10.2139/ssrn.4693148. https://ssrn.com/abstract=4693148
23. Urban, J.E.: Software prototyping and requirements engineering. Arizona State University, Department of Computer Science and Engineering, Tempe, AZ, State-of-the-art Research Report 137 p., prepared for Rome Laboratory (RL/C3CB), Griffiss AFB, NY; Distributed by ITT Systems & Sciences Corporation and Data & Analysis Center for Software (DACS) (1992)
24. Zambrano, P., Torres, J., Anchundia, C., Illicachi, J.: Gender-based violence as a cyberthreat: the impact of social media on digital security. In: Proceedings of [Nombre del Congreso o Revista si aplica], Quito, Ecuador, research Project PIS-23-23 (2025)
25. Modha, S., Mandl, T., Majumder, P., Patel, D.: Overview of the hasoc track at fire 2019: hate speech and offensive content identification in Indo-European languages. In: Working Notes of FIRE 2019 - Forum for Information Retrieval Evaluation, Kolkata, India (2019). https://hasocfire.github.io/2019/

Effects of Technology Acceptance Factors on Satisfaction with Virtual Personal Assistants

Hyeon Jo(✉)

Department of Strategic Planning, RealSecu, Busan, Republic of Korea
sineoriz@gmail.com

Abstract. Artificial intelligence (AI) has evolved steadily, and the market is growing rapidly. AI enhances consumers' lives in the form of virtual personal assistants (VPAs). The purpose of this study is to identify the predictors of user satisfaction with VPAs from the perspective of technology acceptance. This study employed partial least squares structural equation modeling (PLS-SEM) to analyze data from 241 samples. The findings show that satisfaction is significantly influenced by perceived ease of use, perceived usefulness, and perceived enjoyment. It was confirmed that the antecedents of technology acceptance have a significant effect on satisfaction with VPAs. The results of this study will help researchers and practitioners identify new factors that explain satisfaction.

Keywords: artificial intelligence · virtual personal assistant · satisfaction · technology acceptance factor

1 Introduction

Artificial intelligence (AI) contributes considerably to the development of mankind. It plays a significant role in information processing and decision-making in various industries [1–3]. AI is easily accessible in our daily life. It is installed on information technology (IT) objects including speakers, TVs, and navigation. Smartphone users are aware of the presence of mounted personal assistants such as Bixby, Siri, and Alexa. This is usually called virtual personal assistant (VPA). Depending on the context, AI personal assistant (AIPA), smart voice assistant (SVA), and intelligent personal assistant (IPA) are used interchangeably [4, 5]. It has been about 10 years since VPA was born and is still evolving. It is important to investigate the effect on the satisfaction of VPA in this environment. This study aims to identify factors that affect user satisfaction with VPA based on the technology acceptance model (TAM).

TAM explains the process of acceptance of technology by users [6]. Users accept the technology if they perceive it as easy and useful [7]. Perceived ease of use and perceived usefulness have been widely introduced in explanations such as acceptance, continuous use, and satisfaction of technology in various IT domains [8–10]. Previous studies on VPA have adopted perceived ease of use and perceived usefulness to explain the user's behavior [11–13]. Perceived enjoyment plays an important role in the process of acceptance or use of hedonic technology [14]. VPA can now serve as friends to users.

H. R. Arabnia et al. (Eds.): CSCI 2025, CCIS 2968, pp. 316–323, 2027.
https://doi.org/10.1007/978-3-032-28457-0_22

Users may find it fun to communicate with the VPA. Therefore, this study proposes that satisfaction is determined by perceived ease of use, perceived usefulness, and perceived enjoyment.

2 Related Work and Hypotheses

Figure 1 depicts the theoretical framework for clarifying the major factors in shaping users' satisfaction with VPA. This research postulates that satisfaction is significantly influenced by perceived ease of use, perceived usefulness, and perceived enjoyment.

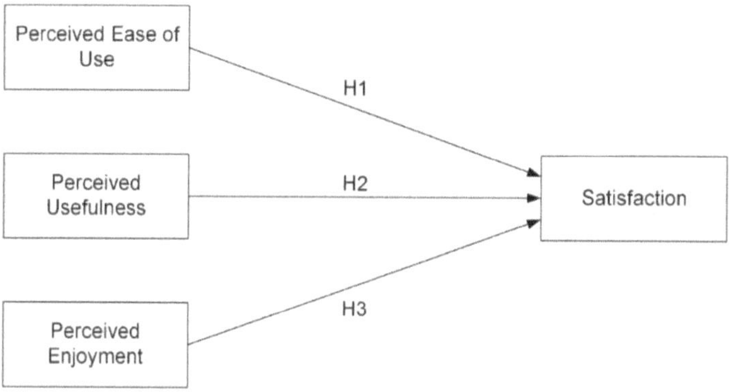

Fig. 1. Research model

2.1 Perceived Ease of Use

Perceived ease of use means the degree to which users believe that using an information system would be free of effort [7]. Previous studies have revealed that perceived ease of use significantly affect users' behavior in several IT contexts [9, 15, 16] Perceived ease of use significantly influences satisfaction with AI chatbot [13], When a VPA is more comfortable to use, users may be more satisfied with it. Thus, perceived ease of use is believed to affect satisfaction.

H1. Perceived ease of use significantly impacts satisfaction.

2.2 Perceived Usefulness

Perceived usefulness represents the extent to which users believe that using an information system would enhance their job performance (Davis 1989). Former works have supported that perceived usefulness serves as the key factor of satisfaction [17–19]. Perceived usefulness significantly determines attitude or satisfaction in the contexts of AI artifacts [12, 13], Therefore, perceived usefulness is expected to significantly raise the level of satisfaction.

H2. Perceived usefulness significantly impacts satisfaction.

2.3 Perceived Enjoyment

Perceived enjoyment is defined as the degree to which an individual perceived an information system as playful, besides any expected performance [20]. It plays a salient role in elevating the level of satisfaction [13]. If VPA provides users with more pleasure, users may experience intrinsic gratification. Hence, perceived enjoyment is suggested to improve satisfaction.

H3. Perceived enjoyment significantly impacts satisfaction.

3 Empirical Methodology

3.1 Instrument

This study surveyed VPA users to identify the affecting factors of satisfaction. All survey items were drawn from extant studies. Measure items were modified to fit the context of VPA. Each item was measured using a 7-point scale (1 = strongly disagree and 7 = strongly agree). The questionnaire was first prepared in English by the author. The English version was translated into Korean by a bilingual researcher in the management information system (MIS) area. Then, the Korean version was back translated into English by a bilingual professional. The two English versions of the questionnaire had only slight differences that were adjusted by the author. Table A1 shows the measurement items and related studies.

3.2 Subject and Data Collection

The survey agency collected the data from registered panels in South Korea. The company has 1.3 million panels. After discarding insincere responses, 241 responses were used for the final analysis. A total of 129 respondents are male and 112 respondents are female, showing an almost uniform distribution. Table 1 details the demographic information of a sample

Table 1 Demographic Information of Sample

Demographics	Item	Subjects ($N = 241$)	
		Frequency	Percentage
Gender	Male	129	53.5%
	Female	112	46.5%
Age	10 s	10	4.1%
	20 s	64	26.6%
	30 s	86	35.7%
	40 s	61	25.3%
	50 s	20	8.3%
VPA Type	Siri	97	40.2%

(*continued*)

Table 1 (*continued*)

Demographics	Item	Subjects ($N = 241$)	
		Frequency	Percentage
	Bixby	117	48.5%
	Alexa	2	0.8%
	Google Assistant	23	9.5%
	Etc.	2	0.8%

4 Analysis and Results

This study carried out structural equation modeling (SEM) to test the measurement model and structural model. The research model was validated by partial least squares (PLS). PLS-SEM offers the advantage of having fewer restrictions on the distribution of sample size and residuals compared to covariance-based SEM [21].

4.1 Measurement Model

This study tested the reliability, convergent validity, and discriminant validity of the measurements. Cronbach's alpha and composite reliability (CR) were calculated to assess reliability. If Cronbach's alpha and CR estimates are higher than 0.70, the reliability is acceptable [22]. As shown in Table 2, the CR scores and Cronbach's alpha value were well above 0.7, indicating a satisfactory level of reliability.

Average variance extracted (AVE) and the factor loadings were examined to evaluate convergent validity. AVE values ranged from 0.690 to 0.875, which are well over the expected threshold of 0.5 [22]. The lowest value (SAT2) of factor loadings was 0.757. Thus, convergent validity was achieved [23]. Discriminant validity was satisfied since the square root value of AVE for each construct was greater than the inter-construct correlation [24]. The analysis results of discriminant validity are illustrated in Table 3.

Table 2 Test Results of Reliability and Convergent Validity

Construct	Items	Mean	St. Dev.	Factor Loading	Cronbach's Alpha	CR	AVE
Satisfaction	SAT1	4.817	1.229	0.869	0.796	0.876	0.704
	SAT2	4.573	1.154	0.757			
	SAT3	4.647	1.214	0.885			
Perceived Ease of Use	PEU1	4.066	1.398	0.822	0.776	0.870	0.690
	PEU2	4.452	1.384	0.821			

(*continued*)

Table 2 (*continued*)

Construct	Items	Mean	St. Dev.	Factor Loading	Cronbach's Alpha	CR	AVE
	PEU3	4.834	1.281	0.848			
Perceived Usefulness	PUS1	4.871	1.251	0.894	0.900	0.937	0.833
	PUS2	4.627	1.346	0.931			
	PUS3	4.423	1.337	0.912			
Perceived Enjoyment	PEN1	3.763	1.516	0.924	0.929	0.955	0.875
	PEN2	3.722	1.426	0.954			
	PEN3	3.979	1.561	0.928			

Table 3 Test Results of Discriminant Validity

Constructs	1	2	3	4
1. Satisfaction	0.839			
2. Perceived Ease of Use	0.558	0.831		
3. Perceived Usefulness	0.593	0.702	0.913	
4. Perceived Enjoyment	0.472	0.517	0.451	0.936

Note: Diagonal values are the square root of AVE

4.2 Structural Model

An SEM was conducted to evaluate the hypothesized paths among the constructs through PLS. A bootstrap resampling method with 500 resamples was performed. The results of the structural model test are summarized in Table 4.

Table 4 Results of PLS-SEM

H	Predictor	Outcome	Coefficient	*t*-value	Hypothesis
H1	Perceived Ease of Use	Satisfaction	0.199	2.527	Supported
H2	Perceived Usefulness	Satisfaction	0.360	4.668	Supported
H3	Perceived Enjoyment	Satisfaction	0.208	3.240	Supported

5 Conclusion

5.1 Summary

This study investigates the precursors of satisfaction with VPAs. It suggested that perceived ease of use, perceived usefulness, and perceived enjoyment significantly influence satisfaction. This study conducted a cross-sectional analysis by surveying Korean VPA users. This study used the PLS-SEM method to validate the measurement model and structural model.

5.2 Implications for Theory and Practice

This study provides some contributions as follows. First, this study applied technology acceptance factors to explain satisfaction, one of the post-adoption scales of VPA users. In the technology acceptance model, preceding factors affect acceptance intention [7]. It has been found that acceptance factors affect continuance intention and attitude in the AI contexts [11, 13]. This study strengthened the theory of technology acceptance by confirming the role of acceptance factors. The results of the study showed that the acceptance factor had a significant effect on the overall process from the adoption of the technology to post-adoption. To improve user satisfaction, practitioners will have to consider determinants from the acceptance stage to the formation process of satisfaction. Second, this study contributes to the existing literature in that it identified the role of perceived enjoyment. VPA was designed to help users make decisions. After that, users also use VPAs for enjoyment. The research results showed that perceived enjoyment improves satisfaction. Therefore, it will be valuable for scholars to further subdivide and explore the aspect of playfulness. Service providers will be able to increase satisfaction at a faster pace by grasping users' disposition toward comedy.

5.3 Limitations and Future Research Directions

The limitations of this study and future research directions are as follows. First, this study did not reflect the type of VPA. The proposed research model may vary depending on the type of VPA. It will be valuable for future research to perform multi-group analysis by dividing the user group according to the type of VPA. Second, this study did not include variables specific to VPA. Researchers are encouraged to include the unique factors (e.g. voice, reaction speed, and answer accuracy) that can only be applied to VPA within the research model.

Appendix A

Table A1 Measurement Items

Construct	Items	Description	Source
Satisfaction		How do you feel about your overall experience of using the AI personal assistant?	[13]
	SAT1	Very dissatisfied — very satisfied	
	SAT2	Very displeased — very pleased	
	SAT3	Very frustrated — very contented	

(continued)

Table A1 *(continued)*

Construct	Items	Description	Source
Perceived Usefulness	PUS1	I find the AI personal assistant useful in my daily life.	[13]
	PUS2	Using the AI personal assistant helps me accomplish tasks more quickly.	
	PUS3	Using the AI personal assistant increases my productivity.	
Perceived Ease of Use	PEU1	My interaction with the AI personal assistant is clear and understandable.	[13]
	PEU2	Interacting with the AI personal assistant does not require much mental effort.	
	PEU3	I find the AI personal assistant easy to use.	
Perceived Enjoyment	PEN1	I enjoy interacting with the AI personal assistant.	[13]
	PEN2	It is fun and pleasant to interact with the AI personal assistant.	
	PEN3	Interacting with the AI personal assistant is exciting.	

References

1. Sahu, C.K., Young, C., Rai, R.: Artificial intelligence (AI) in augmented reality (AR)-assisted manufacturing applications: a review. Int. J. Prod. Res. **59**(16), 4903–4959 (2021)
2. Jia, P., Stan, C.: Artificial intelligence factory, data risk, and VCs' mediation: the case of ByteDance, an AI-powered startup. J. Risk Financial Manag. **14**(5), 203 (2021)
3. Mhlanga, D.: Industry 4.0 in finance: the impact of artificial intelligence (ai) on digital financial inclusion. Int. J. Financial Stud. **8**(3), 45 (2020)
4. Sun, Y., Li, S., Yu, L.: The dark sides of AI personal assistant: effects of service failure on user continuance intention. Electron. Mark., 1–23 (2021)
5. Han, S., Yang, H.: Understanding adoption of intelligent personal assistants: a parasocial relationship perspective. Ind. Manag. Data Syst. **118**(3), 618–636 (2018)
6. Davis, F.D.: A technology acceptance model for empirically testing new end-user information systems: Theory and results. Ph.D. dissertation, MIT Sloan School of Management, Cambridge, MA (1986)
7. Davis, F.D.: Perceived usefulness, perceived ease of use, and user acceptance of information technology. MIS Q. **13**(3), 319–340 (1989)
8. Jo, H.: Understanding the key antecedents of users' continuance intention in the context of smart factory. Technol. Anal. Strateg. Manag. **35**, 1–14 (2021)
9. Jo, H.: Key factors influencing loyalty and satisfaction toward ERP: mediating role of flow. J. Knowl. Econ. **14** (2022)
10. Jo, H.: Examining the key factors influencing loyalty and satisfaction toward the smart factory. J. Bus. Ind. Mark. **38** (2022) ahead-of-print(ahead-of-print)
11. Nguyen, Q.N., Ta, A., Prybutok, V.: An integrated model of voice-user interface continuance intention: the gender effect. Int. J. Hum.–Comput. Interact. **35**(15), 1362–1377 (2019)

12. Nguyen, D.M., Chiu, Y.-T.H., Le, H.D.: Determinants of continuance intention towards banks' chatbot Services in Vietnam: a necessity for sustainable development. Sustainability. **13**(14), 7625 (2021)
13. Ashfaq, M., et al.: I, chatbot: Modeling the determinants of users' satisfaction and continuance intention of AI-powered service agents. Telematics Inform. **54**, 101473 (2020)
14. van der Heijden, H.: User acceptance of hedonic information systems. MIS Q. **28**(4), 695–704 (2004)
15. Lisha, C., et al.: Integrating guanxi into technology acceptance: an empirical investigation of WeChat. Telematics Inform. **34**(7), 1125–1142 (2017)
16. Jo, H.: Effects of perceived factors on the word-of-mouth of SNS. J. Inf. Technol. Serv. **11**(3), 227–240 (2012)
17. Jo, H., Kim, S.-H., Lee, S.-K.: A study on the success factors of EDI information system: focused on medical industry. J. Korean Data Inf. Sci. Soc. **22**(2), 323–333 (2011)
18. Jung, S., Lee, J., Jo, H.: ERP success factors in perspective of IS success model and TAM: focused on shipbuilding and marine engineering industry. J. Internet Electron. Commer. Res. **13**(2), 85–103 (2013)
19. Kim, B.: The diffusion of mobile data services and applications: exploring the role of habit and its antecedents. Telecommun. Policy. **36**(1), 69–81 (2012)
20. Davis, F.D., Bagozzi, R.P., Warshaw, P.R.: Extrinsic and intrinsic motivation to use computers in the workplace. J. Appl. Soc. Psychol. **22**(14), 1111–1132 (1992)
21. Chin, W.W.: The partial least squares approach to structural equation modeling. In: Modern Methods for Business Research, vol. 295, pp. 295–336 (1998)
22. Fornell, C., Larcker, D.F.: Evaluating structural equation models with unobservable variables and measurement error. J. Mark. Res. **18**(1), 39–50 (1981)
23. Bagozzi, R.P., Yi, Y., Phillips, L.W.: Assessing construct validity in organizational research. Adm. Sci. Q. **36**(3), 421–458 (1991)
24. Ashfaq, M., Yun, J., Yu, S., Loureiro, S.M.C.: I, chatbot: Modeling the determinants of users' satisfaction and continuance intention of AI-powered service agents. Telematics Inform. **54**, 101473 (2020)

A New Design of Mars Orbital Station Based on In-situ Resources Utilization and Artificial Intelligence Technology

Robin Zhang[1], Patrick Bruskiewich[2], Kevin Zhang[3], and Chris Cheng Zhang[4](✉)

[1] St. Patrick Regional Secondary School, Vancouver, Canada
[2] Sager Education Inc., Vancouver, Canada
pbruskiewich@sagereducation.com
[3] University of British Columbia, Vancouver, Canada
kzhang47@student.ubc.ca
[4] Canada Youth Robotics Club, Vancouver, Canada
head@roboca.org

Abstract. A recent international call for designs for a Mars Orbiting Station (MOS) large enough to accommodate 2,200 inhabitants was undertaken. The call was set out by Global Future Space Scholars Meet (GFSSM) which is a UK, USA and PRC group of space professionals. The MOS is comparable in mass and complexity to a Nimitz class, nuclear powered aircraft carrier. This short paper outlines the main features of a design for the GFFSM done by a group in Canada. We present details about the MOS overall structure, the different com- ponents, the electrical supply, the propulsion system, the agricultural section as well as aspects of the daily routine for both people and robots. The overall cost of such a system is estimated at 15 Trillion USD, would take nearly twenty years to construct and would be in Mars orbit for about five decades.

Keywords: Mars Orbital Station · In-Situ Resources Utilization (ISRU) · CANDU Reactor

1 A Space Faring Civilization

In the first century of the 21st century, humanity may perhaps finally become a *Space Faring Civilization.* To do this we have to venture permanently out of Low Earth Orbit (LEO). Seventy years ago, in 1953 Dr. Von Braun wrote the first definitive study of a Mars mission in his pioneering book '*Project Mars: A Technical Tale*' [1].

Planning, financing, as well as science and technology cycles, being what they are, there will be a lead up time of at the minimum 25 years and perhaps even twice that length of time, before habitats are in place on or near Mars. In the meantime there may be a handful of brief, tentative crewed exploratory missions to Mars.

H. R. Arabnia et al. (Eds.): CSCI 2025, CCIS 2968, pp. 324–331, 2027.
https://doi.org/10.1007/978-3-032-28457-0_23

The authors recently participated in an international planning exercise, the *Global Future Space Scholars Meet* (GFSSM) that tasked for the design of a Mars Orbital Station (MOS) for 2,200 inhabitants. The Global Future Space Scholars Meet is an international consortium made up of professionals from the UK, US and PRC. [7] A Mars Orbital Station is a structure is comparable in mass and complexity to a Nimitz class, nuclear powered aircraft carrier. This paper is meant to share some of the design features of our Mars orbiting space station proposal.

2 Overview of the Mars Orbiting Station Design

The Mars Orbital Station (MOS) developed by RB-Team is a space city station pro- posed to hold a total of 2,200 occupants at any given time in Martian orbit; 2,000 of the crew onboard are full-time residents that live, work, and operate the space station. The remaining 200 crew are visitors. This structure considers the aspects of physics-related challenges (gravity, momentum) and tackles them in an effective and realistic way. MOS serves a future gateway to interstellar human habitation and comfort amidst rising global populations and uncertainties on earth. This space station concept envisions the potentials of human habitat, incorporated through innovative and smart ways.

The MOS is a space structure that extends 1,310 m in length. The station's foundation is a triangular-shaped rod that extends throughout the entirety of the space station. MOS contains three critical habitat sections: the Residential Wheel, the Agricultural Barrel, and Recreational Wheel which each have their respective purposes. All wheels rotate to mimic artificial gravity and the Recreational Wheel rotates in opposite direction to the Residential Wheel and Agricultural Barrel to cancel out angular momentum from the structure's rotation itself.

The Residential Wheel holds all full-time residents and any additional visitors. The Residential Wheel features two levels with rows of rooms on both floors. Each floor in the Residential Wheel has a floor area of 39,000 m^2.

The Agricultural Barrel is the food source of the entire station; all needs to ensure the survival of our residents are provided from here. The barrel is the largest and heaviest structure of the space station, so we intentionally placed it in the centre of the structure to evenly balance the weight it influences on the station's structure. The agricultural spaces will rely on advanced hydroponics and include vertical stacked gardens. Each barrel floor is over 20 m in height, which is sufficient for the amount of space needed to provide food on a daily occurrence. There are 225,000 m^2 of area of the total combined three levels in the Agricultural Barrel.

The Recreational Wheel is mounted to the front of the foundation through six evenly placed spokes (similar to bicycle wheels) that also individually serve as an elevator to transport between the two levels of the wheel and into the foundation of the space station.

Inside both the Residential Wheel and Recreational Wheel, there are hospitals, clinics, and dentists as well as other medical facilities. The space station is highly automated, which incorporates the use of AI in daily operations.

We propose innovative technology to help address key potential issues onboard large structures (such as our MOS), where saving time, effort, and maximizing practicality are essential. The RB-Bot is our handy helper onboard the MOS to assist with daily needs on the space station. There is one RB-Bot for every resident onboard and hundreds of additional remaining robots that are free. Each robot has a humanoid body with a similar size to that of an average person (Fig. 1).

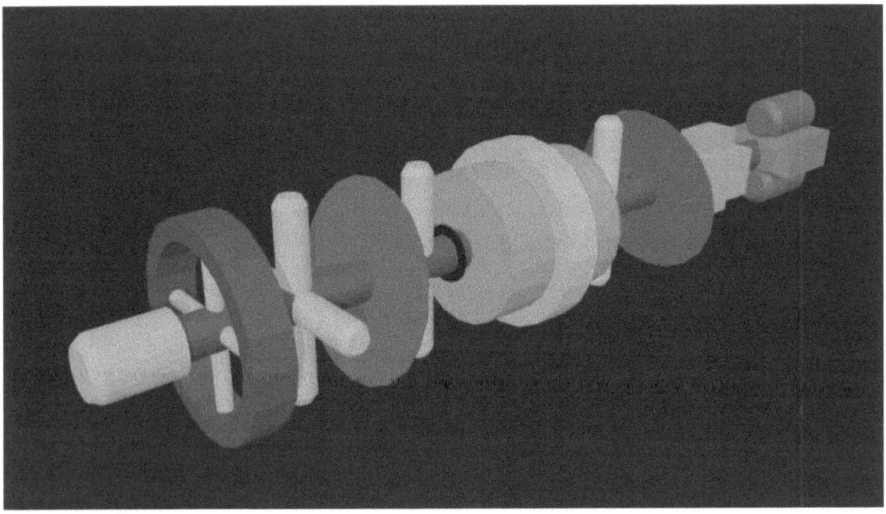

Fig. 1. Schematic representation of MOS. Colors designate different sections.

Sections specified individually from bow to stern along the Central Foundational Rod (dark grey).

- Front Command Section (grey)
- Recreational Wheel (red)
- Front docking ports and evacuation ports (grey)
- Front photovoltaic mirror disk (blue)
- Mid-docking ports and evacuation ports (grey)
- Agricultural Barrel (green)
- Residential Wheel (yellow)
- Rear docking port and evacuation ports (grey)
- Rear photovoltaic mirror disk (blue)
- Fuel tanks (cyan)
- Reactors and propulsion systems (purple)

MOS will be initially built in low-lunar orbit by using resources from the moon wherever possible. The construction will continue during its transfer from lunar to Martian orbit with a minimum human crew coming alongside to command and direct robots actively working on construction. The fitting out and commissioning will be accomplished in Martian orbit. The first harvest of the Agricultural Barrel serves as the first commissioning steps and provides a food source for the human crew that has arrived.

The electrical power system has redundancy in it. The key source of electricity are two CANDU reactors, which will serve to provide electrical power and thermal heat for the space station. In the Martian orbit, such distances receive much less light than Earth, so it is absolutely essential to have enough power and heating to keep the crew warm. After all, Mars only receives 43% of sunlight that earth does! The space station cannot solely rely on its photovoltaic systems to provide electricity. In fact, it is only the secondary system.

The primary system will be the nuclear reactors that are located at the stern of the space station. Adjacent to the nuclear reactors are the primary propulsion systems, which is an arc jet propulsion system that uses water as its propellant. The nuclear re- actors serve a dual-use purpose; in transfer from the Moon to Mars, it's primarily being used by the arc jet propulsion system. Once in orbit, the reactors are providing electrical and thermal power to the station (Fig. 2).

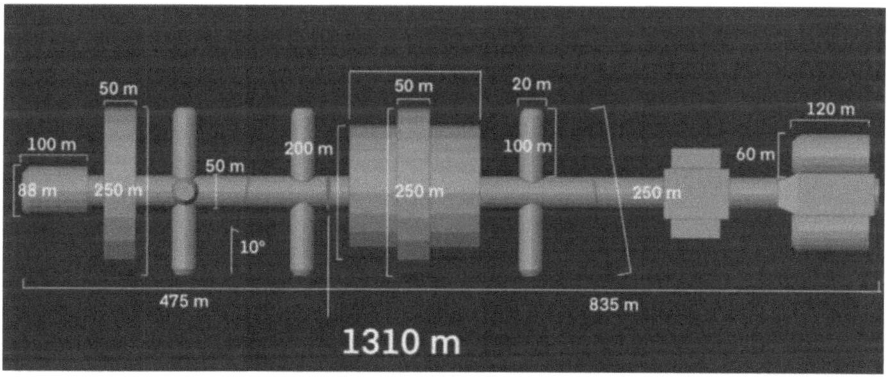

Fig. 2. Schematic of the proportions of section sizes.

The MOS is fitted with a total of 88 lifeboats. We have ensured to evenly distribute the lifeboat evacuation ports across the space station by placing their evacuation ports in total seats more than enough to evacuate all crew on board. Once activated, the lifeboats will deploy in a steady sequence to land on the surface of Mars (assuming that there is a station at the surface). This design plan and evacuation sequence ensures that it is very likely that all crew can evacuate from the station and land on Mars safely. Our lifeboat design takes inspiration from the *Dream Chaser* winged-spacecraft because its design offers notable advantages. A landing field will be required on Mars.

3 Electrical Power is Everything

In space flight electrical power is more fundamental than other consumables even oxygen or water. We are reminded of this in the feature film Apollo 13. "*Electrical Power is Everything*!" All systems aboard a spacecraft operate using electrical power.

In deep space in orbit around the planet Mars, thermal energy is also needed to keep human crewed systems within safe operating parameters. Most of this thermal energy will be provided by the thermal output of the Primary electrical supply (CANDU).

In the MOS design there are five types of electrical supply (Table 1).

Table 1. Five types of electrical supply on the MOS.

Type	Description	MW Electrical
Primary	Two 1,000 MW CANDU Reactors	$2 \times 1{,}000$ MW
Secondary	Photovoltaic systems	12.6 MW
Tertiary	Passive Thermal Ionic Electric	4.3 MW
Quaternary	Beta-emission Electric generators	$>$ 1 MW
Quinary	Batteries and Fly Wheels	$>$ 20 MW

The primary power source, with one reactor online at around 40% capacity would represent about 90% of the normal electrical power supply of MOS, while the Secondary and Tertiary systems providing the remaining 5% and both the Quaternary and Quinary standing by as an emergency draw. In orbit around Mars the MOS would cycle from one CANDU reactor to the other on a scheduled basis. Thorium extracted on the moon would be used to breed fertile U-233 to allow for extended use of the fuel in the CANDU reactors.

During the transfer flight from low lunar orbit the MOS will house a minimal human crew and both reactors will be online to provide power to the Arcject propulsion system. Once in Mars orbit is the CANDU reactors would provide both electrical and thermal power to MOS. The secondary system appears adequate to provide for the crew's daily electricity needs as well as top up the battery supply of electricity.

There are several reason why a CANDU design is considered. First CANDU has high efficiency and high online time. Second is the fact that the Calandria is unpressurized while the fuel tubes are. This greatly increases the safety of the reactor. Third the system is moderated by heavy water which simplifies the operation of the reactor. ISRU heavy water can be separated out of the water at the South Pole of the Moon. Fourth the uranium fuel does not have to be enriched. Fifth, with a high neutron efficiency CANDU can breed U-233 using Th-232. Deposits of Th-232 has been found on the moon. Sixth, it can be maintained simply and in an ongoing fashion. CANDU reactors are the safest power reactors presently in operation [3].

4 MOS Propulsion

To make a MOS viable an efficient and cost effective way must be found to transfer large masses from LEO, or Lunar orbit to Mars. The MOS outlined in this paper has the mass of a Nimitz class aircraft carrier (about 100,000 tonnes). A modified Hohmann Transfer Orbit could be considered for such a system. A modified Hohmann Transfer Orbit could use different propulsion systems than just traditional chemical rockets. There is a third burn that will be necessary for MOS and that will be an orbital entry burn when MOS arrives at Mars. Intermediate between the chemical rocket and ion propulsion is an advanced Arcjet system that could be designed to deliver thrust at a specific impulse of in excess of 1,000 s. This is the system that should be considered for a MOS. The Arcjet system could be of a dual nature where during the big delta v phases it uses water as its propellant, and then during low thrust phases it can use hydrogen as an ion propulsion system with a specific impulse in excess of 4,000 s [4].

MOS would use a high efficient Arcjet propulsion system that would be powered by two 1,000 MW CANDU reactors and could use water as its propellant. In transit between Erath orbit and Mars the propellant tanks would be the Agricultural Barrel on MOS. The agricultural barrel has a volume of nearly 4 million cubic meters which could store nearly 4 billion kg of water.

The minimal structure would be transferred between Earth orbit and Mars representing perhaps 20% of its final mass. Subsequent transfers between the Earth Mars system would fit out the spacecraft once it is placed in a stable circular orbit around Mars. As far as possible ISRU can be used in the fitting out using both Mars and lunar resources. It might be possible to mine one or both of Mar's satellites for the materials needed to complete and commission MOS. With 4 billion kg of water used as propellant and a specific impulse of 2,500 s or more, there is a thrust budget in excess of 10 trillion Newtons available for MOS. There will be required several rendezvous with MOS to provide additional water propellant to meet the 165 trillion Newton thrust budget required to establish MOS.

It appears that, to make MOS plausible, will require propulsion systems with a specific impulse in excess of 10,000 s. It might be contemplated to use a space tug that is based on *Project Orion*, which has a Impulse in excess of 1.25 million s [5].

5 The Agricultural System

The Agricultural Barrel is the most complex section that we have designed in the MOS. The barrel is the most important part of the station for long-term human habitation in Mars orbit and acts as a gateway to smarter interplanetary farming, while also providing the oxygen needed in MOS and the high daily nutritional needs of all of its 2,200 residents. The Agricultural Barrel is a biosphere that needs to operate well for there to be long-term human habitation aboard MOS.

Such placement additionally also offers other pleasing benefits; on the station's foundation, the sections that rotate create angular momentum, but this is balanced by two separate sections that rotate in opposite directions and the balanced centered mass of the Agricultural Barrel. This provides a dynamic stability to the station and allows the structure to be oriented in a fixed direction is space as it orbits Mars. This helps to stabilize the stations average temperature.

The Agricultural Barrel is a cylindrical structure made up of three staggered levels, with the outer barrel having the most floor area. In the Agricultural Barrel all three levels combined have a total area of 225,000 square meters. Each level is 25 m in height. The volume of the Agricultural Barrel is around 5.5 million cubic meters.

To complement the biome's oxygen regeneration via photosynthesis in the Agricultural Barrel, a molecular sieve system is used to extract carbon dioxide from the MOS atmosphere in the residential and recreational regions and fed into the Agricultural Barrel. It is recommended that the Agricultural Barrel is operated at a carbon dioxide rich concentration to optimize food production by the C3/C4 crops. Advanced hydroponics will be used in the Agricultural Barrel. Organic waste from byproducts such as unconsumed vegetable matter, livestock waste and other compostable can be reintroduced into the agricultural biome as fertilizer.

Wherever possible the. Agricultural Barrel will provide both oxygen regeneration, via photosynthesis, as well as foodstuff for the 2,200 inhabitants of the Mars Orbital Station. It is envisaged that supplies may still be needed from off-station, either from Earth or from near Mars sites, however, the inhabitants will focus wherever possible on ISRU principles, which may require a somewhat advanced and disciplined lifestyle.

6 Conclusion

In considering the crew rotation for the MOS it was suggested that a three shift crew might be used to reduce peak demand on infrastructure. A twenty four clock would be fine on MOS given that the rotation of Mars is 24.6 h. It was also suggested that each human member of the crew would be 'matched' with a personal robot who would complement the work needed to provide for the needs of each member of the human crew. The robots would be tasked to assist MOS activity on a 24×6 schedule, with one day in seven for planned maintenance of the robots. While its 'human match' sleeps, the robot might be in the Agricultural Barrel tending to food production, or at work in some custodial function, or helping to tend to logistical needs, or may be tasked to attend to planned maintenance of MOS.

The construction of MOS comes in five phases, first lunar infrastructure development, second basic MOS structure construction in Low Lunar Orbit, third during the Trans-Martian Insertion where power systems and pressurized parts of the ship are brought on stream, fourth with final fitting out and commissioning in Mars orbit and five with the transfer of people and logistics to MOS.

The full construction time is estimated at around 18 years (936 weeks) beginning with project's the inaugural meeting. Under the GFSSM proposal, the Mars Orbiting Station commissioning is slated for 2070. After commissioning MOS is designed to operate for a minimum of 50 years, with a major overhaul each 12 to 15 years. It should be in continual crewed use until 2125, a century hence.

An estimate for the total project costs (in 2025 US$s) is outlined in Table 2.

Table 2. Estimate of Total Project Cost for MOS in 2025 US$s

Stage	(Costs $ Trillions)	Remark	Notes:
1	3.5	Lunar Surface & Orbit	Infrastructure for construction
2	2.5	Lunar Orbit	Basic structure
3	2	Mars Transfer	Power systems and pressurized
4	2	Mars Orbit	Fitting out and Commissioning
5	1.6	Crew Transfer	People and Logistics
Intermediate Total	11.6		
Contingency	3.4		To cover inflation
Gross Total 2025	15		

A budget of 15 trillion USD for MOS would represent perhaps 3% of the world's GDP in 2070. It is expected that a consortium representing perhaps one-fifth of the world's population would work together to make MOS possible. The present day *Artemis Accords* might be used as a template for such an international MOS Consortia.

References

1. Von Braun, W.: refer to https://archive.org/details/ProjectMars (1953)
2. Refer to: https://spacescholars.org/
3. Refer to: https://en.wikipedia.org/wiki/CANDU_reactor
4. Refer to: https://www.matsusada.com/column/ion_engine.html
5. Refer to: Project Orion. https://ntrs.nasa.gov/api/citations/20000096503/downloads/20000096503.pdf

Optimism vs Skepticism in the Use of Intelligent Personal Assistant (IPA)

Hyeon Jo(✉)

Department of Strategic Planning, RealSecu, Busan, Republic of Korea
sineoriz@gmail.com

Abstract. This study aims to investigate the determinants of users' continuance intention in the context of intelligent personal assistants (IPAs), with a particular focus on optimism and skepticism. A cross-sectional survey was conducted among Korean IPA users, and the data were analyzed using partial least squares structural equation modeling (PLS-SEM). The findings indicate that system quality, information quality, and optimism serve as key antecedents of continuance intention. In contrast, skepticism does not exert a significant effect. Based on these results, both theoretical and practical implications are discussed for researchers and service providers.

Keywords: artificial intelligence · intelligent personal assistant · continuance intention · optimism · skepticism

1 Introduction

An intelligent personal assistant (IPA) is an artificial intelligence (AI)-based information processing system that interacts with users through text or voice interfaces. Typical examples include voice-based assistants embedded in smartphones, such as Bixby and Siri. IPAs learn users' speech patterns and improve recognition accuracy through repeated interactions. In the literature, IPAs are also referred to as virtual personal assistants (VPAs), voice-controlled AI systems, or AI-powered digital assistants [1–3].

Users rely on IPAs for various tasks, such as setting alarms, managing schedules, and controlling connected devices. In addition, some users engage in casual or conversational interactions with IPAs. Notably, IPAs have also been shown to support independent living among older adults and, in some cases, contribute to life-saving interventions [4–6]. As the IPA market continues to expand, understanding users' continuance intention has become increasingly important. Accordingly, this study aims to identify the key factors influencing users' intention to continue using voice-based IPAs installed on smartphones.

Prior research suggests that information systems (IS) success factors significantly influence continuance intention in IPA contexts [7, 8]. Specifically, system quality and information quality have been found to affect continuance intention through trust and satisfaction in text-based IPA environments. These relationships are likely to extend to voice-based IPAs. Therefore, this study posits that system quality and information quality are critical determinants of continuance intention.

H. R. Arabnia et al. (Eds.): CSCI 2025, CCIS 2968, pp. 332–340, 2027.
https://doi.org/10.1007/978-3-032-28457-0_24

Optimism, a dimension of technology readiness, reflects an individual's positive propensity toward technology [9]. Previous studies have demonstrated that optimism positively influences the intention to use AI-based robo-advisors [10] and affects shopping intentions through perceived ease of use and usefulness in AI-driven retail environments [11]. Thus, users with higher levels of optimism toward IPAs are expected to exhibit stronger continuance intention.

Skepticism, in contrast, refers to a tendency to question or resist new technologies [12]. Individuals with higher skepticism may initially resist adoption but can become consistent users once they accept the technology [13]. Some studies have reported a positive relationship between skepticism and continuance intention [14]. Given that IPAs are relatively mature technologies, it is plausible that skeptical users may still demonstrate sustained usage after initial adoption.

This paper is organized as follows. Sect. 2 reviews relevant literature and develops the hypotheses. Sect. 3 describes the research methodology, including measurement and data collection. Sect. 4 presents the empirical results. Finally, Sect. 5 discusses conclusions, implications, and limitations.

2 Related Work and Hypotheses

Figure 1 illustrates the research framework for examining the determinants of continuance intention among IPA users. This study proposes that continuance intention is influenced by system quality, information quality, optimism, and skepticism.

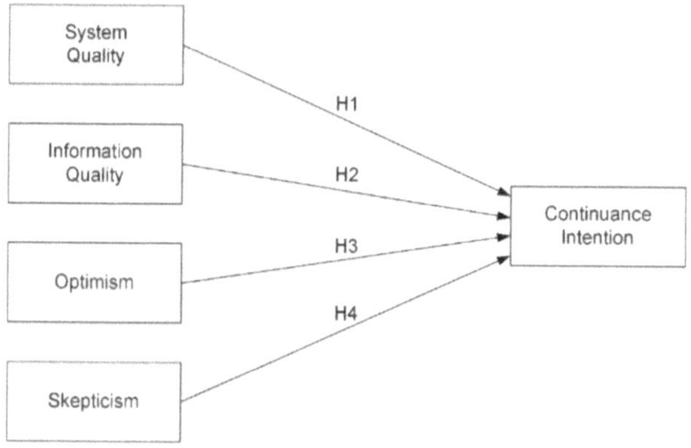

Fig. 1 Research model

2.1 System Quality

System quality is defined as the reliability, ease of use, and response time of an information system [15]. It has consistently been identified as a critical determinant of user behavior across various technological contexts [16–18]. A more stable and responsive IPA is likely to encourage continued usage. Therefore, this study proposes the following hypothesis:

H1. System quality significantly influences continuance intention.

2.2 Information Quality

Information quality refers to the accuracy, timeliness, and completeness of information provided by an information system [19]. Prior studies have shown that information quality indirectly affects continuance intention in IPA contexts [7, 8, 20]. High-quality information is expected to enhance user satisfaction and encourage continued use. Thus:

H2. Information quality significantly influences continuance intention.

2.3 Optimism

Optimism represents a positive belief in the benefits of technology, including improved control, flexibility, and efficiency [13]. It has been found to significantly influence the adoption of AI assistants [10]. Users with a more favorable perception of IPAs are more likely to continue using them. Therefore:

H3. Optimism significantly influences continuance intention.

2.4 Skepticism

Skepticism reflects a cautious attitude toward adopting new technologies or services [12]. Although skeptical users may resist initial adoption, they may exhibit stronger loyalty once the technology is accepted [21]. Prior research suggests that skepticism can positively influence continuance intention in voice assistant contexts. Accordingly:

H4. Skepticism significantly influences continuance intention.

3 Methodology

3.1 Instrument

Measurement items were adapted from established literature in information systems and service research. The items were modified to ensure contextual relevance to IPAs. All constructs were measured using a 7-point Likert scale. The questionnaire was originally developed in English and subsequently translated into Korean by a bilingual expert. Table A1 presents the measurement items and corresponding sources.

3.2 Sample

The survey was administered through Embrain, a leading online research panel in South Korea with approximately 1.3 million registered participants. A quota sampling method was applied based on gender and age. Data collection was conducted from January 18 to January 20, 2022. After excluding incomplete responses, 236 valid responses were retained for analysis. Table 1 summarizes the demographic characteristics of the sample.

Table 1 Profile of the respondents

Demographics	Item	Subjects ($N = 236$)	
		Frequency	Percentage
Gender	Male	125	53.0
	Female	111	47.0
Age	10 s	10	4.2
	20 s	60	25.4
	30 s	86	36.4
	40 s	61	25.8
	50 s	19	8.1
IPA Type	Siri	95	40.3
	Bixby	114	48.3
	Alexa	2	0.8
	Google Assistant	23	9.7
	Etc.	2	0.8

4 Analysis and Results

This study employed partial least squares structural equation modeling (PLS-SEM) to evaluate both the measurement and structural models. PLS-SEM offers advantages in terms of fewer distributional assumptions and greater flexibility regarding sample size and residual distributions compared to covariance-based structural equation modeling [22].

4.1 Measurement Model

The measurement model was assessed in terms of reliability, convergent validity, and discriminant validity. Cronbach's alpha and composite reliability (CR) were calculated to evaluate internal consistency. Values exceeding 0.70 indicate acceptable reliability [23]. As shown in Table 2, all CR and Cronbach's alpha values were well above the threshold of 0.70, confirming satisfactory reliability.

Convergent validity was examined using average variance extracted (AVE) and factor loadings. The lowest AVE value (skepticism) was 0.733, exceeding the recommended threshold of 0.50 [23]. Additionally, the lowest factor loading (SKT2) was 0.842, which is higher than the minimum acceptable level of 0.70 [24], thereby supporting convergent validity.

Discriminant validity was assessed using the Fornell–Larcker criterion. The square root of AVE for each construct exceeded the corresponding inter-construct correlation coefficients, indicating adequate discriminant validity [23]. The results are presented in Table 3.

Table 2 Factor analysis and reliability

Construct	Items	Mean	St. Dev.	Factor Loading	Cronbach's Alpha	CR	AVE
Continuance Intention	COI1	4.932	1.184	0.938	0.915	0.946	0.855
	COI2	4.720	1.281	0.893			
	COI3	4.877	1.210	0.943			
System Quality	SYQ1	4.364	1.250	0.914	0.922	0.951	0.865
	SYQ2	4.398	1.222	0.941			
	SYQ3	4.398	1.246	0.935			
Information Quality	INQ1	4.398	1.341	0.880	0.883	0.928	0.811
	INQ2	4.581	1.224	0.916			
	INQ3	4.525	1.198	0.905			
Optimism	OPT1	4.814	1.214	0.917	0.904	0.940	0.839
	OPT2	5.051	1.138	0.919			
	OPT3	4.873	1.190	0.912			
Skepticism	SKT1	3.788	1.586	0.870	0.636	0.846	0.733
	SKT2	3.758	1.410	0.842			

Table 3 Fornell-Larcker scale results

Constructs	1	2	3	4	5
1. Continuance Intention	0.925				
2. System Quality	0.663	0.930			
3. Information Quality	0.682	0.781	0.901		
4. Optimism	0.674	0.600	0.625	0.916	
5. Skepticism	0.106	0.168	0.172	0.018	0.856

Note: Diagonal values are the square root of AVE

4.2 Structural Model

The structural model was evaluated using bootstrapping with 500 resamples. The results indicate that system quality, information quality, and optimism significantly influence continuance intention, whereas skepticism does not have a significant effect (Table 4).

Table 4 Results of PLS-SEM

H	Predictor	Outcome	β	t	Hypothesis
H1	System quality	Continuance intention	0.228	2.628	Supported
H2	Information quality	Continuance intention	0.273	3.136	Supported
H3	Optimism	Continuance intention	0.366	3.728	Supported
H4	Skepticism	Continuance intention	0.015	0.260	Not Supported

5 Conclusion

5.1 Summary

This study examines the determinants of continuance intention among IPA users. It posited that system quality, information quality, optimism, and skepticism would significantly influence continuance intention. Data were collected from Korean IPA users using a cross-sectional survey design, and PLS-SEM was employed to evaluate both the measurement and structural models.

5.2 Implications for Researchers and Practitioners

This study offers several important implications for both researchers and practitioners. First, it contributes to the literature by integrating the core constructs of the IS success model with individual technology-related dispositions to explain continuance intention in the IPA context. While prior studies have primarily relied on frameworks such as the IS success model, technology acceptance model, and expectation-confirmation theory [7, 8, 20], this study extends existing research by incorporating users' technological propensity alongside system-related factors. Accordingly, future research should further examine the role of individual traits in both adoption and post-adoption stages. From a practical perspective, the findings suggest that service providers should consider users' personal dispositions when promoting IPA services. Understanding differences in users' attitudes toward technology may enable more effective segmentation and targeted strategies. Second, this study contributes by simultaneously examining optimism and skepticism within a single framework. Previous research has typically focused on either optimism, grounded in technology readiness [10, 11], or skepticism, reflecting resistance among late adopters [14]. By integrating both perspectives, this study provides a more comprehensive understanding of user psychology. The findings indicate that only optimism has a significant effect on continuance intention, whereas skepticism does not. This suggests that enhancing positive perceptions of technology may be more critical than mitigating skepticism. For practitioners, conducting qualitative approaches such as focus group interviews segmented by optimism and skepticism levels may provide valuable insights for market targeting and service design.

5.3 Limitations and Future Research

This study has several limitations. First, the data were collected from a single country, South Korea, which may limit the generalizability of the findings. Given that the model includes individual-level psychological traits, the results may vary across cultural contexts. Future studies should incorporate more diverse international samples to enhance external validity. Second, this study considered a limited set of individual technology-related traits. Individual differences in technology adoption are multifaceted, including factors such as innovativeness and self-efficacy. Future research should incorporate a broader range of personal characteristics to develop a more comprehensive and robust research model.

Appendix A

Table A1. Measurement Items

Construct	Items	Survey Item	Source
Continuance Intention	COI1	I intend to continue using the AI personal assistant in the future.	Ashfaq, Yun [8]
	COI2	I intend to increase my use of the AI personal assistant in the future.	
	COI3	I will continue using the AI personal assistant as regularly as I do now.	
System Quality	SYQ1	In terms of system quality, I would rate the AI personal assistant highly.	Nguyen, Ta [20]
	SYQ2	Overall, the AI personal assistant is of high quality.	
	SYQ3	Overall, I would give the AI personal assistant a high quality rating.	
Information Quality	INQ1	This AI personal assistant provides sufficient information.	Ashfaq, Yun [8]
	INQ2	The information provided by this AI personal assistant is up-to-date.	
	INQ3	The information provided by the AI personal assistant is reliable.	
Optimism	OPT1	New information technologies, such as AI personal assistants, improve quality of life.	Pillai, Sivathanu [11]
	OPT2	New information technologies, such as AI personal assistants, are convenient for users.	

(*continued*)

Table A1. (*continued*)

Construct	Items	Survey Item	Source
	OPT3	New information technologies, such as AI personal assistants, help make my life more efficient and effective.	
Skepticism	SKT1	I feel somewhat apprehensive about high-tech products such as AI personal assistants.	Pal, Babakerkhell [14]
	SKT2	I tend to approach new technologies with skepticism and caution.	

References

1. Shao, Z., et al., Anthropomorphism Affordance, Trust and Continuance Intention in Virtual Personal Assistants. 2021.
2. Hasan, R., Shams, R., Rahman, M.: Consumer trust and perceived risk for voice-controlled artificial intelligence: the case of Siri. J. Bus. Res. **131**, 591–597 (2021)
3. Balakrishnan, J., Dwivedi, Y.K.: Conversational commerce: entering the next stage of AI-powered digital assistants. Ann. Oper. Res., 1–35 (2021)
4. Shalini, S., et al.: Development and comparison of customized voice-assistant systems for independent living older adults. In: International Conference on Human-Computer Interaction. Springer (2019)
5. Pradhan, A., Lazar, A., Findlater, L.: Use of intelligent voice assistants by older adults with low technology use. ACM Trans. Comput. Hum. Interact. **27**(4), 1–27 (2020)
6. Picard, C., et al.: Can Alexa, Cortana, Google assistant and Siri save your life? A mixed-methods analysis of virtual digital assistants and their responses to first aid and basic life support queries. BMJ Innov. **6** (2020) bmjinnov-2018-000326
7. Nguyen, D.M., Chiu, Y.-T.H., Le, H.D.: Determinants of continuance intention towards banks' chatbot Services in Vietnam: a necessity for sustainable development. Sustainability. **13**(14), 7625 (2021)
8. Ashfaq, M., et al.: I, chatbot: Modeling the determinants of users' satisfaction and continuance intention of AI-powered service agents. Telematics Inform. **54**, 101473 (2020)
9. Parasuraman, A.: Technology readiness index (tri):a multiple-item scale to measure readiness to embrace new technologies. J. Serv. Res. **2**(4), 307–320 (2000)
10. Flavián, C., et al.: Intention to use analytical artificial intelligence (AI) in services–the effect of technology readiness and awareness. J. Serv. Manag. (2021)
11. Pillai, R., Sivathanu, B., Dwivedi, Y.K.: Shopping intention at AI-powered automated retail stores (AIPARS). J. Retailing Consum. Serv. **57**, 102207 (2020)
12. Jahanmir, S.F., Lages, L.F.: The late-adopter scale: a measure of late adopters of technological innovations. J. Bus. Res. **69**(5), 1701–1706 (2016)
13. Parasuraman, A., Colby, C.L.: An updated and streamlined technology readiness index: TRI 2.0. J. Serv. Res. **18**(1), 59–74 (2015)
14. Pal, D., Babakerkhell, M.D., Zhang, X.: Exploring the determinants of users' continuance usage intention of smart voice assistants. IEEE Access. **9** (2021)
15. Gao, L., Waechter, K.A.: Examining the role of initial trust in user adoption of mobile payment services: an empirical investigation. Inf. Syst. Front. **19**(3), 525–548 (2017)

16. Jo, H.: Understanding the key antecedents of users' continuance intention in the context of smart factory. Technol. Anal. Strateg. Manag., 1–14 (2021)
17. Suzianti, A., Paramadini, S.A.: Continuance intention of E-learning: the condition and its connection with open innovation. J. Open Innov. Technol. Mark. Complex. **7**(1), 97 (2021)
18. Aparicio, M., Bacao, F., Oliveira, T.: Grit in the path to e-learning success. Comput. Hum. Behav. **66**, 388–399 (2017)
19. DeLone, W.H., McLean, E.R.: Measuring e-commerce success: applying the DeLone & McLean information systems success model. Int. J. Electron. Commer. **9**(1), 31–47 (2004)
20. Nguyen, Q.N., Ta, A., Prybutok, V.: An integrated model of voice-user interface continuance intention: the gender effect. Int. J. Hum.–Comput. Interact. **35**(15), 1362–1377 (2019)
21. Jahanmir, S.F., Lages, L.F.: The lag-user method: using laggards as a source of innovative ideas. J. Eng. Technol. Manag. **37**, 65–77 (2015)
22. Chin, W.W., The partial least squares approach to structural equation modeling. Modern Methods for Business Research, 1998. 295(2), 295–336.
23. Fornell, C., Larcker, D.F.: Evaluating structural equation models with unobservable variables and measurement error. J. Mark. Res. **18**(1), 39–50 (1981)
24. Bagozzi, R.P., Yi, Y., Phillips, L.W.: Assessing construct validity in organizational research. Adm. Sci. Q. **36**(3), 421–458 (1991)

Computational Intelligence

Solving a Highly-Constrained Bi-objective Shift Scheduling Problem Using a Specialized NSGA-II

Bryan González-Ramírez(✉), Nicolás Paz-Tralma, and Elizabeth Montero

Departamento de Informática, Universidad Técnica Federico Santa María, Avenida España, 1680 Valparaíso, Chile
{bryan.gonzalezr,nicolas.paz,elizabeth.montero}@usm.cl

Abstract. Highly constrained optimization problems challenge metaheuristics. This work tackles employee shift scheduling as a multi-objective problem balancing satisfaction and coverage. We adapted NSGA-II with problem-specific representation, tailored operators, and hybrid initialization. The model considers ten realistic constraints and two objectives: minimizing preference dissatisfaction and staffing penalties. Six structural indicators characterize instance complexity, and AMPL-Gurobi benchmarks provide reference Pareto fronts for ten problem instances. NSGA-II efficiently generates diverse, high-quality and feasible solutions for small instances, and competitive results for medium-sized ones. Scalability issues arise in larger, highly constrained instances, highlighting the impact of problem structure on performance.

Keywords: Shift scheduling · multi-objective optimization · NSGA-II · instance complexity · employee satisfaction

1 Introduction

Solving hard optimization problems is a complex task for metaheuristic algorithms. Highly constrained optimization problems typically require specialized methods, and multi-objective versions of these problems pose an additional challenge due to the need to find sets of feasible, well-distributed, and high-quality solutions.

In this work, we focus on the Employee Shift Scheduling Problem (ESSP), which involves assigning work shifts to employees while satisfying multiple constraints such as maximum and minimum shifts, working hours, free days, weekends, and consecutive shift rules, aiming to optimize both operational efficiency and employee satisfaction. A literature review of the main ESSP approaches is presented in Sect. 2.

Research Track: CSCI-RTCI.

H. R. Arabnia et al. (Eds.): CSCI 2025, CCIS 2968, pp. 343–358, 2027.
https://doi.org/10.1007/978-3-032-28457-0_25

A specially adapted NSGA-II version is presented, aiming to find high-quality solutions for well-known benchmark problem instances.

The main contributions of this work are as follows. First, we provide exact Pareto front approximations using AMPL-Gurobi for ten benchmark instances, including computed hypervolume values as reference solutions. Second, we propose six structural indicators to characterize instance complexity along dimensions including preference concentration, coverage tension, and constraint pressure. Third, we present a specialized NSGA-II with problem-specific representation, a feasible sequence-based initialization method, five tailored mutation operators and structure-preserving crossovers, designed using a parameter-tuning approach.

Experiments show the approach efficiently solves small-size instances (4–8 employees), outperforming exact methods in hypervolume, and achieves competitive results for medium-size instances (14–20 employees). However, scalability limitations emerge for larger instances (30–40 employees), with large hypervolume gaps revealing effectiveness thresholds for standard NSGA-II configurations.

Section 3 details the problem formulation, Sect. 4 presents the algorithm, Sect. 5 describes the experimental setup, and Sect. 7 discusses results and conclusions.

2 Literature Review

The Employee Shift Scheduling Problem is a problem that can have many variants, one of the best known being the General Employee Shift Scheduling Problem [9]. Its popularity stems from the fact that it is a much more general version of the problem, allowing it to be customized to each organization's needs; however, it is still complex, taking into account, for example, the skills of each employee.

As presented in [3], most research on this problem modifies certain elements, such as constraints or the objective function, to adapt the formulation to specific legal requirements or organizational contexts. Beyond manually designing and adjusting constraints, an alternative approach uses AI to automatically learn them from historical schedules. For instance, in [10], past rosters are modeled as tensors to extract human-readable constraints that capture their underlying structure. Among the main variants is the Stochastic Demand Employee Scheduling Problem [11], which accounts for the uncertainty in employee demand across different shifts. Additionally, the Vehicle Routing Problem with Time Windows (VRPTW) [14] treats the classic VRP as a scheduling problem when including Time Windows. Adding them requires coordinating not just which customers to visit and in what order, but also when each visit occurs to meet time constraints. In essence, the VRPTW combines routing and scheduling into a single integrated problem. In this way, we can also find the study of the schedule of time-dependent ambulance deployment and shifts of the crews [16]. This variant addresses the joint optimization of ambulance locations over time and the assignment of crew shifts.

In [1], authors propose a hybrid algorithm that integrates integer programming (IP) and variable neighborhood search (VNS). After generating an initial solution using a greedy heuristic, the method applies variable neighborhood descent to refine it, and then incorporates IP into a *ruin-and-recreate* scheme to diversify the search. In [12], authors address the challenge posed by restrictions on the number of consecutive assignments, which are known to make solving the problem considerably more difficult. A *reformulation* of these restrictions as subproblems with an efficient internal structure is presented. This reformulation, integrated within a new integer programming model, allows for a significant reduction in computation times to achieve optimal solutions with standard solvers. Later, in [5], authors model the ESSP as a partial weighted maxSAT problem, taking advantage of advances in Boolean satisfiability. This approach allows constraints such as cardinalities and preferences to be formulated naturally and facilitates the use of powerful SAT solvers. Finally, in [13], authors introduce a column generation-based heuristic method, aimed at solving large instances for which traditional exact methods are impractical. Unlike the full branch-and-price approach, this strategy performs an incomplete search that prioritizes good solutions in reduced time, using a fast, low-precision first-order linear programming solver. Beyond minimizing costs and maximizing employee satisfaction, other objective functions include reducing incompatibility among employees sharing a shift [7], lowering fatigue from shift work [1], improving productivity by leveraging employee skills, and promoting fair workload distribution.

3 Problem Definition

The main parameters considered in modeling this problem include employee preferences, employee workload preferences (for example, some employees prefer to work hard over several days and have more days off), employee shift type preferences, legal or company policy limitations on consecutive work days, employee requests for time off, and company requirements for weekend coverage. These preferences are important factors determining employee satisfaction and must be considered when scheduling.

3.1 Mathematical Model

This mathematical model is based on the one presented in [13], however, a multi-objective approach will be considered. In previous work, the penalty for not meeting employee satisfaction and the penalty for underfilling/overfilling shifts were treated as a single objective function. In this model, they are treated as separate objectives, allowing for greater flexibility in prioritizing one objective over another and, in turn, enabling the identification of alternative solutions.

Parameters. We consider a set I of employees, a set D of days of size $|D| = h$, a set W of weekends in the planning horizon, and a set T of shifts. N_i is the set of days the employee i can not be assigned to a shift, while b_i and c_i are the minimum and maximum number of minutes the employee i must/can be assigned. f_i and g_i are the minimum and maximum shifts the employee i must/can work. o_i is the minimum consecutive rest days that the employee i can have, and a_i is the maximum number of weekends that the employee i can work. R_t corresponds to the set of shifts that cannot be assigned after shift t, and l_t is the duration in minutes of shift t. m_{it} correspond to the maximum shifts of type t that can be assigned to employee i, q_{idt} is the penalty if shift t is **not** assigned to employee i on day d, while p_{idt} is the penalty if shift t is assigned to employee i on day d. s_{dt} is the preferred number of employees assigned to shift t on day d, while u_{dt} and v_{dt} are the penalties for under- and over-coverage of shift t in day d.

Decision Variables. We consider binary variables $x_{idt} = 1$, if employee i is assigned to shift t on day d, 0 otherwise, $k_{iw} = 1$ if employee i works the weekend w, 0 otherwise. Also, the under and over-coverages are measured using y_{dt} = under-coverage of shift t on day d, and z_{dt} = over-coverage of shift t on day d.

Constraints. Equation (1) allows at most one shift per employee per day. Equation (2) establishes constraints concerning consecutive shifts. Equation (3) controls the maximum shifts per employee while Eq. (4) controls the minimum and maximum work hours per employee. Equations (5) and (6) control the maximum and minimum consecutive shift sequences per employee. Equation (8) controls the minimum of free consecutive days, and Eqs. (9) and (7) control the maximum worked weekends. Equation (10) forces the free days explicitly asked. Equation (11) controls the minimum coverage per shift.

$$\sum_{t\in T} x_{idt} \leq 1, \quad \forall i \in I, d \in D \tag{1}$$

$$x_{idt} + x_{i(d+1)u} \leq 1, \\ \forall i \in I, d \in \{1, \ldots, h-1\}, t \in T, u \in R_t \tag{2}$$

$$\sum_{d\in D} x_{idt} \leq m_{it}, \quad \forall i \in I, t \in T \tag{3}$$

$$b_i \leq \sum_{d\in D}\sum_{t\in T} l_t x_{idt} \leq c_i, \quad \forall i \in I \tag{4}$$

$$\sum_{j=d}^{d+g_i}\sum_{t\in T} x_{ijt} \leq g_i, \quad \forall i \in I, d \in \{1, \ldots, h-g_i\} \tag{5}$$

$$\sum_{t\in T} x_{idt} + \sum_{j=d+1}^{d+s}\sum_{t\in T} x_{ijt} + \sum_{t\in T} x_{i(d+s+1)t} > 0, \\ \forall i \in I, s \in \{1, \ldots, f_i - 1\}, d \in \{1, \ldots, h-s+1\} \tag{6}$$

$$\sum_{w\in W} k_{iw} \leq a_i, \quad \forall i \in I \tag{7}$$

$$1 - \sum_{t\in T} x_{idt} + \sum_{j=d+1}^{d+s}\sum_{t\in T} x_{ijt} + 1 - \sum_{t\in T} x_{i(d+s+1)t} > 0, \\ \forall i \in I, s \in \{1, \ldots, o_i - 1\}, d \in \{1, \ldots, h-s+1\} \tag{8}$$

$$k_{iw} \leq \sum_{t\in T} x_{i(7w-1)t} + \sum_{t\in T} x_{i(7w)t} \leq 2k_{iw}, \forall i \in I, w \in W \tag{9}$$

$$x_{idt} = 0, \quad \forall i \in I, d \in N_i, t \in T \tag{10}$$

$$\sum_{i\in I} x_{idt} - z_{dt} + y_{dt} = s_{dt}, \quad \forall d \in D, t \in T \tag{11}$$

Objective Functions. Equation (12) minimizes the penalty for unfilled preferred shift assignments while Eq. (13) minimizes the penalty related to under- and over-coverage of shifts. These objective functions correspond to the joint objective function presented in [1]. Here we aim to find sets of trade-off solutions that optimize both, considering different levels of relative importance.

$$\min \quad \sum_{i\in I}\sum_{d\in D}\sum_{t\in T} q_{idt}(1 - x_{idt}) + \sum_{i\in I}\sum_{d\in D}\sum_{t\in T} p_{idt} x_{idt} \tag{12}$$

$$\min \quad \sum_{d\in D}\sum_{t\in T} y_{dt} u_{dt} + \sum_{d\in D}\sum_{t\in T} z_{dt} v_{dt} \tag{13}$$

4 Proposed Approach

This section presents our NSGA-II-based approach. NSGA-II [2] improves the original algorithm with efficient non-dominated sorting and crowding distance [4] to maintain Pareto front diversity. Algorithm 1 outlines its main steps, requiring population size p and crossover/mutation probabilities (c_r, m_r).

The algorithm starts generating the initial population on line 2. The offspring is then generated using selection, crossover, and mutation operators in line 3. At each iteration, in line 5, parent and offspring populations are combined in R_g, which is then classified in non-dominated fronts in line 6 to obtain a new population in line 7. New offspring are obtained on line 8 and the process is repeated until the maximum generations is reached.

Algorithm 1. NSGA-II.

Require: p, c_r, m_r
Ensure: Pareto front F.
1: $g \leftarrow 0$;
2: Generate initial population $P_g(p)$;
3: Generate offspring $Q_g(p)$ (selection, crossover and mutation);
4: **while** $g < MaxGen$ **do**
5: $\quad R_g = P_g \cup Q_g$;
6: $\quad$ Classify non-dominated fronts $F = (F_1, F_2, ...)$;
7: $\quad$ Obtain new population $P_{g+1}(p)$;
8: $\quad$ Generate offspring $Q_{g+1}(p)$ (selection, crossover, mutation);
9: $\quad g \leftarrow g + 1$
10: **end while**
11: $R_g = P_g \cup Q_g$;
12: Classify non-dominated fronts $F = (F_1, F_2, ...)$;
13: **return** F_1

Next, we explain the main components of the NSGA-II algorithms adapted to solve the ESSP.

4.1 Representation

Solutions use a hybrid *genotype-phenotype* encoding.

At the **genotypic** level, each solution is an one-dimension integer array of size $e \times h$ (e employees, h planning horizon), with entries 0 for days off and $[1, s]$ for shifts, allowing standard genetic operators.

At the **phenotypic** level, each employee i maintains sequences $\mathcal{S}_i$ (contiguous shift blocks) and starting days $\mathcal{D}_i$. Each sequence stores shift types, length, and total work time; decoding places sequences at their starting days, filling gaps with days off. This enables precomputed feasible sequences and block-wise local search.

Figure 1 shows examples of solution representations. On the left, a genotypic array for $e = 3$, $h = 3$, $s = 1$ is shown, while the figure on the right depicts its phenotypic decoding, where precomputed sequences $\mathcal{S}_i$ are placed at starting days $\mathcal{D}_i$, filling remaining gaps with days off.

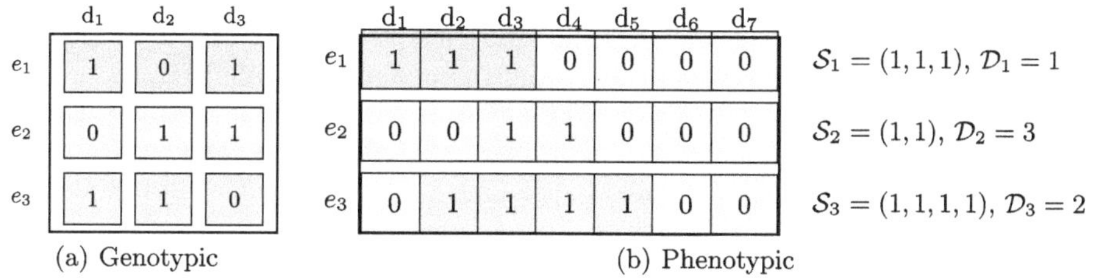

Fig. 1. Solution representation

4.2 Evaluation Functions

In NSGA-II, constrained domination ranks solutions first by feasibility and then by classic domination based on objectives (12) and (13). The constraint violation for all constraints except 1 (handled by the representation) determines the constrained-domination level.

4.3 Mutation Operators

The algorithm implements five different mutation operators, applied in a probabilistically weighted manner. These operators are specifically designed for the problem, seeking to maintain the feasibility of solutions as much as possible while exploring the search space.

1. **Sequence Add mutation:** This mutation inserts a precomputed sequence block into an employee's schedule. It selects an employee e_i, a sequence S of length L, and a starting day d_k, replacing any overlapping assignments. This introduces diversity by exploring new feasible or partially feasible coverage blocks. Figure 2 shows two existing sequences (red) and the added block (green) starting at d_4.
2. **Replace Sequence mutation:** This mutation substitutes an existing sequence with another of equal length. A random employee e_i and one of their sequences S_{ij} are selected. The algorithm searches for a sequence that improves *preference satisfaction* or *coverage balance*; if none is found, a top candidate is chosen randomly to maintain exploration. Figure 3 shows an example of the replace sequence mutation process. Before mutation, employee e_1 has two sequences (red); the second sequence is replaced by a new one of the same length (green), preserving feasibility and potentially improving objectives.

3. **Shift mutation:** This operator enhances local exploration by slightly modifying existing assignments while preserving feasibility. A random employee e_i and one of their sequences are selected, and the sequence is *shifted* forward or backward without overlapping other sequences or violating constraints. Each feasible position is scored based on coverage improvement or preference satisfaction, and a roulette-wheel selection chooses the final placement. This operator introduces fine-grained local adjustments that improve coverage balance and preferences without disrupting the overall schedule (Fig. 4).
4. **Adaptive replace mutation:** This operator introduces adaptive variability by modifying sequence lengths. A random employee e_i and one of their sequences are selected. Instead of replacing it with a sequence of the same length, a shorter feasible length is chosen, reducing workload while maintaining continuity and avoiding gaps. All feasible sequences of the new length are scored for coverage and preference improvement, and a roulette-wheel selection chooses the replacement, favoring sequences that enhance overall schedule quality (Fig. 5).
5. **Replace Employee mutation:** This mutation replaces a randomly selected employee e_i with a feasible substitute e_j, introducing new working patterns while maintaining feasibility. It promotes large-scale exploration, diversifies the population, and helps escape local optima by generating different coverage and preference combinations (Fig. 6).

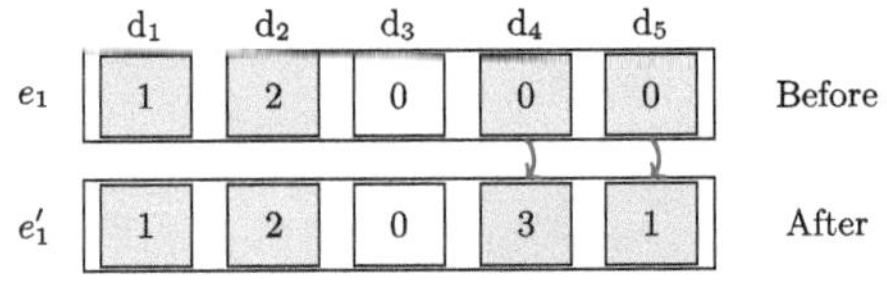

Fig. 2. Sequence Add mutation. A precomputed sequence is inserted for employee e_1, replacing any overlapping blocks. First sequence block is {1, 2}, replaced by {3, 1}.

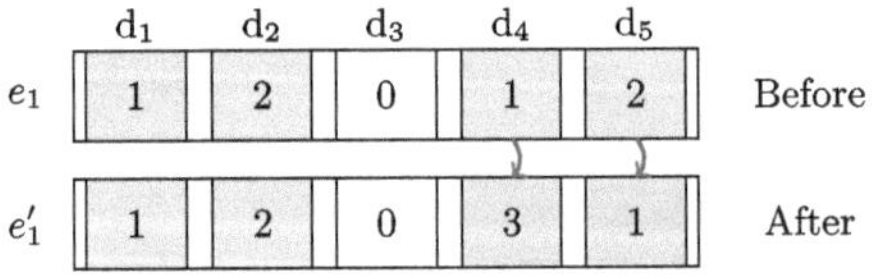

Fig. 3. Replace Sequence mutation. The second sequence of e_1 is replaced by another of equal length to improve preferences or coverage.

Each operator is applied with a normalized probability (p_{mut_i}), allowing control over the relative frequency of use of each mutation type. Operators 2)-5) are specifically designed to preserve constraints such as the total number of assigned turns, while exploring different temporal distributions of these turns.

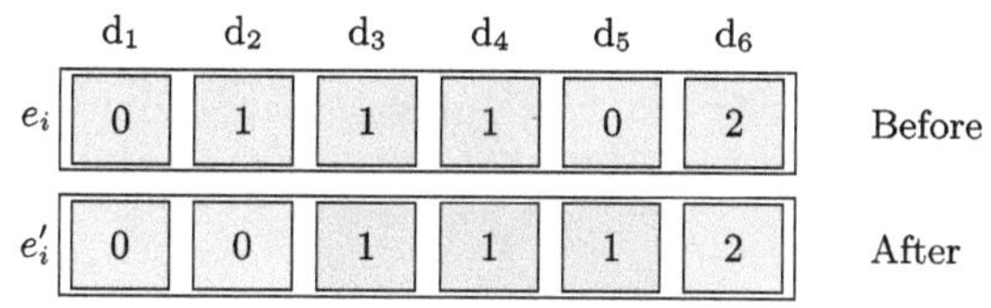

Fig. 4. Shift mutation applied to a single employee's sequence. The red block is shifted to a new feasible position (green), improving coverage or preference quality.

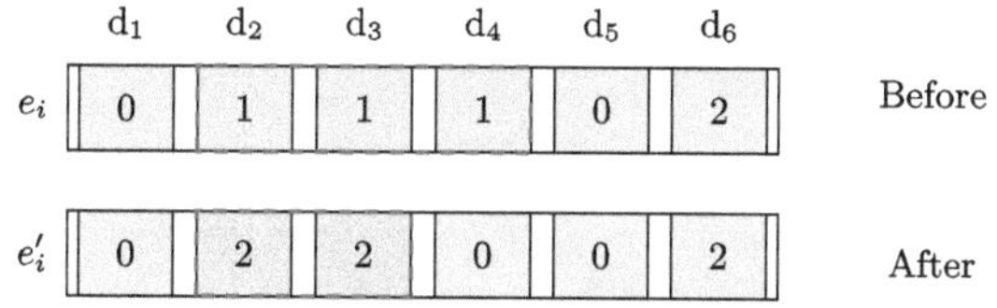

Fig. 5. Adaptive replace mutation. The original sequence (red) is replaced by a shorter, improved sequence (green) that enhances coverage or preferences. (Color figure online)

Crossover Operators. Two crossover operators focused on preserving relevant structures in the solution are implemented. Employee crossover preserves the sequence of shifts per employee, and the day crossover preserves the shift patterns per day. For each parent pair, one of these operators is selected probabilistically.

1. **Employee crossover:** This crossover operator preserves complete shift assignments for selected employees, maintaining the internal consistency of their schedules. For each employee, one of the two parents is randomly chosen to contribute their full shift block to one offspring, while the other parent provides the complementary assignment to the other offspring. This mechanism promotes diversity while retaining feasible and coherent work patterns. Figure 7 illustrates this process for two employees over four days.
2. **Sequence crossover:** This operator combines the precomputed sequence structure of both parents, selecting whole sequences from either parent for each employee. For every employee, all their feasible sequences are considered, and each sequence is inherited by one of the two offspring with equal

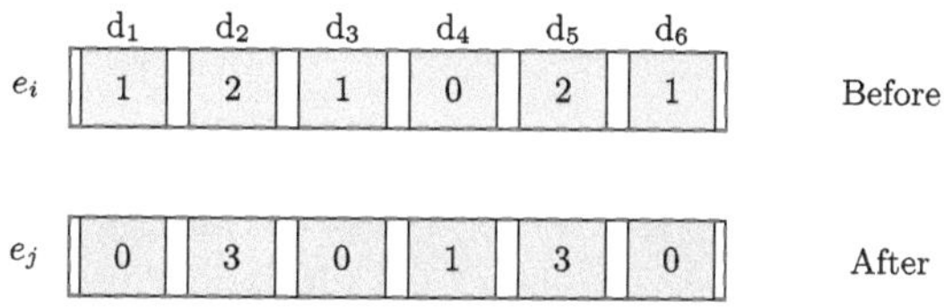

Fig. 6. Replace employee mutation. The entire schedule of employee e_i (red) is replaced by another feasible employee e_j (green), increasing exploration potential. (Color figure online)

probability. This allows offspring to inherit complete, constraint-consistent building blocks (sequences) rather than fragmented daily assignments, thus preserving the structural feasibility learned during evolution.

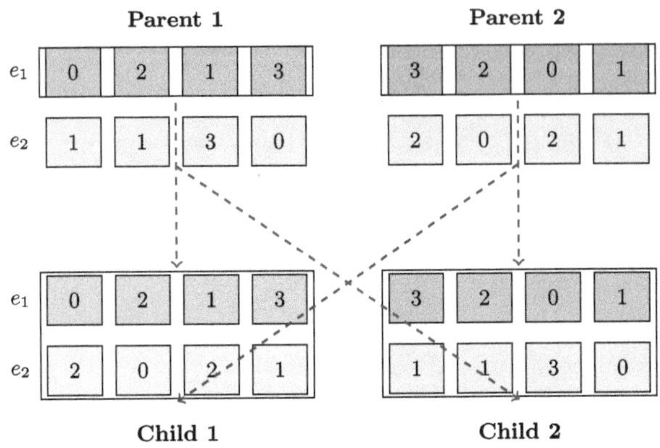

Fig. 7. Employee crossover: complete employee schedules are inherited by alternating between parents. Colors indicate offspring.

4.4 Initialization

The initialization consists of two sequential phases: *Sequence Gather* and *Feasible Employees*, designed to construct high-quality, diverse, and feasible initial individuals.

In the first phase, all feasible work-rest sequences for each employee are generated via backtracking with aggressive pruning, discarding partial assignments that violate constraints such as maximum consecutive shifts, minimum rest days, or forbidden transitions. Each sequence is stored in a *sequence pool*, indexed by length for efficient access during mutations and construction. In the second phase, full feasible schedules are assembled by combining sequences from the pool based on their lengths, reducing the search space. A subsequent *repair stage* replaces conflicting sequences to satisfy global constraints, producing fully feasible employees stored in an *employee pool*. Initial individuals are then created by randomly selecting from this pool, ensuring feasibility.

4.5 Algorithm Parametrization

Table 1 summarizes the parameter settings used in this study. These parameters define the configuration of the NSGA-II algorithm, including population size, number of generations, and the probabilities associated with crossover and mutation operators. In particular, separate probabilities for different mutation and crossover operators allow fine-grained control over exploration and exploitation, while the `mutamount` parameter adjusts the intensity of mutation applications.

Table 1. Parametrization scenario

Parameter	Description	Values
P	population size	[4, ∞]
GENERATIONS	generations	[1, ∞]
p_{cross}	crossover probability	[0, 1]
p_{mut}	mutation probability	[0, 1]
$p_{mut\_1} \cdots p_{mut\_5}$	mutation operators probabilities	[0, 1]
mutamount	proportion of changes in each mutation	[0, 1]
p_{cross_1} and p_{cross_2}	probability of crossover operators	[0, 1]

5 Methodology

In this section, we present our algorithm selection methodology to determine the best parametrization for each problem instance being solved. We used ParamILS [8] to determine the best parameter configuration for each of the 10 problem instances.

All experiments were executed on an Intel(R) Xeon(R) CPU E5-2680 v3, 2.50GHz server.

The 10 problem instances used in this study are available from previous work [13]. To better understand their structure and expected computational difficulty, we computed well-known instance features from [15], such as Total Coverage Constrainedness (TCC), Day and Shift Coverage Distribution (DCD, SCD), and Nurse Preference Distribution (NPD).

Furthermore, we introduce new features specifically designed for the instances used in this work. Here we include features related to positive preference concentrations, tension between negative preferences and required coverage, versatility of staff, demand concentration on weekends, and pressure of successive constraints. Details on these features are presented in Table 2. The computed instance features are summarized in Table 3. These metrics provide a baseline characterization of the structure of problem instances, independent of algorithm performance, and were used to guide the design of representation and mutation operators. Although high employee counts (Instances 8 and 9) correspond to lower Total Coverage Constrainedness (TCC), Instance 10 -despite a similar size- exhibits a TCC comparable to the smallest instances, indicating its relationship with factors beyond the scale of the problem. Other indicators, such as shift diversity (SDI), weekend concentration (WCR_c, WCR_p), and preference-related metrics (PCI, SCP), highlight structural differences that may correlate with algorithmic difficulty, motivating future feature-based analyses and predictive tuning.

We can observe that instances with larger horizons, more employees, or more shifts (e.g., instances 8–10) tend to present slightly higher values in complexity indicators such as Soft Conflict Pressure (SCP), Shift Diversity Index (SDI),

Table 2. Proposed instance features for characterization: PCI (Preference Conflict Index), SCP (Soft Conflict Pressure), SDI (Shift Diversity Index), WCR_c (Weekend Concentration Ratio coverage), WCR_p (Weekend Concentration Ratio preferences), and TIP (Turn Incompatibility Pressure).

Feature	Description	Formula
PCI	Concentration of positive preferences on same shift/day.	$\frac{1}{\lvert D\rvert \lvert T\rvert} \sum_{d,t} \frac{\lvert \{i: p_{idt}=+1\} \rvert^2}{\lvert I\rvert}$
SCP	Tension from negative preferences vs. required coverage.	$\frac{1}{\lvert D\rvert \lvert T\rvert} \sum_{d,t} \frac{\lvert \{i: p_{idt}=-1\} \rvert}{\text{req\_cov}_{dt}}$
SDI	Versatility of staff regarding shifts they can cover.	$\frac{1}{\lvert I\rvert} \sum_{i} \frac{\lvert \{t: \exists d,\ p_{idt} \neq 0\} \rvert}{\lvert T\rvert}$
WCR_c	Demand concentration on weekends.	$\frac{\sum_{(d,t)\in \mathcal{W}\times T} s_{dt}}{\sum_{(d,t)\in D\times T} s_{dt}}$
WCR_p	Concentration of preferences on weekends.	$\frac{\lvert \mathcal{P}_{\mathcal{W}}\rvert}{\lvert \mathcal{P}\rvert}$
TIP	Pressure from succession constraints with few staff.	$\frac{\lvert C\rvert}{\lvert T\rvert^2} \cdot \frac{1}{1+\lvert I\rvert/\lvert T\rvert}$

Table 3. Features for problem instances

Inst.	$\lvert H\rvert$	$\lvert E\rvert$	$\lvert S\rvert$	TCC	DCD	SCD	NPD	SPD	DPD	TIP	PCI	SCP	SDI	WCR_c	WCR_p
1	2	8	1	0.63	0.48	0	0.01	0	0.75	0	0.04	0.32	0.75	0.28	0.15
2	2	14	2	0.55	0.28	0.02	0.01	0.32	0.57	0.03	0.05	0.49	0.57	0.32	0.24
3	2	20	3	0.55	0.19	0.05	0.01	0.23	0.32	0.04	0.04	0.26	0.32	0.32	0.28
4	4	10	2	0.65	0.25	0.04	0.01	0.25	0.65	0.04	0.04	0.31	0.65	0.27	0.24
5	4	16	2	0.64	0.03	0.05	0	0.24	0.75	0.03	0.04	0.28	0.75	0.29	0.26
6	4	18	3	0.59	0.18	0.02	0.01	0.27	0.59	0.05	0.04	0.34	0.59	0.28	0.25
7	4	20	3	0.56	0.1	0.08	0	0.3	0.57	0.04	0.07	0.45	0.57	0.29	0.14
8	4	30	4	0.57	0.06	0.02	0	0.27	0.47	0.04	0.04	0.32	0.47	0.28	0.27
9	4	36	4	0.41	0.09	0.03	0	0.23	0.41	0.04	0.04	0.4	0.41	0.29	0.22
10	4	40	5	0.62	0.03	0.04	0	0.25	0.36	0.04	0.02	0.33	0.36	0.3	0.33

and Weekend Compression Ratio (WCR_p), which can be a sign of the need for larger populations and longer runs.

5.1 Parameter Tuning

ParamILS [8] is an automatic configuration framework that systematically finds high-quality parameter settings for metaheuristics. It combines iterated local search with intensification/diversification strategies to explore a discretized parameter space efficiently. Local moves modify individual parameters, while perturbations escape local optima, under a limited total evaluation budget.

We tune 12 NSGA-II parameters, fixing the total number of evaluations at 1,000,000. The number of generations is set accordingly based on population size. To guide the tuning towards feasible high-quality solutions, we define a combined metric of objective quality and constraint violation:

$$\text{Score} = \text{Norm}(\text{Fo}_1) + \text{Norm}(\text{Fo}_2) + \text{Norm}(\text{Violation}) \tag{14}$$

where each component is normalized by its maximum observed value across runs, ensuring proportional contribution and fair comparison between configurations.

6 Results

NSGA-II performance was evaluated using the tuned parameter configurations. Three independent groups of runs were performed (10×100k, $10 \times 1{,}000$k, and $10 \times 10{,}000$k evaluations). Within each group, Pareto fronts from all runs were aggregated and checked for dominance to assess convergence. The highest-evaluation group consistently produced the best results, which are reported in Table 5 with the number of feasible non-dominated solutions, hypervolume [6] and execution times. The hypervolume indicator (HV) measures the objective-space volume dominated by the obtained Pareto front with respect to a reference point r. Here we use $r_i = \max_{s \in S}(f_i(s)) \times \alpha_i$, using $\alpha_1 = 3$ and $\alpha_2 = 5$. The corresponding aggregated fronts are shown in Fig. 8. For comparison, an exact Pareto front was computed using AMPL-Gurobi.

The AMPL experiments consisted of 11 runs per instance, each using different weight combinations of the two normalized objective functions. This weighted-sum approach allowed us to approximate the exact Pareto front for each instance. Each run had a time limit of 43,200 s (12 h). Table 4 summarizes the AMPL results, including hypervolume values, reference points, number of non-dominated solutions, and computation times.

Table 4. Summary of results for the exact method (AMPL).

Inst.	HV	Ref F1	Ref F2	# sol.	Min t. (s)	Max t. (s)	Av t. (s)	Total t. (s)
1	82 700	3 006	35	4	0	1	0.27	3
2	749 622	5 103	185	6	0	6	1.36	15
3	642 150	13 845	50	4	1	6	3.82	42
4	893 757	6 606	185	5	0	10	3.73	41
5	4 519 457	18 084	275	8	1	70	19.18	211
6	10 378 310	25 644	445	8	3	31	13.45	148
7	16 169 460	25 941	660	8	2	144	42.36	466
8	37 336 341	36 885	1 090	8	17	14 217	2 936.00	32 296
9	244 765	1 503	260	3	7	3 610	3 279.00	36 069
10	33 754 419	49 140	770	7	11	160	50.73	558

NSGA-II and AMPL obtain a comparable number of feasible non-dominated solutions in most cases, with NSGA-II outperforming in some smaller and

Table 5. Performance comparison between NSGA-II and AMPL.

Instance	Solutions		Time (s)		Gap HV (%)	Time ratio
	NSGA-II	AMPL	NSGA-II	AMPL		
1	7	4	81	3	+0.25	27×
2	6	6	110	15	−10.80	7.3×
3	4	4	198	42	−8.56	4.7×
4	12	5	177	41	−12.60	4.3×
5	12	8	247	211	−8.26	1.1×
6	15	8	344	148	−9.40	2.32×
7	31	8	346	466	−7.99	−1.3×
8	38	8	1 046	32 296	−23.45	−30.8×
9	7	3	2 011	36 069	−100	−17.9×
10	31	7	717	558	−27.15	1.28×

medium-sized instances (e.g., Instances 1 and 6). This highlights its ability to find diverse, high-quality solutions efficiently. While AMPL achieves higher precision for certain instances, it requires significantly more computation time. The proposed NSGA-II demonstrates strong performance for small and medium instances, efficiently maintaining feasible and diverse solutions. However, scalability challenges emerge for larger instances (30–40 employees), where solution quality degrades significantly despite reduced computation time compared to exact methods.

NSGA-II performance aligns with the structural insights from Table 3. Instances 8 and 9, which exhibit the poorest convergence for both AMPL and NSGA-II, correspond to high employee counts and lower TCC highlighted above. In contrast, Instance 10, despite having a relatively low TCC similar to smaller instances, achieved considerably shorter execution times, suggesting that multiple structural features beyond TCC influence algorithm difficulty and should be considered when interpreting performance. A convergence analysis shows distinct behaviors across instances. For Instances 1 and 2, the algorithm achieves excellent convergence, even surpassing the number of feasible non-dominated solutions of AMPL–Gurobi. Instances 5 and 10, representing medium and medium-large problems, struggle to approach the Pareto front, though increasing generations improves convergence, indicating the effectiveness of the proposed crossover and mutation operators. Larger problem sizes likely reduce the impact of beneficial mutations, increasing reliance on stochastic effects. Future research could explore (i) adaptive mutation control to reduce randomness in large instances and (ii) rigorous statistical evaluation using multi-seed experiments, aggregated convergence curves, and significance tests.

Figure 8 shows these patterns, highlighting NSGA-II's strong performance on smaller instances and growing challenges with larger, more constrained ones.

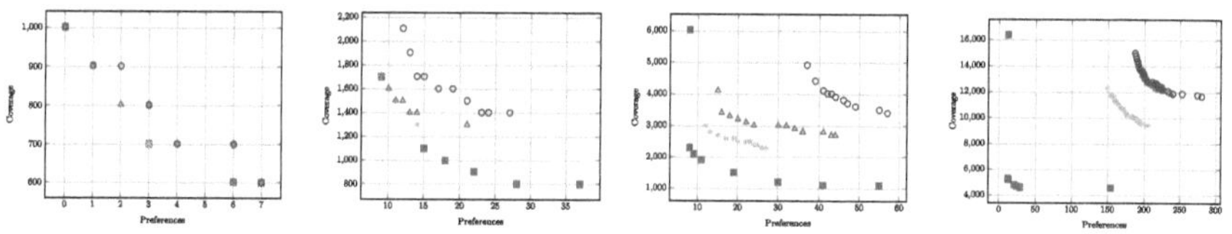

(a) Instance 1 (b) Instance 2 (c) Instance 5 (d) Instance 10

Mono-objetive◆, AMPL ■, NSGA-II (100K) ○ NSGA-II (1000K)△ NSGA-II(10000K) ∗

Fig. 8. Pareto fronts: NSGA-II across different evaluation amounts

7 Conclusions

This work tackles the multi-objective Employee Shift Scheduling Problem using a tailored NSGA-II with hybrid representation, specialized mutations, and two-phase initialization, producing feasible, diverse solutions with good convergence and reasonable computation times. Beyond the optimization framework, this research incorporated a structural characterization of the benchmark instances, introducing both classical and newly defined features to describe their intrinsic difficulty. These indicators provided valuable insight into the relationship between instance properties and algorithmic performance, opening the door to more informed parameter selection strategies. Exact AMPL–Gurobi benchmarks were also developed to provide reference results and validate the effectiveness of the metaheuristic approach. Comparisons showed that the proposed NSGA-II delivers competitive Pareto fronts, particularly on small and medium-sized instances, while maintaining robustness across different problem settings.

Future work should focus on: (i) exploiting the proposed features for predictive algorithm configuration, (ii) developing adaptive mutation strategies for large instances, (iii) hybrid approaches combining NSGA-II with local search or constraint programming for highly constrained cases.

Acknowledgments. This study was funded by FONDECYT project 1230365.

Disclosure of Interests. The authors have no competing interests to declare that are relevant to the content of this article.

References

1. Amindoust, A.: A hybrid genetic algorithm for nurse scheduling problem considering the fatigue factor. J. Healthc. Eng. (2021)
2. Bagchi, T.P.: The nondominated sorting genetic algorithm: NSGA, pp. 171–202. Springer, Boston (1999). https://doi.org/10.1007/978-1-4615-5237-6_8
3. Bruecker, P.D.: Workforce planning incorporating skills: state of the art. Eur. J. Oper. Res. **243**, 1–16 (2014)

4. Deb, K., Pratap, A., Agarwal, S., Meyarivan, T.: A fast and elitist multiobjective genetic algorithm: NSGA-II. IEEE Trans. Evol. Comput. **6**(2), 182–197 (2002). https://doi.org/10.1109/4235.996017
5. Demirović, E., Musliu, N., Winter, F.: Modeling and solving staff scheduling with partial weighted MaxSat. Ann. Oper. Res. **275**(1), 79–99 (2019). https://doi.org/10.1007/s10479-017-2693-y
6. Fonseca, C.M., Paquete, L., López-Ibáñez, M.: An improved dimension-sweep algorithm for the hypervolume indicator. In: Proceedings of the 2006 Congress on Evolutionary Computation (CEC 2006), pp. 1157–1163. IEEE Press, Piscataway (2006). https://doi.org/10.1109/CEC.2006.1688440
7. Hamid, M.: A multi-objective model for a nurse scheduling problem by emphasizing human factors. J. Eng. Med. (2019)
8. Hutter, F., Hoos, H. H., Leyton-Brown, K., Stützle, T.: ParamILS: an automatic algorithm configuration framework. J. Artif. Intell. Res. 36, 267–306 (2009). ISSN: 1076–9757
9. Kletzander, L.: Solving the general employee scheduling problem. Comput. Oper. Res. **113**, 104794 (2020)
10. Kumar, M., Teso, S., De Causmaecker, P., De Raedt, L.: Automating personnel rostering by learning constraints using tensors. In: IEEE 31st International Conference on Tools with Artificial Intelligence (ICTAI), pp. 697–704 (2019). https://doi.org/10.1109/ICTAI.2019.00102
11. Sadjadi, S.: A new nonlinear stochastic staff scheduling model. Sci. Iranica **18**(3), 699–710 (2011)
12. Smet, P.: Constraint reformulation for nurse rostering problems. In: PATAT 2018: Twelfth International Conference on the Practice and Theory of Automated Timetabling, Vienna, Austria, pp. 69–80 (2018)
13. Strandmark, P.: First-order linear programming in a column generation-based heuristic approach to the nurse rostering problem. Comput. Oper. Res. (2020)
14. Truden, C.: Decomposition of the vehicle routing problem with time windows on the time dimension. Transp. Res. Procedia **62**, 131–138 (2022)
15. Vanhoucke, M., Maenhout, B.: On the characterization and generation of nurse scheduling problem instances. Eur. J. Oper. Res. **196**(2), 457–467 (2009). https://doi.org/10.1016/j.ejor.2008.03.044
16. Wiesche, L.: Time-dependent ambulance deployment and shift scheduling of crews. In: Doerner, K.F., Ljubic, I., Pflug, G., Tragler, G. (eds.) Operations Research Proceedings 2015. ORP, pp. 591–596. Springer, Cham (2017). https://doi.org/10.1007/978-3-319-42902-1_80

Solving Large Feature Selection Problems Using an Improved NSGA-II

Nicolás Paz-Tralma(✉), Bryan González-Ramírez, and Elizabeth Montero

Departamento de Informática, Universidad Técnica Federico Santa María, Avenida España, 1680 Valparaíso, Chile
{nicolas.paz,bryan.gonzalezr,elizabeth.montero}@usm.cl

Abstract. This work addresses the feature selection problem as a multi-objective optimization task, motivated by the increasing number of features in datasets used in modern machine learning applications. The goal is to identify a reduced subset of features that improves classification accuracy. We propose using the Chi-square statistical test to initialize the evolutionary algorithm NSGA-II, taking advantage of its population-based nature to find a set of non-dominated, high-quality solutions. Two initialization strategies are proposed and evaluated. Six datasets from different domains were tested, ranging from 16 to 12,600 features. Improvements in accuracy were observed, especially when initializing with 10% of the features. Moreover, the proposed strategies did not significantly increase computational time and achieved competitive results compared to the state-of-the-art in terms of hypervolume.

Keywords: Feature selection · NSGA-II · Chi-square · Multi-objective optimization · Evolutionary algorithms

1 Introduction

The increasing volume and dimensionality of modern datasets pose significant challenges in machine learning, particularly in domains such as biomedicine and image analysis [1]. A crucial preprocessing step in such high-dimensional settings is feature selection, which aims to identify a compact subset of relevant features to improve classification performance while reducing computational complexity.

Feature selection simultaneously pursues two goals: reducing the number of selected features and improving classification accuracy. This makes it suitable to be formulated as a multi-objective optimization problem. However, due to its combinatorial complexity, the problem is NP-hard [13], motivating the use of evolutionary algorithms. Among them, the Non-Dominated Sorting Genetic Algorithm II (NSGA-II) [3] is widely used due to its capability to generate a diverse set of non-dominated solutions, effectively constructing a Pareto front

Research Track: CSCI-RTCI.

H. R. Arabnia et al. (Eds.): CSCI 2025, CCIS 2968, pp. 359–375, 2027.
https://doi.org/10.1007/978-3-032-28457-0_26

that balances the competing objectives. One persistent challenge in NSGA-II, however, is the generation of the initial population, which significantly influences convergence behavior and solution quality.

To address this issue, we propose initializing the population using the Chi-square statistical test [5], selecting features most strongly associated with the target class. We also introduce a hybrid initialization strategy that combines Chi-square-based selection with random feature inclusion, thus promoting both exploitation and exploration. This study aims to evaluate whether such initialization strategies can improve the performance of NSGA-II in feature selection. We examine the impact of initializing with 10% and 50% of the total features, assessing how these strategies influence accuracy, generalization, and computational cost. Our approach is validated using six datasets of varying size and domain. We benchmark our results against CNSGA-II [7], using the hypervolume indicator as the primary performance metric.

The remainder of this paper is organized as follows: Sect. 2 reviews relevant literature, Sect. 3 defines the problem formally, Sect. 4 presents our proposed methods, Sect. 5 details the experimental setup and datasets, Sect. 6 discusses the results and Sect. 7 concludes the study and shows directions for future research.

2 Literature Review

Classical Feature selection algorithm approaches are divided into three categories: filter methods, wrapper models, and embedded models [13].

Filter models evaluate features without using any classification algorithm. These algorithms usually consist of two steps. The first is to classify the characteristics based on a single criterion. The second is to choose the best-ranked features to induce classification models [13]. One of the variants used to address the problem is the correlation-based fast filter [18]. This discusses the concept of *predominant correlation* and proposes an efficient method to analyze redundancy between features without a full pairwise analysis. The algorithm calculates symmetric uncertainty values for each feature, selecting relevant features based on a predefined threshold, and sorting them in descending order. After that, it processes the sorted list and removes the redundant features, keeping only the predominant ones. Tests showed a more efficient approach able to reduce dimensionality and improve precision compared to other algorithms. The problem with filter models is that they completely ignore the performance of the subset of features in the induction algorithm. For this reason, wrapper models use a specific classifier to evaluate the quality of the selected features, thus offering a simple and powerful way to approach the problem. First, the feature set is searched, then the selected subset is evaluated, and finally, the previous steps are repeated until the desired quality. This is why the wrapper model has a higher computational cost compared to the filter model [13]. Embedded models address feature selection through the construction of classifiers. There are three types of integrated methods. Pruning methods use all features to train a model and then try to eliminate some features by setting the corresponding coefficients to zero to maintain the performance of the model [13]. In the second type are models with a built-in mechanism for feature selection, such as ID3 [9] and C4.5 [10] that build decision trees by selecting in each node the attribute that maximizes the information gain. Finally, regularization models aim to minimize fitting errors and force the coefficients to be as small as possible. Features that reach zero are removed [6].

The feature selection problem has also been tackled using multi-objective search algorithms. In [14] the Flexible Cut-Point Particle Swarm Optimization is presented, which provides the possibility of choosing a flexible number of cut points to discretize the characteristics. Applying the latter means that continuous data will become discrete data. The approach allows discretization and selection to be done in a single structure, reducing complexity in large volumes of data. In [12], authors propose a Differential Multi-Objective Evolution (DEMOFS). This method uses a representation of real numbers between 0 and 1, where each value corresponds to the probability of choosing a given feature. This method makes use of the NSGA-II structure for the part of the non-dominated ordering of individuals and crowding distance. DEMOFS is compared with two single-objective approaches DEFS and DEFS2, also proposed in [12], demonstrating its ability to solve multi-objective versions of the problem. The Compact NSGA-II (CNSGA-II) [7], unlike the classical algorithm, uses probability vectors to represent solutions. This allows Pareto front solutions to be improved by more efficiently exploring the search space. Five data sets were used in the experimentation, which were related to image and face recognition. The results showed that CNSGA-II achieved a better reduction of selected features and lower minimum classification error in three of the five instances used, outperforming NSGA-II in terms of hypervolume. In [16] NSGA-II is combined with ReliefF to show the internal information of the data set and assign to each characteristic a weight based on its importance. During the initialization of the population and the transformation process of NSGA-II, the characteristics with greater weight are more likely to be chosen.

In conclusion, although filter, wrapper, embedded approaches, and multi-objective evolutionary algorithms have demonstrated significant advances in feature selection, their proper combination remains an open challenge for large feature selection problems. This highlights the need to explore evolutionary algorithm strategies based on filter algorithms, which justifies and supports the present research as a logical contribution to the field.

3 Problem Definition

The feature selection problem primarily arises when working with machine learning models that handle large datasets with thousands of features. From such vast amounts of data, it is common to identify irrelevant or redundant features that add complexity to the classification process without improving the results. This defines the challenge of reducing the dimensionality of the data without compromising the performance of the model, aiming to improve the efficiency and effectiveness of machine learning approaches in real-world applications [13].

Feature selection is a complex engineering problem that seeks to identify an optimal subset of relevant features, thus helping to reduce complexity when handling large datasets. This problem involves two conflicting objectives: minimizing the number of selected features and maximizing the classification accuracy of the model. Consequently, feature selection can be approached as a single-objective or multi-objective optimization problem, depending on the adopted methodology. Feature selection does not modify the original data, but rather selects a subset of the existing features [1].

The multi-objective optimization problem involves the simultaneous optimization of two or more conflicting objectives. This type of problem can be addressed by either transforming it into a single-objective problem, for example, by applying weights to each objective, or by obtaining a set of solutions that meet the established requirements and constraints. The key challenge is that generally, no single solution can optimize all objectives at the same time, as they tend to conflict with one another [15].

In this work, we propose an improved NSGA-II approach, where initialization strategies are enhanced to solve feature selection problems by minimizing the number of features and maximizing the model's accuracy.

4 Algorithm Proposal

NSGA-II is a well-known evolutionary multi-objective algorithm in the literature [3]. NSGA-II incorporates a complexity-efficient non-dominated front definition approach and the crowding distance to compare equivalent solutions. The algorithm 1 shows the main structure of the NSGA-II approach. The algorithm requires setting the population size, the crossover and mutation probabilities, and the selection probability.

Algorithm 1. NSGA-II

Require: λ, p_c, p_m, h
Ensure: Pareto front F.
$g = 0$
P_g = Generate initial population(λ)
Q_g = Generate offspring using selection, crossover and mutation(λ, p_c, p_m)
while $g < MaxGen$ **do**
    R_g = Combine $P_g \cup Q_g$
    F = Classify R_g on $(F_1, F_2, ...)$
    P_{g+1} = Obtain new population(λ)
    Q_{g+1} = Generate offspring using selection, crossover and mutation(λ, p_c, p_m)
    $g = g + 1$
end while
R_g = Combine $P_g \cup Q_g$
F = Classify R_g on $(F_1, F_2, ...)$
return F_1

NSGA-II begins with the initialization of the population in line 2. Then, genetic selection, crossover, and mutation operators are applied to generate the offspring solutions in line 3. Between lines 4 and 10, an elitist choice is made to create the population of the next generation. This selection is made considering the parent and offspring populations. This procedure is implemented between lines 5 and 7. Infeasibility is incorporated in the non-dominance analysis. In this sense, given two solutions i and j, i is preferred if (1) the solution i is feasible, and solution j is not, (2) both solutions are infeasible, but solution i has a lower level of violation of constraints than j, and (3) both solutions are feasible and i dominates j based on objective functions. At the end of each generation, new offspring are generated, and the process is repeated. In the end, the first best non-dominated front obtained from the last parent and offspring population is returned in line 13.

In our approach, we represent solutions using binary arrays of size D, with D is the total number of features. The selection and mutation operators used are binary tournament selection and bit-flip mutation, as described in [3]. For the crossover operator, a two-point crossover is implemented.

Equations (1) and (2) show the two objective functions implemented. In Eq. (1), C corresponds to the total number of classes, FP_i represents the values that are false positives in class C, and S_i represents all values in class C. On the other hand, in Eq. (2), #selected_features corresponds to the number of features used. This number can be lower than the number of selected features indicated by the chromosome representation.

$$F_1 = 1 - \frac{\sum_{i=1}^{C} FP_i}{\sum_{i=1}^{C} S_i} \tag{1}$$

$$F_2 = \#\text{selected\_features} \tag{2}$$

Next, we explain in detail the initialization processes proposed.

4.1 Initialization

Classic Initialization. The initialization of the NSGA-II algorithm generates random binary solutions. In this case, for each solution, a $h\%$ probability of selecting a feature is set. NSGA-II algorithm uses a 50% probability [3]. In this case, each individual in the initial population is expected to possess half of the problem's active features.

Chi-Square Initialization. To improve the population initialization in the NSGA-II algorithm, a strategy based on the Chi-square statistical test (χ^2) [5] is implemented. This technique allows for identifying the features that have a greater dependence on the target variable. This would be useful to reduce the number of features without affecting the accuracy of the model.

The Chi-square statistical test (χ^2) is a nonparametric test for analyzing differences between categorical variables in a population. The null hypothesis states that the characteristics being compared are independent (unrelated) [17]. The objective here is to compare the observed frequencies (O), which correspond to the actual data, with the expected frequencies (E), which correspond to the expected values. If the observed values are similar to the expected ones, the hypothesis is not rejected, and if they differ from each other, then the hypothesis is rejected, indicating that a significant relationship may exist. In the context of machine learning, it is a tool used in classification problems to measure the dependence between each characteristic and the target class [5,17]. It is applied in cases where the data are categorical or discretized, and allows for the identification of characteristics that exhibit a significant relationship with the target variable.

The initialization process is presented in Algorithm 2, where n corresponds to the number of intervals in which the data is discretized, j the percentage of features to be selected, $prob1$ the probability of mutating a bit, *percent* the percentage of features with better Chi-squared values, and P the initial population. In line 1, the algorithm discretizes the features X. This procedure converts continuous features into discrete values for the Chi-square test. Then, in line 2, the χ^2 test scores are computed for each feature. These scores are then normalized to facilitate selection. In line 3, the $j\%$ of features with the highest χ^2 scores are selected. In line 4, an individual is created containing the features selected above, and in line 5, a vector with the feature indices ordered from highest to lowest χ^2 value is generated. In line 6, the first individual in the population is initialized with M_0. Subsequently, in lines 7 and 8, the number of features that can be mutated in M_0 is computed, as well as the number of times a bit will be mutated in each individual. The values obtained in the two previous lines are used in the cycle between lines 9 and 12, where the idea is to generate a loop to have different individuals in the population. To do this, the first thing to do is copy the first individual to another in line 10. Subsequently, in line 11, individuals are varied by mutating a total of *var* bits, which are randomly chosen, ensuring that the mutated characteristics are not repeated and limiting the possible characteristics selected from the vector M based on the variable *limit*. Once the cycle is finished, the initial population is returned in line 13.

Hybrid Initialization. We also evaluate a hybrid approach that follows the same first steps as the Chi-square strategy (lines 1–5), selecting the top $j/2\%$ of features. Then, an additional $j/2\%$ is randomly chosen from the remaining features, ensuring no overlap with the first half of the selected features.

5 Experimental Setup

In this section, we present the main components of our experimental setup, including the features of the computational test environment, the NSGA-II parameter settings, and the learning approach and datasets used for the experiments. All experiments were run in an Intel(R) Xeon(R) CPU E5-2680 v3 at 2.50GHz with Ubuntu 22.04.3.

Algorithm 2. Chi-square-based initialization

Require: X, $c0$, j, D, λ, $prob$, $percent$
Ensure: Initial population P
X_{Disc} = Discretize data(X)
Chi = Compute and normalize scores $\chi^2(X_{Disc}, y)$
Chi_j = Select best scored features(Chi, j)
M_0 = Binary vector (Chi_j)
M = Sorted list features (Chi_j)
$P_0 = M_0$
$limit = D * percent$
var = Compute mutation for bits ($limit$, $prob$)
while ($i < \lambda - 1$) **do**
    $P_i = P_0$
    P_i = Mutate random different positions (var, $limit$, M)
end while
return P

Experiments were executed using the datasets listed in Table 1. The first two datasets were used in [11] and are available in this GitHub repository. The *warpAR10P* and *warpPIE10P* datasets were used in [7] and are available at [11]. The *penbased* and *texture* datasets were used in [4] and are available in the *KEEL Dataset Repository*. The datasets were selected for their variety in size, domain, and classification types (binary and multiclass), ranging from 16 to 12,600 features and 77 to 10,992 samples.

Table 1. Datasets

Name	Features (D)	Samples	Classes
DLBCL	5,469	77	2
Lung Cancer	12,600	203	5
warpAR10P	2,400	130	10
warpPIE10P	2,420	210	10
Penbased	16	10,992	10
Texture	40	5,500	11

To evaluate the proposed approach, the k-NN algorithm [2] is used together with Stratified K-Fold Cross-Validation (SKCV) [8]. It should be noted that the value of k associated with k-NN is different from the value of K associated with SKCV. For k-NN, $k = 3$ is used, as in [16].

To evaluate the performance, each dataset is randomly split in every run: 80% for training the algorithm and 20% for testing the results. Each algorithm is executed 10 times per data set. For all of our experiments, we used a mutation probability $p_m = 1/D$, and a crossover probability $p_c = 0.9$. Mutation and crossover probabilities were obtained as recommendations from literature [3,16]. Moreover, from preliminary experiments, a population size of 20 individuals was chosen to reduce computational costs per iteration, and 500 generations were used to achieve a total effort equivalent to that presented in the literature [7]. Two values were tested in our experiments for h and j: 10% and 50%, corresponding to the percentage of features initially selected in each strategy. The mutation probability $prob$ takes two values depending on the dataset dimensionality: 10% for datasets with fewer than 1,000 features, and 0.1% otherwise. This setting ensures a comparable number of mutations across datasets and initialization methods. Finally, $select$ was fixed at 50% to ensure that the same subset of top-ranked features was considered in the Chi-square initialization, regardless of whether 10% or 50% of initial features were ultimately selected. These values were also derived from preliminary experiments.

6 Results

Tables 2, 3, and 4 show the results obtained by the NSGA-II algorithm using the three initialization approaches. The minimum values, averages, and the highest standard deviation in each row are bold. The results indicate that Chi-square-based initialization consistently achieves lower average errors in both training and testing in most instances, which highlights its elitist character by prioritizing relevant features from the outset. However, this also limits its exploratory capacity, as evidenced by its higher minimum number of features and lower standard deviation. In contrast, random initialization exhibits greater variability in both error and the number of features selected, suggesting a broader exploration but with less consistency, especially on challenging datasets such as warpAR10P. The hybrid strategy achieves a better balance, reaching competitive minimum error rates, lower average number of features in most cases, and broader exploration of the solution space, making it a valuable alternative for improving the effectiveness of NSGA-II in both accuracy and dimensionality reduction.

Tables 5, 6 and 7 show the results found by NSGA-II algorithm for a 50% of initial features. The minimum values, averages, and the highest standard deviation in each row are bold. The results obtained with a 50% of the initial features

Table 2. Minimum and av. ± st. dev. of percentage error in training set - Scen. 10%

Instance	Random		Chi-square		Hybrid	
	Min	ave ± sd	Min	ave ± sd	Min	ave ± sd
DLBCL	**0.00**	1.76 ± 0.89	**0.00**	2.86 ± **1.59**	**0.00**	**0.46** ± 0.87
Lung Cancer	**0.00**	**2.84** ± **2.41**	3.68	5.11 ± 0.91	3.72	4.92 ± 0.81
warpAR10P	22.10	32.67 ± **4.86**	**5.00**	**14.32** ± 3.50	12.39	18.87 ± 3.45
warpPIE10P	1.30	2.89 ± 1.02	**0.00**	**2.23** ± 0.82	1.09	2.69 ± **1.06**
penbased	0.70	13.56 ± 19.55	**0.60**	**13.22** ± 19.00	0.61	14.42 ± **20.07**
texture	**0.67**	**9.72** ± 15.82	0.76	**9.72** ± 15.53	0.80	11.66 ± **18.12**

Table 3. Minimum and av. ± st. dev. of percentage error in testing set - Scen. 10%.

Instance	Random		Chi-square		Hybrid	
	Min	ave ± sd	Min	ave ± sd	Min	ave ± sd
DLBCL	**0.00**	**10.13** ± **6.54**	6.25	11.07 ± 4.79	14.29	17.08 ± 4.39
Lung Cancer	**2.56**	15.14 ± **9.69**	7.26	**11.05** ± 3.46	5.41	16.21 ± 7.53
warpAR10P	62.14	71.19 ± 4.96	**37.54**	**50.22** ± 7.68	39.58	58.88 ± **9.63**
warpPIE10P	39.39	65.54 ± **11.28**	37.07	**53.73** ± 8.59	**36.62**	53.75 ± 7.29
penbased	1.48	15.97 ± **19.81**	**1.46**	**15.73** ± 19.44	1.55	16.12 ± 19.54
texture	2.49	11.60 ± 14.26	**2.36**	**11.36** ± 14.59	2.49	12.63 ± **16.15**

Table 4. Minimum and av. ± st. dev. of number of features in training set - Scen. 10%

Instance	Random		Chi-square		Hybrid	
	Min	ave ± sd	Min	ave ± sd	Min	ave ± sd
DLBCL	270	295.83 ± 12.85	329	333.92 ± 5.37	**265**	**288.03** ± **16.54**
Lung Cancer	932	1051.47 ± **120.06**	1027	1032.25 ± 3.73	**930**	**955.36** ± 16.47
warpAR10P	**87**	**111.88** ± 15.43	115	134.10 ± **25.90**	112	128.02 ± 8.94
warpPIE10P	68	93.34 ± **11.47**	73	91.47 ± 9.59	**58**	**76.52** ± 6.83
penbased	**1**	**6.45** ± 3.56	**1**	6.79 ± 3.89	**1**	6.58 ± **3.93**
texture	**1**	7.47 ± **4.82**	**1**	6.96 ± 4.37	**1**	**6.49** ± 4.18

Table 5. Minimum and av. ± st. dev. of percentage error in training set - Scen. 50%

Instance	Random		Chi-square		Hybrid	
	Min	ave ± sd	Min	ave ± sd	Min	ave ± sd
DLBCL	3.10	4.52 ± 0.52	**1.43**	**3.70** ± **0.89**	6.00	8.17 ± 0.77
Lung Cancer	**1.96**	**4.01** ± 1.17	3.62	5.01 ± 1.02	4.29	6.02 ± **1.20**
warpAR10P	39.65	41.83 ± **1.60**	27.91	31.07 ± 1.40	**21.05**	**24.64** ± 1.31
warpPIE10P	**1.50**	4.23 ± 0.97	1.82	**3.19** ± **1.40**	2.68	4.00 ± 0.73
penbased	0.60	**13.34** ± 18.91	0.63	14.59 ± **20.46**	**0.58**	14.22 ± 20.12
texture	**0.73**	**9.19** ± 15.07	**0.73**	9.91 ± 15.98	0.75	10.06 ± **16.32**

present a behavior similar to that observed with 10%, but with a lower overall performance. This is due to the higher presence of irrelevant features from the beginning, which raises errors and makes dimensionality reduction more difficult. The Chi-square strategy maintains its elitist character, but with greater dispersion; the random strategy continues to stand out for its exploratory capacity, and the hybrid strategy retains its balance between quality and diversity. Overall, the previous patterns are maintained, but the limitations are intensified by increasing the proportion of initial characteristics.

Table 6. Minimum and av. ± st. dev. of percentage error in testing set - Scen. 50%

Instance	Random		Chi-square		Hybrid	
	Min	ave ± sd	Min	ave ± sd	Min	ave ± sd
DLBCL	16.67	21.76 ± **4.92**	17.00	23.56 ± 2.87	**12.96**	**16.14** ± 2.07
Lung Cancer	**2.50**	13.70 ± 6.57	6.07	**12.59** ± 4.58	5.59	18.15 ± **6.85**
warpAR10P	60.61	72.44 ± 7.39	**45.77**	68.24 ± **12.06**	49.21	**63.81** ± 10.68
warpPIE10P	53.44	67.29 ± **11.06**	53.40	**66.75** ± 7.95	**49.50**	68.21 ± 7.34
penbased	**1.27**	15.66 ± **19.64**	1.48	**15.16** ± 19.19	1.60	15.35 ± 19.47
texture	**2.37**	**11.43** ± 14.38	2.52	12.04 ± 15.37	2.52	12.49 ± **16.04**

Table 7. Minimum and av. ± st. dev. of number of features in training set - Scen. 50%

Instance	Random		Chi-square		Hybrid	
	Min	ave ± sd	Min	ave ± sd	Min	ave ± sd
DLBCL	**1892**	**1912.87** ± 18.19	2003	2041.77 ± 19.59	1939	1967.08 ± **26.33**
Lung Cancer	**5277**	**5377.11** ± 43.65	5531	5557.71 ± 16.78	5362	5416.06 ± **46.47**
warpAR10P	574	592.81 ± 12.77	618	648.06 ± **22.62**	**551**	**584.92** ± 16.23
warpPIE10P	**534**	**580.53** ± **32.18**	586	609.25 ± 11.16	555	589.76 ± 20.90
penbased	**1**	6.70 ± 3.91	**1**	**6.27** ± 3.52	**1**	6.67 ± **3.93**
texture	**1**	7.60 ± 5.08	**1**	7.70 ± **5.25**	**1**	**7.37** ± 4.73

Figures 1, 2, 3 and 4 show the Pareto fronts found in scenarios 10% and 50% for training and testing sets. It is clear from these results that the last two sets are easily solvable for the three initialization approaches in both scenarios. In these cases, all methods obtained well-distributed and high-quality non-dominated solutions. In the hardest cases, the Chi-square approach focuses on low-error solutions, while the random approach tends to generate more diverse fronts with fewer features. The hybrid strategy achieved the best trade-offs in several instances under the 10% scenario, but the random initialization performed better overall when using 50% of initial features.

Tables 8, 9, 10 and 11 show the hypervolume obtained in scenarios 10% and 50% for training and testing sets. The highest average hypervolume per instance is shown bold. To ensure consistency across all evaluations, hypervolume was computed using a normalized version of function 2 was adapted to use the reference point (1, 1).

Tables 12 and 13 show the hypervolume obtained in scenarios 10% and 50% for training and testing sets compared to CNSGA-II [7]. The highest average hypervolume per instance is shown bold. The comparison with CNSGA-II shows that, in the training set with 50% of initial features, only the hybrid initialization outperforms the state of the art in the first instance, while both proposed strategies improve the average performance in the other case. In testing,

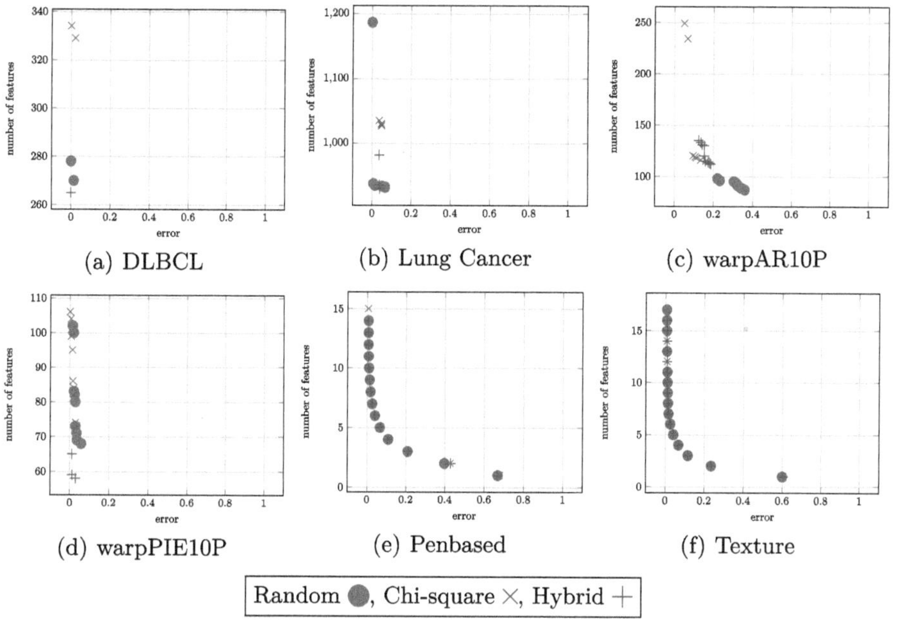

Fig. 1. Training Pareto Fronts - Scenario 10%.

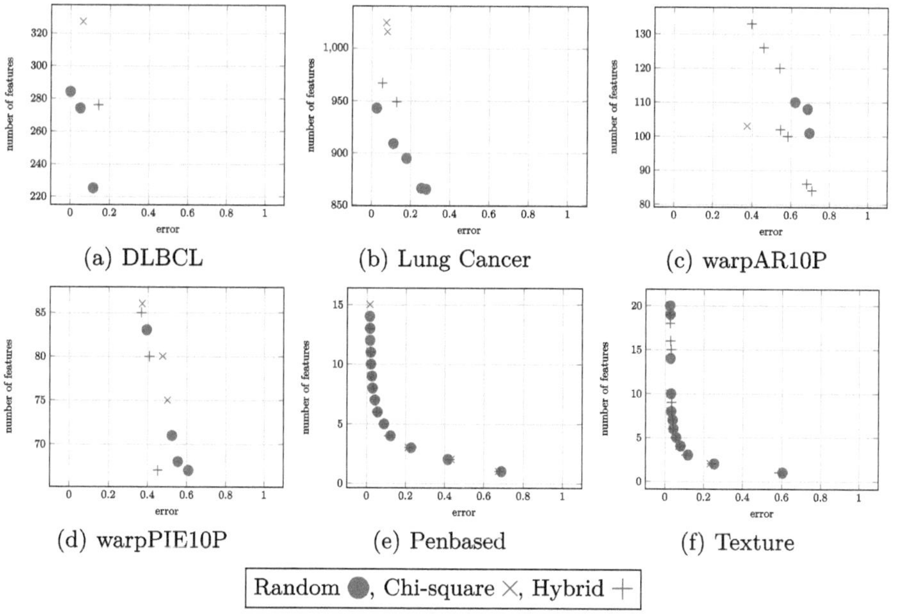

Fig. 2. Testing Pareto Fronts - Scenario 10%.

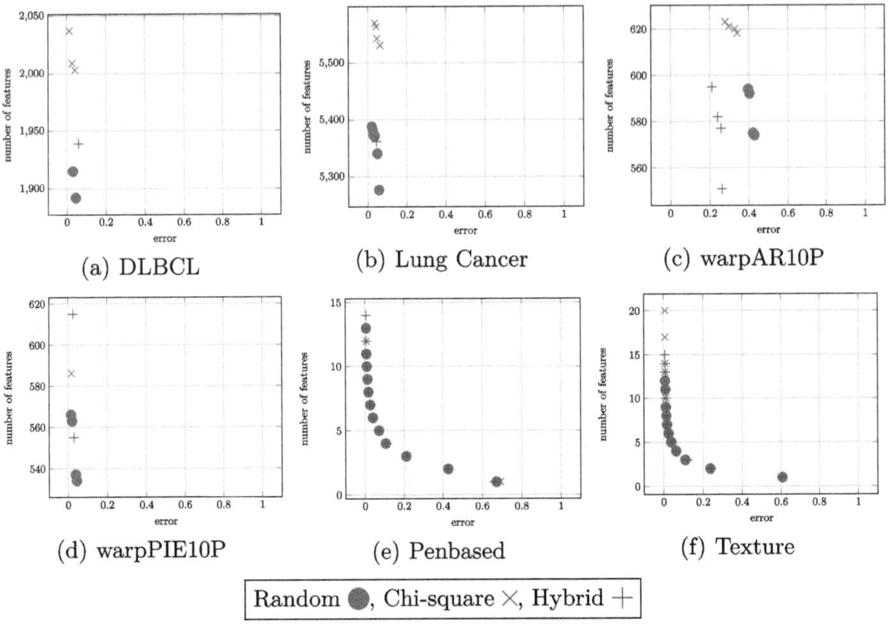

Fig. 3. Training Pareto Fronts - Scenario 50%.

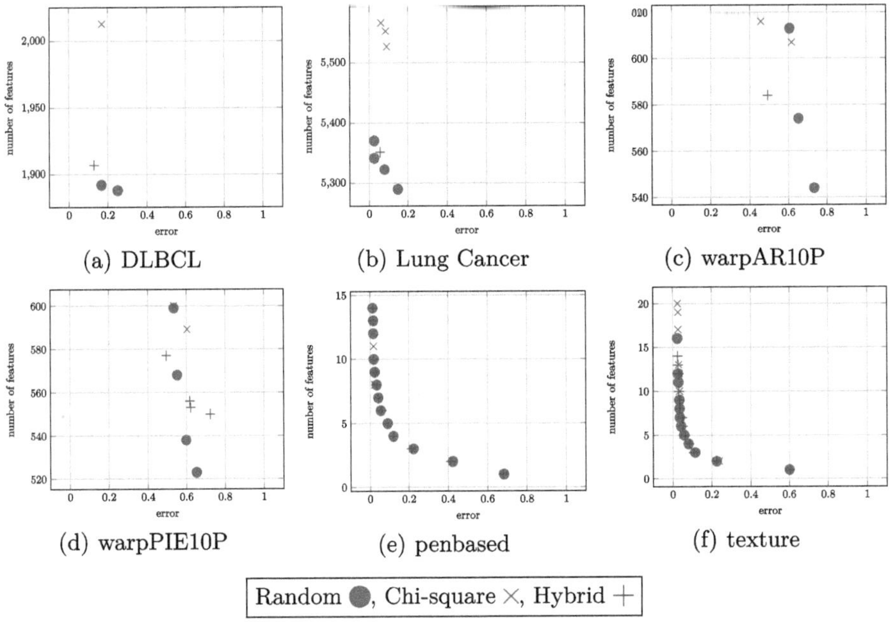

Fig. 4. Testing Pareto Fronts - Scenario 50%.

Table 8. Best and average ± st. deviation Hypervolume in training set - Scen. 10%.

Instance	Random		Chi-square		Hybrid	
	Best	ave ± sd	Best	ave ± sd	Best	ave ± sd
DLBCL	0.7717	**0.7159 ± 0.0406**	0.6518	0.6011 ± 0.0253	0.5882	0.5521 ± 0.0203
Lung Cancer	0.9258	**0.9203 ± 0.0051**	0.8841	0.8707 ± 0.0096	0.8901	0.8798 ± 0.0097
warpAR10P	0.7478	0.6638 ± 0.0480	0.8619	**0.8288 ± 0.0287**	0.8311	0.7817 ± 0.0329
warpPIE10P	0.9472	0.9375 ± 0.0091	0.9591	0.9444 ± 0.0074	0.9648	**0.9494 ± 0.0103**
penbased	0.8360	**0.8343 ± 0.0010**	0.8338	0.8331 ± 0.0005	0.8344	0.8329 ± 0.0008
texture	0.9402	**0.9386 ± 0.0010**	0.9401	0.9384 ± 0.0010	0.9396	0.9384 ± 0.0006

Table 9. Best and average ± st. dev. Hypervolume in testing set - Scenario 10%.

Instance	Random		Chi-square		Hybrid	
	Best	ave ± sd	Best	ave ± sd	Best	ave ± sd
DLBCL	0.9481	**0.8512 ± 0.0658**	0.8814	0.8399 ± 0.0466	0.8138	0.7627 ± 0.0516
Lung Cancer	0.9015	0.7679 ± 0.0748	0.8520	**0.7988 ± 0.0451**	0.8733	0.7551 ± 0.0575
warpAR10P	0.3613	0.2853 ± 0.0565	0.5978	**0.4826 ± 0.0937**	0.5707	0.3954 ± 0.0984
warpPIE10P	0.5853	0.3468 ± 0.1325	0.6070	**0.4918 ± 0.0863**	0.6127	0.4660 ± 0.0857
penbased	0.8204	0.8180 ± 0.0020	0.8207	**0.8186 ± 0.0015**	0.8238	0.8184 ± 0.0033
texture	0.9218	0.9145 ± 0.0053	0.9245	**0.9174 ± 0.0054**	0.9218	0.9147 ± 0.0044

Table 10. Best and average ± st. dev. Hypervolume in training set - Scen. 50%.

Instance	Random		Chi-square		Hybrid	
	Best	ave ± sd	Best	ave ± sd	Best	ave ± sd
DLBCL	0.6297	**0.6210 ± 0.0046**	0.6185	0.6048 ± 0.0081	0.6068	0.5895 ± 0.0089
Lung Cancer	0.5612	**0.5524 ± 0.0060**	0.5377	0.5290 ± 0.0051	0.5498	0.5365 ± 0.0080
warpAR10P	0.4546	0.4334 ± 0.0094	0.5344	0.5075 ± 0.0145	0.5938	**0.5707 ± 0.0101**
warpPIE10P	0.7559	**0.7321 ± 0.0148**	0.7441	0.7255 ± 0.0141	0.7476	0.7262 ± 0.0119
penbased	0.8346	0.8333 ± 0.0007	0.8337	**0.8337 ± 0.0001**	0.8348	0.8329 ± 0.0011
texture	0.9407	**0.9387 ± 0.0013**	0.9405	0.9385 ± 0.0013	0.9399	0.9383 ± 0.0008

Table 11. Best and average ± st. dev. Hypervolume in testing set - Scen. 50%.

Instance	Random		Chi-square		Hybrid	
	Best	ave ± sd	Best	ave ± sd	Best	ave ± sd
DLBCL	0.5451	0.4957 ± 0.0333	0.5245	0.4827 ± 0.0225	0.5669	**0.5392 ± 0.0165**
Lung Cancer	0.5613	**0.5032 ± 0.0476**	0.5243	0.4871 ± 0.0321	0.5430	0.4812 ± 0.0414
warpAR10P	0.2933	0.1992 ± 0.0649	0.4031	0.2470 ± 0.0951	0.3843	**0.2624 ± 0.0860**
warpPIE10P	0.3504	0.2499 ± 0.0828	0.3505	**0.2607 ± 0.0596**	0.3846	0.2486 ± 0.0631
penbased	0.8211	0.8183 ± 0.0022	0.8223	**0.8190 ± 0.0022**	0.8211	0.8184 ± 0.0024
texture	0.9230	0.9163 ± 0.0050	0.9198	0.9160 ± 0.0036	0.9227	**0.9165 ± 0.0046**

CNSGA-II obtains better results, indicating stronger generalization. However, with 10% of initial features, the proposed methods, especially Chi-square, outperform CNSGA-II in all cases, confirming their effectiveness under more restrictive scenarios and the importance of a good initialization strategy. These findings suggest that the benefit of using prior information becomes more significant when the number of selected features is small. Even with fewer initial features, the proposed strategies guided NSGA-II toward better solutions than CNSGA-II, which used a much larger portion of the feature space. This highlights the crucial role that initialization plays in balancing performance and generalization in feature selection tasks. It should also be clarified that the CNSGA-II results were obtained using 50% of the initial features, while our methods were evaluated under both 10% and 50% scenarios. Therefore, the comparison between 10% and 50% should not be interpreted as a direct equivalence in the amount of prior information available, but rather as an analysis of how initialization size affects convergence and generalization.

Table 12. Average ± st. dev. Hypervolume in train. and test. sets – Scen. 50%.

Instance	Set	Random	Chi-square	Hybrid	CNSGA-II
warpAR10P	Train	0.4334 ± 0.0094	0.5075 ± 0.0145	**0.5707 ± 0.0101**	0.54 ± 0.03
	Test	0.1992 ± 0.0649	0.2470 ± 0.0951	0.2624 ± 0.0860	**0.33 ± 0.04**
warpPIE10P	Train	**0.7321 ± 0.0148**	0.7255 ± 0.0141	0.7262 ± 0.0119	0.65 ± 0.01
	Test	0.2499 ± 0.0828	0.2607 ± 0.0596	0.2486 ± 0.0631	**0.43 ± 0.04**

Table 13. Average ± st. dev. Hypervolume in train. and test. sets – Scen. 10%.

Instance	Set	Random	Chi-square	Hybrid	CNSGA-II
warpAR10P	Train	0.6638 ± 0.0480	**0.8288 ± 0.0287**	0.7817 ± 0.0329	0.54 ± 0.03
	Test	0.2853 ± 0.0565	**0.4826 ± 0.0937**	0.3954 ± 0.0984	0.33 ± 0.04
warpPIE10P	Train	0.9375 ± 0.0091	0.9444 ± 0.0074	**0.9494 ± 0.0103**	0.65 ± 0.01
	Test	0.3468 ± 0.1325	**0.4918 ± 0.0863**	0.4660 ± 0.0857	0.43 ± 0.04

Table 14 presents the average execution time (in seconds) per dataset and initialization strategy, considering scenarios with 10% and 50% of initial features. Execution times were similar across all initialization strategies, showing that Chi-square and Hybrid methods do not add significant overhead compared to Random. However, times increased consistently in the 50% scenario due to the higher cost of evaluating larger feature subsets. Thus, using 10% of initial features improves NSGA-II efficiency without affecting the fairness of comparisons.

Table 14. Average execution time (in seconds) per strategy and dataset

Instance	Scenario	Random	Chi-square	Hybrid
DLBCL	10%	3,334	3,347	3,302
	50%	5,082	5,353	5,317
Lung Cancer	10%	7,076	7,116	7,124
	50%	13,480	11,740	11,766
warpAR10P	10%	3,412	3,413	3,391
	50%	4,150	4,222	4,088
warpPIE10P	10%	3,607	3,643	3,602
	50%	4,675	4,664	4,597
penbased	10%	8,589	8,555	8,679
	50%	8,815	8,793	8,828
texture	10%	6,831	6,944	6,927
	50%	6,780	6,601	6,757

7 Conclusions and Future Work

The experimental results indicate that the Chi-square-based initialization improves both the average and minimum error, especially when using 10% of the features. The hybrid strategy achieved a better balance between exploration and exploitation, particularly in complex instances. Furthermore, initializing with 10% of the features led to lower error rates, reduced dimensionality, and improved hypervolume values, outperforming in some cases the classic approach and the CNSGA-II baseline. Importantly, these improvements did not incur a significant increase in execution time. The strategies performed consistently across diverse datasets. Overall, integrating statistical measures into the initialization process of evolutionary algorithms like NSGA-II appears to be a promising direction for improving performance in feature selection tasks.

Future work could focus on refining the initialization and variation mechanisms by integrating statistical criteria into mutation or crossover operators, and exploring alternative filter-based methods. Additionally, it would be relevant to evaluate the proposed strategies on more diverse, large-scale, or imbalanced datasets to assess their scalability and robustness. Finally, further comparisons with other state-of-the-art multi-objective algorithms would help validate the competitiveness and generalization capacity of the proposed approaches.

Acknowledgments. This study was funded by FONDECYT project 1230365.

Disclosure of Interests. The authors have no competing interests to declare that are relevant to the content of this article.

References

1. Cai, J., Luo, J., Wang, S., Yang, S.: Feature selection in machine learning: a new perspective. Neurocomputing **300**, 70–79 (2018). https://doi.org/10.1016/j.neucom.2017.11.077
2. Cunningham, P., Delany, S.J.: k-nearest neighbour classifiers - a tutorial. ACM Comput. Surv. **54**(6) (2021). https://doi.org/10.1145/3459665
3. Deb, K., Pratap, A., Agarwal, S., Meyarivan, T.: A fast and elitist multiobjective genetic algorithm: NSGA-II. IEEE Trans. Evol. Comput. **6**(2), 182–197 (2002). https://doi.org/10.1109/4235.996017
4. Gong, H., Li, Y., Zhang, J., Zhang, B., Wang, X.: A new filter feature selection algorithm for classification task by ensembling Pearson correlation coefficient and mutual information. Eng. Appl. Artif. Intell. **131**, 107865 (2024). https://doi.org/10.1016/j.engappai.2024.107865
5. Jin, X., Xu, A., Bie, R., Guo, P.: Machine learning techniques and chi-square feature selection for cancer classification using SAGE gene expression profiles. In: Li, J., Yang, Q., Tan, A.-H. (eds.) BioDM 2006. LNCS, vol. 3916, pp. 106–115. Springer, Heidelberg (2006). https://doi.org/10.1007/11691730_11
6. Ma, S., Huang, J.: Penalized feature selection and classification in bioinformatics. Brief. Bioinform. **9**(5), 392–403 (2008)
7. Miyandoab, S.Z., Rahnamayan, S., Bidgoli, A.A.: Compact NSGA-II for multi-objective feature selection. In: 2023 IEEE International Conference on Systems, Man, and Cybernetics (SMC). IEEE (2023). https://doi.org/10.1109/smc53992.2023.10394458
8. Prusty, S., Patnaik, S., Dash, S.: SKCV: stratified k-fold cross-validation on ml classifiers for predicting cervical cancer. Frontiers in Nanotechnology **4**, 972421 (2022). https://doi.org/10.3389/fnano.2022.972421
9. Quinlan, J.R.: Induction of decision trees. Mach. Learn. **1**(1), 81–106 (1986). https://doi.org/10.1007/BF00116251
10. Salzberg, S.L.: C4.5: Programs for machine learning by J. Ross Quinlan. Morgan Kaufmann Publishers, Inc., 1993. Mach. Learn. **16**(3), 235–240 (1994)
11. Statnikov, A., Aliferis, C.F., Tsamardinos, I., Hardin, D., Levy, S.: A comprehensive evaluation of multicategory classification methods for microarray gene expression cancer diagnosis. Bioinformatics **21**(5), 631–643 (2004). https://doi.org/10.1093/bioinformatics/bti033
12. Storn, R., Price, K.: Differential evolution - a simple and efficient heuristic for global optimization over continuous spaces. J. Glob. Optim. **11**(4), 341–359 (1997). https://doi.org/10.1023/A:1008202821328
13. Tang, J., Alelyani, S., Liu, H.: Feature Selection for Classification: A Review, pp. 37–64. CRC Press (2014). https://doi.org/10.1201/b17320
14. Tran, B., Xue, B., Zhang, M.: A new representation in PSO for discretization-based feature selection. IEEE Trans. Cybern. **48**(6), 1733–1746 (2018). https://doi.org/10.1109/TCYB.2017.2714145
15. Xu, Y., Yan, C., Liu, H., Wang, J., Yang, Z., Jiang, Y.: Smart energy systems: a critical review on design and operation optimization. Sustain. Cities Soc. **62**, 102369 (2020). https://doi.org/10.1016/j.scs.2020.102369
16. Xue, Y., Zhu, H., Neri, F.: A feature selection approach based on NSGA-II with ReliefF. Appl. Soft Comput. **134**, 109987 (2023). https://doi.org/10.1016/j.asoc.2023.109987
17. Yang, Y., Pedersen, J.O.: A comparative study on feature selection in text categorization. In: Proceedings of the 14th International Conference on Machine Learning (ICML), pp. 412–420 (1997)

18. Yu, L., Liu, H.: Feature selection for high-dimensional data: a fast correlation-based filter solution. In: Proceedings of the 20th International Conference on Machine Learning (ICML), pp. 856–863 (2003)

A Multi-objective Optimization Approach for Tuning Multivariable DMC Controllers

Víctor Huilcapi[1(✉)], Ricardo Cajo[2], Juan Manuel Herrero[3], and Xavier Blasco[3]

[1] Universidad Politécnica Salesiana, Guayaquil, Ecuador
vhuilcapi@ups.edu.ec

[2] Facultad de Ingeniería en Electricidad y Computación, Escuela Superior Politécnica del Litoral, ESPOL, Campus Gustavo Galindo Km 30.5 Vía Perimetral, P.O. Box 09-01-5863, 090150 Guayaquil, Ecuador
rcajo@espol.edu.ec

[3] Instituto Universitario de Automática e Informática Industrial, Universitat Politècnica de València, 46022 Valencia, Spain
{juaherdu,xblasco}@isa.upv.es

Abstract. This paper proposes a posteriori multi-objective optimization framework for tuning multivariable Dynamic Matrix Control (DMC) controllers. To demonstrate the application of this method, a classical multivariable control system from the literature with two inputs and two outputs (2×2 MIMO) is used, and the multi-objective evolutionary algorithm *ev*-MOGA was applied to tune the DMC parameters. The proposed approach offers relevant information for a designer regarding the effects of DMC tuning parameters on the system's output performance, as well as the aggressiveness of the control efforts. To visualize Pareto fronts, the *Level Diagram* (LD) tool is used, allowing for the observation and quantification of trade-offs between the design objectives of the formulated multi-objective problem (MOP). The proposed methodology offers designers relevant information to make informed decisions on tuning DMC controllers and selecting the optimal parameters to achieve satisfactory performance in the multivariable process. Performance metrics were used to evaluate the quality of the DMC controller tuned using the proposed approach, showing satisfactory performance. The DMC controller was compared with two decentralized controllers (PI controllers), and the DMC showed better performance than both controllers.

Keywords: Multivariable system · DMC controller · Multi-objective optimization · Pareto front · Multi-objective evolutionary algorithm

1 Introduction

Among the most notable modern control strategies is Model Predictive Control (MPC), as it has proven effective in controlling complex multivariable processes

H. R. Arabnia et al. (Eds.): CSCI 2025, CCIS 2968, pp. 376–390, 2027.
https://doi.org/10.1007/978-3-032-28457-0_27

with hard constraints. Within this domain, Dynamic Matrix Control (DMC), a subclass of MPC, has gained widespread adoption in industrial practice due to its explicit model-based formulation and the capacity to predict system behavior over a finite horizon [3]. However, despite its advantages, DMC faces significant challenges in tuning its parameters, especially in multivariable contexts where multiple performance indicators must be balanced. This challenge opens the door to applying multi-objective optimization (MOO) methodologies, which provide a structured framework for managing trade-offs between conflicting objectives.

An approach that explores global optimization techniques to tune DMC controllers in multivariable processes (MIMO) is shown in [21]. By employing methods like PSO, the Firefly Algorithm, the Grey Wolf Optimizer, and the Jaya algorithm, they demonstrated the potential of global optimization to enhance convergence and DMC controller performance. In addition, [20] provides an extensive review of metaheuristic multi-objective optimization approaches in intelligent control, aimed at offering a broader understanding of the field. They emphasized the challenges of managing conflicting objectives, such as accuracy, energy efficiency, and cost, and highlighted the strength of evolutionary and swarm-based algorithms in navigating complex, multimodal tuning landscapes.

In parallel, structural modifications to the DMC framework were proposed in [19] to improve tuning flexibility and control performance, particularly in nonlinear systems. Their contributions opened new avenues for adapting DMC to the needs of industrial environments of higher complexity. Expanding the industrial relevance of MOO, [18] demonstrates the application of an economical MPC for drinking water transport networks. Through Pareto-based decision-making, they replaced traditional trial-and-error tuning with a systematic approach, demonstrating improved efficiency in a real-world case study in Barcelona. Concurrently, in [24] the application of MOO in the mineral processing industry was addressed. Their work on grinding circuits-systems known for slow dynamics and long dead times-demonstrated how compromise programming can translate high-level operational goals into practical tuning decisions. In line with this trend, [6] introduced a gain-free MPC approach that eliminates the need for weighting factors. Applied to a nine-level ANPC converter, their method achieved robust performance and reduced sensitivity to parameter mismatch.

In [14], an advanced multi-objective optimization algorithm (MOEA/D) was applied to a real-time temperature control system, demonstrating that multi-objective methods perform effectively in practical applications. Finally, in one of the most comprehensive comparative studies, 77 evolutionary algorithms were evaluated for PID tuning [1]. Their results highlighted that the performance of each algorithm varies depending on the metric and that MOEADAWA emerges as a strong candidate for future use in complex controller architectures such as DMC.

This paper proposes a structured framework for tuning multivariable Dynamic Matrix Control controllers using multi-objective optimization. Unlike heuristic or single-objective approaches, it formulates the tuning as a multi-objective problem balancing tracking performance and control effort. A benchmark 2 × 2 MIMO system is used, and the DMC parameters are optimized using the *ev−MOGA* algorithm [7,8,11]. Trade-offs are visualized via *Level Diagrams* [2,12], helping designers understand how tuning affects system behavior and control aggressiveness. This approach offers insights that are not possible with single-objective methods.

The proposed methodology addresses several key limitations in the existing literature:

- **Scalability:** applicable to complex multivariable systems with strong couplings.
- **Informed trade-offs:** uses Pareto fronts and Level Diagrams for guided parameter selection.
- **A multidimensional design approach:** where the designer can select the number of design objectives based on the depth of analysis required for the multivariable process.

The remainder of this paper is organized as follows. Section 2 explains the basics of multi-objective optimization and DMC controllers. Section 3 describes the system used. Section 4 details the tuning method and configuration of the *ev−MOGA* algorithm. Section 5 shows and discusses the results. Finally, Sect. 6 presents the conclusions.

2 Preliminary Conceptual Framework

2.1 Multi-objective Optimization Fundamentals

A multi-objective optimization design (MOOD) approach consists of three main phases. Phase 1 poses the multi-objective optimization problem. Phase 2 is the multi-objective optimization process, which consists of searching for Pareto fronts and Pareto sets. Phase 3 is the multi-criteria decision-making stage, where the designer must select one or more optimal solutions for the MOP [4,10].

In the **first phase**, a MOP can be posed as shown in (1)–(5).

$$\min_{\mathbf{x}\in D} \boldsymbol{F}(\mathbf{x}, \boldsymbol{\vartheta}) \tag{1}$$

$$\boldsymbol{F}(\mathbf{x}, \boldsymbol{\vartheta}) = \{f_1(\mathbf{x}, \boldsymbol{\vartheta}), f_2(\mathbf{x}, \boldsymbol{\vartheta}), ..., f_k(\mathbf{x}, \boldsymbol{\vartheta})\} \tag{2}$$

$$P(\mathbf{x}) \leq 0 \tag{3}$$

$$Q(\mathbf{x}) = 0 \tag{4}$$

$$\mathbf{x}_{min} \leq \mathbf{x}_i \leq \mathbf{x}_{max}; i = [1, ..., n] \tag{5}$$

where $\mathbf{x} = (x_1, x_2, \ldots x_n) \in \mathbb{R}^n$ is the decision vector; D is the search space; $\boldsymbol{F}(\mathbf{x}, \boldsymbol{\vartheta}) \in \mathbf{R}^m$ is a vector containing the MOP design objectives; $\boldsymbol{\vartheta}$ represents possible uncertainties in the system; $P(\mathbf{x})$ and $Q(\mathbf{x})$ are the system constraints; $\mathbf{x}_{min}$ and $\mathbf{x}_{max}$ are the boundaries of the search space.

In the **second phase**, a multi-objective optimization algorithm must be appropriately selected to obtain the Pareto fronts of the MOP. Currently, there are several modern optimization approaches for this task, for example, machine learning algorithms, metaheuristic algorithms with swarm intelligence, evolutionary algorithms, genetic algorithms, and others [22,23]. In this paper, the multi-objective evolutionary algorithm $ev - MOGA$ [7,8,11] was used. After obtaining the Pareto fronts, it is necessary to analyze the characteristics of the solutions and the trade-offs among them. Based on this analysis and according to their preferences, the designer can propose an optimal solution to the MOP. To help a designer choose the best solutions (according to the designer's preferences) from the Pareto front, various visualization tools have been developed as shown in [13,17]. It is important to emphasize that all Pareto front solutions are optimal and that the choice between one solution or another depends on the designer's preferences and the characteristics of the MOP.

In the Pareto front visualization stage (**third phase**), the visualization tool called *Level Diagram* (LD) was used [2,12]. LD enables a multidimensional analysis of the solutions on the Pareto front. It is based on the classification of the Pareto front $\boldsymbol{F}(\mathbf{x}, \boldsymbol{\vartheta})$ according to its proximity to the ideal solution. The ideal solution (utopian solution) is the point with the minimum values for each design objective in the MOP. LD uses different norms ($1 - norm, 2 - norm, \infty - norm$) to represent a Pareto front. Each norm provides a designer with relevant information about the trade-offs between solutions on the Pareto front. A traditional representation of a Pareto front for a two-dimensional MOP is shown in Fig. 1 and it is also represented using LD with $\infty - norm$ in Fig. 2.

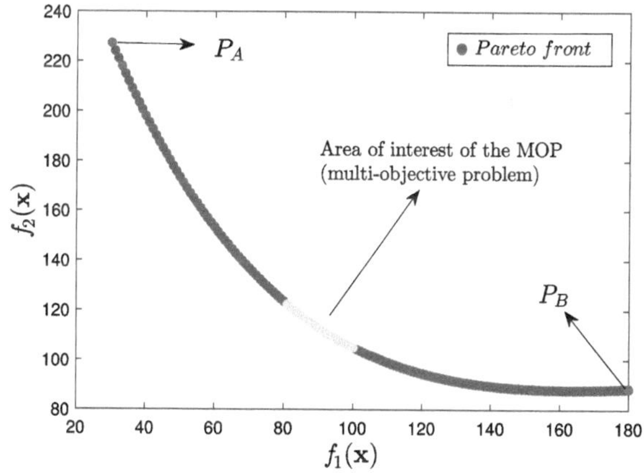

Fig. 1. Conventional visualization of a bi-dimensional Pareto front.

2.2 Fundamentals of Dynamic Matrix Control (DMC)

Model predictive control (MPC) is a widely accepted methodology for controlling complex industrial processes. The MPC control strategy calculates future

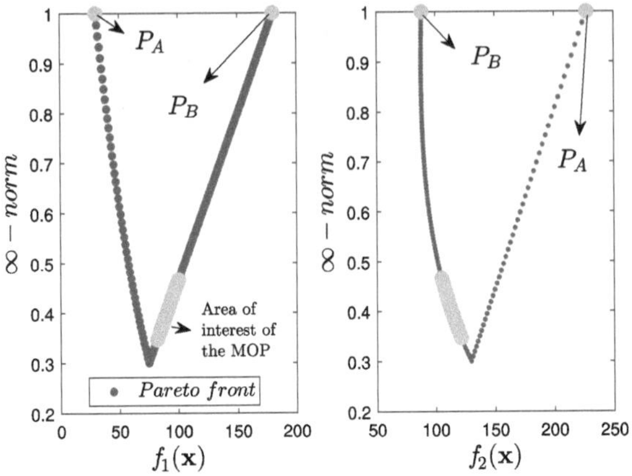

Fig. 2. Bi-dimensional Pareto front obtained using LD with $\infty - norm$.

process outputs for a given prediction horizon. MPC calculates a set of future control signals using an optimization criterion that minimizes the tracking error of a set of input references. The MPC algorithm uses the receding horizon approach and applies only the first control signal calculated for the current period to the process. In each subsequent control period, it repeats the optimization process with new information available, which allows it to continuously update the control strategy [3,5,16]. One of the most widely used MPC algorithms in the industry is the Dynamic Matrix Control (DMC) algorithm duo to its easy implementation and high effectiveness. The DMC algorithm uses linear models of stable open-loop processes, uses the system's step response to calculate predictions, and considers constant disturbances throughout the prediction horizon. In a DMC, future outputs over the prediction horizon p (from $t+1$ to $t+p$) are adjusted to follow an input reference sp by calculating a sequence of future control moves c (from t to $t+c-1$) as shown in (6).

$$\hat{y}(t+j) = \sum_{i=1}^{j} g_i \Delta u(t+j+i) + y_0 + \sum_{i=j+1}^{M} g_i \Delta u(t+j-i) + \hat{n}(t+j) \tag{6}$$

where, $\hat{y}(t+j)$ represents the future output of the system at time $t+j$; t is the control period; y_0 is the initial condition of the output; $\Delta u(t)$ are the system inputs at different time intervals (control actions); $\hat{n}(t+j)$ are non-measurable disturbances or unmodeled dynamics of the system; g_i are the coefficients of the system response to a unit step input; M is the number of intervals required for the system to stabilize. The free response of the system does not depend on future control actions and is shown in (7).

$$\hat{y}_f(t+j) = \sum_{i=j+1}^{M} g_i \Delta u(t+j-i) \tag{7}$$

The calculation of future outputs (considering past inputs up to the current time) over the entire prediction horizon p for the time intervals from $t+1$ to $t+p$ is shown in (8).

$$\begin{bmatrix} \hat{y}(t+1) \\ \vdots \\ \hat{y}(t+p) \end{bmatrix} = \begin{bmatrix} \hat{y}_f(t+1) \\ \vdots \\ \hat{y}_f(t+p) \end{bmatrix} + \mathbf{G} \cdot \begin{bmatrix} \Delta u(t) \\ \vdots \\ \Delta u(t+c-1) \end{bmatrix} + \begin{bmatrix} \hat{n}(t+1) \\ \vdots \\ \hat{n}(t+p) \end{bmatrix} \quad (8)$$

where $\mathbf{G}$ is known as the ***dynamic matrix*** of the system and can be calculated as shown in (9).

$$\mathbf{G} = \begin{bmatrix} g_1 & 0 & 0 \cdots & 0 \\ g_2 & g_1 & 0 \cdots & 0 \\ \vdots & \vdots & \ddots \ddots & \vdots \\ g_c & g_{c-1} & \cdots\cdots & g_1 \\ \vdots & \vdots & \ddots \ddots & \vdots \\ g_p & g_{p-1} & \cdots\cdots & g_{p-c+1} \end{bmatrix} \quad (9)$$

For future outputs (within the prediction horizon) to reach the input reference sp, the DMC algorithm must calculate the future movements c of the inputs. To achieve this, a DMC creates a residue r between the predicted output and the reference by minimizing the squared deviations from this residue. The optimization problem is shown in (10).

$$\begin{aligned} \min_{\Delta \mathbf{u}} \mathbf{J}(\Delta \mathbf{u}) &= r^T r = \sum_{j=1}^{p} [sp(t+j) - \hat{y}(t+j)]^2 \\ &= [\mathbf{sp} - \hat{\mathbf{y}}_f + \mathbf{G}\Delta \mathbf{u}]^T [\mathbf{sp} - \hat{\mathbf{y}}_f + \mathbf{G}\Delta \mathbf{u}] \end{aligned} \quad (10)$$

where: $\mathbf{sp} = [sp(t+1), \cdots, sp(t+p)]$; $\hat{\mathbf{y}}_f = [\hat{y}_f(t+1) + \hat{n}(t), \cdots, \hat{y}_f(t+p) + \hat{n}(t)]$; $\Delta \mathbf{u} = [\Delta u(t), \cdots, \Delta u(t+c-1)$.

It is possible to add weighting factors (λ and α) to the optimization problem, which allows penalizing the amplitude of the input movements to obtain different closed-loop system performances. The reformulated optimization problem is defined in (11).

$$\begin{aligned} \min_{\Delta \mathbf{u}} \mathbf{J}(\Delta \mathbf{u}) &= [\mathbf{sp} - \hat{\mathbf{y}}_f + \mathbf{G}\Delta \mathbf{u}]^T \alpha [\mathbf{sp} - \hat{\mathbf{y}}_f + \mathbf{G}\Delta \mathbf{u}] + \Delta \mathbf{u}^T \lambda \Delta \mathbf{u} \\ \Delta \mathbf{u} &= (\mathbf{G}^T \alpha \mathbf{G} + \lambda)^{-1} \mathbf{G}^T \alpha (\mathbf{sp} - \hat{\mathbf{y}}_f) \end{aligned} \quad (11)$$

The coefficients λ and α allow us to obtain desirable behavior from the DMC controller. The control horizon c and the prediction horizon p also play a significant role in optimizing the performance index $\mathbf{J}$. To illustrate the challenges associated with selecting DMC tuning parameters, this paper proposes a multi-objective framework that provides designers with detailed information to make informed decisions regarding these parameters.

3 System Description

To demonstrate the methodological approach proposed in this paper for tuning multivariable DMC controllers using a multi-objective optimization approach, the multivariable system defined in [15] is employed, as shown in (12).

$$\begin{bmatrix} y_1(s) \\ y_2(s) \end{bmatrix} = \mathbf{G(s)} \begin{bmatrix} u_1(s) \\ u_2(s) \end{bmatrix} = \begin{bmatrix} \frac{5e^{-40\,s}}{100\,s+1} & \frac{1e^{-4\,s}}{10\,s+1} \\ \frac{-5e^{-4\,s}}{10\,s+1} & \frac{5e^{-40\,s}}{100\,s+1} \end{bmatrix} \begin{bmatrix} u_1(s) \\ u_2(s) \end{bmatrix} \quad (12)$$

$$\Lambda_{RGA} = \begin{bmatrix} 0.8333 & 0.1667 \\ 0.1667 & 0.8333 \end{bmatrix}; \Lambda_{DRGA} = \begin{bmatrix} 0.25 & 0.75 \\ 0.75 & 0.25 \end{bmatrix} \quad (13)$$

where: $y_1(s)$, $y_2(s)$ and $u_1(s)$, $u_2(s)$ represent the outputs and inputs of the system, respectively. Furthermore, the relative gain array (Λ_{RGA}) and the dynamic relative gain array (Λ_{DRGA}) are obtained in (13). These arrays show that there is an ambiguity when trying to control the system with PI controllers (these types of controllers are shown in detail in [9]).

In this paper, a new *centralized control* approach is proposed for this multivariable plant, applying a DMC control structure and comparing its performance with the controllers proposed in [9,15].

4 Methodology

To tune the DMC controller shown in Sect. 2.2, a multi-objective optimization problem is posed with four design objectives and a decision vector containing five DMC tuning parameters (input and output weighting parameters) as shown in (14)–(22).

$$\min_{\mathbf{x}^{DMC}} J(\mathbf{x}^{DMC}) \quad (14)$$

$$J(\mathbf{x}^{DMC}) = \{J_1(\mathbf{x}^{DMC}), J_2(\mathbf{x}^{DMC}), J_3(\mathbf{x}^{DMC}), J_4(\mathbf{x}^{DMC})\} \quad (15)$$

$$J_1(\mathbf{x}^{DMC}) = \int_0^{t_f} e_1^2\big|_{sp_1=1}^{sp_2=0}\,dt + \int_0^{t_f} e_1^2\big|_{sp_1=0}^{sp_2=1}\,dt \quad (16)$$

$$J_2(\mathbf{x}^{DMC}) = \int_0^{t_f} e_2^2\big|_{sp_1=1}^{sp_2=0}\,dt + \int_0^{t_f} e_2^2\big|_{sp_1=0}^{sp_2=1}\,dt \quad (17)$$

$$J_3(\mathbf{x}^{DMC}) = \int_0^{t_f} u_1^2\big|_{sp_1=1}^{sp_2=0}\,dt + \int_0^{t_f} u_1^2\big|_{sp_1=0}^{sp_2=1}\,dt \quad (18)$$

$$J_4(\mathbf{x}^{DMC}) = \int_0^{t_f} u_2^2\big|_{sp_1=1}^{sp_2=0}\,dt + \int_0^{t_f} u_2^2\big|_{sp_1=0}^{sp_2=1}\,dt \quad (19)$$

$$t_f = 1000\ seconds$$

$$\mathbf{x_{min}}^{DMC} \leq \mathbf{x}^{DMC} \leq \mathbf{x_{max}}^{DMC} \quad (20)$$

$$J_1(\mathbf{x}^{DMC}) < 250, J_2(\mathbf{x}^{DMC}) < 110, J_3(\mathbf{x}^{DMC}) < 30, J_4(\mathbf{x}^{DMC}) < 60 \quad (21)$$

$$\mathbf{x}^{DMC} = [c, \alpha_1, \alpha_2, \lambda_1, \lambda_2] \tag{22}$$

The MOP proposes four design objectives to analyze the performance independently, that is, without combining errors e_1 and e_2 into a single objective. The effect of control efforts u_1 and u_2 is also evaluated independently (without mixing u_1 and u_2). This approach allows us to perform a multidimensional analysis of the MOP without losing information relevant to the designer.

The design objectives J_1 and J_2 evaluate the performance of the system outputs y_1 and y_2 in tracking the input references sp_1 and sp_2, respectively. The Integral of the Squared Error (ISE) was used as a performance index. The design objectives J_3 and J_4 evaluate the performance of control efforts (DMC controller aggressiveness) using the Integral of the Squared Control action (ISU). To establish the bounds of the decision vector $\mathbf{x}^{DMC}$, a reference DMC controller was tuned to reduce the search space and lower the computational cost of the optimization process. In this paper, an unconstrained multivariable DMC controller was tuned (the constraints are stated in the MOP). The bounds of the decision vector are shown in Table 1.

Table 1. Decision vector description and bounds for DMC controller tuning.

Parameter	*Description*	Type	$[\mathbf{x_{min}}^{DMC} - \mathbf{x_{max}}^{DMC}]$
c	DMC control horizon	Integer	[1–10]
α_1	Error weighting for y_1	Real	[0.1–10]
α_2	Error weighting for y_2	Real	[0.1–10]
λ_1	Weighting of movements for u_1	Real	[1–10]
λ_2	Weighting of movements for u_2	Real	[1–5]

Algorithm 1 presents the $ev-MOGA$ configuration. The *Level Diagram* (LD) tool was used in the Pareto front visualization stage.

5 Results and Discussion

In the multi-objective optimization process, Pareto fronts and Pareto sets were obtained, as shown in Figs. 3 and 4, respectively. To represent the Pareto fronts and analyze the trade-offs between design objectives, LD with ∞-norm was used since an increase in this norm shows how much one objective can be worsened compared to another.

The objectives $J_1(\mathbf{x}^{DMC})$ and $J_2(\mathbf{x}^{DMC})$ evaluate the performance of the system outputs y_1 and y_2, respectively, in tracking a unit step input reference. The objectives $J_3(\mathbf{x}^{DMC})$ and $J_4(\mathbf{x}^{DMC})$ assess the aggressiveness of the control efforts for the system's control signals u_1 and u_2, respectively. The brown solutions ($\bullet$) for objectives $J_1(\mathbf{x}^{DMC})$ and $J_2(\mathbf{x}^{DMC})$ (see Fig. 3) show that there is a conflict in controlling outputs y_1 and y_2. That is, the best-performing solutions for the output y_1 correspond to the worst-performing ones for the output

Algorithm 1. *ev – MOGA* algorithm configuration (DMC controller tuning)

$eMOGA.objfun = fcoste\ multi-objective\ DMC\ controller$
$eMOGA.objfun\_dim = 4$
$eMOGA.search\ space\ UB = [10\ 10\ 10\ 10\ 5\ ]$
$eMOGA.search\ space\ LB = [1\ 0.1\ 0.1\ 0.1\ 1\ ]$
$eMOGA.Nind\_P = 500$
$eMOGA.Generations = 500$
$eMOGA.n\_div = 50$
$eMOGA.Nind\_GA = 8$
$[pfront, pset, eMOGA] = evMOGA(eMOGA)$
$function: \ J = fcoste\ multi - objective\ DMC(model, param)$
$c = model(1); \alpha_1 = model(2); \alpha_2 = model(3); \lambda_1 = model(4); \lambda_2 = model(5)$
$simulink\ (MIMO\_DMC\ controller)$
$J_1(\mathbf{x}^{DMC}) = min(\Delta e_1^2);\ J_2(\mathbf{x}^{DMC}) = min(\Delta e_2^2)$
$J_3(\mathbf{x}^{DMC}) = min(\Delta u_1^2);\ J_4(\mathbf{x}^{DMC}) = min(\Delta u_2^2);$
$J = [J_1\ J_2\ J_3\ J_4]$

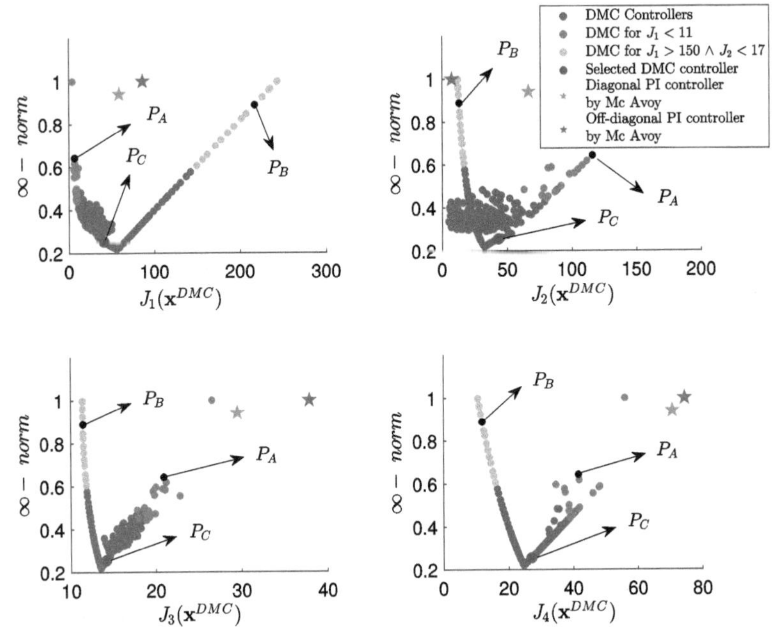

Fig. 3. Pareto fronts for tuning DMC controllers.

y_2 (since the brown regions are on opposite sides of the Pareto fronts). Likewise, the yellow solutions (●) are located on opposite sides of objectives $J_1(\mathbf{x}^{DMC})$ and $J_2(\mathbf{x}^{DMC})$, which implies that the solutions with the best performance to control the output y_2 correspond to those with the worst performance to control the output y_1. Regarding the DMC controller's control actions, objective $J_3(\mathbf{x}^{DMC})$ shows that the solutions with the best performance for the output

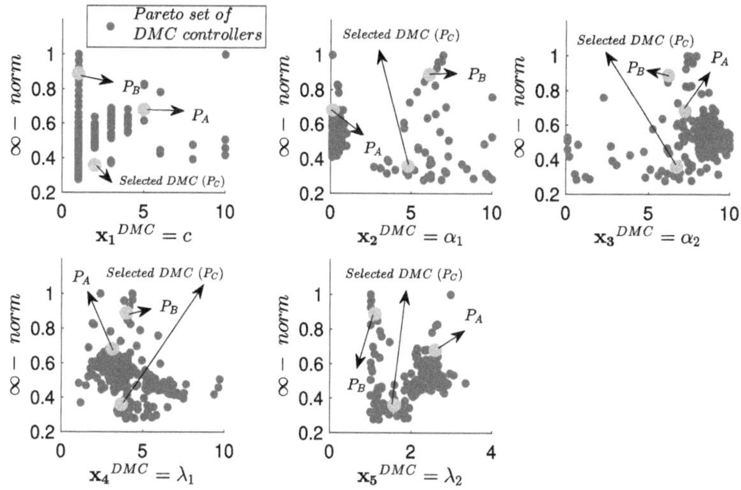

Fig. 4. Pareto set for tuning DMC controllers.

y_1 (brown region) require greater control effort (more aggressive controllers). The objective $J_4(\mathbf{x}^{DMC})$ shows the opposite, that is, the solutions with the best performance for the output y_2 (yellow region) require less control effort (less aggressive controllers).

To evaluate the trade-offs between the solutions on the Pareto fronts (DMC controllers), three controllers were selected: P_A, P_B and P_C. The solution in P_A represents the best controller for output y_1, as it has the lowest objective $J_1(\mathbf{x}^{DMC})$ compared to P_B and P_C, but it is also the worst controller for output y_2, as the objective $J_2(\mathbf{x}^{DMC})$ is higher compared to P_B and P_C (see Table 2). On the other hand, the front solution in P_B represents the best controller for output y_2 because it has the lowest objective $J_2(\mathbf{x}^{DMC})$ compared to P_A and P_C, but it is also the worst controller for output y_1 since $J_1(PB) \gg J_1(PC) \gg J_1(PA)$. It can also be observed in Table 2 that controllers with better performance for output y_1 require greater control effort because $J_3(PA) > J_3(PC) > J_3(PB)$. Furthermore, the controllers with the best performance for output y_2 require less control effort since $J_4(PB) > J_4(PC) > J_4(PA)$. It is possible to observe in Table 2 that the DMC controller tuned with the methodology proposed in this paper has a better performance in three design objectives ($J_1(PC)$, $J_3(PC)$, $J_4(PC)$) than the two controllers (PI diagonal PI and PI off-diagonal) tuned in [15]. Also, this multi-objective methodology allows a multidimensional analysis of the Pareto fronts when tuning the DMC controller, giving a main role to the designer when choosing a solution for the MOP. For example, a designer may prefer a DMC controller with better performance for the output y_1 and choose a solution from the brown region. Another designer may prefer a controller with better performance for the output y_2 and should select a solution from the yellow region. In this paper, the final solution chosen

Table 2. Performance of DMC controllers in PA, PB, and PC versus PI controllers. The performance is shown for each design objective $J(\mathrm{x})$.

DMC controllers				PIs controllers		
$Objectives$	P_A	P_C	P_B	$objectives$	$Diagonal$	$Off-diagonal$
$J_1(\mathbf{x}^{DMC})$	**6.29**	39.87	216.8	$J_1(\mathbf{x}^{PI})$	58.33	85.84
$J_2(\mathbf{x}^{DMC})$	115.9	43.21	**13.18**	$J_2(\mathbf{x}^{PI})$	66.33	7.43
$J_3(\mathbf{x}^{DMC})$	20.93	14.14	11.4	$J_3(\mathbf{x}^{PI})$	29.51	37.87
$J_4(\mathbf{x}^{DMC})$	41.58	26.98	11.96	$J_4(\mathbf{x}^{PI})$	70.70	74.47

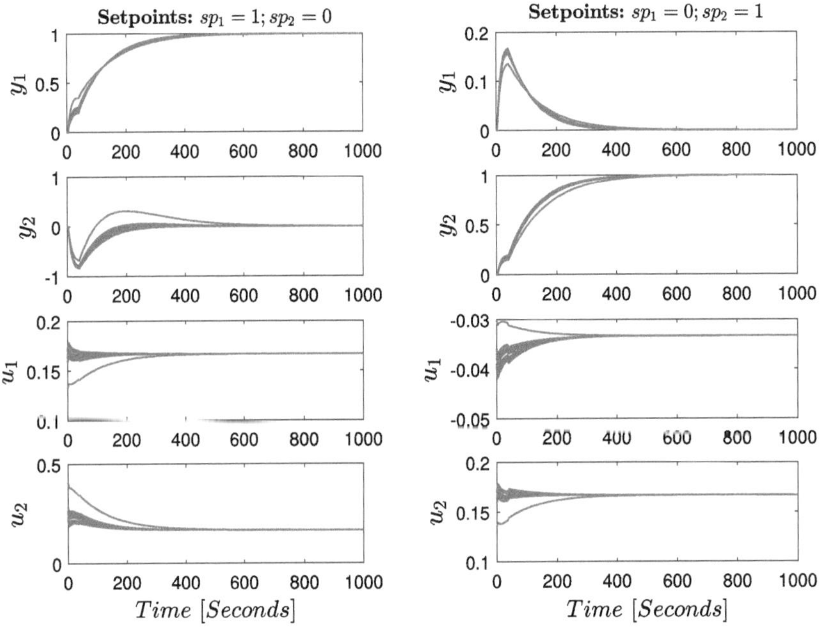

Fig. 5. Responses of DMC controllers: $J_1 < 11$ on the Pareto front.

for the MOP is the DMC controller in P_C, which satisfactorily controls both outputs y_1 and y_2 (to obtain balanced performance in each output, without favoring aggressive control in either of them). The tuned parameters for the DMC controller in P_C are: $c = 2$, $\alpha_1 = 4.8$, $\alpha_2 = 6.7$, $\lambda_1 = 3.7$ y $\lambda_2 = 1.6$.

Figure 5 shows the responses of the best-performing DMC controllers for output y_1 ($J_1(\mathbf{x}^{DMC}) < 11$). The output y_1 stabilizes at approximately 300 s and the output y_2 at approximately 400 s. Likewise, Fig. 6 shows the responses of the DMC controllers with the best performance for output y_2 ($J_2(\mathbf{x}^{DMC}) < 17$ and $J_1(\mathbf{x}^{DMC}) > 150$). In this case, output y_2 stabilizes in approximately 300 s, while output y_1 stabilizes in approximately 450 s.

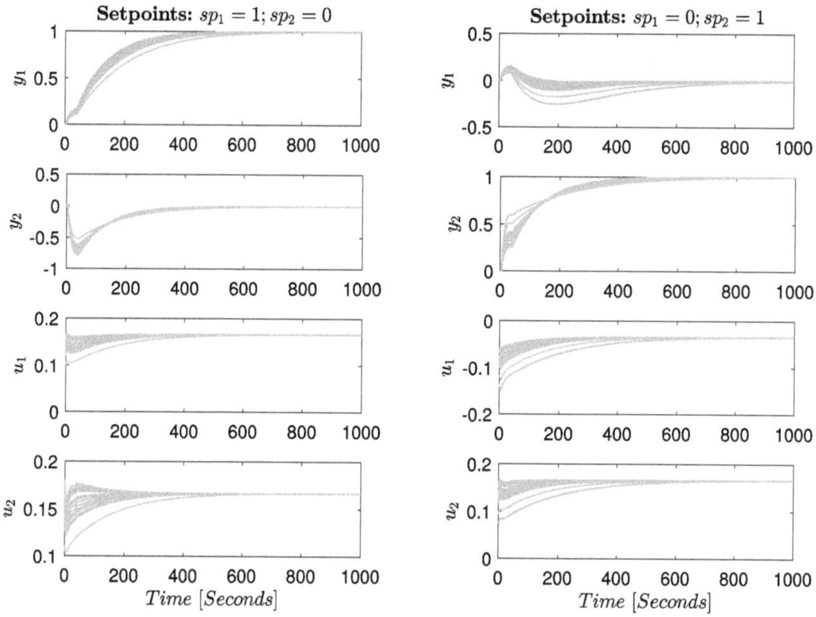

Fig. 6. Responses of DMC controllers: $J_2 < 17 \wedge J_1 > 150$ on the Pareto front.

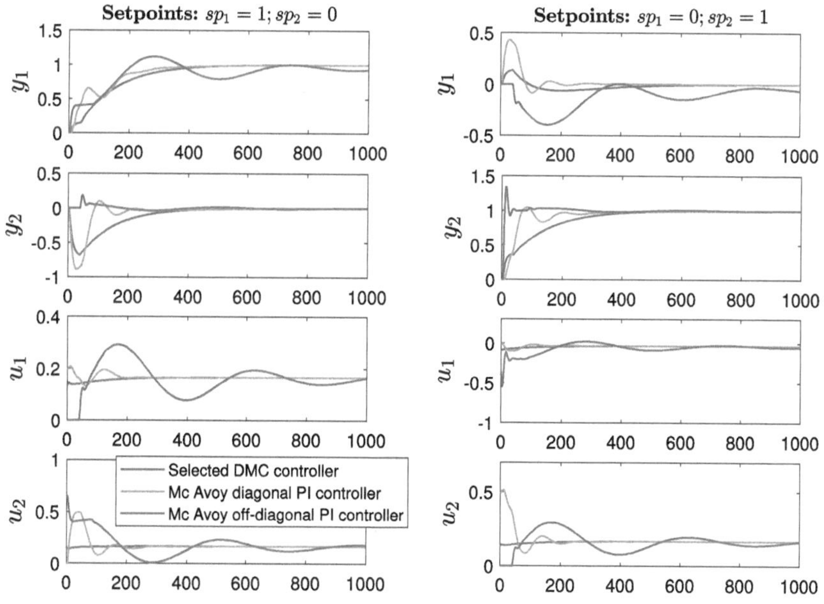

Fig. 7. Responses of the selected DMC controller vs. Mc Avoy controllers.

Finally, Fig. 7 presents the DMC controller selected in P_C and compares its performance with the PI controllers tuned by the methodology proposed in [15].

It is observed that the DMC controller has a better performance in controlling the output y_1 than the diagonal and off-diagonal PI controllers since it has a shorter settling time and fewer oscillations (in addition, the off-diagonal PI controller has a steady-state error). This can also be verified in Table 2 since $J_1(\mathbf{x}^{DMC})$ in $P_C < J_1(\mathbf{x}^{PI})$ for both PI controllers. For output y_2 the DMC controller in P_C exhibits better performance than the diagonal PI controller since it has fewer oscillations, and also $J_2(\mathbf{x}^{DMC})$ in $P_C < J_2(\mathbf{x}^{PI}\ diagonal)$. For output y_2 the off-diagonal PI controller performs better than the DMC controller because it manages to stabilize rapidly ($J_2(\mathbf{x}^{PI}\ off-diagonal) < J_2(\mathbf{x}^{DMC})(P_C)$), However, this has the disadvantage that the control of output y_1 is very oscillatory and presents a steady-state error. The control actions of the DMC controller in P_C are much smoother than those of the diagonal and off-diagonal PI controllers as can be seen in Fig. 7. Table 2 also shows that the design objectives $J_3(\mathbf{x}^{DMC})$ and $J_4(\mathbf{x}^{DMC})$ are much lower than $J_3(\mathbf{x}^{PI})$ and $J_4(\mathbf{x}^{PI})$, respectively.

6 Conclusions

This paper proposes a framework for tuning multivariable DMC controllers using a multi-objective optimization approach. The method was applied to a system with two inputs and two outputs (2×2 MIMO), allowing for the extraction of relevant information about the trade-offs between the outputs controlled by the DMC (MOP design objectives: $J_1(\mathbf{x}^{DMC})$ and $J_2(\mathbf{x}^{DMC})$). For example, it shows that the system output y_1 conflicts with the output y_2. That is, the Pareto front solutions with the best performance for output y_1 correspond to the solutions with the worst performance for output y_2 and vice versa. The trade-offs of the control efforts for the multivariable system are also shown to illustrate how aggressive a DMC controller is compared to another. This information can only be obtained using a multi-objective design approach such as the one proposed in this paper. Furthermore, it was possible to observe in the Pareto set containing the DMC tuning parameters how these directly affect the quality of the DMC controller. With all the information provided by the proposed methodology and based on the designer's preferences, a DMC controller was selected for the multivariable system. The selected DMC controller showed satisfactory performance in stabilizing the system by applying non-aggressive control efforts. Finally, the performance of the selected DMC controller was compared with two decentralized controllers (PI controllers), showing that the DMC controller outperforms both PI controllers.

Acknowledgements. The authors thank the Research Group in Control Systems and Robotics (GISCOR). This work has been supported by the Universidad Politécnica Salesiana through project No. 063-004-2025-06-27 and by the grant PID2024-158941OB-I00 founded by MCIN/AEI/10.13039/501100011033, Spain.

References

1. Optimizing Control Systems: A Comparative Analysis of 77 Multi-Objective Evolutionary Algorithms, ASME International Mechanical Engineering Congress and Exposition, Volume 5: Dynamics, Vibration, and Control (2024). https://doi.org/10.1115/IMECE2024-140634
2. Blasco, X., Herrero, J.M., Reynoso-Meza, G., Iranzo, M.A.M.: Interactive tool for analyzing multiobjective optimization results with level diagrams. In: Proceedings of the Genetic and Evolutionary Computation Conference Companion, pp. 1689–1696 (2017)
3. Camacho, E.F., Bordons, C., Camacho, E.F., Bordons, C.: Constrained Model Predictive Control. Springer (2007)
4. Cao, J., Yang, Y., Zhang, J., Chen, Z., Liu, Z.: A multi-task optimization algorithm via reinforcement learning for multimodal multi-objective optimization. Expert Syst. Appl. **283**, 127862 (2025)
5. Cutler, C.R.: Dynamic matrix control-a computer control algorithm. In: Proceedings of the Joint Automatic Control Conference, 1979 (1979)
6. Harbi, I., et al.: Multi-objective control of nine-level ANPC converters: a robust and gain-free MPC method. In: 2023 IEEE International Conference on Predictive Control of Electrical Drives and Power Electronics (PRECEDE), pp. 1–6. (2023). https://doi.org/10.1109/PRECEDE57319.2023.10174579
7. Herrero, J., Blasco, X., Martínez, M., Ramos, C., Sanchis, J.: Non-linear robust identification of a greenhouse model using multi-objective evolutionary algorithms. Biosyst. Eng. **98**(3), 335–346 (2007)
8. Huilcapi, V., Blasco, X., Herrero, J.M., Pajares, A., Reynoso-Meza, G.: Application of an input-output pairings selection methodology to control multivariable systems based on multi-objective optimization. In: 2021 International Conference on Computational Science and Computational Intelligence (CSCI), pp. 443–449. IEEE (2021)
9. Huilcapi, V., Blasco, X., Herrero, J.M., Reynoso-Meza, G.: A loop pairing method for multivariable control systems under a multi-objective optimization approach. IEEE Access **7**, 81994–82014 (2019)
10. Huilcapi, V., Blasco, X., Herrero, J.M., Reynoso-Meza, G.: A loop pairing method for non-linear multivariable control systems under a multi-objective optimization approach. IEEE Access **8**, 41262–41281 (2020)
11. Huilcapi, V., Castillo, C., Sanchez, D., Cajo, R.: Optimization of an air pressure system: a multi-objective control and modeling approach. IEEE Access (2024)
12. Huilcapi, V., García, G., Ghia, E., Soto, B.: Modeling and control of a peltier thermoelectric system applying a multi-objective optimization approach. In: Berrezueta-Guzman, S., Torres, R., Zambrano-Martinez, J.L., Herrera-Tapia, J. (eds.) TICEC 2024. CCIS, vol. 2273, pp. 303–317. Springer, Cham (2024). https://doi.org/10.1007/978-3-031-75431-9_20
13. Li, G., Xiao, R.: A radial visualization method based on knee point information for many-objective optimization. Appl. Soft Comput. **175**, 113064 (2025)
14. Li, M., Liu, R., Wu, P., Wang, J.: Multipoint temperature model predictive control with improved MOEA/D. Trans. Inst. Meas. Control. **47**(4), 771–780 (2025). https://doi.org/10.1177/01423312241260107
15. Mc Avoy, T., Arkun, Y., Chen, R., Robinson, D., Schnelle, P.D.: A new approach to defining a dynamic relative gain. Control. Eng. Pract. **11**(8), 907–914 (2003)

16. Morato, M.M., Normey-Rico, J.E., Sename, O.: Model predictive control design for linear parameter varying systems: a survey. Annu. Rev. Control. **49**, 64–80 (2020)
17. Nagar, D., Ramu, P., Deb, K.: Visualization and analysis of pareto-optimal fronts using interpretable self-organizing map (ISOM). Swarm Evol. Comput. **76**, 101202 (2023)
18. Ocampo-Martinez, C., Toro, R., Puig, V., Van Impe, J., Logist, F.: Multi-objective-based tuning of economic model predictive control of drinking water transport networks. Water **14**(8) (2022). https://doi.org/10.3390/w14081222, https://www.mdpi.com/2073-4441/14/8/1222
19. Parrott, E., Chevarie, B., Ellis, K., Versloot, J., Dubay, R.: A survey of advanced dynamic matrix control algorithms for improved performance. In: Progress in Canadian Mechanical Engineering, vol. 3. University of Prince Edward Island. Robertson Library, Charlottetown, P.E.I. (2020)
20. Rodríguez-Molina, A., Mezura-Montes, E., Villarreal-Cervantes, M.G., Aldape-Pérez, M.: Multi-objective meta-heuristic optimization in intelligent control: a survey on the controller tuning problem. Appl. Soft Comput. **93**, 106342 (2020). https://doi.org/10.1016/j.asoc.2020.106342
21. Sawulski, J., Lawrynczuk, M.: Optimisation-based tuning of dynamic matrix control algorithm for multiple-input multiple-output processes. In: 2018 23rd International Conference on Methods & Models in Automation & Robotics (MMAR). IEEE (2018)
22. Van-Canh, N., Anh-Thang, N., Ngoc-Linh, P., Thuy-Duong, N.: Multi-objective optimization of SUS430C steel turning process using hybrid machine learning and evolutionary algorithm approach. Results Eng. 104233 (2025)
23. Xu, X., Xie, C., Ma, L., Yang, L., Zhang, T.: Multi-objective evolutionary algorithm with two balancing mechanisms for heterogeneous UAV swarm path planning. Appl. Soft Comput. **173**, 112927 (2025)
24. Yamashita, A.S., Martins, W.T., Pinto, T.V.B., Raffo, G.V., Euzébio, T.A.M.: Multiobjective tuning technique for MPC in grinding circuits. IEEE Access **11**, 43041–43054 (2023). https://doi.org/10.1109/ACCESS.2023.3269559

A Benchmark Suite of Reddit-Derived Datasets for Mental Health Detection

Khalid Hasan and Jamil Saquer(✉)

Missouri State University, Springfield, MO, USA
{kh597s,jamilsaquer}@missouristate.edu

Abstract. The growing availability of online support groups has opened up new windows to study mental health through natural language processing (NLP). However, it is hindered by a lack of high-quality, well-validated datasets. Existing studies have a tendency to build task-specific corpora without collecting them into widely available resources, and this makes reproducibility as well as cross-task comparison challenging. In this paper, we present a uniform benchmark set of four Reddit-based datasets for disjoint but complementary tasks: (i) detection of suicidal ideation, (ii) binary general mental disorder detection, (iii) bipolar disorder detection, and (iv) multi-class mental disorder classification. All datasets were established upon diligent linguistic inspection, well-defined annotation guidelines, and human-judgmental verification. Inter-annotator agreement metrics always exceeded the baseline agreement score of 0.8, ensuring the labels' trustworthiness. Previous work's evidence of performance on both transformer and contextualized recurrent models demonstrates that these models receive excellent performances on tasks (F1 ≈93–99%), further validating the usefulness of the datasets. By combining these resources, we establish a unifying foundation for reproducible mental health NLP studies with the ability to carry out cross-task benchmarking, multi-task learning, and fair model comparison. The presented benchmark suite provides the research community with an easy-to-access and varied resource for advancing computational approaches toward mental health research.

Keywords: Mental Disorder detection · Benchmark datasets · Reddit · Suicidal ideation · Bipolar disorder

1 Introduction

Mental illness disorders such as depression, anxiety, bipolar disorder, and post-traumatic stress disorder (PTSD) are a worldwide issue of concern to hundreds of millions of individuals worldwide [20]. Though online support groups for mental health are ubiquitous, particularly on platforms such as Reddit, automated mental health language detection and analysis remain hindered by a fundamental bottleneck: the inability to leverage known, standardized datasets. Model

H. R. Arabnia et al. (Eds.): CSCI 2025, CCIS 2968, pp. 391–403, 2027.
https://doi.org/10.1007/978-3-032-28457-0_28

building is isolated without these datasets, and reproducibility between studies is poor.

Recent improvements in natural language processing (NLP), particularly transformer-based models [5,16], have shown promise in detecting mental health markers from web content. Their performance, however, largely relies upon the quality and consistency of the underlying dataset. While prior work has proposed tasks such as suicidal ideation identification, disorder-specific identification, and multi-class classification, no end-to-end benchmark set is available to facilitate fair comparison, reproducibility, and generalizability across tasks.

To bridge this gap, we introduce a universal benchmark dataset from Reddit mental illness communities. The benchmark is a consolidation of four datasets that we originally introduced in task-focused experiments: Suicidal Ideation Detection [7], General Mental Disorder Detection [10], Bipolar Disorder Detection [9], and Multi-Class Mental Disorder Classification [11]. The linguistic analysis process, annotation guidelines, and judgmental validation carefully crafted the datasets to ensure high reliability and replicability. In earlier research, we demonstrated the utility of these datasets separately by training and testing models such as RoBERTa, BERT, and LSTM variants. In this paper, however, we do not seek to present new models but to combine these datasets into a benchmark suite that highlights their overall consistency, linguistic diversity, and utility to future study. We make the datasets available for other researchers at Zenodo [8].

We make three contributions:

1. **Dataset Resource Consolidation:** We introduce four empirically supported Reddit-based datasets as a standard benchmark set, where all the datasets contain one task in mental health detection.
2. **Empirical and Human Validation:** We characterize dataset strength through linguistic analysis, annotation guidelines, inter-annotator agreement, and reported benchmark performance by prior work.
3. **Benchmark Potential:** We stake a claim for future use of these datasets for cross-task comparison, multi-task learning, and standardized benchmarking, and thus making them a foundation for reproducible mental health NLP research.

With this benchmark set, we hope to provide the research community with a reproducible, reliable, and heterogeneous foundation for equitable model comparison, facilitating multi-task methods, and simplifying further development of NLP applications to mental health.

2 Related Work

The intersection of natural language processing (NLP) and mental health research has grown exponentially over the last few years, driven by large amounts of user-posted text available on websites like Reddit, Twitter (now X), and dedicated support sites. Previous research has established that linguistic markers are effective predictors of psychological distress, suicidality, and some disorders such as depression, anxiety, or bipolar disorders [3,4]. These studies established the feasibility of social media data usage in mental health detection, but were often plagued by small sample sizes or task-specific data.

Several datasets have since been made available to examine online discussion on mental disorders. Initial datasets such as the CLPsych shared tasks [19] focused on depression and suicide risk prediction based on tweets, while others, such as DAIC-WOZ [6], explored multi-modal signals from clinical interviews. More recently, Reddit has been a treasure trove for large datasets [24] with its topic-centered specialist communities and unstructured, anonymous posts. These projects tend to be strongly specialized in a single activity or application, however, and hence are of lesser general use.

Benchmark suites such as GLUE [23], SuperGLUE [22], and MMLU [12] have accelerated NLP research by offering standardized testing for a range of tasks. Mental health NLP lacks such benchmark suites. While individual datasets have been shown to support good model performance, the absence of a common resource causes reproducibility, cross-task generalizability, and comparing approaches on an equal footing to be difficult. Several recent studies [13] highlight this limitation and call for the development of standardized benchmarks for health-targeted NLP.

In contrast to previous efforts that introduced single-task corpora or collaborative challenges, our contribution unifies four empirically established Reddit-derived datasets into a comprehensive benchmark covering suicidal ideation detection, bipolar disorder detection, general mental health condition detection, and multi-class mental condition classification. Each dataset has been validated in prior work, both empirically and linguistically, with inter-annotator agreement consistently exceeding Cohen's $\kappa = 0.8$. By consolidating these resources, we provide a robust and diverse benchmark suite that can serve as a reference for future research in mental health detection, reproducibility studies, and multi-task learning.

3 Datasets

Our dataset spans four classification tasks:

3.1 Suicidal Ideation

The suicidal ideation data were collected from Reddit for three months (October-December 2022). The suicidal ideation-related subreddit, r/SuicideWatch, and non-suicidal related communities (e.g., r/socialanxiety, r/TrueOffMyChest, r/bipolar, r/confidence, r/geopolitics) were the source of dataset posts. Subreddit selection was based on two considerations: (i) community size and activity, to guarantee sufficient data coverage, and (ii) existence of posts that could reasonably be classified as suicidal or non-suicidal. After this filtering, we collected 37,821 posts, which we proceeded to annotate into the two categories: suicidal and non-suicidal.

Linguistic Review. To find typical suicidal communication phrases, we used the TextRank algorithm [18] through the PyTextRank library[1]. This picked up highly pertinent expressions such as "so much pain", "no hope", "suicidal thoughts", "my suicide note", "no other way", "just a burden", and "my own death". These phrases are strongly indicative of suicidal ideation and align with established clinical descriptions of emotional distress.

Table 1. An overview of linguistic analysis of the suicidal dataset

Linguistic Attributes	Values	
	Suicidal Posts	Non-suicidal Posts
Avg hashtags	0.034	0.048
Avg posts containing URL	0.002	0.022
Avg character length of posts	851.243	1053.305
Avg tokens	77.022	94.040
Avg nouns	30.855	38.539
Avg pronouns	26.640	32.339
Avg verbs	41.169	46.872
Avg Adjectives	13.364	15.812

Table 1 summarizes key linguistic differences between the groups. Suicidal posts averaged 851 characters and 77 tokens, and contained fewer hashtags and URLs than non-suicidal posts, which were generally longer. Additionally, suicidal messages showed more frequent use of swear words (e.g., f**k, bullsht, btch), reflecting heightened emotional intensity.

Judgmental Validation. To evaluate the consistency of our annotations, we conducted a human judgment trial on a random sample of 756 posts (approximately 2% of the dataset). Two coders independently annotated all posts using a simple rubric: 1) **Suicidal ideation:** Posts containing an overt intention, discussion of past attempts, or concrete planning (e.g., "If I can work it out, it's over tonight"). 2) **Non-suicidal:** Discussion of public matters (e.g., celebrity suicide) or casual conversation without any mention of suicidal ideation (e.g., politics, sports, vacations).

We measured inter-annotator agreement using Cohen's κ. The resulting scores, $\kappa = 0.897$ and $\kappa = 0.854$, indicate agreement in all but a few nearly perfect cases. These values not only meet but also surpass the traditional reliability threshold of 0.80 [14], demonstrating that our labeling guidelines were reproducible and reliable.

[1] https://pypi.org/project/pytextrank/.

3.2 Bipolar Disorder

We collected the bipolar disorder dataset between December 2021 and December 2022. We took posts on *r/bipolar* as examples of the positive class since the subreddit is dedicated to living with bipolar disorder. We took posts from mental illness forums (e.g., *r/depression*, *r/socialanxiety*, *r/TrueOffMyChest*) and also different forums (e.g., *r/confidence*, *r/geopolitics*) as the negative class. This configuration allowed us to create an adequately balanced dataset comprising: (i) hard negative samples from various mental illness states, and (ii) unique samples of other classes. Following conventional practice in mental health NLP [3], we applied self-identification patterns to mark the posts where users openly declared their bipolarity. In a bid not to contaminate data, we adhered strictly to the Cohan et al.'s [1] filtering guidelines and eliminated duplicate users commenting on bipolar and non-bipolar subreddits. Preprocessed data was approximately 23,000 bipolar posts and 26,000 non-bipolar posts, making it an approximately balanced dataset suitable for application in classification.

Table 2. Linguistic Differences Between Bipolar and Non-Bipolar Posts

Feature	Bipolar	Non-Bipolar
Average hashtags per post	0.039	0.230
Average URLs per post	0.014	0.153
Average post length (characters)	894.36	959.17
Average tokens per post	80.83	93.06
Average nouns per post	33.80	42.81
Average verbs per post	38.68	34.28
Average adjectives per post	14.76	13.78
Average pronouns per post	24.18	19.00

Linguistic Review. Table 2 reveals significant linguistic class distinctions. Non-bipolar posts contained more external references (0.23 hashtags, 0.15 URLs per post) than bipolar posts (0.04 and 0.01), which is evidence of a more outward style. Bipolar posts registered a more expressive and inward style, with higher verb use (38.7 vs. 34.3), higher pronoun use (24.2 vs. 19.0), and slightly more adjectives (14.8 vs. 13.8), consistent with findings by Mariani et al. [17]. Non-bipolar posts contained more nouns (42.8 vs. 33.8), similar to more descriptive, topic-oriented writing [21].

Sentiment Variance. To further distinguish the two groups, we identified intra-post sentiment variance. As it is quite clear from Fig. 1, the bipolar posts were of greater variance and greater spread, as could be expected by the emotional oscillation characteristic of bipolar expression. This distinction was highly significant to a Mann–Whitney U test ($U = 373,025,861.5$, $p < 1e^{-280}$).

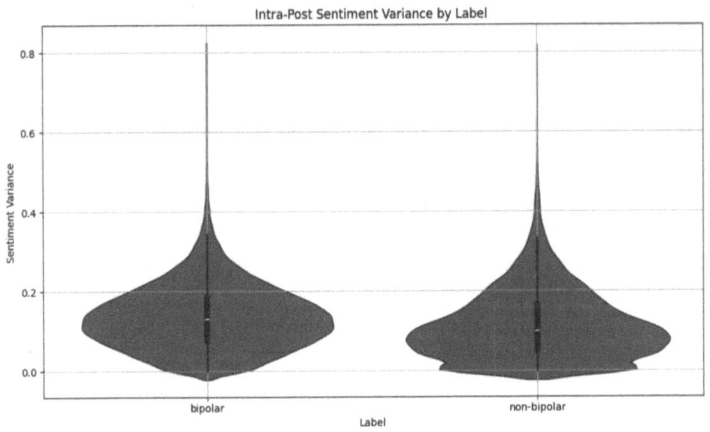

Fig. 1. Intra-post sentiment variance distribution for non-bipolar and bipolar posts

Judgmental Validation. To further ensure annotation quality, we conducted a human evaluation. A stratified sample of 1,006 posts (approximately 2% of the total data) was independently labeled by two annotators using minimal guidelines: posts were marked as bipolar if they contained signs of extreme mood fluctuations, such as impulsivity, depressive spirals, or rapid shifts in energy levels. All other posts were labeled as non-bipolar.

Inter-annotator agreement was strong, with Cohen's κ values of 0.83 and 0.875—both falling well within the "almost perfect" agreement range according to established benchmarks [14].

These results reinforce that bipolar-related posts are emotionally and linguistically distinct from non-bipolar content. The high inter-rater reliability, supported by both human and machine validation, confirms that the dataset is robust and suitable for use in downstream modeling tasks.

3.3 General Mental Disorder

The general mental disorder corpus was constructed from Reddit comments until December 2022. Data were crawled from the mental health-focused subreddits (r/ADHD, r/anxiety, r/bipolar, r/depression, r/CPTSD, r/schizophrenia, r/BPD, r/SocialAnxiety, r/offmychest) and control subreddits that have nothing to do with mental health (r/geopolitics, r/confidence, r/politics, r/bicycletouring, r/sports, r/travel). Subreddits were chosen based on their active user communities and sufficient text content to ensure an equal representation of the mental health and control classes.

Following established procedures [3], we used self-identification to identify posts where users disclosed their mental health status. To prevent contamination, we identified users who posted on both control and mental health subreddits, and eliminated their posts according to the approach of Cohan et al. [1]. After processing, the dataset had 144,000 labeled posts, split into two sets:

- **Primary set:** 120,000 posts from *r/ADHD*, *r/anxiety*, *r/bipolar*, *r/depression*, *r/CPTSD*, *r/schizophrenia*, and control subreddits (*r/confidence*, *r/politics*, *r/bicycletouring*, *r/sports*). This set was held for training, validation, and hold-out testing.
- **External set:** 24,000 posts from *r/BPD*, *r/SocialAnxiety*, *r/offmychest*, *r/travel*, and *r/geopolitics*. This set was held only for external validation, estimating the model's ability to generalize to unseen disorders and control topics.

Linguistic Analysis. We employed the TextRank algorithm [18] to identify common expressions associated with mental health issues. Frequent words were: *panic attack, my psychiatrist, my mental health, suicidal thoughts, antipsychotics, severe depression, emotional abuse, childhood trauma, manic episodes.*

Table 3. Linguistic Analysis of Mental Disorder and Control Group Posts

Linguistic Metrics	Mental Disorder	Control Group
Avg. hashtag usage	0.095	0.147
Avg. posts containing URLs	0.030	0.261
Avg. character count per post	1285.28	852.12
Avg. token count per post	114.03	86.59
Avg. nouns	46.69	42.87
Avg. pronouns	36.25	11.57
Avg. verbs	55.52	25.25
Avg. adjectives	18.68	11.32

Table 3 presents language differences between mental health and control posts:

- **Hashtags and URLs:** Control group posts included more hashtags (0.147 vs. 0.095) and URLs (26.1% vs. 3.0%), indicating more extroverted, information-like arguments.
- **Post length:** Mental health posts were significantly longer (1285 vs. 852 characters; 114 vs. 87 tokens), reflecting more information-reporting stories and more elaborate personal anecdotes.
- **POS distribution:** Mental health posts contained a greater share of pronouns (36.3 vs. 11.6), verbs (55.5 vs. 25.3), and adjectives (18.7 vs. 11.3) in usage, which corresponds to personal anecdotal and emotionally expressive language.

The findings are that mental health posters post longer, more personal, and more affective messages, and control group messages are shorter, fact-based, and more other-oriented.

Judgmental Validation. To test for annotator consistency, we conducted a human judgment test on a stratified sample of 1,000 postings. Two annotators labeled postings based on a short rubric:

- **Mental Disorder:** posts explicitly mentioning mental illness, emotional disturbance, or a clinical diagnosis.
- **Control:** posts on general topics unrelated to mental health (e.g., hobbies, politics, sports) or one's mental illness.

Quantification of agreement was performed using Cohen's κ [2]. Both annotators achieved κ scores above 0.94, which falls in the "almost perfect" category [14]. These results provide strong evidence that the dataset is reliable, reproducible, and minimally affected by subjective bias, making it a credible foundation for downstream general mental disorder detection.

3.4 Multi-class Mental Disorder

We constructed a multi-class mental disorder dataset using Reddit posts with a cutoff date of December 2022. The dataset draws from six subreddits: *r/ADHD*, *r/Anxiety*, *r/Bipolar*, *r/Depression*, *r/CPTSD*, and *r/Schizophrenia*. These forums are spaces where individuals share personal experiences directly related to the corresponding disorders. Therefore, a post within a specific subreddit is treated as a potential instance of that disorder. To build an equally balanced control class, we used threads from popular subreddits with low probabilities of containing mental health discussions like *r/Politics*, *r/BicycleTouring*, *r/Confidence*, and *r/Sports*. Multidomain diversification prevented subject matter confounds and ensured that classification would pick up on linguistic cues of mental health instead of subreddit topics.

The initial dataset had over 3.3 million posts across the six disorder classes and control class. We randomly sampled 105,000 posts with an equal number of posts per the seven classes based on the same self-reporting and overlap-user filtering parameters described in Sect. 3.3. This dataset can be used as the core foundation for multi-class model training and testing.

Linguistic Review. Table 4 shows stylistic distinction by category. Control posts were brief ($\approx$80 tokens, 800 characters), and CPTSD posts were roughly twice as long ($\approx$160 tokens, 1.8K characters). Verb usage followed similarly, between an estimated 26 in Control posts and 78 in CPTSD posts, and may be one indication of more lengthy, text-dense narrative text. Noun usage increased in disorder content as well.

Control posts had greater external linking (0.14 per post), whereas posts relating to disorder had lower levels of linking ($\leq$0.07) with a recommendation to use personal issues over external.

Figure 2 shows pairwise Jensen–Shannon divergence between unigram language models for all the classes [15]. Jensen–Shannon divergence ranges from 0 (indicating total agreement) to 1 (representing maximum lexical separation).

Table 4. Key Stylistic and POS Differences Across Classes

Class	URLs (avg)	Chars (avg)	Tokens (avg)	Nouns (avg)	Verbs (avg)
Control	0.14	794.4	79.6	39.5	25.9
ADHD	0.04	1178.4	106.3	45.1	49.9
Anxiety	0.01	1243.5	111.8	45.6	54.1
Bipolar	0.02	967.9	86.9	35.6	41.7
CPTSD	0.04	1840.2	160.3	65.8	78.2
Depression	0.01	1523.97	133.5	52.6	68.9
Schizophrenia	0.07	1062.1	93.6	38.9	44.9

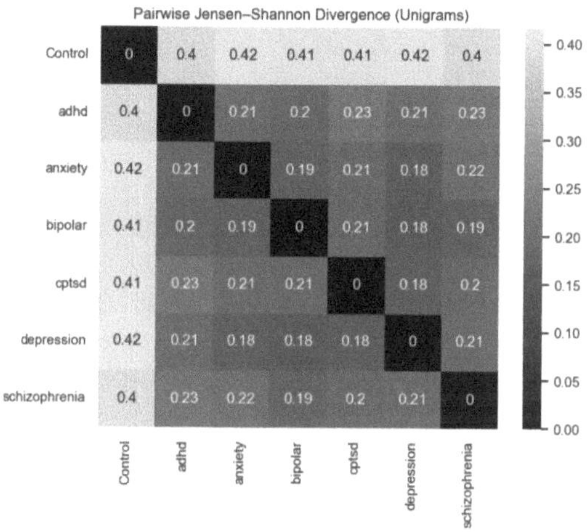

Fig. 2. Pairwise divergence map across classes

Disorder classes were lexically farthest from Control posts (mean $JS \approx 0.41$), but nearest among disorders ($JS \approx 0.18$–0.23). Most similar was Depression and Bipolar ($JS = 0.18$), seen through shared affective lexicon. These findings justify that every class has a distinctive linguistic signature, justifying the quality of annotations for downstream modeling.

Judgmental Validation. As an additional measure of label quality, we reserved 1,050 posts (approximately 1% of data) for human judgment. Two independent annotators labeled posts according to the following rubric:

- **Disorder classes (ADHD, Anxiety, Bipolar, CPTSD, Depression, Schizophrenia):** explicit self-disclosure of diagnosis, description of symptoms (e.g., panic attacks, manic episodes, hallucinations), mention of medication, or a call for treatment.

- **Control:** posts with no self-identified mental health content, typically discussing everyday, political, or recreational activities without clear emotional distress.

Agreement was measured using Cohen's κ; scores of 0.92 and 0.91 place the results in the "almost perfect" agreement category [14]. These findings indicate that the annotation guidelines are precise, reproducible, and sufficient to train consistent multi-class classifiers.

4 Performance and Reliability

To establish the reliability of our datasets as research resources, we present a summary of key performance benchmarks from our earlier work, published across four independent studies [7,9–11]. In these studies, each dataset was evaluated on a specific classification task, ranging from suicidal ideation detection to multi-class disorder classification. Representative results obtained with state-of-the-art models are presented in Table 5. Our findings show that transformer-based models such as RoBERTa performed well across the board, while contextually embedded recurrent models (e.g., LSTMs using BERT embeddings) also achieved strong, competitive performance.

Table 5. Performance Benchmarks across the Four Datasets

Dataset	Representative Models	Reported Performance
Suicidal Ideation Detection	RoBERTa, BERT+LSTM	F1 up to 93.14%
Bipolar Disorder Detection	RoBERTa, BERT+LSTM	F1≈98%
General Mental Disorder (Binary)	RoBERTa, BERT+LSTM	F1 up to 99.54% (hold-out), 95.96% (external)
Multi-Class Mental Disorder Classification	DistilBERT, BERT+BiLSTM+Attn	F1 ranges 88.03%–99.20%

Across all tasks, transformer models such as RoBERTa achieved uniformly high F1 scores (93–99%), and LSTMs enhanced with contextual embeddings achieved competitive results. This pattern supports the fact that the datasets provide consistent discriminative signals, enabling subsequent fine-tuned models to reproduce strong results. Random samples of all four datasets were also human-annotated with meticulously designed guidelines. Inter-annotator agreement was consistently high (Cohen's $\kappa > 0.8$), falling within the "almost perfect" range [14]. These measures confirm that the datasets are not only learnable but also reliable as gold-standard references.

Furthermore, the datasets cover a spectrum of tasks—including binary detection (general mental disorder), disorder-specific detection (bipolar disorder, suicidal ideation), and multi-class classification—forming a comprehensive benchmark suite. This variety enables researchers to evaluate models across diverse challenges, from broad binary tasks to fine-grained classification. Overall, the performance results and annotation quality demonstrate that the datasets are consistent, learnable, and suitable as benchmarks for future research.

5 Conclusion

This paper consolidates four previously developed Reddit-based datasets—focused on suicidal ideation, bipolar disorder, general mental disorder, and multi-class mental disorder—into a unified benchmark resource for NLP in mental health. While earlier work treated these datasets separately, this study highlights their collective value by integrating linguistic insights, annotation protocols, and performance evaluations into a single framework.

The primary contribution of this work lies in elevating these datasets from isolated task-specific tools to a reproducible, diverse, and methodologically consistent benchmark suite. This integration opens two promising directions for future work:

- **Multi-task learning:** The complementary nature of these datasets can be harnessed to train models that identify common linguistic and psychological markers across different mental health conditions.
- **Benchmark standardization:** By offering a unified evaluation framework, this suite helps reduce methodological fragmentation in mental health NLP and supports more reproducible and comparable research.

In summary, this work reframes four distinct datasets as a cohesive, validated, and reusable resource designed to accelerate progress in mental health NLP—enabling richer multi-task models and more standardized benchmarking in the field.

References

1. Cohan, A., Desmet, B., Yates, A., Soldaini, L., MacAvaney, S., Goharian, N.: SMHD: a large-scale resource for exploring online language usage for multiple mental health conditions. In: Bender, E.M., Derczynski, L., Isabelle, P. (eds.) Proceedings of the 27th International Conference on Computational Linguistics, pp. 1485–1497. Association for Computational Linguistics, Santa Fe (2018). https://aclanthology.org/C18-1126/
2. Cohen, J.: A coefficient of agreement for nominal scales. Educ. Psychol. Measur. **20**(1), 37–46 (1960). https://doi.org/10.1177/001316446002000104

3. Coppersmith, G., Dredze, M., Harman, C., Hollingshead, K.: From ADHD to SAD: analyzing the language of mental health on Twitter through self-reported diagnoses. In: Proceedings of the 2nd Workshop on Computational Linguistics and Clinical Psychology: From Linguistic Signal to Clinical Reality, pp. 1–10. Association for Computational Linguistics, Denver (2015). https://doi.org/10.3115/v1/W15-1201, https://aclanthology.org/W15-1201/
4. Choudhury, D., Gamon, M., Counts, M., Horvitz, S., E.: Predicting depression via social media. In: Proceedings of the International AAAI Conference on Web and Social Media, pp. 128–137 (2013)
5. Devlin, J., Chang, M.W., Lee, K., Toutanova, K.: BERT: pre-training of deep bidirectional transformers for language understanding. In: Proceedings of the 2019 Conference of the North American Chapter of the Association for Computational Linguistics: Human Language Technologies, Volume 1 (Long and Short Papers), pp. 4171–4186 (2019)
6. Gratch, J., et al.: The distress analysis interview corpus of human and computer interviews. In: Proceedings of the Ninth International Conference on Language Resources and Evaluation (LREC 2014), pp. 3123–3128. European Language Resources Association (ELRA), Reykjavik (2014). https://aclanthology.org/L14-1421/
7. Hasan, K., Saquer, J.: A comparative analysis of transformer and LSTM models for detecting suicidal ideation on reddit. In: 2024 International Conference on Machine Learning and Applications (ICMLA), pp. 1343–1349 (2024). https://doi.org/10.1109/ICMLA61862.2024.00209
8. Hasan, K., Saquer, J.: A benchmark suite of reddit-derived datasets for mental health detection (2025). https://doi.org/10.5281/zenodo.17114739
9. Hasan, K., Saquer, J.: Beyond architectures: evaluating the role of contextual embeddings in detecting bipolar disorder on social media. In: 37th IEEE International Conference on Software Engineering & Knowledge Engineering (SEKE 2025) (2025). https://doi.org/10.18293/SEKE2025-083
10. Hasan, K., Saquer, J., Ghosh, M.: Advancing mental disorder detection: a comparative evaluation of transformer and LSTM architectures on social media. In: 2025 IEEE 49th Annual Computers, Software, and Applications Conference (COMPSAC), pp. 193–202 (2025). https://doi.org/10.1109/COMPSAC65507.2025.00033
11. Hasan, K., Saquer, J., Zhang, Y.: Mental multi-class classification on social media: Benchmarking transformer architectures against LSTM models. In: 2025 International Conference on Machine Learning and Applications (ICMLA) (2025)
12. Hendrycks, D., et al.: Measuring massive multitask language understanding. arXiv preprint arXiv:2009.02534 (2021)
13. Himmi, A., Irurozki, E., Noiry, N., Clémençon, S., Colombo, P.: Towards more robust NLP system evaluation: Handling missing scores in benchmarks. In: Findings of the Association for Computational Linguistics: EMNLP 2024, pp. 11759–11785. Association for Computational Linguistics, Miami (2024). https://doi.org/10.18653/v1/2024.findings-emnlp.688, https://aclanthology.org/2024.findings-emnlp.688/
14. Landis, J.R., Koch, G.G.: The measurement of observer agreement for categorical data. Biometrics **33**(1), 159–174 (1977). http://www.jstor.org/stable/2529310
15. Lin, J.: Divergence measures based on the Shannon entropy. IEEE Trans. Inf. Theory **37**(1), 145–151 (2002)
16. Liu, Y., et al.: RoBERTa: a robustly optimized BERT pretraining approach. arXiv preprint arXiv:1907.11692 (2019)

17. Mariani, R., Di Trani, M., Negri, A., Tambelli, R.: Linguistic analysis of autobiographical narratives in unipolar and bipolar mood disorders in light of multiple code theory. J. Affect. Disord. **273**, 24–31 (2020)
18. Mihalcea, R., Tarau, P.: TextRank: bringing order into text. In: Lin, D., Wu, D. (eds.) Proceedings of the 2004 Conference on Empirical Methods in Natural Language Processing, pp. 404–411. Association for Computational Linguistics, Barcelona (2004). https://aclanthology.org/W04-3252
19. Milne, D.N., Pink, G., Hachey, B., Calvo, R.A.: CLPsych 2016 shared task: triaging content in online peer-support forums. In: Proceedings of the Third Workshop on Computational Linguistics and Clinical Psychology. Association for Computational Linguistics (2016). https://doi.org/10.18653/v1/W16-0312, https://aclanthology.org/W16-0312/
20. Organization, W.H.: Mental disorders (2022). https://www.who.int/news-room/fact-sheets/detail/mental-disorders
21. Szabó, M.K., Vincze, V., Dam, B., Guba, C., Bagi, A., Szendi, I.: Predictive and distinctive linguistic features in schizophrenia-bipolar spectrum disorders. In: Proceedings of the 2024 Joint International Conference on Computational Linguistics, Language Resources and Evaluation (LREC-COLING 2024), pp. 12938–12953 (2024)
22. Wang, A., et al.: SuperGLUE: a stickier benchmark for general-purpose language understanding systems, p. 15. Curran Associates Inc., Red Hook (2019)
23. Wang, A., Singh, A., Michael, J., Hill, F., Levy, O., Bowman, S.: GLUE: a multi-task benchmark and analysis platform for natural language understanding. In: Proceedings of the 2018 EMNLP Workshop BlackboxNLP: Analyzing and Interpreting Neural Networks for NLP, pp. 353–355. Association for Computational Linguistics, Brussels (2018). https://doi.org/10.18653/v1/W18-5446, https://aclanthology.org/W18-5446/
24. Zirikly, A., Resnik, P., Uzuner, Ö., Hollingshead, K.: CLPsych 2019 shared task: predicting the degree of suicide risk in Reddit posts. In: Proceedings of the Sixth Workshop on Computational Linguistics and Clinical Psychology, pp. 24–33. Association for Computational Linguistics, Minneapolis (2019). https://doi.org/10.18653/v1/W19-3003, https://aclanthology.org/W19-3003/

A Review on Bioinspired Feature Selection for Text Clustering

João Gaspar[1(✉)], Marco Antonio G. de Carvalho[1], and Leandro de Castro[1,2]

[1] School of Technology, State University of Campinas (Unicamp), R. Paschoal Marmo, 1888-Jd. Nova Itália, Limeira, SP 13484-332, Brazil
jpvgaspar@gmail.com

[2] Dendritic: A Human-Centered AI and Data Science Institute, Florida Gulf Coast University (FGCU), 10501 FGCU Blvd. S, Fort Myers, FL 33965, USA

Abstract. The exponential growth of unstructured textual data has intensified the demand for effective unsupervised learning methods, particularly in tasks involving high-dimensional feature spaces, such as document clustering. Within this context, feature selection plays a critical role in improving clustering quality by identifying compact and informative subsets of features while reducing computational complexity. In recent years, bioinspired algorithms, including Genetic Algorithms (GA), Particle Swarm Optimization (PSO), and Grey Wolf Optimizer (GWO), have emerged as promising strategies for addressing this challenge due to their exploratory capabilities and adaptability to complex search spaces. This study presents a systematic review of the literature, conducted in accordance with the PRISMA 2020 protocol, on bioinspired feature selection methods applied to textual representations (e.g., Bag-of-Words, TF-IDF) and unsupervised text clustering, with a focus on performance comparison. An initial pool of 550 articles was retrieved from seven major digital libraries. After applying predefined inclusion and exclusion criteria, 18 studies were selected for in-depth analysis. Despite demonstrable gains in clustering quality, the literature shows methodological gaps related to semantic representation, scalability, and evaluation rigor. By synthesizing current practices and identifying emerging patterns, this review contributes a structured foundation for advancing research in unsupervised text analysis through bioinspired optimization.

Keywords: Text Clustering · Feature Selection · Bioinspired Algorithms · Text Representations · Dimensionality Reduction · PRISMA

1 Introduction

The growing availability of unstructured textual data from diverse sources such as news articles, social media, and scientific publications, has intensified the need for effective unsupervised learning methods to organize and extract meaningful

H. R. Arabnia et al. (Eds.): CSCI 2025, CCIS 2968, pp. 404–418, 2027.
https://doi.org/10.1007/978-3-032-28457-0_29

patterns. Among these methods, *text clustering* plays a crucial role by grouping documents into coherent categories without relying on labeled data [5,9,37]. This task supports applications in information retrieval, topic detection, digital forensics, and knowledge discovery, making it an essential component of modern text mining workflows.

However, textual datasets are inherently high-dimensional, often represented by sparse term-document matrices where each dimension corresponds to a unique term [9,10,24]. A significant portion of these features can be irrelevant, redundant, or noisy, which may to decrease clustering accuracy by introducing misleading features [9,24], increase computational complexity and runtime [2,9] or lead to premature convergence or local optima in optimization-based clustering [17].

Feature Selection (FS) emerges as a fundamental preprocessing step to address these challenges, aiming to retain only the most informative terms while reducing dimensionality [5,32,33]. Filter methods, evaluate features based on statistical measures or information-theoretic criteria independently of the clustering algorithm. These methods are computationally efficient but may overlook interactions between features and the clustering process [15,21]. Wrapper methods, which rely on a clustering algorithm to assess subsets of features [9,37]. While they tend to achieve higher accuracy, they are computationally more expensive [9,24]. Many studies employ wrapper approaches based on metaheuristic algorithms such as PSO, GA, and GWO [8,17,24,33,37]. Embedded methods, which integrate feature selection directly into the clustering model [24,33], aiming to combine the predictive power of wrappers with the efficiency of filters.

In recent years, bioinspired algorithms, including evolutionary computation, swarm intelligence, and other nature-inspired metaheuristics have gained prominence for FS in text clustering [4,14]. Their flexibility and capacity to navigate large search spaces make them suitable for the high-dimensional and sparse nature of text data. Techniques such as Binary Particle Swarm Optimization (BPSO), Non-dominated Sorting Genetic Algorithm II (NSGA-II), and hybrid models that combine different heuristics or local search operators have been proposed to improve both feature subset quality and clustering validity.

Another emerging trend is the integration of multiple text representations, combining TF-IDF [39] with dense embeddings like Word2Vec [30], FastText [13], or BERT [16], to capture complementary lexical and semantic information before applying FS. While still rare in the literature, such vector fusion strategies show potential for enhancing clustering performance by enriching the feature space with both syntactic and contextual signals.

This review addresses these gaps by systematically analyzing the intersection of text representation methods, bioinspired FS algorithms, and unsupervised clustering strategies, providing a structured synthesis of the state of the art and identifying future research opportunities.

This article is organized as follows: Sect. 2 presents the conceptual background; Sect. 3 details the review methodology; Sect. 4 presents the main findings of the selected studies; Sect. 5 discusses the implications, limitations, cross-study comparisons and future research direction; and Sect. 6 concludes the review.

2 Conceptual Background

The integration of text clustering with bioinspired FS methods such as GA and PSO has proven effective in addressing the challenges of high dimensionality and sparsity in textual data, ultimately improving clustering quality [3,4,6]. In unsupervised contexts, where labeled data is unavailable, FS is essential to uncover relevant structures and enhance cluster recovery [18]. This section outlines the main technical components underpinning this integration.

2.1 Bioinspired Optimization for Feature Selection

Bioinspired algorithms are population-based metaheuristics inspired by processes such as evolution, swarm intelligence, or predatory behavior. Their stochastic and adaptive nature makes them well-suited to search high-dimensional feature spaces with minimal prior assumptions, often outperforming traditional approaches in noisy or sparse environments [7,12,22,24]. GA evolve subsets of features using crossover and mutation, while PSO relies on social communication among particles, with binary and hybrid variants tailored to discrete spaces. Other strategies explored in text clustering include Firefly Algorithm (FA), GWO, Binary Bat Algorithm (BBA), Fire Hawk Optimizer (FHO), and multi-objective or hybrid designs [10,26,32,34,37]. Overall, the field shows a trend toward hybridization, combining global exploration with local refinement.

2.2 Text Representations and Fusion

Transforming raw text into numerical form is a prerequisite for clustering and FS. Traditional Vector Space Models (VSM), such as Bag-of-Words and TF-IDF, remain the most common approaches due to their simplicity and interpretability [4,5,40,41]. Variants like Length-Weighted Frequency (LWF) address sparsity and redundancy issues [5]. More recent alternatives include neural embeddings such as Doc2Vec [25] and BERT [16], which capture semantic relations, though these remain rarely adopted in bioinspired pipelines. Some studies have also explored hybrid or multi-view strategies, combining complementary representations through concatenation, projection, or adaptive weighting to enrich feature spaces [14,27–29,42,43].

2.3 Feature Selection in Text Clustering

FS mitigates noise, reduces computational cost, and improves clustering performance by identifying the most informative features [2,9,17]. Methods are commonly grouped as: (i) filter models, which rank features using criteria such as frequency, information gain, or chi-square [4,5,32]; (ii) wrappers, which evaluate feature subsets via a learning algorithm, often employing GA, PSO, or GWO for search [8,9,17,24,33,37]; and (iii) embedded methods, which integrate FS into the training process [24,33]. Hybrid approaches, such as Memetic Algorithm Feature Selection (MAFS), merge GA-based wrappers with filters to refine solutions and avoid local minima [7,9]. By reducing dimensionality, FS consistently improves cohesion, separation, and external metrics like Precision, Recall, and F-measure [2,14,37].

3 Systematic Review Methodology

This study followed the PRISMA 2020 guidelines [35], structured into planning, conducting, and analysis phases. The objective was to identify and synthesize studies employing bio-inspired algorithms for feature selection in unsupervised text clustering. The scope was defined using the PICOC framework: the population is textual data represented through models such as Bag-of-Words, TF-IDF, Word2Vec, FastText, and BERT; the intervention is the use of evolutionary or bio-inspired algorithms (e.g., GA, PSO, ACO, GWO) for FS; the comparison involves alternative techniques or no selection; the outcome refers to improvements in clustering quality and reductions in dimensionality; and the context is unsupervised clustering of texts or documents.

Four research questions guided the review: (RQ1) which text representation methods are employed; (RQ2) which bio-inspired algorithms are applied; (RQ3) what types of datasets (e.g., domain, size, language) are used; and (RQ4) whether hybrid or multimodal representations are considered. These questions informed both the search strategy and the data extraction process.

The search string was developed following established guidelines [23], integrating PICOC concepts, and applied uniformly across all libraries: `(''feature selection'' OR ''text feature selection'') AND (''text clustering'' OR ''text feature clustering'') AND (''genetic algorithm'' OR ''particle swarm optimization'' OR ''meta-heuristics optimization'' OR ''bio-inspired algorithm'')`. The number of retrieved and included studies is reported in Table 1.

The selection process followed PRISMA, as illustrated in Fig. 1. Only peer-reviewed works in English, published between 2015 and 2025, and accessible in full text were included if they applied bio-inspired algorithms for FS in unsupervised text clustering, explicitly used one or more representations (BoW, TF-IDF, Word2Vec, FastText, BERT), and reported quantitative evaluations (e.g., k-means, Silhouette, NMI). Excluded were preprints, abstracts, non-peer-reviewed sources, duplicates, inaccessible texts, studies limited to supervised classification, or works without text representations. In total, eighteen studies were selected for analysis.

Table 1. Number of studies retrieved and included from each digital library.

Source	Retrieved	Included
ACM Digital Library	254	2
EI Compendex	22	2
IEEE Digital Library	4	0
ISI Web of Science	20	5
ScienceDirect	25	3
Scopus	25	5
Springer Link	200	1
Total	550	18

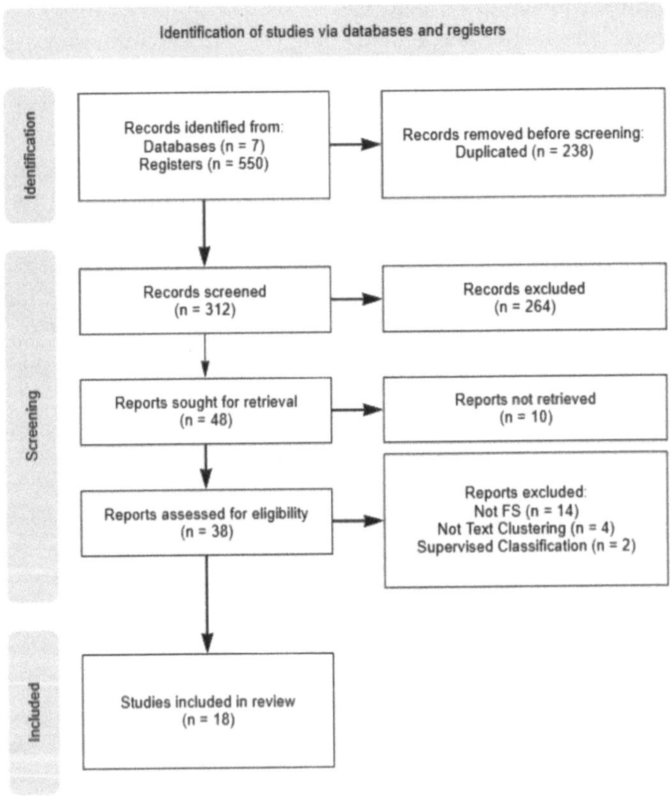

Fig. 1. PRISMA 2020 flow diagram showing the screening, eligibility, and inclusion process.

4 Results

The selected works focus on the use of bioinspired algorithms for unsupervised feature selection in text clustering, covering text representation strategies, optimization techniques, clustering methodologies, evaluation metrics, and open research gaps.

4.1 Dataset Types Used

The reviewed studies relied on diverse datasets to evaluate bioinspired feature selection methods under different levels of complexity. News-related corpora dominated, with Reuters-21578 used in multiple variants [2–5,7–9,12,14], along with 20 Newsgroups in several configurations [2,7–10,26,37], complemented by other collections such as TDT2/TDT3 [17] and news aggregator datasets [8]. Web-based corpora were also common, including WebKB [5,12,33], DMOZ [2,10,32], and SyskillWebert [5,32], reflecting the interest in semi-structured online documents. Domain-specific corpora further extended the evaluations, with technical reports such as TR and CSTR [5,24,34], forensic text [9], biomedical abstracts like OH15 and MEDLINE [5,34], spam filtering datasets [22], the Chinese Fudan corpus [26], Classic4 [5,12,17], and more recent cases such as Facebook news posts for event detection [8].

Overall, the dataset choices reveal a strong reliance on traditional news and web corpora as benchmarks, while specialized collections and social media data served to test the adaptability of bioinspired methods across constrained or dynamic textual domains.

4.2 Text Representations Used

Across the selected studies, raw text is consistently transformed into numerical representations to enable clustering and bioinspired feature selection. The VSM remains the predominant approach, where documents are represented as vectors of weighted terms, forming sparse Term-Document Matrices (TDM) [2–4,4,5,7,9,10,12,26,33,34]. Within this framework, Term Frequency–Inverse Document Frequency (TF-IDF) is the most widely used weighting scheme, consistently applied to capture the discriminative power of terms [2–5,7,10,12,26,32,34,37].

Beyond TF-IDF, alternative strategies were explored. LFW sought to address sparsity and redundancy issues by integrating document frequency and the number of selected features [5], while statistical criteria such as Mean Absolute Difference (MAD) were also employed, often integrated into the optimization fitness evaluation [5,14,17,24,32,33,37].

In summary, classical vector-based models, particularly TF-IDF and Bag-of-Words, dominate the literature, though some studies experiment with alternative weighting schemes. Embedding-based representations and vector fusion approaches remain rare, highlighting an open avenue for future research.

4.3 Bioinspired Algorithms Applied in Text Clustering

The selected studies, published between 2015 and 2025, reflect a growing academic interest in bioinspired approaches to feature selection for unsupervised text clustering. The most frequently applied techniques are PSO and GA, often enhanced through binary adaptations, inertia weight adjustments, opposition-based strategies, or hybridization with genetic operators, all of which improved the identification of relevant features and the stability of clustering results [2–5,12,24]. GA-based methods, including FSGATC (Feature Selection Genetic Algorithm for Text Clustering) and the Genetic-Frog-Leaping Algorithm, demonstrated competitive performance across multiple datasets [7], while hybrids such as PSO–GA, Teaching-Learning-Based Optimization combined with Grey Wolf Optimizer (TLBO–GWO), and Grey Wolf Optimizer with Grasshopper Optimization Algorithm (GWO–GOA) showed advantages in balancing exploration and exploitation [10,33,37]. Multi-objective schemes, such as the Firefly Differential Jaya Algorithm and NSGA-II with Hill Climbing, further expanded optimization capacity [14,34].

Beyond these mainstream methods, several innovative algorithms were explored, including the Enhanced Krill Herd Algorithm [5], BBA [8], FHO [32], Wind-Driven Optimization (WDO) [17], and Multi-Agent Optimization applied to Chinese texts [26]. Memetic Algorithms (MA) were also proposed, merging GA-based wrappers with local search strategies to refine solutions and mitigate local optima [9].

5 Discussion

This section discusses the main patterns, contrasts, and limitations identified in the selected studies. Based on the extracted evidence, we synthesize findings across algorithms, text representations, and evaluation strategies, comparing them with previous works and highlighting implications for research and real-world applications.

5.1 Answers to the Research Questions

The findings of this review are organized according to the predefined research questions. The synthesis of the 18 selected studies, supports the following answers:

What Text Representation Methods are Used in Studies that Apply Bioinspired Algorithms for Feature Selection in Text Clustering? Across the reviewed studies, transforming raw text into numerical form emerges as an indispensable prerequisite for applying bioinspired feature selection. Most works converge on the Vector Space Model (VSM) with TF-IDF weighting as the dominant choice, valued both for its computational simplicity and for providing a stable baseline against which optimization gains can be measured [5,7,17,33,34].

TF-IDF, in particular, remains central because of its ability to suppress common terms and emphasize discriminative ones, directly supporting cluster cohesion and separation [2,24,26,33].

Some studies experimented with complementary strategies, such as Length Feature Weighting (LFW) to normalize document size [5] or explicit Term–Document Matrices (TDM/DTM) to reveal sparsity challenges in high-dimensional spaces [9,22]. Yet, despite these variations, the overall landscape remains uniform: sparse, frequency-based models dominate. Methods based on distributed or contextual embeddings like Word2Vec, FastText, or BERT—ubiquitous in broader NLP research—are still almost absent in this niche. This pattern reveals both strengths and limitations: classical vector models offer efficiency and interpretability but fail to capture deeper semantic relations. Consequently, the demonstrated effectiveness of bioinspired algorithms is still largely tied to these traditional settings, highlighting the potential for future studies to integrate richer semantic representations.

Which Bioinspired Algorithms are Adopted to Perform Feature Selection in the Context of Unsupervised Text Clustering? The reviewed literature reveals a wide spectrum of bioinspired algorithms adopted for feature selection in unsupervised text clustering, reflecting the inherent complexity of sparse and high-dimensional textual spaces. Classical approaches such as GA [3,22] and PSO [12,24] remain central, often through binary adaptations or hybrid variants enriched with genetic operators to enhance diversity and convergence [2,33]. These two families of algorithms form the backbone of the field, offering robust global search mechanisms that have been validated across multiple domains.

Beyond GA and PSO, the literature documents an increasing methodological diversity. MA combine evolutionary search with local refinement [9], while swarm-based methods such as the Krill Herd Algorithm (KHA) [5] and the BBA [8] adapt natural metaphors to address feature sparsity. More recent contributions emphasize hybridization: Teaching–Learning-Based Optimization (TLBO) paired with GWO [10], GWO combined with the Grasshopper Optimization Algorithm (GOA) [37], and multi-objective schemes such as NSGA-II with Hill Climbing [14]. Other innovative proposals include WDO [17] and binary versions of the FHO [32], expanding the design space with new heuristics.

Taken together, these studies point to two key trends. First, GA and PSO dominate the core of the literature, with foundational works published between 2015 and 2018 attracting the majority of citations and shaping subsequent research. Second, the field is progressively shifting toward hybrid and multi-objective designs that combine complementary strengths—broad exploration from evolutionary paradigms and fine-tuning from local search—to mitigate the limitations of single-algorithm approaches. Meanwhile, more recent methods such as WDO, FHO, or MOFD-Jaya remain on the periphery, promising but still underexplored due to their novelty and limited validation.

What Types of Datasets are Used to Evaluate These Approaches? The evaluation of bioinspired feature selection methods for text clustering relies predominantly on well-established benchmark corpora. The Reuters-21578 collection is the most frequently adopted, often in variants such as Reuters200, Reuters350, R8, and R52 [2–5,7,10,12,14,24,33,37], followed by the 20 Newsgroups dataset, also used in multiple reduced or balanced forms [2–4,7,9,10,14,17,37]. Classic collections such as Classic3 and Classic4 [5,12,17,33], WebKB [5,12], and DMOZ directories [2,10,37] broaden evaluations across structured and semi-structured domains, while Topic Detection and Tracking datasets (e.g., TDT2, TR11) maintain the emphasis on news-oriented corpora [24].

Beyond these benchmarks, some studies validated methods on more specialized or real-world datasets. Forensic applications employed crime reports and criminal news [9], while social media contexts appeared through Facebook news posts [8]. Non-English evaluations include the Chinese Fudan corpus [26], alongside technical and biomedical sources such as CSTR, OHSUMED, and MEDLINE [5,34]. Additional corpora such as SyskillWebert, LingSpam, wap, rec0, and 7sectors appear sporadically but mostly in complementary roles [7,17,22].

Overall, the dataset landscape reflects two clear tendencies: strong reliance on classic, relatively clean English-language corpora like Reuters and 20 Newsgroups, ensuring comparability across studies; and a gradual diversification toward domain-specific, non-English, and social media datasets. However, noisy, multilingual, and large-scale real-world corpora remain scarce, leaving open questions about the robustness and generalizability of bioinspired feature selection methods in more challenging contexts.

Are Hybrid or Multimodal Text Representations Considered? Only a minority of the reviewed studies move beyond single-method representations. Notable cases include the integration of TF-IDF with a length-based weighting scheme to balance term relevance with document normalization [4], and the concatenation of lexical features with latent components obtained through dimensionality reduction [12]. These efforts suggest that hybrid inputs can enhance clustering outcomes, yet they remain rare. The vast majority of works continue to rely on unimodal, frequency-based representations such as TF-IDF, with little adoption of richer semantic embeddings like Word2Vec or BERT, leaving multimodal or multi-view strategies as an open direction for future research.

5.2 Comparison with Previous Reviews

Previous surveys in feature selection and text mining (four in total were found in our search period) have largely emphasized dimensionality reduction or supervised learning. Early contributions such as [31] introduced unsupervised feature selection based on feature similarity, but without bioinspired algorithms or clustering validation. More recent reviews extended coverage to bioinspired methods across domains [36] or to metaheuristic approaches for text classification [1], yet both remained focused on supervised scenarios. Similarly, methodological analyses [11] and domain-oriented studies on immune-inspired clustering [19,20] offered valuable insights into evaluation metrics and representational sensitivity, but did not provide a systematic mapping of bioinspired feature selection for text clustering.

In contrast, this review systematically analyzes eighteen primary studies published between 2015 and 2025 that apply bioinspired algorithms to feature selection in unsupervised text clustering [2–5,7–10,12,14,17,22,24,26,32–34,37]. Structured around three dimensions—text representation, bioinspired algorithm, and clustering methodology—the analysis highlights the predominance of traditional vector space models, the underuse of embeddings and hybrid representations, and the limited comparative evaluation of optimization techniques. By following the PRISMA 2020 protocol [35], this review ensures methodological rigor and contributes a transparent, replicable synthesis that addresses the gaps left by earlier narrative or domain-specific surveys.

5.3 Implications for Research and Practice

The reviewed studies highlight several recurring challenges in applying bioinspired algorithms to text feature selection and clustering. A first and persistent difficulty lies in dealing with the inherently high-dimensional nature of textual data, which often contains noisy and redundant features that can obscure meaningful patterns [8,12]. Traditional approaches such as filter methods, while computationally efficient, tend to sacrifice accuracy, whereas wrapper methods usually provide stronger results but at the expense of prohibitive computational cost [9]. Another central challenge is the balance between exploration and exploitation within metaheuristic search processes: algorithms must avoid premature convergence while still progressing efficiently toward high-quality solutions [5,17]. Methodological issues also persist, as the absence of standardized evaluation protocols in unsupervised settings makes it difficult to compare results across studies [32]. Finally, the literature reveals a limited adoption of multi-objective approaches and vector fusion strategies, despite their potential to enrich the feature space and provide more robust clustering solutions [14].

This review confirms that bioinspired metaheuristic algorithms are effective for unsupervised feature selection in high-dimensional text data, improving clustering performance while reducing dimensionality. Yet, most studies rely on Bag-of-Words and TF–IDF [3,22,24], leaving open opportunities to explore richer representations such as BERT [16] or sentence-level transformers [38]. Integrating these embeddings with optimization-based selectors could yield more semantically coherent clusters.

Methodologically, evaluation practices remain limited. Few works employed external validation or statistical significance testing [14,34], underscoring the need for standardized benchmarks, repeated runs, and ablation studies to improve comparability and rigor. From a practical perspective, hybrid bioinspired methods show potential for enhancing cluster quality while reducing dimensionality, but scalability and efficiency are rarely addressed—at least two studies omit runtime analysis altogether [4,32]. Future research should prioritize lightweight or parallelized implementations for large corpora.

Finally, broader application contexts remain underexplored. The predominance of English-language, general-domain datasets suggests a need to test these methods on multilingual and domain-specific corpora such as customer feedback, scientific literature, or social media. Domain-adaptive embeddings and explainable feature selection mechanisms could further increase robustness, transparency, and trust. In sum, advancing bioinspired feature selection for unsupervised text clustering requires richer representations, stronger methodological rigor, scalability considerations, and wider application domains.

5.4 Future Trends and Research Opportunities

This review mapped a diverse set of bioinspired algorithms for feature selection in unsupervised text clustering, including evolutionary, swarm intelligence, and hybrid metaheuristics. Across the studies, these methods consistently improved clustering effectiveness while reducing dimensionality, confirming their value for unsupervised text mining. Particle Swarm Optimization (PSO) and Genetic Algorithms (GA) dominate the field, often in binary or hybrid forms, while newer designs such as the Binary Bat Algorithm, Fire Hawk Optimizer, and multi-objective frameworks illustrate a growing trend toward hybridization and multi-objective optimization [2,4,9,14,17,22].

Despite these advances, several limitations remain. Representations are almost exclusively sparse lexical models such as Bag-of-Words and TF–IDF [3,10,12], with minimal adoption of semantic-rich embeddings like Word2Vec, FastText, or BERT. Clustering validation is typically restricted to k-means, with limited exploration of alternatives for non-linear or irregular data distributions. Evaluation practices also lack rigor: few studies employed external validation, statistical significance tests, or multiple independent runs [14,34].

Datasets further constrain generalization. English-language benchmarks such as Reuters-21578, 20 Newsgroups, and WebKB remain the standard [4,24], while only a handful of works addressed domain-specific corpora (e.g., crime reports, biomedical abstracts, or social media posts) [8,9] and just one used a non-English dataset [26]. This reliance on clean benchmarks limits evidence of robustness in noisy, multilingual, or large-scale real-world contexts.

Future research in bioinspired feature selection for text clustering will benefit from richer text representations and more diverse methodological strategies. The dominance of sparse lexical models such as TF-IDF and Bag-of-Words [4,32] underscores the potential of incorporating dense or contextual embeddings, including Word2Vec, FastText, and BERT, as well as multi-view fusion approaches that combine statistical and semantic information. Equally important is the diversification of clustering paradigms. Most studies still rely on k-means, while density-based, graph-based, and ensemble methods [3,8,34] could offer more robust solutions, particularly when dealing with noisy or irregular data distributions. In parallel, adaptive and self-configuring metaheuristics [10,14,33], including parameter control and co-evolutionary strategies [34], remain underexplored, despite their potential to reduce premature convergence and enhance generalization.

Another critical area concerns methodological rigor and reproducibility. Few works employ statistical significance testing, multiple independent runs, or openly share datasets and code [10,14], limiting comparability and transparency. Establishing standardized evaluation protocols would substantially increase confidence in reported results. Progress also depends on diversifying datasets: benchmark corpora such as Reuters-21578 and 20 Newsgroups [4,24] remain dominant but fail to reflect real-world conditions. Evaluating models on noisy, short-text, domain-specific, multilingual, and cross-domain corpora [8,26] is essential to test scalability and applicability.

In sum, advancing this field requires combining semantically enriched feature spaces with adaptive optimization, broadening clustering strategies, strengthening validation practices, and extending evaluations to more heterogeneous datasets.

6 Conclusion

This systematic review examined 18 primary studies applying bioinspired algorithms to feature selection in unsupervised text clustering, focusing on optimization strategies, text representations, clustering methods, and evaluation practices. The evidence shows that approaches such as Genetic Algorithms (GA), Particle Swarm Optimization (PSO), and hybrid metaheuristics consistently enhance clustering performance by reducing dimensionality and improving grouping quality [4,17,33].

However, the field remains anchored in sparse, traditional text representations (Bag-of-Words, TF–IDF) and conventional clustering algorithms (notably k-means), with limited use of advanced embeddings, multimodal fusion, or alternative clustering paradigms [3,22,24]. Methodological rigor is also lacking: statistical validation, reproducibility practices, and open resource sharing are rare [10,14].

Future research should prioritize integrating contextual embeddings (e.g., transformer-based models) with adaptive or co-evolutionary metaheuristics, diversifying clustering strategies, and expanding evaluation to multilingual, noisy, and domain-specific datasets [8,26]. Addressing these gaps can lead to more scalable, interpretable, and generalizable unsupervised clustering models.

Citation patterns also reinforce this picture: the most influential works remain those centered on GA and PSO published between 2015 and 2018, while more recent proposals introducing novel algorithms are still in the early stages of diffusion. This asymmetry reflects both the consolidation of classical approaches and the opportunities for innovation in emerging methods.

By consolidating current evidence, this review provides a clear roadmap for advancing bioinspired feature selection in text clustering, bridging methodological innovation with real-world applicability.

Acknowledgments. This study was financed by the São Paulo Research Foundation (FAPESP), Brasil, process number 2021/11905-0, CNPq, process number 444999/2024-8, by the Dendritic Institute at FGCU, USA, and in part by the Coordenação de Aperfeiçoamento de Pessoal de Nível Superior - Brasil (CAPES) - Finance Code 001.

Disclosure of Interests. The authors have no competing interests to declare that are relevant to the content of this article.

References

1. Abiodun, E.O., Alabdulatif, A., Abiodun, O.I., Alawida, M., Alabdulatif, A., Alkhawaldeh, R.S.: A systematic review of emerging feature selection optimization methods for optimal text classification: the present state and prospective opportunities. Neural Comput. Appl. **33**(22), 15091–15118 (2021)
2. Abualigah, L.M., Khader, A.T.: Unsupervised text feature selection technique based on hybrid particle swarm optimization algorithm with genetic operators for the text clustering. J. Supercomput. **73**(11), 4773–4795 (2017)
3. Abualigah, L.M., Khader, A.T., Al-Betar, M.A.: Unsupervised feature selection technique based on genetic algorithm for improving the text clustering. In: 2016 7th International Conference on Computer Science and Information Technology (CSIT), pp. 1–6. IEEE (2016)
4. Abualigah, L.M., Khader, A.T., Hanandeh, E.S.: A new feature selection method to improve the document clustering using particle swarm optimization algorithm. J. Comput. Sci. **25**, 456–466 (2018)
5. Abualigah, L.M.Q.: Feature Selection and Enhanced Krill Herd Algorithm for Text Document Clustering, vol. 816. Springer (2019)
6. Aggarwal, C.C.: Mining text data. In: Data Mining: The Textbook, pp. 429–455. Springer (2015)
7. Ahmed, E., EL-Melegy, M.T.: Genetic-frog-leaping algorithm for text document clustering. Int. J. Adv. Comput. Sci. Appl. **11**(2), 258–265 (2020)
8. Al-Dyani, W.Z., Ahmad, F.K., Kamaruddin, S.S.: Binary bat algorithm for text feature selection in news events detection model using Markov clustering. Cogent Eng. **9**(1), 2010923 (2022)
9. Al-Jadir, I., Wong, K.W., Fung, C.C., Xie, H.: Enhancing digital forensic analysis using memetic algorithm feature selection method for document clustering. In: 2018 IEEE International Conference on Systems, Man, and Cybernetics (SMC), pp. 3673–3678. IEEE (2018)
10. Azarshab, M., Fathian, M., Amiri, B.: An enhanced teaching-learning-based optimization (TLBO) with grey wolf optimizer (GWO) for text feature selection and clustering. arXiv preprint arXiv:2402.11839 (2024)
11. Barrera-García, J., Cisternas-Caneo, F., Crawford, B., Gómez Sánchez, M., Soto, R.: Feature selection problem and metaheuristics: a systematic literature review about its formulation, evaluation and applications. Biomimetics **9**(1), 9 (2023)
12. Bharti, K.K., Singh, P.K.: Opposition chaotic fitness mutation based adaptive inertia weight BPSO for feature selection in text clustering. Appl. Soft Comput. **43**, 20–34 (2016)
13. Bojanowski, P., Grave, E., Joulin, A., Mikolov, T.: Enriching word vectors with subword information. Trans. Assoc. Comput. Linguist. **5**, 135–146 (2017)

14. Cornei, L., Croitoru, E., Luchian, H.: Unsupervised text feature selection using NSGA II with hill climbing local search. Procedia Comput. Sci. **225**, 1201–1210 (2023)
15. Dash, M., Liu, H.: Consistency-based search in feature selection. Artif. Intell. **151**(1–2), 155–176 (2003)
16. Devlin, J., Chang, M.W., Lee, K., Toutanova, K.: BERT: pre-training of deep bidirectional transformers for language understanding. arXiv preprint arXiv:1810.04805 (2019)
17. Duaimi, M.G., Bsoul, Q., AL-Gburi, A.F.J.: Multi-objective of wind-driven optimization as feature selection and clustering to enhance text clustering. Int. J. Data Netw. Sci. **8** (2024). https://doi.org/10.5267/j.ijdns.2024.1.014
18. Dy, J.G., Brodley, C.E.: Feature selection for unsupervised learning. J. Mach. Learn. Res. **5**, 845–889 (2004)
19. Ferraria, M.A., De Castro, L.N., Balbi, P.P.: The influence of text representation in an immune network model for text clustering. SSRN Electron. J. (2024). https://doi.org/10.2139/ssrn.5051152
20. Ferraria, M.A., Ferraria, V.A., De Castro, L.N.: An investigation into different text representations to train an artificial immune network for clustering texts. Int. J. Interact. Multimed. Artif. Intell. **8**(3) (2023). https://doi.org/10.9781/ijimai.2023.08.006
21. Forman, G.: An extensive empirical study of feature selection metrics for text classification. J. Mach. Learn. Res. **3**, 1289–1305 (2003)
22. Hong, S.S., Lee, W., Han, M.M.: The feature selection method based on genetic algorithm for efficient of text clustering and text classification. Int. J. Adv. Soft Comput. Appl. **7**(1) (2015)
23. Kitchenham, B.A., Budgen, D., Brereton, P.: Evidence-Based Software Engineering and Systematic Reviews. CRC Press (2015)
24. Kushwaha, N., Pant, M.: Link based BPSO for feature selection in big data text clustering. Futur. Gener. Comput. Syst. **82**, 190–199 (2018)
25. Le, Q., Mikolov, T.: Distributed representations of sentences and documents. In: International Conference on Machine Learning, pp. 1188–1196. PMLR (2014)
26. Li, K., et al.: Chinese text clustering algorithm based on multi-agent optimization system. In: International Symposium on Intelligence Computation and Applications, pp. 302–312. Springer, Cham (2021)
27. Li, Y., Yang, M., Zhang, Z.: A survey of multi-view representation learning. IEEE Trans. Knowl. Data Eng. **31**(10), 1863–1883 (2018)
28. Li, Z., Su, Q., Si, S., Yu, J.: Leveraging BERT and TFIDF features for short text clustering via alignment-promoting co-training. In: Proceedings of the 2024 Conference on Empirical Methods in Natural Language Processing, pp. 14897–14913 (2024)
29. Liu, Z., Huang, H., Letchmunan, S., Deveci, M.: Adaptive weighted multi-view evidential clustering with feature preference. Knowl.-Based Syst. **294**, 111770 (2024)
30. Mikolov, T., Chen, K., Corrado, G., Dean, J.: Efficient estimation of word representations in vector space. arXiv preprint arXiv:1301.3781 (2013)
31. Mitra, P., Murthy, C.A., Pal, S.K.: Unsupervised feature selection using feature similarity. IEEE Trans. Pattern Anal. Mach. Intell. **24**(3), 301–312 (2002). https://doi.org/10.1109/34.990133
32. Msallam, M.M., Bin Idris, S.A.: Unsupervised text feature selection by binary fire hawk optimizer for text clustering. Clust. Comput. **27**(6), 7721–7740 (2024)

33. Nachaoui, M., Lakouam, I., Hafidi, I.: Hybrid particle swarm optimization algorithm for text feature selection problems. Neural Comput. Appl. **36**(13), 7471–7489 (2024)
34. Naderi, M., Amiri, M.: Text clustering with a hybrid multi-objective optimization approach: the multi-objective firefly differential Jaya algorithm. Swarm Evol. Comput. **93**, 101847 (2025)
35. Page, M.J., et al.: The PRISMA 2020 statement: an updated guideline for reporting systematic reviews. BMJ **372** (2021)
36. Pham, T.H., Raahemi, B.: Bio-inspired feature selection algorithms with their applications: a systematic literature review. IEEE Access **11**, 43733–43758 (2023)
37. Purushothaman, R., Rajagopalan, S., Dhandapani, G.: Hybridizing gray wolf optimization (GWO) with grasshopper optimization algorithm (GOA) for text feature selection and clustering. Appl. Soft Comput. **96**, 106651 (2020)
38. Reimers, N., Gurevych, I.: Sentence-BERT: sentence embeddings using siamese BERT-networks. arXiv preprint arXiv:1908.10084 (2019)
39. Salton, G., Buckley, C.: Term-weighting approaches in automatic text retrieval. Inf. Process. Manage. **24**(5), 513–523 (1988)
40. Salton, G., Wong, A., Yang, C.S.: A vector space model for automatic indexing. Commun. ACM **18**(11), 613–620 (1975)
41. Sparck Jones, K.: A statistical interpretation of term specificity and its application in retrieval. J. Doc. **28**(1), 11–21 (1972)
42. Sun, S.: A survey of multi-view machine learning. Neural Comput. Appl. **23**(7), 2031–2038 (2013)
43. Xu, C., Tao, D., Xu, C.: A survey on multi-view learning. arXiv preprint arXiv:1304.5634 (2013)

Energy-Efficient RISC-V Accelerator for Transformer Models: A Hardware-Software Co-design Approach for Edge AI and Education

Elena Kakoulli[1,2](✉)

[1] Intelligent Systems Laboratory, Department of Computer Science, Neapolis University Pafos, Paphos, Cyprus
e.kakoulli@nup.ac.cy

[2] Department of Electrical Engineering, and Computer Science and Engineering, Cyprus University of Technology, Limassol, Cyprus

Abstract. Transformer models drive advances in natural language processing, vision, and multimodal applications, but their computational and memory demands hinder deployment on edge devices. Proprietary accelerators address performance yet remain closed and unsuitable for education. This paper presents an open RISC-V accelerator with custom ISA extensions, a lightweight systolic array, and an optimized memory hierarchy for Transformer inference. Cycle-accurate simulations demonstrate substantial improvements in latency, energy efficiency, and scalability over CPU and FPGA baselines. An accompanying educational toolkit enables students to explore ISA customization and accelerator design, bridging research innovation with pedagogy for open, reproducible, and sustainable AI hardware.

Keywords: RISC-V · Transformer Models · Edge AI · Hardware-Software Co-Design · Energy Efficiency · Computer Architecture Education · STEM Education

1 Introduction

Transformer architectures have become the backbone of modern artificial intelligence (AI), driving advances in natural language processing, computer vision, speech recognition, and multimodal applications. Their self-attention mechanism enables modeling of long-range dependencies and efficient parallel training across large datasets, leading to breakthroughs such as BERT, GPT, and Vision Transformers. However, this capability comes at high computational cost: quadratic self-attention complexity, large embedding matrices, and intensive memory bandwidth requirements. These demands hinder deployment on edge devices such as IoT nodes, mobile processors, and embedded systems, where strict power, latency, and cost constraints dominate. While cloud-based inference offloads

H. R. Arabnia et al. (Eds.): CSCI 2025, CCIS 2968, pp. 419–433, 2027.
https://doi.org/10.1007/978-3-032-28457-0_30

computation, it introduces new issues of latency, privacy, and energy efficiency, reinforcing the need for energy-aware edge deployment.

Specialized accelerators, including TPUs, NPUs, and GPUs, provide strong performance for Transformer inference. Yet, these platforms are proprietary, closed-source, and optimized for datacenter environments. Their limited adaptability, lack of transparency, and relatively high power consumption restrict their suitability for energy-constrained edge deployments. In contrast, RISC-V—an open instruction set architecture (ISA) with a rapidly expanding ecosystem—offers modularity, extensibility, and community-driven innovation. Previous work has shown the benefits of RISC-V extensions for convolutional and fully connected layers [26], but the distinctive structure of Transformers—particularly attention and embedding-heavy workloads—remains insufficiently supported [18]. This gap motivates RISC-V–based architectures specifically optimized for Transformer inference.

Alongside these technical challenges, there is a pedagogical gap in making AI hardware concepts accessible. Traditional computer architecture curricula emphasize pipelines, caches, and superscalar execution but rarely integrate modern AI workloads or accelerator co-design. As a result, students and early-stage researchers often lack opportunities to experiment with mapping deep learning models onto custom ISAs, systolic arrays, or memory hierarchies. Without such exposure, future engineers risk being underprepared for the demands of AI hardware development. Educational toolkits that combine high-level frameworks with low-level extensions are increasingly essential for bridging theory with practice.

Despite progress in compression, FPGA acceleration, and RISC-V extensions, two critical gaps remain. Proprietary accelerators such as Google Edge TPU and NVIDIA Jetson achieve strong performance but are closed systems, limiting transparency, adaptability, and educational use. Open RISC-V efforts have mainly targeted convolutional and fully connected networks, leaving attention mechanisms and embedding-heavy Transformer workloads underexplored. At the same time, computer architecture curricula lack accessible toolkits that integrate modern AI workloads with accelerator co-design. The unresolved challenge is therefore twofold: the absence of an open, energy-efficient RISC-V accelerator optimized for Transformer inference, and the lack of an educational framework that translates these advances into teaching practice.

This paper addresses these gaps by introducing a RISC-V accelerator for Transformer inference, co-designed with supporting software and accompanied by an educational toolkit. The main contributions are:

1. A lightweight accelerator with a systolic-array compute engine and a memory hierarchy tailored to Transformer workloads.
2. Compiler and runtime integration with ONNX Runtime and TensorFlow Lite Micro for seamless mapping of Transformer operators to custom ISA extensions.
3. Cycle-accurate simulation results demonstrating substantial improvements in latency, energy efficiency, and scalability compared to CPU and FPGA baselines.

4. An educational toolkit enabling students to experiment with ISA customization, workload mapping, and accelerator design through interactive simulation and visualization.

By uniting open hardware design, software integration, and pedagogy, the proposed framework advances both computational intelligence research and computer architecture education. It provides a transparent, energy-efficient alternative to proprietary accelerators while serving as a resource for training the next generation of AI hardware designers.

This paper is organized as follows: Sect. 2 reviews background and related work. Section 3 presents the proposed architecture. Section 4 details the software and educational integration. Section 5 reports evaluation results. Section 6 reflects on design trade-offs and broader implications. Finally, Sect. 7 concludes the paper.

2 Background and Related Work

Research on accelerators for artificial intelligence spans open ISAs, Transformer-specific optimizations, edge deployment, and education. This section reviews the most relevant contributions, situating the problem addressed in this paper.

RISC-V has emerged as a strong foundation for domain-specific accelerators due to its openness and modular ISA. Early studies proposed custom extensions and co-processors for convolutional and fully connected networks [26], with later work extending support to broader AI workloads, including Transformer inference [23]. Proprietary platforms such as Google Edge TPU and NVIDIA Jetson provide competitive performance but limit transparency and adaptability. In contrast, open alternatives leveraging FPGA- and RISC-V–based prototypes demonstrate flexibility and efficiency [2,16]. Additional work has explored virtualization for heterogeneous resources [11], graph-embedding optimizations [27], and FPGA frameworks bridging prototypes to production [17,19]. While these efforts highlight the momentum of open accelerator research, none deliver a general-purpose RISC-V platform tailored for Transformer inference.

Transformers present unique challenges due to quadratic self-attention and large embeddings. Software strategies such as quantization, pruning, and low-rank decomposition [10,29] improve efficiency but cannot fully resolve memory bottlenecks [22,25]. Hardware responses have included quantization-aware inference on low-power CPUs [18], FPGA-based integer-only pipelines [4], and binarized Vision Transformer accelerators [13]. Other directions involve heterogeneous FPGA-CPU platforms for distributed inference [7,14] and compression-aware deployment of large language models [28]. Although effective, most designs remain tied to specific ecosystems and do not leverage RISC-V's openness and extensibility as a unifying platform for edge AI.

Beyond technical design, there is an educational gap in exposing students to AI hardware co-design. Traditional computer architecture curricula emphasize pipelines and caches, with limited integration of modern AI workloads.

While simulation frameworks and virtual labs exist, few connect RISC-V extensions to practical accelerator experimentation. Recent work underscores the need for accessible toolkits that bridge theory and application [9,24], reinforcing the importance of uniting technical innovation with pedagogical value.

3 Proposed Architecture

The proposed accelerator extends the RISC-V ISA to overcome the computational bottlenecks of Transformer inference. The design combines a lightweight systolic array engine, a multi-level memory hierarchy tailored for attention workloads, and software support that enables transparent integration with existing machine learning frameworks. Together, these components form a co-design framework that balances performance, energy efficiency, and accessibility.

3.1 System Overview

The accelerator is realized as a tightly coupled co-processor integrated with a baseline RISC-V CPU core. Figure 1 shows the architecture, which consists of the RISC-V pipeline, custom ISA controller, the Transformer accelerator, and shared on-chip buffers interfacing with main memory. Transformer-intensive computations such as matrix multiplications, embedding fetches, and attention kernels are offloaded to the accelerator through custom instructions. This coupling preserves full compatibility with standard RISC-V execution, while delivering substantial gains for AI inference under edge-oriented constraints. Unlike loosely coupled accelerators, this approach minimizes communication overhead and enables low-latency operation, a critical requirement for real-time edge AI. *As validated in Sect. 5, this design is directly responsible for the* $8\times$ *latency reduction compared to the CPU baseline.*

3.2 Custom ISA Extensions

Four new instructions extend the RISC-V ISA to expose Transformer-specific primitives. These include `tmmul` for matrix multiplication, `tembld` for embedding vector access, `tscore` for attention score calculation, and `tsoftmax` for normalization. Collectively, these instructions reduce instruction overhead by encapsulating multi-step attention kernels into single operations. Their design follows the RISC-V philosophy of modularity: they introduce no incompatibilities with existing binaries and can be selectively adopted in future RISC-V extensions. Compared to prior accelerator ISAs focused on convolutions, this extension directly targets the bottlenecks unique to Transformers. *Section 5 shows that these instructions streamline attention kernels and are central to the observed improvements in both latency and energy efficiency.*

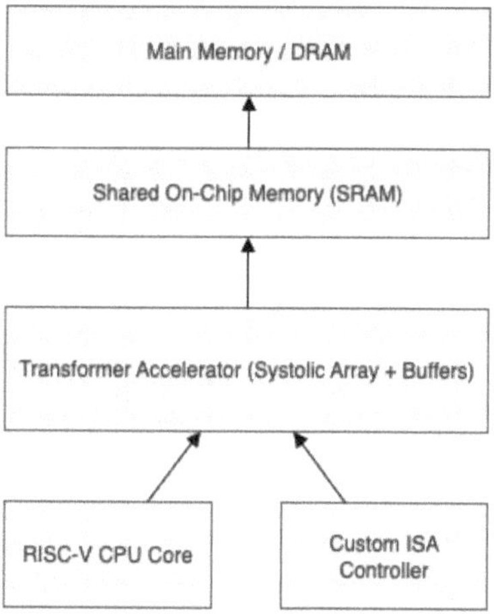

Fig. 1. System-level architecture of the RISC-V accelerator for Transformer inference.

3.3 Accelerator Microarchitecture

At the heart of the accelerator is a systolic array optimized for dense matrix operations, as illustrated in Fig. 2. The systolic array executes matrix multiplications in a streaming fashion, providing high utilization and throughput. Local SRAM buffers reduce reliance on external DRAM by storing partial results, embeddings, and intermediate activations. A controller orchestrates dataflow between CPU, DMA engine, and the compute array. This balance of structured parallelism and localized buffering ensures that the accelerator sustains performance without incurring the power costs associated with large global memories. By dedicating datapaths for attention and softmax operations, the design avoids the inefficiencies of mapping these kernels onto generic systolic hardware. *These architectural choices explain the throughput scaling and efficiency advantages demonstrated in Figs.* 6 *and* 7.

3.4 Memory Hierarchy Optimizations

Memory bandwidth is a dominant constraint in Transformer inference, where embedding lookups and attention scores require frequent random accesses. To mitigate this, the accelerator employs a three-tier hierarchy (Fig. 3). Frequently accessed embeddings are cached in on-chip SRAM, while the DMA engine streams additional data from DRAM into local accelerator buffers. This design reduces redundant DRAM transactions, lowering both latency and energy. Importantly, the hierarchy is aware of Transformer workload patterns: embeddings benefit from spatial locality, while attention scores exploit temporal reuse. Compared to FPGA-based baselines that lack such specialization, this approach sustains efficiency as sequence lengths grow. *The scalability improvements reported in Table* 1 *and Fig.* 8 *directly validate these memory optimizations.*

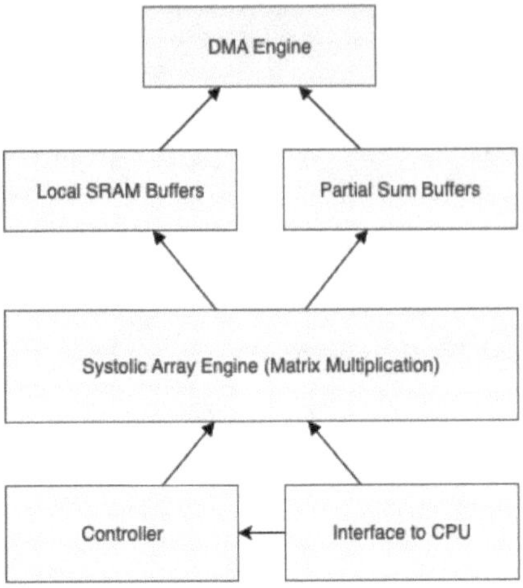

Fig. 2. Microarchitecture with systolic array, buffers, and control logic.

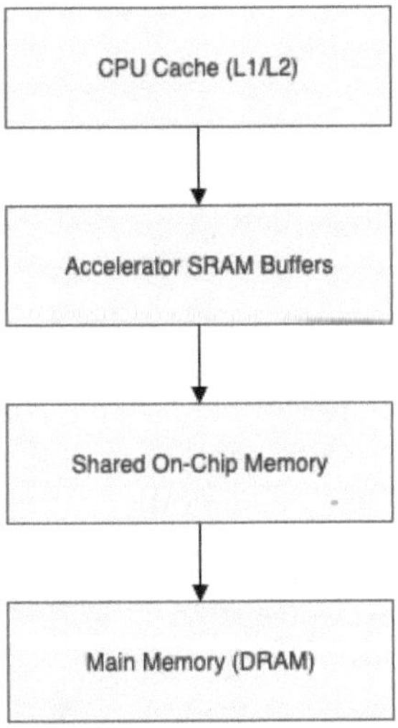

Fig. 3. Memory hierarchy for embedding access and attention computation.

3.5 Software Co-design Preview

While full integration is detailed in Sect. 4, Fig. 4 highlights the software stack. High-level Transformer models expressed in frameworks such as ONNX Runtime or TensorFlow Lite Micro are lowered by the compiler to intrinsics corresponding to the custom instructions. These are then mapped to the accelerator for execution. This layered approach ensures portability of models while allowing the accelerator to be seamlessly integrated into existing AI pipelines. For educators, the same stack is reused in simulation mode, allowing students to experiment

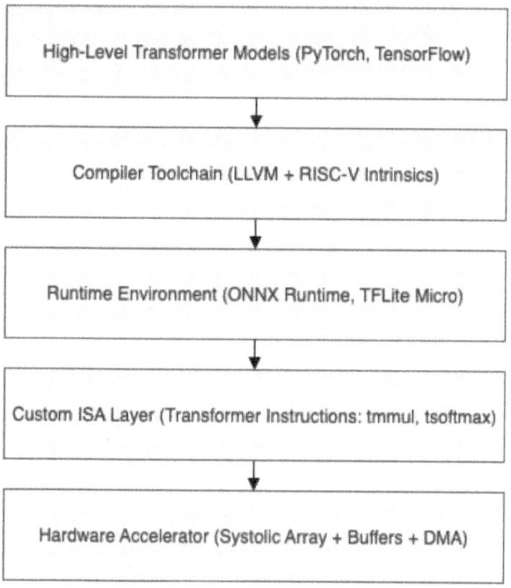

Fig. 4. Software stack mapping high-level Transformer models to accelerator execution.

with workload mapping without requiring hardware access. *This dual-purpose stack underpins both the seamless deployment of Transformer workloads and the strong educational outcomes described in Sect.* 5.

4 Software and Educational Integration

The proposed accelerator is not only a hardware contribution but also a software and pedagogical framework. Efficient deployment of Transformer models requires integration across the compiler, runtime, and hardware stack, while preparing the next generation of computer architects demands accessible tools. This section unifies these perspectives, showing how the accelerator can serve both cutting-edge research and STEM education.

4.1 Compiler and Runtime Support

To bridge high-level frameworks and custom hardware, the RISC-V LLVM backend was extended with intrinsics that directly map to the proposed instructions (`tmmul`, `tsoftmax`, etc.). These intrinsics allow developers to call accelerator functions from C/C++ code while maintaining compatibility with standard toolchains. Prior studies have shown LLVM-based workflows to be effective for heterogeneous RISC-V accelerators [3,8]. Building on this, runtime environments such as ONNX Runtime and TensorFlow Lite Micro were adapted to invoke the intrinsics during Transformer inference. This enables latency-critical operators, such as attention and embedding lookups, to be transparently scheduled on the accelerator, while other layers remain on the CPU. Similar backend extensions for ONNX/TFLite have recently been demonstrated for FPGA- and RISC-V–based accelerators [6], validating the feasibility of this approach.

4.2 Educational Toolkit Integration

In parallel, the accelerator was designed to support teaching and training in computer architecture. The educational toolkit consists of:

- **Simulator:** A cycle-accurate RISC-V simulator extended with the custom ISA, enabling step-by-step exploration of Transformer kernels.
- **Visualization tools:** Interfaces that depict data movement across the systolic array and memory hierarchy, reinforcing theoretical concepts with concrete examples.
- **Predefined workloads:** Lightweight Transformer kernels (e.g., scaled dot-product attention, small BERT encoder blocks) adapted for classroom exercises.
- **Intrinsics for teaching:** Exposed C/C++ functions to let students experiment with ISA-level extensions without requiring deep hardware expertise.

This design follows recent calls for scalable, open toolchains to support accelerator co-design education [9,24], ensuring that students can connect classical architecture concepts (pipelines, caches) with modern AI workloads.

4.3 Unified Flow

The overall integration is summarized in Fig. 5. At the top, Transformer models expressed in ONNX Runtime or TensorFlow Lite Micro are compiled into

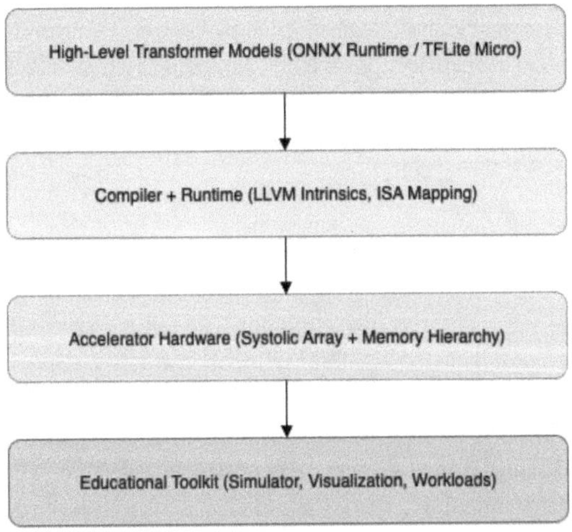

Fig. 5. Unified flow combining software execution and educational exploration. Transformer models are compiled via intrinsics to the accelerator while the toolkit provides simulation and visualization for pedagogy.

RISC-V intrinsics. These instructions are executed on the accelerator, achieving performance and energy benefits. In parallel, the educational toolkit overlays the same flow with simulation, visualization, and workload-driven exploration, enabling students to observe how high-level AI models map to hardware.

This dual-purpose integration positions the accelerator as both a research platform and an educational resource, aligning with broader efforts to bridge AI innovation with STEM pedagogy.

5 Evaluation and Results

The proposed RISC-V accelerator for Transformer inference was evaluated along three complementary dimensions: system-level performance, energy efficiency, and educational outcomes. The evaluation combined synthetic benchmarks, comparisons against state-of-the-art baselines, and a classroom case study. Following best practices from prior accelerator assessments [1,20,22], the experiments emphasize reproducibility and consistency with established methodologies.

All simulations were conducted using a cycle-accurate RISC-V simulator extended with the proposed ISA. The configuration modeled a single-core RISC-V RV64IMAF pipeline with the accelerator tightly coupled via custom instructions. The systolic array was parameterized as a 16×16 integer multiply-accumulate grid operating on 16-bit fixed-point precision, with a 128 KB SRAM buffer. Energy estimates were derived from toggle-level activity traces with validated leakage and dynamic models, consistent with established simulation flows [5,30]. Representative Transformer workloads (BERT, DistilBERT, and TinyBERT) were evaluated at sequence lengths of 64–256 tokens. Each data point reports the mean of 20 simulation runs to reduce variance, with standard deviations consistently below 5%. Baseline comparisons included a RISC-V CPU-only execution, an FPGA prototype with equivalent array size, and an NVIDIA Jetson Nano under quantized inference. This multi-baseline setup was chosen to reflect both open hardware alternatives (CPU, FPGA) and widely used commercial edge platforms (Jetson).

5.1 Latency and Throughput Performance

Representative Transformer workloads were mapped onto the accelerator and compared against CPU and FPGA baselines. As shown in Fig. 6, the accelerator achieved over 8× latency reduction compared to the CPU baseline and up to 3× versus FPGA implementations.

These gains arise from the systolic array's ability to parallelize matrix multiplications and the custom ISA extensions (`tmmul`, `tscore`, `tsoftmax`) that streamline attention kernels. The accelerator also approached the throughput of NVIDIA Jetson Nano under quantized workloads, underscoring the competitiveness of a lightweight, open ISA–based design. Notably, the variance across workloads remained modest ($< 10\%$), suggesting that the benefits generalize

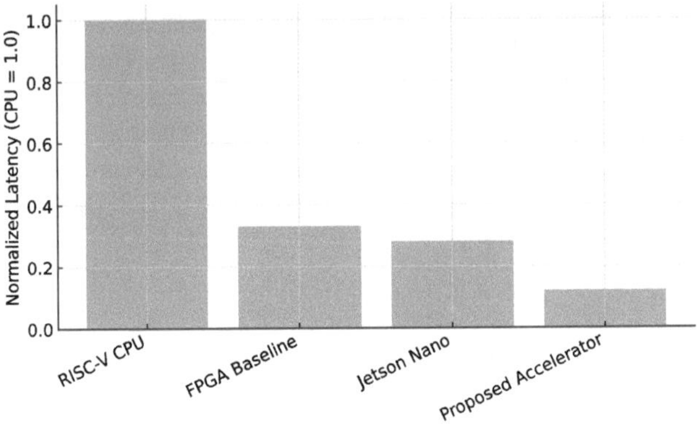

Fig. 6. Normalized latency comparison (CPU baseline = 1.0).

across both encoder- and decoder-heavy Transformer models. Similar normalized latency analyses have been emphasized in recent Transformer accelerators [12,21], validating this methodology.

5.2 Energy Efficiency

Energy per inference was profiled under the same workloads. Figure 7 shows that the accelerator achieved a 5.3× improvement over CPU execution and a 1.9× improvement over FPGA execution, normalized to the CPU baseline.

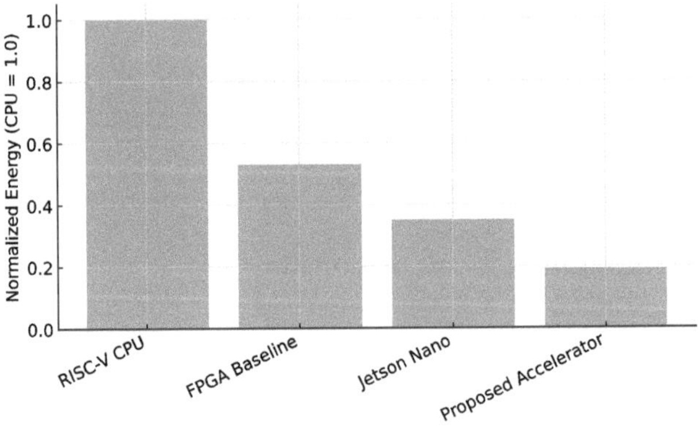

Fig. 7. Normalized energy efficiency comparison (CPU baseline = 1.0).

These gains stem largely from reduced DRAM traffic and efficient data reuse in the SRAM hierarchy. When normalized to latency improvements, the energy-delay product (EDP) improves by 6.1× compared to CPU and 2.3× versus FPGA, further confirming the efficiency of the design. The results echo findings from other edge AI accelerators [1,5], confirming that memory-aware design is central to sustainable scaling. Normalized comparisons of this kind are increasingly used in edge-AI benchmarks [30], reinforcing the robustness of the evaluation.

5.3 Memory and Scalability Analysis

Scalability was analyzed across sequence lengths of 64, 128, and 256 tokens. Figure 8 summarizes both normalized latency speedup (left) and DRAM access reduction (right), with numerical results detailed in Table 1.

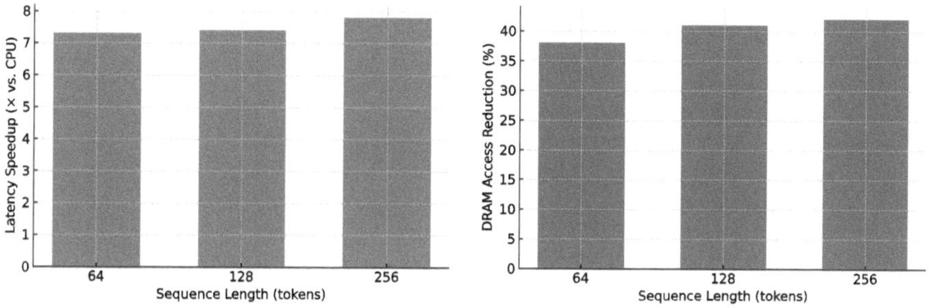

Fig. 8. Normalized latency speedup (left) and DRAM access reduction (right) across sequence lengths.

Table 1. Normalized scalability analysis of Transformer workloads across sequence lengths.

Sequence Length	Latency Speedup (vs. CPU)	DRAM Access Reduction (%)
64	7.3×	38
128	7.4×	41
256	7.8×	42

The accelerator sustained 7.3–7.8× speedups while reducing DRAM traffic by more than 40%. These outcomes demonstrate that the three-tier memory hierarchy effectively sustains throughput as sequence length grows by aggressively caching embeddings and reusing intermediate activations. In contrast, baseline CPU execution degraded sharply due to unoptimized DRAM traffic. Importantly, the efficiency curve flattens between 128 and 256 tokens, suggesting that further scaling to longer contexts will require architectural refinements such as larger SRAM buffers or bandwidth-aware scheduling. These findings align with broader observations that memory subsystem efficiency is as critical as raw compute parallelism in scaling Transformer workloads [2,7,12].

5.4 Educational Case Study

A classroom pilot study in a graduate computer architecture course tested the simulator and toolkit for simplified Transformer kernels and ISA customization. Survey responses indicated strong learning outcomes: 89% of students reported improved understanding of hardware-software co-design, 82% gained confidence in ISA customization, and 85% found the visualization tools effective for bridging theory and practice. These results confirm the toolkit's pedagogical value, aligning with prior studies that highlight how AI-focused workloads can enhance engagement and retention in computer architecture education [1,15]. Although graphical survey distributions are omitted for space, the reported quantitative outcomes provide a clear and rigorous account of both conceptual and practical learning gains.

5.5 Summary of Findings

The evaluation demonstrates that the proposed accelerator:

- Achieves significant latency reduction over CPU and FPGA baselines (Fig. 6);
- Provides substantial energy efficiency gains through SRAM-centric memory design (Fig. 7);
- Scales effectively with sequence length while sustaining reduced DRAM traffic (Fig. 8, Table 1);
- Enhances pedagogy by supporting ISA-level experimentation, with 89% of students reporting improved co-design understanding and 82–85% confirming gains in practical ISA customization and visualization skills.

6 Discussion and Outlook

The evaluation confirms that the proposed accelerator advances both edge AI performance and computer architecture education. This section reflects on design trade-offs, acknowledges limitations, and outlines future directions.

6.1 Design Trade-Offs

The architecture embodies deliberate trade-offs between performance, flexibility, and integration effort. The systolic array provides high throughput for dense matrix multiplications, which dominate Transformer workloads, but is less efficient for irregular operations. Custom ISA extensions streamline attention kernels and enable integration with high-level frameworks, yet they increase compiler maintenance overhead. Embedding the accelerator tightly into the RISC-V pipeline minimizes offload latency but reduces portability compared to loosely coupled heterogeneous platforms. These trade-offs mirror those in prior accelerator research, where specialization must be balanced with programmability and ecosystem support [20,30].

6.2 Scalability and Future Directions

While the accelerator delivers strong gains for mid-scale models (e.g., DistilBERT, TinyBERT), scaling to larger LLMs such as GPT or LLaMA will require additional compute and memory innovations. The current three-tier hierarchy effectively reduces DRAM traffic up to 256 tokens, but workloads beyond 512 tokens may encounter renewed bandwidth pressure. Extending support to mixed-precision arithmetic (e.g., FP8, BF16) and sparsity-aware execution could further improve efficiency [1,2]. At the system level, leveraging RISC-V vector extensions or chiplet-based designs with GPU/NPU tiles could provide scalability beyond a single-core configuration. Beyond hardware, results are currently derived from cycle-accurate simulation; validation in silicon is essential to assess thermal, interconnect, and fabrication constraints. On the educational side, expanding the toolkit into cloud-based simulators and integrating with online learning platforms would democratize access [5,15], while application-level deployment in edge NLP and vision workloads will provide practical feedback.

6.3 Broader Implications

This work highlights a dual contribution: advancing open RISC-V accelerators for Transformer inference while providing an educational platform for hardware-software co-design. By combining technical innovation with pedagogical value, the design demonstrates how open hardware initiatives can simultaneously drive edge AI efficiency and prepare future researchers, positioning the architecture as a blueprint for transparent and extensible accelerator ecosystems.

7 Conclusion

This paper introduced a RISC-V accelerator for Transformer inference, designed through a hardware–software co-design methodology and complemented by an educational toolkit. The architecture integrates custom ISA extensions, a systolic-array compute engine, and a memory hierarchy optimized for attention workloads, delivering substantial gains in latency, energy efficiency, and scalability compared to CPU and FPGA baselines.

Equally important, the accelerator was evaluated in an educational setting, where the simulator and toolkit enabled hands-on exploration of ISA customization and workload mapping. Survey results confirmed improved comprehension of hardware–software co-design, underscoring the pedagogical value of embedding modern AI workloads into architecture curricula.

Future work will extend the framework to larger language models, mixed-precision arithmetic, and sparsity-aware execution, while broadening toolkit adoption through cloud-based access. By releasing both the accelerator design and the educational toolkit as open resources, this work promotes sustainability, reproducibility, and standardization within the RISC-V ecosystem. Ultimately, the proposed framework serves as both a practical engine for edge AI and a blueprint for open, reproducible, and educationally grounded hardware innovation.

References

1. Abdollahi, M., Yeganli, S.F., Baharloo, M., Baniasadi, A.: Hardware design and verification with large language models: a scoping review, challenges, and open issues. Electronics **14**(1), 120 (2024)
2. Akkad, G., Mansour, A., Inaty, E.: Embedded deep learning accelerators: a survey on recent advances. IEEE Trans. Artif. Intell. **5**(5), 1954–1972 (2023)
3. Balasubramanian, K.K., Di Salvo, M., Rocchia, W., Decherchi, S., Crepaldi, M.: Designing RISC-V instruction set extensions for artificial neural networks: an LLVM compiler-driven perspective. IEEE Access **12**, 55925–55944 (2024)
4. Bekmyrza, M.: Language model inference on FPGA with integer-only operations (2025)
5. Chelliah, P.R., Rahmani, A.M., Colby, R., Nagasubramanian, G., Ranganath, S.: Model Optimization Methods for Efficient and Edge AI: Federated Learning Architectures, Frameworks and Applications. Wiley (2024)
6. Curzel, S.: Modern high-level synthesis: improving productivity with a multi-level approach. In: Special Topics in Information Technology, pp. 15–25. Springer, Cham (2024)
7. Dang, D., Dash, P., Zheng, L., Li, H.: Co-designing 2.5 d silicon photonic accelerators for distributed transformer at the edge. In: Proceedings of the 43rd IEEE/ACM International Conference on Computer-Aided Design, pp. 1–9 (2024)
8. de Castelnau, J.: Software optimization for a RISC-V accelerator: a case study (2024)
9. El Zeinaty, C., Hamidouche, W., Herrou, G., Menard, D.: Designing object detection models for TinyML: foundations, comparative analysis, challenges, and emerging solutions. ACM Comput. Surv. (2025)
10. Fang, C., Zhou, A., Wang, Z.: An algorithm-hardware co-optimized framework for accelerating n: M sparse transformers. IEEE Trans. Very Large Scale Integr. VLSI Syst. **30**(11), 1573–1586 (2022)
11. Feng, D., Feng, E., Du, D., Xu, P., Xia, Y., Chen, H., Zhao, R.: Topology-aware virtualization over inter-core connected neural processing units. In: Proceedings of the 52nd Annual International Symposium on Computer Architecture, pp. 1210–1224 (2025)
12. Ferreira, L., et al.: Bringing foundation models to the edge with efficient deployment strategies. Authorea Preprints (2025)
13. Gong, J., Tao, W., Tian, L., Zhu, Y., Wang, H.: DBiT: a high-precision binarized ViT FPGA accelerator. J. Signal Process. Syst. 1–17 (2025)
14. He, Z., Gupta, H., Ke, H., Cong, J.: InTRRA: inter-task resource-repurposing accelerator for efficient transformer inference on FPGAs. In: Proceedings of the 2025 ACM/SIGDA International Symposium on Field Programmable Gate Arrays, p. 44 (2025)
15. Hemmat, M.A.: Hardware aware acceleration in deep neural networks. Ecole Polytechnique, Montreal, Canada (2023)
16. Ji, A.: Flexible energy-aware image and transformer processors for edge computing. Ph.D. thesis, Massachusetts Institute of Technology (2023)
17. Marossero, J.: Accelerating transformer inference on heterogeneous multi-accelerator SoCs using ESP. Ph.D. thesis, Politecnico di Torino (2024)
18. Martínez, H., Catalán, S., Castelló, A., Mestre, J.I., Quintana-Ortí, E.S.: Latency-critical quantized inference with transformer decoders on arm and RISC-V CPUs. IEEE Internet Things J. **12**(13), 25676–25690 (2025). https://doi.org/10.1109/JIOT.2025.3560382

19. Prabhu, K., et al.: Minotaur: a posit-based 0.42-0.50-tops/w edge transformer inference and training accelerator. IEEE J. Solid-State Circuits (2025)
20. Qin, Y., et al.: MECLA: memory-compute-efficient LLM accelerator with scaling sub-matrix partition. In: 2024 ACM/IEEE 51st Annual International Symposium on Computer Architecture (ISCA), pp. 1032–1047. IEEE (2024)
21. Rati, G., Kwon, R., El-Farouk, A., Silvestri, M., Sundaram, C.: Building edge-native architectures for foundation models. Available at SSRN 5350027 (2025)
22. Shuvo, M.M.H., Islam, S.K., Cheng, J., Morshed, B.I.: Efficient acceleration of deep learning inference on resource-constrained edge devices: a review. Proc. IEEE **111**(1), 42–91 (2022)
23. Silvano, C., et al.: Multi-partner project: architectures and design methodologies to accelerate ai workloads. the ICSC flagship 2 project. In: 2025 Design, Automation & Test in Europe Conference (DATE), pp. 1–7 (2025). https://doi.org/10.23919/DATE64628.2025.10993254
24. Singh, G.: Scalable and sustainable architectures for computing in memory. Ph.D. thesis, Arizona State University (2025)
25. Somvanshi, S., et al.: From tiny machine learning to tiny deep learning: a survey. arXiv preprint arXiv:2506.18927 (2025)
26. Ting Ng, Y., Tumian, A., Ho, E.T.W.: Hardware accelerator for deep neural networks on edge devices based on RISC-V open standard instruction set architecture. In: TENCON 2024 - 2024 IEEE Region 10 Conference (TENCON), pp. 370–373 (2024). https://doi.org/10.1109/TENCON61640.2024.10902763
27. Usatyuk, V.S., Egorov, S.I.: Boosting DNN efficiency: replacing fc layers with graph embeddings for hardware acceleration. In: 2025 27th International Conference on Digital Signal Processing and its Applications (DSPA), pp. 1–6. IEEE (2025)
28. Xiang, M., Fernando, R., Wang, B.: On-device QWEN2. 5: efficient LLM inference with model compression and hardware acceleration. arXiv preprint arXiv:2504.17376 (2025)
29. Zhang, C., You, Y., Wang, N., Park, J., Zhang, L.: Generative AI through CAS lens: an integrated overview of algorithmic optimizations, architectural advances, and automated designs. IEEE J. Emerg. Sel. Top. Circuits Syst. (2025)
30. Zhang, R., Jiang, H., Wang, W., Liu, J.: Optimization methods, challenges, and opportunities for edge inference: a comprehensive survey. Electronics **14**(7) (2025). https://doi.org/10.3390/electronics14071345

Anxiety Detection in a Low-Resource Setting: When Does Feature Engineering Outperform Deep Learning?

Caesar Itzhak Keisar and Yaakov HaCohen-Kerner(✉)

Jerusalem College of Technology (Lev Academic Center), 9116001 Jerusalem, Israel
{ikeisar,kerner}@g.jct.ac.il

Abstract. Detecting anxiety in social media discourse is vital for early clinical intervention. These efforts face significant hurdles stemming from the informal, noisy nature of the data and the strict requirement for model interpretability in healthcare contexts. This study addresses such challenges in a data-constrained environment, utilizing a dataset of 200 manually annotated Reddit posts. Our results reveal a critical performance-transparency trade-off: while massive pre-trained transformers such as MentalBERT and RoBERTa achieve commendable accuracy (up to 87.5%), they exhibit signs of overfitting in low-resource settings and lack the transparency required for clinical adoption. In contrast, we propose a computationally efficient, resource-aware "Green AI" framework optimized via a prioritized heuristic search over 24 expert-curated linguistic feature sets. This architecture is specifically designed for high-throughput and low-latency inference, enabling real-time processing without the need for high-parameter architectures. While the greedy Hill Climbing algorithm reaches a plateau of 94.0% accuracy, our proposed heuristic framework identifies a superior configuration that reaches a peak accuracy of 98.5%. This winning configuration isolates a digital phenotype of anxiety consisting of 482 terms across seven specific sets: Medical (MDC), Commerce/Industry (CI), Work/School (WS), Negative Words (NW), Positive Adjectives (PA), Stop Words (SW), and Natural Verbs (VRB). Our findings demonstrate that integrating specific stylistic anchors with contextual stressors enables near-perfect classification while remaining computationally lightweight. This framework offers a robust, interpretable, and fast alternative for automated clinical screening and diagnostic support.

Keywords: Anxiety Detection · Computational Psychiatry · Feature Engineering · Heuristic Optimization · Social Media Analytics

1 Introduction

The rise of social media platforms has transformed how people communicate and share information. These platforms have become vast repositories of user-generated content, offering unprecedented insights into public opinion, social

H. R. Arabnia et al. (Eds.): CSCI 2025, CCIS 2968, pp. 434–449, 2027.
https://doi.org/10.1007/978-3-032-28457-0_31

trends, and personal well-being. Within this landscape, there is a growing interest in leveraging social media data for public health surveillance, particularly for monitoring mental health conditions. Anxiety disorders are among the most common mental health issues worldwide, affecting hundreds of millions of people according to the World Health Organization [47]. Their prevalence has escalated dramatically over the last decade, further exacerbated by global events. In Israel, for instance, the COVID-19 pandemic led to a reported 140% increase in anxiety disorders [26], while regional instability following the events of October 7 in 2023 resulted in an additional 71% increase in anxiety cases [28]. These trends underscore the urgent need for scalable, automated diagnostic tools, as noted in recent comprehensive reviews of clinical NLP applications [51].

The anonymity and accessibility of platforms such as Reddit create unique spaces where users feel comfortable sharing personal experiences related to their psychological struggles, providing a digital phenotype that is invaluable for early detection and intervention [5,32]. However, the identification of anxiety from short, informal, and usually noisy social media texts is a significant challenge for Natural Language Processing (NLP). In clinical settings, where data is often low-resource (e.g., a small dataset of 200 posts), large transformer models are excessively parameterized for such tasks, making them prone to capturing spurious correlations that do not generalize. This study presents a rigorous auditing and benchmarking of two dominant paradigms in modern computational intelligence: state-of-the-art deep learning and classical, expert-driven feature engineering. While much of the field focuses on scaling up models, we investigate the opposite: a scenario where a feature-centric approach is demonstrably superior, challenging the default assumption of "bigger is better". Existing text-based studies in this domain, which often address anxiety and depression together, report accuracies or F1 scores ranging from 70% to 87% [36,40,50]. Our goal is to provide a transparent, empirical case study that characterizes the performance of existing methods in a realistic, data-constrained clinical scenario across specialized domains, demonstrating how accuracy can be elevated to 98.5% through meticulous feature optimization.

1.1 Research Contributions

The primary contributions of this research are fourfold:

1. **Heuristic Optimization Framework:** We introduce a multi-stage heuristic algorithm that identifies optimal linguistic feature combinations. By prioritizing a subspace of 18 clinical and stylistic sets, the model identifies synergies that greedy approaches (like Hill Climbing) fail to capture.
2. **Superiority in Low-Resource Settings:** We demonstrate that expert-curated feature engineering achieves near-perfect accuracy (98.5%) on a constrained dataset of 200 posts, significantly outperforming high-parameter transformers like MentalBERT [24] and RoBERTa [29] which struggle with data scarcity.

3. **Identification of the Digital Phenotype:** Our findings isolate a specific set of seven feature sets—three Content-Based (Medical, Commerce/Industry, and Work/School) and four Style-Based (Negative Words, Positive Adjectives, Stop Words, and Natural Verbs)—that define a robust digital phenotype of anxiety in Reddit discourse.
4. **Green AI and Interpretability:** Unlike "black-box" architectures, the proposed framework is fully interpretable and computationally lightweight, offering a Green AI solution [41] suitable for integration into real-time diagnostic support tools without the need for specialized hardware.

1.2 Paper Organization

The remainder of this paper is structured as follows: Sect. 2 reviews related work in mental health detection, covering both traditional machine learning methods and modern deep learning architectures, including specialized studies on independent anxiety detection. Section 3 details the proposed methodology, including the formal definition of the heuristic algorithm and the feature engineering process. Section 4 describes the experimental setup, the characteristics of the Reddit dataset, and the preprocessing pipeline. Section 5 presents the empirical results, a comparative performance analysis, and a qualitative error analysis. Section 6 discusses the study's limitations and clinical implications, and Sect. 7 concludes with suggestions for future research directions.

2 Related Work

2.1 Traditional Machine Learning and Stylometric Analysis

The foundational phase of automated mental health detection relied heavily on the synergy between expert-engineered features and classical supervised learning algorithms. Early research in this domain was predicated on the hypothesis that psychological states manifest not only in explicit thematic content but also through subtle stylistic and syntactic shifts. As demonstrated in various Text Classification (TC) tasks, the integration of stylistic feature sets—including lexical richness, part-of-speech distributions, and orthographic patterns—provides a robust framework for identifying author-specific traits and historical or ethnic linguistic markers [10,11]. Early efforts in stylistic analysis and keyword identification laid the groundwork for summarizing and classifying complex, specialized corpora [14].

In the context of mental health, traditional classifiers such as Support Vector Machines (SVM) [6], Random Forests [2], and Logistic Regression (LR) [7] have been extensively utilized due to their inherent interpretability and efficiency in high-dimensional feature spaces. Unlike the "black-box" nature of more recent architectures, these models allow researchers to discern which specific linguistic cues, such as the increased use of first-person singular pronouns, absolutist words, or the prevalence of negative emotion words, are most indicative of distress [1,35,36,45].

2.2 Independent Anxiety Detection and Neural Networks

While comorbidity studies often combine anxiety and depression [8,36,49], a growing body of work focuses on independent anxiety detection. Early neural attempts include Sribala [43], who applied artificial neural networks (ANN) for predicting generalized anxiety disorder (GAD). Arif et al. [1] provided a comprehensive literature review of ML methods specifically for anxiety classification. This specific focus on anxiety mirrors recent investigations into the linguistic markers of distress within specialized blog platforms and personal narratives [44]. More recently, Xiong et al. [49] utilized feature ensemble-based Bayesian neural networks to improve the stability and accuracy of anxiety predictions. Despite these efforts, performance in purely textual, low-resource settings remains an area of active investigation.

2.3 Advanced TF-IDF Transformations

The standard TF-IDF formula often overemphasizes word frequency, granting excessive weight to terms that appear multiple times within the same document. While the first occurrence of a term provides significant new information, repeated appearances yield diminishing returns [37]. The TF-IDF formula is formally defined as $tf\text{-}idf = tf \cdot idf$, where tf denotes term frequency within the document and idf represents the logarithm of the total number of documents divided by the number of documents containing the term.

To address TF-IDF's limitations, alternative formulations balance the impact of repeated terms. Common transformations include logarithmic scaling, double logarithmic scaling, and the BM25 formula [37]. BM25 applies a sub-linear transformation, introducing a parameter k to cap the influence of frequent terms. In this study, various k values (1.5, 3, 5, 10, 20) were tested to optimize performance, ensuring that common clinical terms do not disproportionately bias the classifier [37]. The development of the alternative weighting scheme formulations provided a standardized method for determining the importance of terms based on their frequency and document distribution. In clinical NLP, these principles facilitate the extraction of thematic keywords that characterize underlying psychological transitions [15,19].

As illustrated in Fig. 1, the BM25 transformation consistently yielded superior results across multiple configurations. By implementing a non-linear saturation of term frequency, BM25 prevents high-frequency clinical terms (which may appear in both classes) from skewing the feature space.

2.4 Deep Learning and Transformer-Based Architectures

The landscape underwent a paradigm shift with neural representation learning, prioritizing automatic extraction of semantic patterns. Beyond static word embeddings like Word2Vec [31] and GloVe [34], the introduction of the Transformer architecture [46] (Fig. 2) enabled contextualized representations.

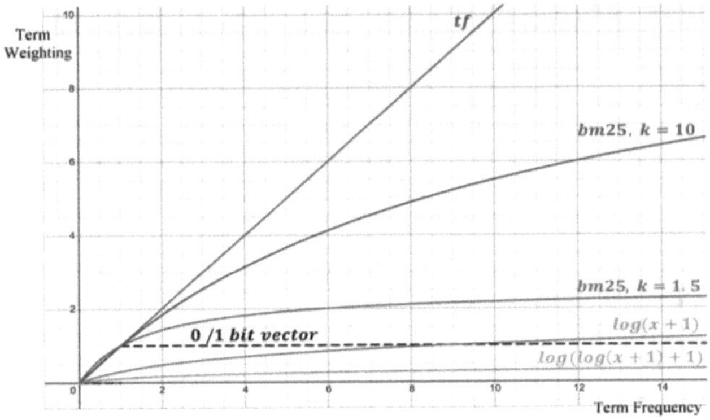

Fig. 1. Comparative analysis of term weighting behavior across different TF-IDF configurations and the BM25 transformation.

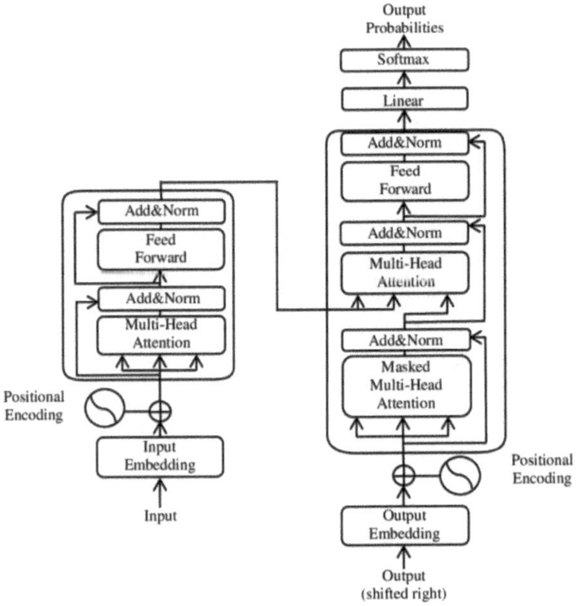

Fig. 2. The Transformer model architecture, utilizing multi-head self-attention mechanisms for contextual representation [46].

As illustrated in Fig. 2, the architecture fundamentally departs from sequential processing (typical of RNNs) in favor of a parallelized attention mechanism. The core component, multi-head attention, allows the model to jointly attend to information from different representation subspaces at various positions. In the context of mental health text mining, this capability is crucial for resolving long-range dependencies, such as linking a distressing event mentioned at the beginning of a post with an emotional reaction described several sentences later. The encoder stack processes the input sequence (e.g., a social media post) to generate contextualized embeddings, which can then be fine-tuned for classification tasks.

Despite the success of general-purpose Transformers, the subtle and highly specialized lexicon of clinical psychology necessitated the development of domain-specific models. This led to the emergence of MentalBERT [24], a variant of BERT [9] pre-trained on large-scale corpora of mental health-related discourse. MentalBERT and its derivatives are specifically tuned to the linguistic markers of anxiety, depression, and suicidal ideation, providing a more nuanced understanding of the speaker's internal state than models trained on general web data.

However, the transition to these multi-billion parameter architectures introduces a new set of challenges. While they offer unprecedented accuracy, they often function as "black boxes", lacking the inherent interpretability of the traditional stylometric models discussed in Sect. 2.1. Moreover, the computational cost associated with fine-tuning and deploying these models has sparked a growing interest in balancing depth with efficiency. This tension highlights the importance of the optimization strategies discussed in the following section, where the goal is to leverage the power of contextual embeddings while maintaining the parsimony required for real-time clinical monitoring.

In this study, we applied three state-of-the-art models, including DistilBERT [39], RoBERTa [29], and a specialized model, MentalBERT [24], which is specifically pre-trained for mental health detection tasks.

Hundreds of configurations were evaluated for these deep learning models. Fine-tuning involved the integration of additional architectural layers, such as Multi-Layer Perceptrons (MLPs) [38], Long Short-Term Memory (LSTM) networks [23], and Convolutional Neural Networks (CNNs) [27]. We explored various activation functions (e.g., ReLU [33], Softmax [3]), dropout rates to prevent overfitting, and batch sizes (2, 4, 8, 16). Learning rates of 1e-3, 1e-4, and 1e-5 were tested across different model states: fully frozen, fully trainable, or with only the final block trainable. Output layers were configured with either 2 or 768 neurons depending on the specific head architecture.

2.5 Preprocessing and Low-Resource Optimization

The inherent volatility of social media discourse poses a significant challenge for supervised TC. Within this context, the strategic selection of preprocessing techniques—such as stemming, lemmatization, and stop-word removal—often exerts a more profound influence on model performance than the selection of the classifier itself [18–20]. This domain-dependent sensitivity is paramount when addressing nuanced clinical manifestations, as evidenced in the detection of anorexia in blog posts [20,21], or in the identification of hate speech spreaders [45].

As established in earlier investigations, preprocessing serves as a critical quality-enhancement phase, effectively purifying the dataset to amplify the signal-to-noise ratio. Basic interventions, such as the normalization of letter casing and the systematic replacement of HTTP hyperlinks with symbolic wildcards, establish a baseline of structural uniformity. Beyond these fundamental steps, rigorous linguistic cleaning—including the disambiguation of acronyms and the correction of chronic spelling errorsensures that morphological variety does not obscure the semantic core of the users intent [16–19,30].

2.6 Character N-Gram Analysis

Research indicates that character-based features can outperform word-based features in certain TC scenarios, particularly when dealing with noisy social media data [12,42]. This study examined character sequences of lengths 3 to 10 (excluding spaces) across the datasets. Classifier performance was assessed using different numbers of features, combined with the four TF-IDF transformations and variations in stop-word inclusion.

3 Methodology

3.1 Problem Formulation and Objective

The task of detecting anxiety in social media discourse is defined as a supervised binary classification problem. Formally, let $D = \{d_1, d_2, \ldots, d_n\}$ be a corpus consisting of n unique Reddit posts. Each document d_i is mapped to a ground-truth label $y_i \in \{0, 1\}$, where $y_i = 1$ indicates the presence of clinical anxiety symptoms and $y_i = 0$ represents a control instance.

Our primary objective is to learn a mapping function $f : D \to Y$ that minimizes the binary cross-entropy loss $\mathcal{L}(\theta)$:

$$\mathcal{L}(\theta) = -\frac{1}{n}\sum_{i=1}^{n} [y_i \log(P(y_i = 1|d_i; \theta)) + (1 - y_i) \log(1 - P(y_i = 1|d_i; \theta))] \quad (1)$$

To achieve this in a low-resource setting, we prioritize a feature-centric approach, ensuring clinical interpretability and computational efficiency in line with Green AI principles [41].

3.2 Linguistic Feature Engineering: The 24 Clinical Lexicons

A central contribution of this methodology is the curation of 24 distinct linguistic feature sets, specifically engineered to capture the "digital phenotype" of anxiety. To ensure a comprehensive representation, the features are evenly divided into 12 style-based (SB) sets and 12 content-based (CB) sets.

The Style-Based sets focus on structural and grammatical patterns (e.g., intensifiers, stop words, and verb distributions), while many Content-Based sets target thematic stressors (e.g., medical symptoms, workplace concerns, and financial pressures). Table 1 provides the comprehensive list of these 24 feature sets. The curation process utilized insights from automated keyword extraction methodologies developed for specialized linguistic domains [13].

Table 1. Comprehensive list of the 24 linguistic feature sets.

Set Code	Feature Set	# Words	Set Code	Feature Set	# Words
ANX	Anxiety, stress, restless	36	NTR	Nature	61
CI	Commerce and industry	49	NV	Negative verbs	36
COM	Community and social	70	NW	Negative words	86
DTH	Death	16	PA	Positive adjectives	36
FD	Food	24	PW	Positive words	66
GEN	General words	89	SF	Self-feeling	68
INF	Intensifiers	29	SLP	Sleep	17
LMT	Limiters and diminutives	16	SW	Stop words	56
MD	Mental disorders	30	TC	Technology	15
MDC	Medical and health	119	TS	Times	55
NB	Negative behavior	57	VRB	Natural verbs	80
NF	Negative feeling	46	WS	Work and school	56

3.3 Preprocessing Auditing: The 128 Configurations

We evaluate 128 unique preprocessing configurations to identify the optimal environment for feature extraction. These configurations involve the combinatorial evaluation of text normalization (casing and hyperlink removal), linguistic reduction (Porter Stemming vs. Lemmatization), and various transformations of TF-IDF. This exhaustive auditing ensures that the linguistic signals are not obscured by suboptimal data cleaning, as noted in previous studies [18,19].

3.4 Optimization Strategies: Hill Climbing Vs. Heuristic Algorithm

To navigate the combinatorial space of 2^{24} (16,777,216) possible feature set combinations, we implemented and compared two distinct search strategies:

1. Hill Climbing (Baseline): The Hill Climbing algorithm operates as a greedy local search, iteratively adding the feature set that provides the highest immediate accuracy gain. In our experiments, it reached a plateau of **94.0% accuracy** using 4 sets (MDC, DTH, ANX, and NW), totaling 257 words.

2. The Proposed Heuristic Optimization Framework: To overcome the local optima of the greedy approach, we implemented a heuristic model based on the methodologies developed in [18,20]. This model utilizes a prioritized subspace filtering logic:

First, we identified the top 17 performing sets that achieved a standalone accuracy of at least 58.5%: MDC, SF, VRB, TS, GEN, SW, NF, NW, PW, WS, COM, NTR, NB, MD, CI, PA, and ANX. We then integrated the core feature sets identified by the Hill Climbing model: MDC, DTH, ANX, and NW. After deduplicating, we arrived at a refined pool of 18 candidate feature sets. By reducing the feature space, we were able to perform an exhaustive combinatorial evaluation of all 2^{18} (262,144) possible combinations.

4 Experimental Setup

4.1 Dataset Description and Source

The empirical foundation of this study is a specialized dataset curated from the Reddit social media platform, specifically targeting the *r/Anxiety* community. The raw data was initially collected via web-scraping from the subreddit, which serves as a global hub for peer-to-peer mental health support [5,50].

The experimental dataset consists of 200 posts manually annotated: 112 (56.0%) tagged as "anxiety" and 88 (44.0%) as "non-anxiety". This distribution provides a relatively balanced class ratio, which is essential for stable training and evaluation in low-resource environments, a common challenge in specialized clinical NLP [22].

4.2 Statistical Data Analysis

A statistical auditing was conducted to identify structural differences between the classes. As summarized in Table 2, anxiety-related posts exhibit distinct length and variance patterns compared to the control group.

Table 2. Statistical distribution and corpus details (N=200 posts).

Class	Docs	% Docs	Words	Chars	Avg. WPD	Avg. CPD	Words Std.	Chars Std.
Anxiety	112	56.0%	14,329	76,976	127.94	687.28	106.16	564.78
Not Anxiety	88	44.0%	10,763	56,974	122.31	647.43	132.08	675.89
Total	**200**	**100%**	**25,092**	**133,950**	**125.46**	**669.75**	**118.02**	**615.11**

4.3 Baselines and Training Protocols

To establish a benchmark, the heuristic model was compared against classical machine learning classifiers and state-of-the-art transformer architectures. The classical baselines include LR [7], Support Vector Classification (SVC) [6], and RF [2].

5 Experimental Results and Discussion

5.1 Baseline and N-Gram Analysis

The initial phase of our evaluation focused on establishing a robust baseline using standard word unigrams. As shown in Table 3, the best baseline performance (67.0% accuracy) was achieved using an LR classifier [7] with a vocabulary of 750 words.

The experiments presented in Table 4 incorporate the mathematical transformations applied to the TF-IDF weighting scheme. By utilizing the BM25 probabilistic relevance framework [37] and sub-linear log-scaling, we observed varying degrees of improvement. Notably, the highest accuracy in this set, 74.0%, was achieved using 750 word unigrams with a Log-transformed TF-IDF scheme.

Table 3. Main baseline accuracy results for word unigrams.

Number of Words	Best Classifier	Best Result
250	SVC	60.5%
500	SVC	65.0%
750	LR	67.0%
1,000	RF, LR	63.0%
1,178	LR	63.5%

Table 4. Classification results using word unigrams and character n-grams.

n-grams	Stop Words	# of n-grams	Classifier	Formula	Accuracy
Word unigrams	With	750	MNB	log	74.0%
Char. 3-grams	With	2,000	LR	BM25, $k = 1.5$	69.5%
Char. 4-grams	With	2,000	LR	BM25, $k = 3$	68.0%
Char. 5-grams	Without	500	LR	Log(log(...))	68.0%
Char. 6-grams	Without	1,500	LR	Tf-idf	67.5%
Char. 7-grams	With	1,500	MLP	Log(log(...))	73.0%
Char. 10-grams	Without	750	RF	BM25, $k = 1.5$	64.5%

5.2 Evaluation of Individual Feature Sets

We evaluated each of the 24 feature sets independently. As shown in Table 5, the *Medical and Health* (MDC) set emerged as the most potent individual predictor (73.5%), followed by *Self-feeling* (SF) at 72.0%.

Table 5. Accuracy of individual feature sets (Selected List).

Set Code	Feature Set	Acc.	Set Code	Feature Set	Acc.
MDC	Medical and health	73.5%	NF	Negative feeling	64.0%
SF	Self-feeling	72.0%	NW	Negative words	63.0%
VRB	Natural verbs	69.0%	ANX	Anxiety, stress, restless	58.5%
TS	Times	69.0%	DTH	Death	54.5%

5.3 Heuristic Optimization Vs. Deep Learning Benchmarks

Table 6 presents the final performance hierarchy. The results demonstrate a clear superiority for the heuristic feature engineering approach. While MentalBERT [24] reaches 87.5%, our heuristic algorithm achieves **98.5% accuracy**.

Table 6. Final comparative performance: Deep Learning vs. Heuristic Optimization.

Model Type	Architecture	Accuracy
Transformer	RoBERTa	77.5%
Transformer	DistilBERT	85.0%
Transformer	MentalBERT	87.5%
Greedy Search	Hill Climbing (4 sets)	94.0%
Heuristic Algorithm	**Optimal Combination (7 sets)**	**98.5%**

5.4 Discussion: Why Does Feature Engineering Outperform Transformers?

The performance gap in this clinical setting ($N = 200$) can be attributed to:

1. **Data Scarcity vs. Parameter Density:** Large-scale transformers like DistilBERT [39], RoBERTa [29], and MentalBERT [24] are prone to overfitting in constrained environments. Our model avoids this pitfall. This aligns with broader observations that even state-of-the-art few-shot learners like GPT-3 [4] often require massive computational overhead and struggle with specialized precision compared to the parsimony of expert features in narrow domains.
2. **Linguistic Nuance:** Anxiety is manifested through subtle stylistic patterns that may be "smoothed out" by neural embeddings but are captured by our seven optimized feature sets [42].

6 Limitations

The primary limitation of this study is the reliance on a relatively small, pre-existing dataset from a single platform (Reddit). While the $N = 200$ subset was manually verified, the limited data volume particularly affected the "data-hungry" transformer models, providing a compelling explanation for the performance gap. This context allowed us to demonstrate that in practical, data-constrained settings, thoughtful feature engineering remains paramount. These results are consistent with the limitations noted in virtual screening tools for clinical anxiety risk, where platform-specific biases can influence model generalizability [48].

Furthermore, our analysis is constrained by the "static" nature of the corpus. Mental health is a dynamic state, and linguistic markers may evolve over time or vary across demographic groups. Future studies should explore longitudinal data [32] to track symptoms over time and ensure that the 24 feature sets remain effective across different social media platforms and digital cultures [20].

7 Conclusions and Future Work

This research demonstrates that in clinical NLP, complex models do not always yield superior results. By integrating domain-specific heuristics with classical machine learning, we achieved 98.5% accuracy, surpassing deep learning benchmarks while providing a computationally efficient solution for high-throughput text processing.

7.1 Conclusions

Our findings challenge the "bigger is better" paradigm, showing that for specialized, low-resource tasks, optimized classical ML can be superior. Achieving near-perfect accuracy using seven specific feature sets underscores the vital role of domain knowledge in detecting mental disorders. Beyond performance, the lightweight architecture ensures low-latency inference, a critical requirement for real-time screening and social media monitoring. We conclude that clinical applications must prioritize interpretability, speed, and precision, as the linguistic nuances of mental distress require a more targeted, resource-efficient approach than general-purpose models currently provide [17,21,38,42,43].

7.2 Future Research Directions

Building upon this study, we propose four key directions: (1) **Cross-Linguistic Validation:** evaluating the generalizability of our heuristic sets across languages like Hebrew or Arabic to determine if the digital phenotype of anxiety remains robust [20,25]; (2) **Scalability Thresholds:** identifying the data volume where neural models might regain an advantage over engineered features; (3) **Active Learning:** implementing human-in-the-loop systems for clinicians to refine the feature sets in real-time; and (4) **Longitudinal Tracking:** monitoring linguistic changes over time to identify early warning signs of clinical relapse [32].

In conclusion, this research reaffirms that human-guided AI remains critical in mental health detection, where resource efficiency ("Green AI") and interpretability are as vital as raw predictive power.

References

1. Arif, M., et al.: Classification of anxiety disorders using machine learning methods: a literature review. Insights Biomed. Res. **4**(1), 95–110 (2020). https://doi.org/10.36959/584/455
2. Breiman, L.: Random forests. Mach. Learn. **45**(1), 5–32 (2001). https://doi.org/10.1023/A:1010933404324
3. Bridle, J.S.: Probabilistic interpretation of feedforward classification network outputs, with relationships to statistical pattern recognition. In: Soulié, F.F., Hérault, J. (eds) Neurocomputing. NATO ASI Series, vol. 68, pp. 227–236. Springer, Heidelberg (1990). https://doi.org/10.1007/978-3-642-76153-9_28
4. Brown, T.B., et al.: Language models are few-shot learners. In: Advances in Neural Information Processing Systems (NeurIPS 2020), vol. 33, pp. 1877–1901 (2020)
5. Coppersmith, G., Dredze, M., Harman, C.: Quantifying mental health signals in Twitter. In: Proceedings of the Workshop on Computational Linguistics and Clinical Psychology (CLPsych), pp. 51–60 (2014). https://doi.org/10.3115/v1/W14-3207
6. Cortes, C., Vapnik, V.: Support-vector networks. Mach. Learn. **20**(3), 273–297 (1995). https://doi.org/10.1007/BF00994018
7. Cox, D.R.: The regression analysis of binary sequences. J. Roy. Stat. Soc.: Ser. B (Methodol.) **20**(2), 215–232 (1958). https://doi.org/10.1111/j.2517-6161.1958.tb00292.x
8. DeMartini, J., Patel, G., Fancher, T.L.: Generalized anxiety disorder. Ann. Intern. Med. **170**(7), ITC49–ITC64 (2019). https://doi.org/10.7326/AITC201904020
9. Devlin, J., et al.: BERT: pre-training of deep bidirectional transformers for language understanding. In: Proceedings of NAACL-HLT, pp. 4171–4186 (2019). https://doi.org/10.18653/v1/N19-1423
10. HaCohen-Kerner, Y., et al.: Cuisine: classification using stylistic feature sets and/or name-based feature sets. JASIST **61**(8), 1644–1657 (2010). https://doi.org/10.1002/asi.21350
11. HaCohen-Kerner, Y., Kass, A., Peretz, A.: HAADS: a Hebrew Aramaic abbreviation disambiguation system. JASIST **61**(9), 1923–1932 (2010). https://doi.org/10.1002/asi.21367
12. HaCohen-Kerner, Y., et al.: Stylistic feature sets as classifiers of documents according to their historical period and ethnic origin. Appl. Artif. Intell. **24**(9), 825–843 (2010). https://doi.org/10.1080/08839514.2010.514197
13. HaCohen-Kerner, Y.: Automatic extraction of keywords from abstracts. In: Palade, V., Howlett, R.J., Jain, L. (eds.) KES 2003. LNCS (LNAI), vol. 2773, pp. 843–849. Springer, Heidelberg (2003). https://doi.org/10.1007/978-3-540-45224-9_112
14. HaCohen-Kerner, Y., Malin, E., Chasson, I.: Summarization of Jewish law articles in Hebrew. In: Proceedings of CAINE 2003, pp. 172–177 (2003)

15. HaCohen-Kerner, Y., Gross, Z., Masa, A.: Automatic extraction and learning of keyphrases from scientific articles. In: Gelbukh, A. (ed.) CICLing 2005. LNCS, vol. 3406, pp. 657–669. Springer, Heidelberg (2005). https://doi.org/10.1007/978-3-540-30586-6_74
16. HaCohen-Kerner, Y., Kass, A., Peretz, A.: Initialism disambiguation: man versus machine. JASIST **64**(10), 2133–2148 (2013). https://doi.org/10.1002/asi.22909
17. HaCohen-Kerner, Y., Dilmon, R., Hone, M., Ben-Basan, M.A.: Automatic classification of complaint letters according to service provider categories. Inf. Process. Manage. **56**(6), 102102 (2019). https://doi.org/10.1016/j.ipm.2019.102102
18. HaCohen-Kerner, Y., Yigal, Y., Miller, D.: The impact of preprocessing on the classification of mental disorders. In: Proceedings of IEEE ICDMW, pp. 52–66 (2019)
19. HaCohen-Kerner, Y., Miller, D., Yigal, I.: The influence of preprocessing on text classification using a bag-of-words representation. PLoS ONE **15**(5), e0232525 (2020). https://doi.org/10.1371/journal.pone.0232525
20. Hacohen-Kerner, Y., Manor, N., Goldmeier, M., Bachar, E.: Detection of anorexic girls-in blog posts written in Hebrew using a combined heuristic AI and NLP method. IEEE Access **10**, 34800–34814 (2022). https://doi.org/10.1109/ACCESS.2022.3162817
21. Hacohen-Kerner, Y., Manor, N., Goldmeier, M., Bachar, E.: Early detection of anorexia in blog posts written in English. ACM TKDD **19**(1), Article 15 (2025). https://doi.org/10.1145/3779415
22. Hedderich, M.A., Lange, L., Adel, H., Strötgen, J., Klakow, D.: A survey on recent approaches for natural language processing in low-resource scenarios. In: Proceedings of the 2021 Conference of the North American Chapter of the Association for Computational Linguistics: Human Language Technologies (NAACL-HLT), pp. 2545–2568 (2021). https://doi.org/10.18653/v1/2021.naacl-main.201
23. Hochreiter, S., Schmidhuber, J.: Long short-term memory. Neural Comput. **9**(8), 1735–1780 (1997). https://doi.org/10.1162/neco.1997.9.8.1735
24. Ji, S., Zhang, T., Ansari, L., Fu, J., Tiwari, P., Cambria, E.: MentalBERT: publicly available pre-trained language models for mental healthcare. arXiv preprint arXiv:2110.15621 (2021)
25. Keinan, R., HaCohen-Kerner, Y.: JCT at SemEval-2023: sentiment analysis for tweets written in low-resource African languages. In: Proceedings of SemEval-2023, pp. 365–378 (2023). https://doi.org/10.18653/v1/2023.semeval-1.49
26. Kimhi, S., Marciano, H., Eshel, Y., Adini, B.: Resilience and demographic characteristics predicting distress during the COVID-19 crisis. Soc. Sci. Med. **265**, 113389 (2020). https://doi.org/10.1016/j.socscimed.2020.113389
27. LeCun, Y., et al.: Gradient-based learning applied to document recognition. Proc. IEEE **86**(11), 2278–2324 (1998). https://doi.org/10.1109/5.726791
28. Levi-Belz, Y., et al.: PTSD, depression, and anxiety after the October 7, 2023 attack in Israel. EClinicalMedicine **68**, 102418 (2024). https://doi.org/10.1016/j.eclinm.2024.102418
29. Liu, Y., Ott, M., Goyal, N., Du, J., Joshi, M., Chen, D., Stoyanov, V.: RoBERTa: a robustly optimized BERT pretraining approach. arXiv preprint arXiv:1907.11692 (2019)
30. McCallum, A., Nigam, K.: A comparison of event models for Naive Bayes text classification. In: AAAI Workshop on Learning for Text Categorization, vol. 752, pp. 41–48 (1998)

31. Mikolov, T., Sutskever, I., Chen, K., Corrado, G.S., Dean, J.: Distributed representations of words and phrases and their compositionality. In: Advances in Neural Information Processing Systems (NIPS 2013), pp. 3111–3119 (2013)
32. Mullatahiri, P.: Detecting depression and anxiety on social media using selective masking. Doctoral dissertation, TU Wien (2024). https://doi.org/10.34726/hss.2024.116631
33. Nair, V., Hinton, G.E.: Rectified linear units improve restricted Boltzmann machines. In: Proceedings of ICML-10, pp. 807–814 (2010)
34. Pennington, J., Socher, R., Manning, C.D.: GloVe: global vectors for word representation. In: Proceedings of EMNLP, pp. 1532–1543 (2014). https://doi.org/10.3115/v1/D14-1162
35. Preoţiuc-Pietro, D., Sap, M., Schwartz, H.A., Ungar, L.: Mental illness detection at the world well-being project for the CLPsych 2015 shared task. In: Proceedings of 2nd CLPsych Workshop, pp. 40–45 (2015). https://doi.org/10.3115/v1/W15-1207
36. Resnik, P., Garron, A., Resnik, R.: Using topic modeling to improve prediction of neuroticism and depression in college students. In: Proceedings of EMNLP, pp. 1348–1353 (2013)
37. Robertson, S., Zaragoza, H.: The probabilistic relevance framework: BM25 and beyond. Found. Trends Inf. Retr. **3**(4), 333–389 (2009). https://doi.org/10.1561/1500000019
38. Rumelhart, D.E., Hinton, G.E., Williams, R.J.: Learning representations by back-propagating errors. Nature **323**(6088), 533–536 (1986). https://doi.org/10.1038/323533a0
39. Sanh, V., Debut, L., Chaumond, J., Wolf, T.: DistilBERT, a distilled version of BERT: smaller, faster, cheaper and lighter. arXiv preprint arXiv:1910.01108 (2019)
40. Sarkar, S., Alhamadani, A., Alkulaib, L., Lu, C.T.: Predicting depression and anxiety on Reddit: a multi-task learning approach. In: Proceedings of IEEE/ACM ASONAM, pp. 427–435 (2022). https://doi.org/10.1145/3551272.3551280
41. Schwartz, R., Dodge, J., Smith, N.A., Etzioni, O.: Green AI. Commun. ACM **63**(12), 54–63 (2020). https://doi.org/10.1145/3381831
42. De Souza, V.B., Nobre, J.C., Becker, K.: DAC stacking: a deep learning ensemble to classify anxiety, depression, and their comorbidity from Reddit texts. IEEE J. Biomed. Health Inform. **26**(7), 3303–3311 (2022). https://doi.org/10.1109/JBHI.2022.3151589
43. Sribala, M.: An approach of artificial neural networks for prediction of generalized anxiety disorder. IJRCAR **3**(3), 118–124 (2015)
44. Tyshchenko, Y.: Automatic detection of depression and anxiety from blog posts. Master's thesis, University of Tartu, Estonia (2018)
45. Uzan, M., HaCohen-Kerner, Y.: Detecting Hate speech spreaders on twitter using LSTM and BERT. In: CLEF 2021 Working Notes, CEUR-WS, vol. 2936, pp. 2178–2185 (2021)
46. Vaswani, A., et al.: Attention is all you need. In: Advances in Neural Information Processing Systems (NIPS 2017), pp. 5998–6008 (2017)
47. World Health Organization: Anxiety Disorders Fact Sheet (2024). https://www.who.int/news-room/fact-sheets/detail/anxiety-disorders
48. Wright-Berryman, J., Cohen, J., Haq, A., Black, D.P., Pease, J.L.: Virtually screening adults for depression, anxiety, and suicide risk using machine learning. Front. Psych. **14**, 1143175 (2023). https://doi.org/10.3389/fpsyt.2023.1143175

49. Xiong, H., Berkovsky, S., Romano, M., Sharan, R.V., Liu, S., Coiera, E., McLellan, L.F.: Prediction of anxiety disorders using a feature ensemble-based Bayesian neural network. J. Biomed. Inform. **123**, 103921 (2021). https://doi.org/10.1016/j.jbi.2021.103921
50. Zarate, D., Ball, M., Prokofieva, M., Kostakos, V., Stavropoulos, V.: Identifying self-disclosed anxiety on Twitter: a natural language processing approach. Psychiatry Res. **330**, 115579 (2023). https://doi.org/10.1016/j.psychres.2023.115579
51. Zhang, T., Schoene, A.M., Ji, S., Ananiadou, S.: Natural language processing applied to mental illness detection: a narrative review. NPJ Digit. Med. **5**(1), 46 (2022). https://doi.org/10.1038/s41746-022-00589-7

Vehicle Routing Optimization Using Tunicate Swarm Algorithm

Leen Alakhras, Soud Khaled, Sana Abouelnour, Mohammed Azmi Al-Betar(✉), and Ghazi Al-Naymat

Artificial Intelligence Research Center (AIRC), College of Engineering and Information Technology, Ajman University, Ajman, United Arab Emirates
m.albetar@ajman.ac.ae

Abstract. A well-known NP-hard combinatorial optimization problem, the Vehicle Routing Problem (VRP) consists of finding the best set of routes for a fleet of vehicles to deliver products to a specified set of consumers. To solve the VRP, we run and contrast three evolutionary computation algorithms—Genetic Algorithm (GA), Particle Swarm Optimization (PSO), and the Tunicate Swarm Algorithm (TSA). The performance of these methods is evaluated using the Solomon benchmark dataset, which represents realistic vehicle routing scenarios involving customer time windows and capacity constraints and the 23 widely used benchmark functions. Every algorithm's advantages and disadvantages in terms of convergence speed, solution quality, and computing economy are meant to be assessed in this work. The results provide insights into the applicability of modern swarm-based metaheuristics such TSA in solving practical routing problems.

Keywords: Genetic Algorithm (GA) · Particle Swarm Optimization (PSO) · Tunicate Swarm Algorithm (TSA) · Vehicle Routing Problem with Time Windows (VRPTW) · Combinatorial Problem

1 Introduction

With the rapid growth of e-commerce and logistics services, effective delivery and transportation systems have become crucial. One of the most frequently studied optimization problems in this area is the Vehicle Routing Problem (VRP), which seeks to minimize the total distance driven by a fleet of vehicles while satisfying various delivery requirements. Logistics firms can save money considerably and enhance customer satisfaction by solving the VRP quickly.

For large-scale VRPs, conventional accurate techniques sometimes become computationally impossible. Hence, heuristics and metaheuristics inspired by natural phenomena have gained popularity for addressing such challenges. This study explores the application of evolutionary computation techniques, specifically Tunicate Swarm Algorithm (TSA), Particle Swarm Optimization (PSO), and Genetic Algorithm (GA), to address the VRP.

H. R. Arabnia et al. (Eds.): CSCI 2025, CCIS 2968, pp. 450–468, 2027.
https://doi.org/10.1007/978-3-032-28457-0_32

The TSA algorithm, which draws inspiration from the swarm behavior and jet propulsion of tunicates, is relatively new, whereas the GA and PSO have been utilized extensively and demonstrated in the literature. By incorporating TSA into the study, this effort aims to investigate its potential as a potent alternative to the VRP solution. The algorithms are evaluated using the Solomon benchmark dataset, which includes realistic problem instances involving time windows, customer locations, and vehicle constraints. The algorithmic efficacy on VRP datasets and benchmark functions is analyzed through a comparative analysis.

1.1 Problem Statement

The Vehicle Routing Problem with Time Windows (VRPTW) seeks to identify the optimal delivery routes for a fleet of cars serving consumers considering capacity and time limits. Particularly with big data sets, this NP-hard problem calls for advanced optimization methods. The three evolutionary computation algorithms we employ in this work to solve VRPTW instances from the Solomon benchmark dataset are Tunicate Swarm Algorithm (TSA), Particle Swarm Optimization (PSO), and Genetic Algorithm (GA). The goal is to evaluate their capacity to produce better, computationally efficient routing solutions.

1.2 Research Objectives

- To implement and validate Genetic Algorithm (GA), Particle Swarm Optimization (PSO), and Tunicate Swarm Algorithm (TSA) for addressing the Vehicle Routing Problem.
- To compare the convergence behavior, solution quality, and computing efficiency of GA, PSO, and TSA.
- To analyze the performance of these algorithms using 23 conventional benchmark test functions and a VRP dataset.
- To assess the appropriateness of the relatively new TSA for combinatorial optimization problems like the VRP.

1.3 Organization of Research

The remainder of this paper is organized as follows. Section 2 reviews previous studies on the application of metaheuristic algorithms to the Vehicle Routing Problem with Time Windows (VRPTW). Section 3 provides the theoretical background of the three optimization methods used in this study—GA, PSO, and the Tunicate Swarm Algorithm (TSA)—along with their mathematical formulations. Section 4 introduces the problem formulation and mathematical model of the VRPTW, describing its objectives and constraints. Section 5 outlines the proposed methodologies for solving the VRPTW using GA, PSO, and TSA, including the implementation process and parameter configuration. Section 6 presents the experimental setup, benchmark datasets, and comparative performance results of the algorithms. Finally, Sect. 7 concludes the study and suggests directions for future research.

2 Literature Review

Genetic Algorithms have gained a lot of traction in the Vehicle Routing Problem with Time Windows (VRPTW). A GA often represents routes as chromosomes and uses a population of possible solutions that is evolved using selection, crossover, and mutation. To maintain feasible solutions, feasibility is ensured via penalty terms that are added to those solutions that violate constraints or with specialized repair operators. For example, Nagata et al. [6] proposed using a memetic GA that penalizes a late arrival with some significant cost added to the fitness function. In general, GA implementations dealing with the VRPTW either penalized time-window or capacity constraint violations or simply did not allow them. Consequently, after running the algorithm, the final reported solutions have no time-window or capacity violations. Most GAs made additional problem-specific modifications to the general algorithm (for example, Solomon's insertion heuristics for seeding the initial population or a local search between population generations) to enhance solution quality.

Soloman's C101 instance is a clustered dataset with relatively tight time windows. GA methods in the literature always find a known optimal solution for C101. For example, Wang, et al. [11] report that a GA reached the best known solution of 828.94 total distance in 10 vehicles on C101 – the benchmark optimum 828.94 (10 vehicles). Likewise, Gocken and Yaktubay [2] also found minimum 828.94 distance and 10 vehicle solutions on C101 by initializing the GA with K-means clustered routes, resulting in a feasible solution with no/zero constraint violations. This indicates that C101, because it is, is relatively easier to solve, GA can always achieve the optimal solution. In the case of instance R103, the customer set is randomly distributed with tight windows which is considerably more difficult. Traditional pure GA approaches struggle on instance R103 relative to the best-known solution and the distance gap indicates that the GA probably produced a good solution but struggled to optimize total travel distance and possibly smaller vehicle counts for this more difficult R103 instance. Liu et al. [5] note that the clustered longest-common-substring GA (CLCS-GA) with embedded local search generally matches best distances on R1 instances. To summarize, traditional GAs not only produced a higher distance on R103 but generally lower quality in terms of distance. However, hybrid approaches can much greater improve travel distances improving exploration and optimize locally.

Like GA, PSO has also demonstrated competitive results on easier problems. On C101, PSO approaches consistently matched the optimal distance of 828.94 with 10 vehicles, often converging faster than GA because of the collective learning behaviour of the swarm [9,11]. For example, R103, however, typical PSO usually produces solutions that are not optimal. The distance reported was 1213.62 when applying 14 vehicles [10], which still did not reach the best known value. On the even more constrained RC106 instance, PSO was reported to be even further away still with distances of around 1384.92 and 12 vehicles [9,10]. Despite this, recent developments around PSO approaches like hybrid PSO, or neighbourhood based PSO, did show these approaches have potential to come closer in competition to GA [10], indicating that while pure PSO has weaknesses with complex VRPTW cases, in some cases modified PSO can be viable alternatives to traditional heuristics.

3 Background

This section discusses the basic theories of the three evolutionary computation algorithms employed in this study: Genetic Algorithm (GA), Particle Swarm Optimization (PSO), and Tunicate Swarm Algorithm (TSA). These algorithms are recognized for their ability to solve challenging optimization problems, especially those that are non-linear, non-convex, and combinatorial in nature.

Evolutionary computation algorithms, a class of optimization techniques inspired by nature, simulate biological and social processes to address difficult real-world problems. Their ability to balance search space exploration and exploitation makes them excellent at combinatorial optimization problems such as the Vehicle Routing Problem (VRP). This study makes use of three popular evolutionary algorithms: the Tunicate Swarm Algorithm (TSA), Particle Swarm Optimization (PSO), and Genetic Algorithm (GA). Each of these algorithms offers unique mechanisms based on natural phenomena, and they differ in their encoding strategies, convergence behaviors, and problem-solving paradigms.

3.1 Genetic Algorithm

Holland introduced the Genetic Algorithm (GA) in 1975 [7], and it is based on Darwinian evolution. It operates on a population of chromosomes, or possible solutions, that undergoes generational changes as a result of biologically inspired operators such as selection, crossover, and mutation. In the context of VRP, solutions are usually displayed as permutations of customer nodes, with optional delimiters to display vehicle route separations. Each solution's fitness is evaluated using a cost function, which typically includes the total route distance and penalties for breaking constraints such as exceeding vehicle capacity or exceeding customer time windows. The population's variety is preserved using genetic operators such as Partially Mapped Crossover (PMX) and mutation (such as swap mutation). Because GA is flexible and can search widely over the solution space, it has shown good performance on VRP.

$$\text{Fitness} = \sum_{v=1}^{k} \left(\sum_{i=0}^{n_v} d_{i,i+1}^{(v)} + \text{Penalty}_{\text{violations}} \right) \tag{1}$$

where k is the number of vehicles, n_v is the number of customers in route v, $d_{i,i+1}^{(v)}$ is the distance between consecutive customers in the route, and $Penalty_{violations}$ accounts for any constraint violations.

The selection probability in GA, often used is roulette wheel selection, calculated as:

$$P_i = \frac{f_i}{\sum_{j=1}^{N} f_j} \tag{2}$$

where f_i is the fitness value of individual i, and N is the population size.

3.2 Particle Swarm Optimization

Kennedy and Eberhart devised Particle Swarm Optimization (PSO) in 1995 [4], which draws influence from the social dynamics evident in fish schools and bird flocks. A swarm of particles, each of which represents a possible solution moves about the solution space by updating its position and velocity in response to both its personal experience and the best solution the swarm discovers globally. Adaptations are required for VRP, a discrete problem, where particle positions are permutations rather than continuous vectors and velocity is interpreted as a series of operations (e.g., swaps). PSO is a good solution when suitably adjusted for combinatorial spaces due of its quick convergence, ease of implementation, and comparatively low parameter sensitivity. The following are the common formulae for updating location and velocity:

$$v_i^{t+1} = wv_i^t + c_1 r_1 (p_i - x_i^t) + c_2 r_2 (g - x_i^t) \tag{3}$$

$$x_i^{t+1} = x_i^t + v_i^{t+1} \tag{4}$$

Here, x_i^t and v_i^t represent the position and velocity of particle i at iteration t, respectively. The terms p_i and g denote the personal best and global best positions, while w is the inertia weight, c_1 and c_2 are acceleration coefficients, and $r_1, r_2 \sim U(0, 1)$ are random variables.

3.3 Tunicate Swarm Algorithm

The bio-inspired Tunicate Swarm Algorithm (TSA) is originally put forward by Kaur et al. in 2020 [3] based on the locomotory and collective behavior of tunicates, which are a class of swimming aquatic animals with swarm coordination behavior. Through the integration of concepts of arbitrary exploration with dynamic attraction to optimum solutions, it precisely balances the local search as well as global search. TSA is updated to use permutation-based encoding analogous to GA for VRP solution. Movement in the discrete domain is redefined by swap, inversion, or segment-based transformations, enabling TSA to adapt route topologies effectively. As a relatively young technology, TSA has shown competitive performance in global optimization tasks and offers fascinating promise for confined and multimodal concerns such as the VRP.

When operating under the jet–propulsion mechanism, each tunicate in the swarm must satisfy three behavioral principles: (i) preventing overlap or collision with neighboring individuals, (ii) navigating efficiently toward the most promising region of the search domain, and (iii) maintaining proximity to the current best individual. The subsequent subsections reformulate its mathematical description.

Collision Avoidance in the Search Region. To prevent mutual interference among tunicates, a directional update vector $\boldsymbol{\Delta}$ is employed to compute each agent's next position:

$$\boldsymbol{\Delta} = \frac{\boldsymbol{\Psi}}{\boldsymbol{\Lambda}}, \tag{5}$$

where

$$\boldsymbol{\Psi} = \eta_2 + \eta_3 - \boldsymbol{\Phi}, \tag{6}$$

$$\boldsymbol{\Phi} = 2\,\eta_1. \tag{7}$$

Here η_1, η_2, and η_3 are independent random values drawn from $\mathcal{U}(0,1)$. The term $\boldsymbol{\Psi}$ models the combined hydrodynamic forces acting on a tunicate, whereas $\boldsymbol{\Phi}$ reflects ocean-flow advection. The social influence term $\boldsymbol{\Lambda}$, representing neighborhood interaction strength, is evaluated as

$$\boldsymbol{\Lambda} = \lfloor \lambda_{\min} + \eta_1(\lambda_{\max} - \lambda_{\min}) \rfloor, \tag{8}$$

where $\lambda_{\min}$ and $\lambda_{\max}$ denote the lower and upper bounds of communication intensity, typically set to 1 and 4, respectively as in the original paper [3].

Choosing the Direction Toward the Optimal Region. Once inter-agent collisions are avoided, each tunicate must orient its motion toward the current global best solution. The guidance distance $\boldsymbol{d}_t$ between a tunicate and the food source $\boldsymbol{z}^*$ at iteration t is defined as

$$\boldsymbol{d}_t = \left| \boldsymbol{z}^* - \eta_4 \times \boldsymbol{z}_t \right|, \tag{9}$$

where $\boldsymbol{z}_t$ is the present position of the tunicate and $\eta_4, \sim \mathcal{U}(0,1)$ is a uniformly distributed random scalar.

Maintaining Proximity to the Best Individual. To ensure collective convergence, tunicates continually reposition themselves around the elite solution according to

$$\boldsymbol{z}_t = \begin{cases} \boldsymbol{z}^* + \boldsymbol{\Delta} \times \boldsymbol{d}_t, & \text{if } u \leq 0.5, \\ \boldsymbol{z}^* - \boldsymbol{\Delta} \times \boldsymbol{d}_t, & \text{if } u > 0.5, \end{cases} \tag{10}$$

Equation (10) keeps each agent oscillating near the leading tunicate, promoting exploitation of promising zones.

Modeling Collective Swarm Dynamics. The cooperative behavior of the population is captured by averaging the best two current solutions. The new position of an agent is then estimated as

$$\boldsymbol{z}_{t+1} \leftarrow \frac{\boldsymbol{z}_t + \boldsymbol{z}_{t+1}}{2 + \eta_1}, \tag{11}$$

where $\boldsymbol{z}_t$ and $\boldsymbol{z}_{t+1}$ correspond to the positions of the two top-ranked tunicates. This adjustment constrains the swarm within a conical region surrounding the elites, thereby preserving diversity while refining exploitation.

4 Problem Mathematical Modelling

4.1 VRPTW Description

A center for distribution O with the maximum carrying capacity of each vehicle C exists in order to define the VRPTW problem. The task demand is at client location a, and there are N client locations. The demands of each client do not exceed the vehicle's maximum carrying capacity. The associated service time window is $[i_a f_a]$, with i_a being the initial service time and f_a being the most recent. The needed service time at the customer's location is s_a. From customer location a to customer point b, a car v goes straight. Vehicle v will wait and the time to begin the service at customer point b will be $begin_b = max[i_b, begin_a + S_a + t_{ab}]$ if it arrives too early. If it arrives later than that, it will not be able to finish the service within the allotted time window and will be penalized. The goal of optimization is to determine the shortest transportation path that maximizes benefits at the lowest possible cost (including waiting times, penalties, and traveling time) while meeting service demands at all client points [1].

4.2 Mathematical Model

Some assumptions based on VRPTW problem description:

- For every route, the overall demand at the consumer location is less than or equal to the maximum load of the vehicle C.
- Every vehicle will leave distribution center O and return to the distribution center after their work is finished.
- Only one car will be accepted for distribution at each customer location.
- The load and speed of every delivery truck are the same.
- Vehicles arriving early or late will be subject to equivalent penalty costs during the soft time frame that corresponds to the demand of all customer points.

The decision variables are specified as follows:

$$x_{abv} = \begin{cases} 1, & \text{if vehicle } v \text{ drives from } a \text{ to } b \\ 0, & \text{otherwise} \end{cases} \tag{12}$$

$$y_{av} = \begin{cases} 1, & \text{if } a\text{'s task is performed by vehicle } v \\ 0, & \text{otherwise} \end{cases} \tag{13}$$

A mathematical model is formulated with the following objective function:

$$\min Z = Z_1 + Z_2 \tag{14}$$

where

$$Z_1 = \sum_{a=1}^{N}\sum_{b=1}^{N}\sum_{v=1}^{M} x_{abv} \cdot d_{ab}, \quad \forall a \neq b \tag{15}$$

$$Z_2 = \sum_{b=1}^{M} P_b \tag{16}$$

where the distance cost is represented by $Z1$, the cost of penalizing the vehicle for breaking the time window by $Z2$, and the penalty for breaking the time window is represented by Pb.

The following are the corresponding constraints:

1. Vehicle capacity constraint: the maximum carrying capacity of the vehicle is not exceeded by the demand of client points on each path.

$$\sum_{a=1}^{N} y_{av} \cdot D_a \leq C, \quad \forall v \in \{1, 2, 3, \ldots, M\} \tag{17}$$

2. Vehicle beginning and ending restrictions, which state that the vehicle must begin at the distribution facility and return there

$$\sum_{a=1}^{N} x_{0av} = \sum_{b=1}^{N} x_{b0v} = 1, \quad \forall v \in \{1, 2, 3, \ldots, M\} \tag{18}$$

3. Constraint on client point visits: every client point is only visited once.

$$\sum_{v=1}^{M} y_{av} = 1, \quad \forall a \in \{1, 2, 3, \ldots, N\} \tag{19}$$

4. Time window restrictions, in which the service duration is inside the appropriate time frame:

$$i_a \leq begin_a \leq f_a, \quad \forall a \in \{1, 2, 3, \ldots, N\} \tag{20}$$

where

$$begin_a = \max\left(i_a, \sum_{b=1}^{N} x_{abv} \cdot (begin_b + S_b + t_{ab})\right), \quad \forall a \neq b, \forall v \in \{1, 2, 3, \ldots, M\} \tag{21}$$

5 Proposed Methodologies

5.1 Genetic Algorithm

To address the Vehicle Routing Problem with Time Windows (VRPTW), we created a customized Genetic Algorithm (GA) specifically tailored to manage the problem's combinatorial complexity and the constraints established in the

Solomon benchmark dataset. Each solution (chromosome) is a probable set of vehicle routes, where each route is a sequence of customer IDs served by a single vehicle, starting and ending at the depot. This encoding enables numerous routes and automatically accommodates time frame and capacity limits. The initial population is generated using a hybrid approach: a portion of the individuals is constructed using a greedy insertion heuristic that prioritizes customers with earlier time windows and proximity to the depot, while the remaining individuals are initialized randomly to encourage solution diversity. This ensures both early exploitation and comprehensive investigation of the solution space.

Fitness evaluation utilizes a composite objective function that minimizes the total distance traveled while including penalties for time window violations and optionally for exceeding vehicle capacity. Selection is accomplished using a tournament technique, which selects parents based on relative fitness among randomly chosen subgroups of the population, balancing selection pressure and variety. The crossover operator is route-based: routes from both parents are combined, and duplicates are deleted. Any unassigned customers are reinserted using a greedy manner to maintain feasibility. Mutation happens with low probability and involves operations such as customer swaps, inter-route moves, and route segment reversals—each aimed at boosting solution exploration and avoiding premature convergence.

After crossover and mutation, a 2-opt local search is performed in every route to reduce intra-route distances and enhance the solutions. In order to ensure that the best solution is maintained, a steady-state replacement strategy with elitism is used to keep the superior individuals and eliminate the poor ones with newly produced offspring. After a certain number of generations, the algorithm terminates. Fitness improvement and population unrest are tracked to observe algorithm performance during the process.

5.2 Particle Swarm Optimizer

In our Particle Swarm Optimization (PSO) approach, each particle encodes a candidate solution as a set of vehicle routes, where each route consists of a sequence of customer IDs, with the depot implicitly positioned at both the start and end. The entire swarm collectively searches across the space of plausible route configurations, adjusting its position over time to discover optimal or near-optimal solutions. Given the discrete nature of VRPTW, typical PSO velocity updates employing numerical vectors are substituted with discrete operators. Specifically, particle "velocity" is characterized as a sequence of swap and move operations: a move transfers a customer between routes, while a swap exchanges the positions of consumers inside a route. These procedures are designed to transition the current solution toward both the particle's personal best and the global best, and the revised solution becomes the new particle position.

Fitness evaluation analyzes solution quality using a composite objective function that combines the overall distance traveled, fines for violations of customer time windows, and optional penalties for capacity violations. The goal is to reduce this aggregate fitness value while ensuring the feasibility of each route.

In order to increase the quality of solutions, on each position change a 2-opt local search is performed. Optimization between routes decreases journey lengths and adds realism. As a prevention for premature convergence, the algorithm features a diversity-guaranteeing mechanism: if, after 20 iterations, inactive particles—those not currently improving—still exist, these may be randomly modified or reinitialized to promote exploration and prevent entrapment in local optima.

Initial particles are generated using a hybrid method: while most are initialized randomly to assure coverage of the solution space, every fifth particle is constructed using a greedy heuristic that emphasizes early time windows and shorter travel distances. This strategy injects high-quality solutions into the initial population from the outset. The optimization process continues for a set number of iterations (e.g., 100). If no improvement in the global best fitness over numerous repetitions, the mutation mechanism is initiated to maintain diversity and promote further exploration of the solution space.

5.3 Tunicate Swarm Algorithm

In our Tunicate Swarm Optimization (TSA) approach, each tunicate in the swarm represents a possible solution encoded as a collection of vehicle routes. Each route is an ordered list of customer IDs, implicitly starting and terminating at the depot to match realistic dispatch and return limits. The TSA algorithm draws inspiration from the natural behaviors of tunicates—marine animals that move in response to propulsion, food sources, and swarm behavior—to lead the search toward effective routing designs. Typical TSA procedures are adapted utilizing permutation-based operators instead of continuous vector updates. Tunicate movement is modeled by three actions. First, jet propulsion randomly perturbs consumer sequences within routes by swapping short segments, permitting local exploration. Second, food seeking allows tunicates to learn from peers by adopting parts of their route architectures. Third, swarm behavior promotes individuals to follow the global best solution through probability-based route alignment and inter-route swaps, promoting the blending of features between solutions.

Fitness evaluation employs a composite objective function that combines total distance traveled, penalties for violating customer time restrictions, capacity overload penalties, and severe penalties for unvisited or duplicate trips to customers. This stimulates the evolution of feasible and logistically efficient routing designs. In order to further enhance solutions, a 2-opt local search is selectively applied to the tours of every tunicate. Local optimization reverses customer subsequences to remove inefficiencies such as detours or backtracking, compacting routes and reducing overall travel distance.

To avoid premature convergence, the algorithm incorporates a diversity-preserving mutation strategy. When the global best fitness stagnates for numerous iterations (e.g., every 20 steps), a subset of tunicates is partially reinitialized or subjected to random customer swaps. This introduces novel variations, enabling the swarm escape local optima and encourage greater exploration.

At the beginning of the algorithm a hybrid initialization is employed. Every fifth tunicate is made greedily, but all the others are initiated randomly, ensuring diversity among initial randomly created tunicates. To start the swarm with good initial locations of the tunicates to help converge in the beginning of the algorithm, the local search algorithm creates routes by giving preference to customers with early time windows and will also limit the distance traveled.

The swarm continues to evaluate candidate solutions and update the global best until the TSA has executed its set number of iterations (e.g., 100). The final result will state the best solution that was obtained. Figure 1 shows the TSA flowchart.

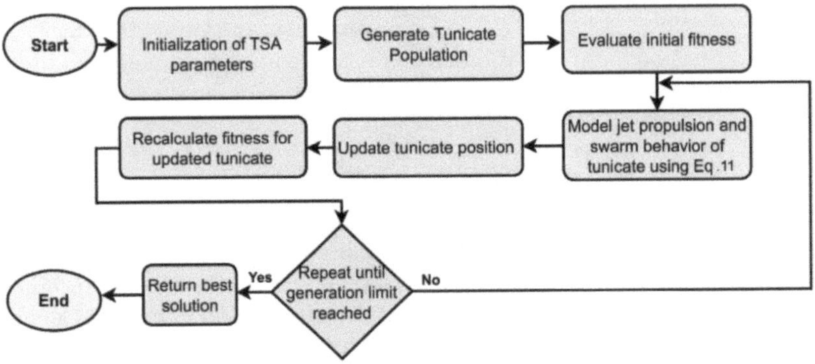

Fig. 1. Flowchart of the Tunicate Swarm Algorithm (TSA).

6 Experiments and Results

6.1 Experimental Design

The Solomon dataset is a common benchmark for evaluating algorithms that solve Vehicle Routing Problem with Time Windows (VRPTW). The dataset contains instances of customers, and those instances are sorted into three types, namely: C - clustered customers, R- randomly distributed customers, and RC- both clustered and randomly distributed customers [8]. For instance, instance C101 would represent a set of customers that were closely clustered together while R103 would have customers that were distributed randomly throughout the map and RC106 would include a randomized and clustered approach which is more difficult for the algorithm to understand. The purpose of the instances is to be able to evaluate an algorithm's capability to deal with spatial and temporal constraints while optimizing a route. Table 1 shows the characteristics of the tested instances.

The performance measures evaluated for each one of the instances were: total distance traveled (shortest path quality), speed of the algorithm or execution time, number of vehicles used, and all constraints violated such as time windows and or vehicle capacity. In addition, in order to assess optimization algorithms, 23 benchmark functions, known as F1 to F23, were used to asses their behaviors with respect to the most important aspects of optimization algorithms, which are convergence behavior, robustness, and ability to escape local optima from various distinct external landscapes. Lastly, the measures provided a clear overview of the performance for each algorithm regarding both routing, and the performance of the algorithms overall.

Table 1. Characteristics of Solomon Benchmark Instances

Instance	Customers	Type	Time Window	Vehicle Capacity
C101	100	Clustered	Narrow	200
R103	100	Random	Narrow	200
RC106	100	Mixed (Random + Clustered)	Narrow	200

6.2 Comparison with GA and PSO Algorithms

The parameter settings for the three metaheuristic algorithms—Genetic Algorithm (GA), Particle Swarm Optimization (PSO) and Tunicate Swarm Algorithm (TSA)—are presented in Table 2. All algorithms were tuned and therefore allowed for fair comparisons. GA had a population of 100 and standard evolutionary parameters, PSO had 30 particles and used an inertia driven movement, while TSA had 30 tunicates and a hybrid initialization and local search probability to promote more exploitation. The consistent and reasonable choice of

Table 2. Algorithm's Parameter Settings

Algorithm	Parameter	Setting
GA	Population size	100
	Generations	100
	Crossover rate	0.8
	Mutation rate	0.2
	Elitism	0.1
PSO	Number of particles	100
	Maximum iterations	100
	Inertia weight (w)	0.5
	Cognitive coefficient (c1)	1.5
	Social coefficient (c2)	1.5
TSA	Number of tunicates	100
	Maximum iterations	100
	Initialization strategy	Greedy every 5th tunicate
	Local search probability	0.1

Table 3. Performance Comparison of Algorithms

Instance	Algorithm	Total Distance	Vehicles Used	Time Window Violations	Capacity Violations	Execution Time (s)
C101	GA	**828.94**	10	0	0	3658
	PSO	880.48	10	0	0	749
	TSA	880.48	10	0	0	**600**
R103	GA	**1489.16**	22	0	0	1996
	PSO	1926.93	20	0	0	360
	TSA	1827.85	20	0	0	**470**
RC106	GA	**1676.38**	17	0	0	2160
	PSO	2049.63	16	0	0	360
	TSA	1995.49	16	0	0	**470**

parameters ensured that differences in performance were due to the behavior of the algorithms, rather than unfair tuning.

Table 3 presents performance comparisons among GA, PSO and TSA on three Solomon VRPTW instances, C101, R103 and RC106. TSA provided feasible solutions with zero violations for time windows and capacity constraints in all performed runs. For the instance C101, TSA was the fastest (159 s), but not as good as GA in terms of total distance. For instances R103 and RC106, TSA's solutions are comparable in quality and speed to the PSO and GA runs—producing fewer vehicles than the GA runs and faster execution times than PSO runs. TSA showed considerably lower execution times when solving the VRPTW instances (e.g. 133 s on RC106 vs. 2160 s on GA). The experimental results obtained between TSA, GA and PSO competence pose TSA as a comparison method to approach complex VRPTW instances, especially, when execution time is paramount. Figures 2, 3 and 4 shows the solution plots for the VRPTW using the three algorithms.

6.3 Comparison with Other Algorithms

Table 4 shows the performance of GA, PSO and TSA on three test Solomon instances (C101, R103, RC106) against reported results in the literature. For example, all three algorithms have satisfied both capacity and time window constraints on the clusterable instance C101. The GA presented in this work and the other GA studies reported in the literature find the known optimal solution (828.94 distance using 10 vehicles), while PSO and TSA fell slightly behind in route cost but there were still feasible solutions. However, the more difficult 'R' and 'RC' instances C101 and RC106, showed a sharper delineation. The proposed GA performed poorly on R103, producing a total distance of 1489.16 with 22 vehicles, whereas literature-reported GA and PSO variants achieved fewer vehicles and lower costs. TSA did not match the best (known) TS results but consistently performed better than PSO and GA on the difficult instances. On RC106, a highly constrained and mixed-instance, TSA once again was superior to base GA and PSO in terms of distance, although it is less than the best-known TS solutions developed previously. Overall, the table illustrates the relative competitiveness TSA has with conventional GA and PSO methods especially in complex

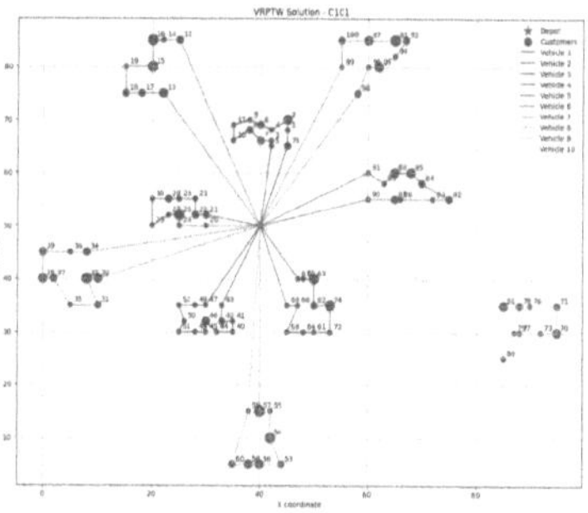

(a) GA

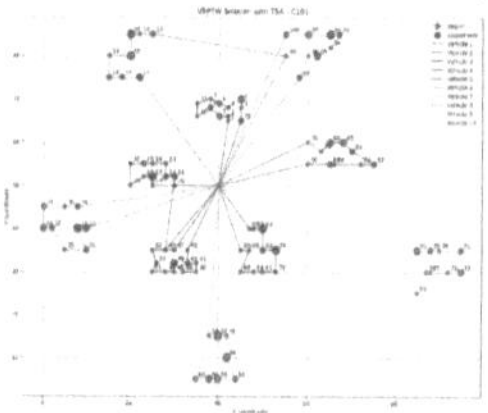

(b) TSA

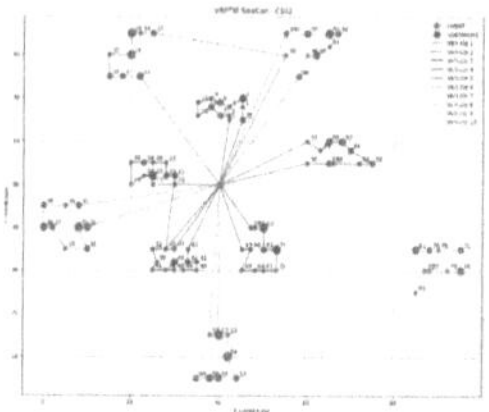

(c) PSO

Fig. 2. C101 instance solution plot for the three algorithms

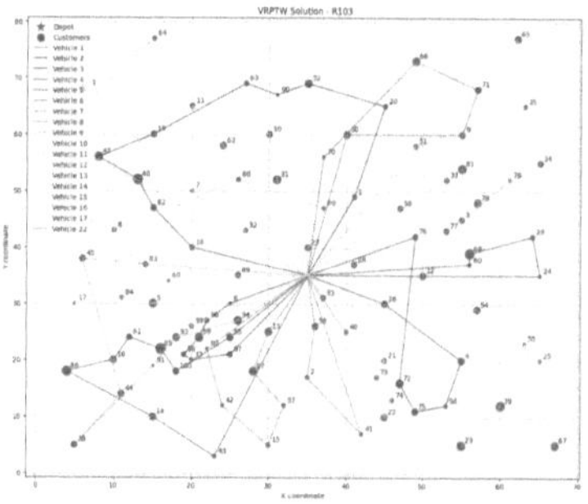

(a) GA

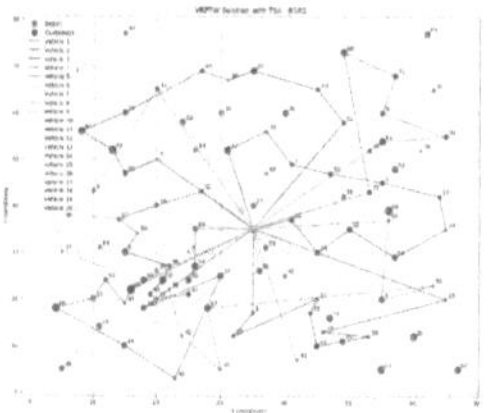

(b) TSA

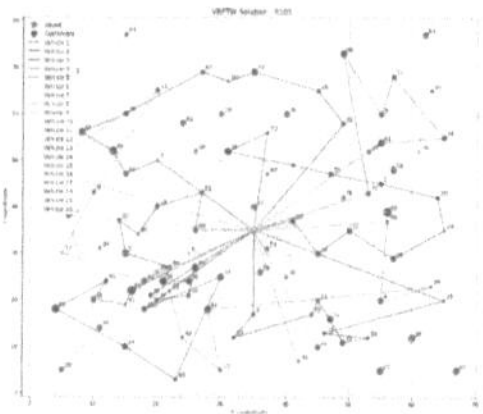

(c) PSO

Fig. 3. R103 instance solution plot for the three algorithms

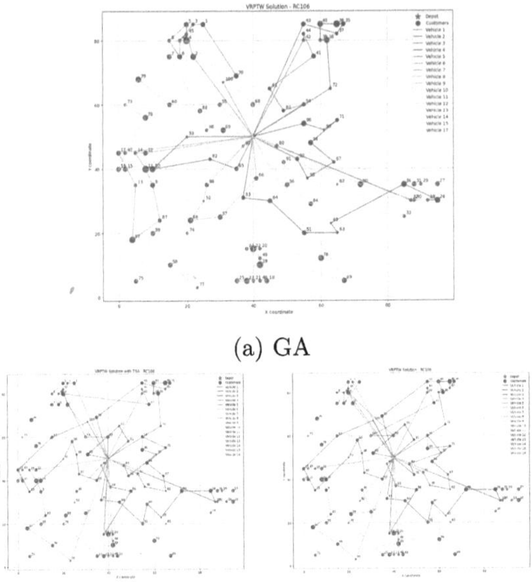

(a) GA

(b) TSA (c) PSO

Fig. 4. RC106 instance solution plot for the three algorithms

instances, and highlights the performance gap relative to highly-tuned TS-based heuristics.

Table 4. Comparison of GA, PSO, and TSA Results on Solomon Instances with Literature Benchmarks

Instance	Algorithm	Total Distance	Vehicles	TW Violations	Cap. Violations
C101	GA (Our Model)	828.94	10	0	0
	PSO (Our Model)	880.48	10	0	0
	TSA (Our Model)	880.48	10	0	0
	GA [11]	828.94	10	0	0
R103	GA (Our Model)	1489.16	22	0	0
	PSO (Our Model)	1926.93	20	0	0
	TSA (Our Model)	1827.85	20	0	0
	GA [5]	1292.85	15	0	0
	PSO [10]	1213.62	14	0	0
RC106	GA (Our Model)	1676.38	17	0	0
	PSO (Our Model)	2049.63	16	0	0
	TSA (Our Model)	1995.49	16	0	0
	GA [11]	1377.35	13	0	0
	PSO [9]	1384.92	12	0	0

6.4 Statistical Evaluation

The performance of TSA is also evaluated against GA and PSO through 23 standard benchmark functions F_1 to F_{23}, the settings were unified throughout all functions and algorithms, with a maximum of 30 runs and 1000 iterations. The average of the best values of the 30 runs and their standard deviations are reported in Tables 5, 6, and 7.

Table 5. Performance of TSA against GA and PSO on Unimodal Functions

Function	Metric	F1	F2	F3	F4	F5	F6	F7
TSA	**Avg.**	**4.34E−47**	**8.79E−29**	**1.29E−10**	**4.24E−03**	**2.82E+01**	3.42E+00	**4.96E−03**
	Std.	**8.28E−47**	**8.98E−29**	**4.13E−10**	**4.82E−03**	**9.86E−01**	5.37E−01	**2.40E−03**
GA	**Avg.**	6.32E+02	1.05E+01	2.15E+03	1.11E+01	2.65E+04	5.92E+02	1.54E−01
	Std.	1.51E+02	1.63E+00	7.23E+02	1.84E+00	1.18E+04	1.43E+02	7.37E−02
PSO	**Avg.**	3.18E−06	1.00E+00	1.33E+03	4.13E+00	9.17E+03	**1.03E−05**	2.67E−02
	Std.	1.41E−05	3.00E+00	2.33E+03	1.09E+00	2.70E+04	**4.37E−05**	1.02E−02

Table 5 highlights the performance of the algorithms against the unimodal functions, where TSA was able to highlight its true competitive nature by outperforming both GA and PSO in 6 out of 7 functions, despite the values being very close to each other for all algorithms. This indicates that although the GA and PSO algorithms also have competitive attributes, TSA was able to outperform them in convergence. As seen in Fig. 5, TSA converged faster than the other algorithms, and the competitive convergence of all algorithms is also clearly highlighted.

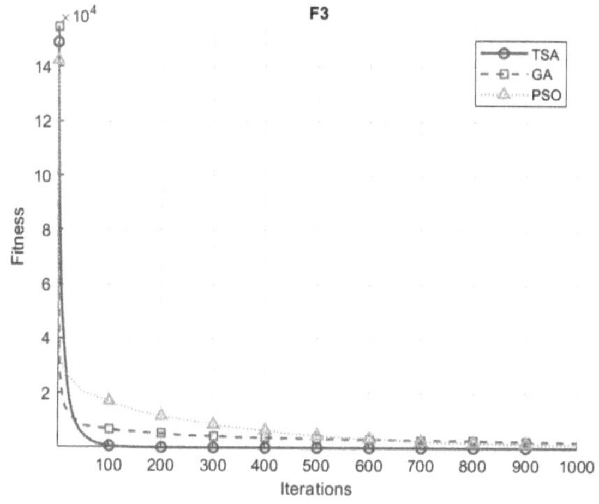

Fig. 5. The convergence behavior of TSA against the other comparative algorithms on F_3

In the analysis of multimodal functions, TSA was able to identify the most optimal solutions in one of the standard multimodal functions and 4 of the complex multimodal functions. Despite not achieving the best optimal value on most the standard multimodal functions, TSA ranked second in almost all of them, as shown in Table 6, demonstrating its ability to reach optimal solutions if slightly modified or hybridized, and therefore can realize its true potential across diverse problem sets.

Table 6. Performance of TSA against GA and PSO on Standard Multimodal Functions

Function	Metric	F8	F9	F10	F11	F12	F13
TSA	**Avg.**	−5.29E+03	1.64E+02	1.91E+00	**7.27E−03**	6.93E+00	2.75E+00
	Std.	5.54E+02	4.56E+01	1.47E+00	**6.77E−03**	3.79E+00	4.67E−01
GA	**Avg.**	−8.61E+03	1.05E+02	6.81E+00	6.39E+00	7.04E+00	7.49E+01
	Std.	**3.89E+02**	**1.34E+01**	6.52E−01	1.47E+00	2.60E+00	1.68E+02
PSO	**Avg.**	**−8.86E+03**	**5.11E+01**	**1.60E−01**	8.04E−03	**5.18E−02**	**4.41E−03**
	Std.	4.68E+02	1.82E+01	**4.27E−01**	1.00E−02	**9.56E−02**	**5.37E−03**

Table 7. Performance of TSA against GA and PSO on Complex Multimodal Functions

Function	Metric	TSA	GA	PSO
F14	**Avg.**	7.34E+00	**9.98E−01**	**9.98E−01**
	Std.	5.05E+00	7.15E−16	**0.00E+00**
F15	**Avg.**	6.14E−03	**5.72E−04**	1.80E−03
	Std.	1.68E−02	**7.36E−05**	4.78E−03
F16	**Avg.**	**−1.03E+00**	**−1.03E+00**	**−1.03E+00**
	Std.	9.49E−03	4.49E−12	**4.33E−16**
F17	**Avg.**	**3.98E−01**	**3.98E−01**	**3.98E−01**
	Std.	5.93E−05	4.37E−08	**0.00E+00**
F18	**Avg.**	1.05E+01	**3.00E+00**	**3.00E+00**
	Std.	2.18E+01	1.16E−10	**5.32E−16**
F19	**Avg.**	**−3.86E+00**	**−3.86E+00**	**−3.86E+00**
	Std.	2.36E−03	1.32E−09	**8.88E−16**
F20	**Avg.**	−3.28E+00	**−3.30E+00**	−3.27E+00
	Std.	6.33E−02	**4.76E−02**	6.87E−02
F21	**Avg.**	−5.88E+00	**−8.24E+00**	−5.40E+00
	Std.	3.25E+00	**3.20E+00**	3.25E+00
F22	**Avg.**	−6.72E+00	−7.87E+00	**−7.93E+00**
	Std.	3.55E+00	**3.35E+00**	3.52E+00
F23	**Avg.**	**−8.19E+00**	−7.89E+00	**−8.19E+00**
	Std.	**3.29E+00**	3.49E+00	3.59E+00

Interestingly, TSA, GA and PSO achieved an equal optimal value in functions F_{17} and F_{19}, this not only highlights the competitive nature of the algorithms but may also be attributed to the complexity of the function, where all algorithms may get stuck in local optima.

Figure 6 does however show how models may at first struggle to reach convergence, and very uniquely, despite TSA initially struggling with its fitness to the function, was able to realize the best optimal value out of the three algorithms, further indicating its ability to reach optimal solutions in complex problem spaces.

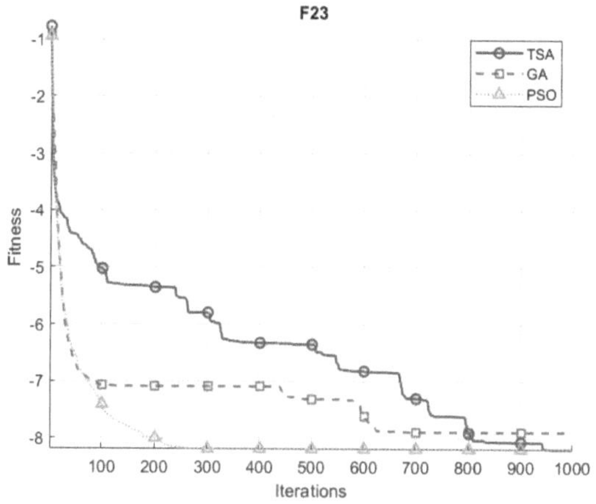

Fig. 6. The convergence behavior of TSA against the other comparative algorithms on F_{23}

7 Conclusion and Future Work

In this project we evaluated and compared the performance of three metaheuristic algorithms: Genetic Algorithm (GA), Particle Swarm Optimization (PSO), and Tunicate Swarm Algorithm (TSA); for solving the Vehicle Routing Problem with Time Windows (VRPTW) using Solomon's benchmark datasets. The TSA was specifically adapted for the discrete nature of the problem to incorporate permutation based solution representations, and TSA was further improved by using a 2-opt local search operator to improve the efficiency of the routes. The comparative results show that although the GA had the shortest total distance for every instance we tested (C101, R103, and RC106 dataset); the TSA provided very good compromise between quality of solution while reducing computation time. For example, TSA had as good as PSO for distance for C101; in addition the TSA was able to out-perform the PSO for both R103 and RC106 with much shorter routes than both other algorithms with fewer vehicles to perform the routes. More importantly, the TSA obtained the shortest execution times for all of the scenarios we tested; the execution times for the TSA was significantly shorter when compared to GA and PSO where the GA was able to execute time blocks while still being feasible (no time window violations or vehicle capacity violations). These results demonstrate the potential of TSA to be a viable and effective method for VRPTW, with future applications focusing on the fast and efficient approach of time critical applications. While TSA provides an effective solution to VRPTW, the performance of TSA is very much dependent on the tuning of parameters and that the implementation of the local search component currently functions in a static context. Future refinements could be made to the local search mechanism to better represent adaptive or self-tuning local search capabilities and hybrid structures that could take better advantage of a combination of algorithms. Future studies could also look to increase the size and dynamism of the VRPTW for TSA, to evaluate the scalability of this approach, and identify how TSA develops ownership in solving VRPTW in dynamic scenarios.

References

1. Chen, G., Gao, J., Chen, D.: Research on vehicle routing problem with time windows based on improved genetic algorithm and ant colony algorithm. Electronics (2079-9292) **14**(4) (2025)
2. Göcken, T., Yaktubay, M.: Comparison of different clustering algorithms via genetic algorithm for VRPTW. In: Proceedings of the 30th DAAAM International Symposium, pp. 574–585 (2019)
3. Kaur, S., Awasthi, L.K., Sangal, A.L., Dhiman, G.: Tunicate swarm algorithm: a new bio-inspired based metaheuristic paradigm for global optimization. Eng. Appl. Artif. Intell. **90**, 103541 (2020)
4. Kennedy, J., Eberhart, R.: Particle swarm optimization. In: Proceedings of ICNN'95-International Conference on Neural Networks, vol. 4, pp. 1942–1948. IEEE (1995)
5. Liu, J., Tong, L., Xia, X.: An improved genetic algorithm for vehicle routing problem with time windows based on cluster of geographic and time windows. Appl. Soft Comput. (in press) (2023). https://doi.org/10.2139/ssrn.4466179. (preprint)
6. Nagata, Y., Bräysy, O., Dullaert, W.: A penalty-based edge assembly memetic algorithm for the vehicle routing problem with time windows. Comput. Oper. Res. **37**(4), 724–737 (2010). https://doi.org/10.1016/j.cor.2009.06.022
7. Sampson, J.R.: Adaptation in Natural and Artificial Systems (John H. Holland) (1976)
8. Solomon, M.M.: Algorithms for the vehicle routing and scheduling problems with time window constraints. Oper. Res. **35**(2), 254–265 (1987)
9. Tang, K., Wen, Z., Tang, X.: Research on route optimization of an express vehicle based on particle swarm optimization algorithm. Int. J. Softw. Hardw. Res. Eng. **8**(8), 1–10 (2020)
10. Wu, Q., et al.: A neighborhood comprehensive learning particle swarm optimization for the vehicle routing problem with time windows. Swarm Evol. Comput. **80**, 100721 (2024). https://doi.org/10.1016/j.swevo.2023.100721
11. Xu, S.H., Liu, J.P., Zhang, F.H., Wang, L., Sun, L.J.: A combination of genetic algorithm and particle swarm optimization (PSO) for vehicle routing problems with time windows. Sensors **15**(9), 21033–21053 (2015). https://doi.org/10.3390/s150921033

Optimizing FSO Links Using AI

Maziar Sayeh(✉) and Peter LoPresti

University of Tulsa, Tulsa, OK 74104, USA
{mas1720,peter-lopresti}@utulsa.edu

Abstract. Free-space optical (FSO) links are essential for global communication because of their high bandwidths and data rates, but their performance can be significantly reduced by atmospheric conditions such as turbulence. This study investigates the use of artificial intelligence (AI) machine learning to optimize FSO parameters under varying atmospheric conditions. Monte-Carlo simulations were run in MATLAB to model beam propagation between a ground-based transmitter and a satellite-based receiver through varying turbulence levels. The simulated data was then used to train two deep learning models: a multilayer perceptron (MLP) and TabNet. Both models were then used to optimize the system parameters, and the best model was validated by re-simulating the optimized inputs. Results showed that both models performed well, with the MLP showing slightly better accuracy and reliability than the TabNet model. The optimization was also found to perform better under low turbulence cases but worse under high turbulence, which may have been due to the design of the loss function setup or due to the complexity of the neural network architectures. Overall, the findings highlighted the potential use of AI and machine learning to improve FSO link performance. This could motivate future work using other machine learning methods such as support vector regression (SVR) and Gaussian process regression (GPR), as well as testing with physical FSO system hardware for real-world validation.

Keywords: Monte Carlo simulation · Neural network machine learning and optimization · Deep learning models · Free-space optical communication. SCI-RTCI

1 Introduction

Free-space optical (FSO) links, which use laser beam propagation between ground-based transmitters and satellite-based receivers, have gained a lot of interest for global communication. FSO transmissions can achieve high bandwidths and data rates. However, they are often crippled by atmospheric conditions such as weather conditions and turbulence that reduce their optical signal-to-noise ratio (OSNR) and other link performance metrics.

Several approaches have been proposed to reduce these impairments and improve overall performance. One common technique is utilizing adaptive optics (AO), which adjusts the optical wavefront to correct distortions caused by turbulence. Studies have shown that AO improves the received power of FSO links as well as reduces the bit

H. R. Arabnia et al. (Eds.): CSCI 2025, CCIS 2968, pp. 469–477, 2027.
https://doi.org/10.1007/978-3-032-28457-0_33

error rates (BER) [1, 2]. Another approach to improve link performance is to develop hybrid systems consisting of both optical and radio frequency (RF) communication. The hybrid FSO/RF links operate by either implementing a switching mechanism that would fall back to RF when conditions were too poor for optical [3], or by simultaneously transmitting both optical and RF and combining the signals, which improved reliability but also increased system complexity [4].

More recently, artificial intelligence (AI) and machine learning have been considered to model and predict FSO link performance. Some researchers have applied neural networks to learn correlations between RF signal strength and optical signal attenuation for hybrid FSO/RF links, which allowed for predictions of link availability [5]. The results suggest the potential of supervised learning in modeling complex environments.

While most approaches rely on detailed measurements such as wavefront sensors, this study investigates the use of AI models trained using a simulated dataset to learn and then optimize FSO link parameters in an environment where such detailed measurements are not available or reliable. Monte-Carlo simulations are used to generate a realistic dataset under different turbulence levels and pointing jitter conditions, which are then used to train deep learning neural network models. The trained models are then used to optimize the FSO link's performance, and the results are analyzed to determine the potential benefit of implementing AI.

2 Methods

2.1 Simulation Setup

A Monte-Carlo simulation was created in MATLAB to model an FSO transmission link between a ground-based transmitter station and a satellite receiver. The transmitter consisted of a single-mode fiber coupled to a three-lens telescope. Gaussian beam propagation was modeled using the ABCD matrix method to calculate the beam waist (W_T) and radius of curvature (R_T) at the telescope output. Adjusting the spacing between the second and third lenses allowed control over these parameters.

The atmosphere was modeled in layers, where each layer was characterized by a constant refractive index structure parameter (C_n^2) [6]. Five phase screens were generated for each layer using the modified Von Kármán power spectral density, and Fresnel propagation was used to simulate how the optical beam moved through the different layers [7]. Minimal turbulence was represented by $C_n^2 = 10^{-20}\ m^{-2/3}$. The received optical power was calculated by integrating the beam intensity across a 75 mm collecting lens, assuming no loss within the receiver optics. A reference power (P_{ref}) was also calculated under minimal turbulence for comparison. For each combination of atmospheric conditions and parameters W_T and R_T, the simulation was run and repeated over 1000 trials to capture the distribution of the received power statistics (average power, OSNR, etc.) for future analysis.

Pointing jitter effects were also modeled using a Beta distribution model [8]. By multiplying the Beta distribution with the received power distribution, the combined effects of turbulence and pointing jitter created a joint power distribution that provided information on how different laser beam choices impacted the received power statistics. Using the probability density function (PDF) fitter application in MATLAB, the statistics were then used to compute BER as a function of OSNR and to provide performance metrics, including the average and variance of received power when the transmitted power was 5 mW (P_{avg} and P_{var}). These values provided crucial metrics for model training and optimization.

$$P_{avg} = a' * b' * P_{ref} \tag{1}$$

$$P_{var} = a' * \left(b'\right)^2 * \left(P_{ref}\right)^2 \tag{2}$$

a′ and b′ are Gamma distribution fitting parameters obtained from the PDF.

2.2 Network Setup

Machine learning models were trained with supervised learning to learn the relationship between the FSO link's parameters and performance metrics. The training dataset was generated by the Monte-Carlo simulations. Five input parameters were used: (1) turbulence strength at the surface layer (first 2 km), (2) turbulence strength at the next atmospheric layer (next 2 km), (3) beam waist at the transmitter side ($\boldsymbol{W_T}$), (4) beam waist at the receiver side ($\boldsymbol{W_L}$), and (5) divergence angle at the transmitter side ($\boldsymbol{\theta_D}$). Then five output parameters were used: (1) reference power *($\boldsymbol{P_{ref}}$)*, (2) OSNR at BER = 10^{-3}, (3) OSNR at BER = 10^{-6}, and (4–5) the Gamma distribution fitting parameters ($\boldsymbol{a'}$) and ($\boldsymbol{b'}$).

An initial multilayer perceptron (MLP) model was tested, consisting of five input neurons, four hidden layers with 64, 128, 128, then 64 neurons, respectively, and five output neurons. A Tanh activation function was implemented, with inputs normalized to [−1,1] to fit within Tanh's range. For training, the model used an Adam optimizer with a learning rate of 0.001 and a mean squared error (MSE) loss. To analyze the model's performance, scatter plots of predicted vs. actual values and residual error histograms were generated. The scatter plots illustrated the accuracy of test data (data points on the dashed line were accurate) and the histograms illustrated the difference in error between the actual and predicted values (a histogram centered primarily at zero meant the model was well-fit; otherwise, the model had bias or was underfit) (Figs. 1 and 2).

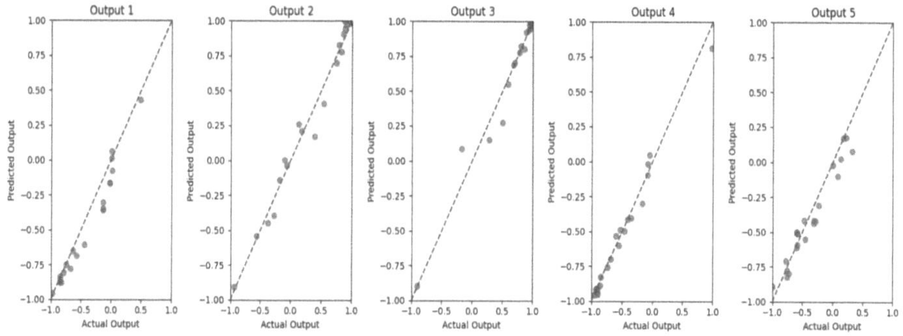

Fig. 1. Scatter plots of predicted vs. actual values for each output parameter using the MLP model.

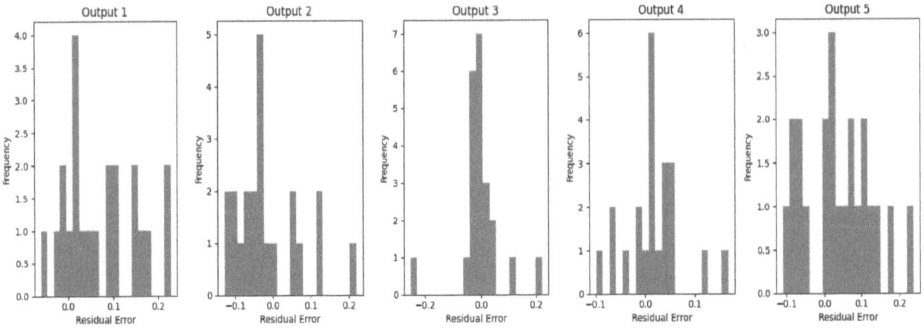

Fig. 2. Residual error histograms for each output parameter using the MLP model.

Despite the MLP achieving good accuracy, it showed signs of overfitting which meant the model had a reduced generalization to new data. To reduce overfitting, a revised MLP architecture was created, instead consisting of two hidden layers with 64 then 32 neurons, respectively, ReLU activation, dropout layers, and batch normalization layers. The data was also modified to z-score standardization instead of being normalized to $[-1,1]$. Training still used the Adam optimizer, but with a weight decay of 10^{-4} and with Huber loss instead of MSE. five-fold cross-validation and early stopping was also added in hopes of improving generalization. A second model, TabNet, was also trained with the same data for comparison with the new MLP because of how TabNet's architecture is designed for tabular data. The TabNet model was designed with five sequential steps in its architecture, a width of 64 for both decision and attention steps, two independent layers and two shared layers, a sparsemax feature selection strategy, an Adam optimizer with a learning rate of 0.001, and five-fold cross-validation and early stopping.

2.3 Optimization Process

After training, both models were used for optimization. Model weights were frozen, and instead of updating the weights during training, the input parameters were updated to minimize custom-defined loss functions. The Adam optimizer learning rate was increased to 0.05. Each optimization trial produced a set of unstandardized input and output values along with performance metrics ($\boldsymbol{P_{avg}}, \boldsymbol{P_{var}}, \boldsymbol{P_{var}/P_{avg}}$), which were saved for later analysis.

Lastly, the optimized inputs of the best model were re-evaluated in the Monte-Carlo simulation to verify that the predicted optimization values produced similar results to the simulation. This validation confirmed that the optimization results were consistent with the FSO link model.

3 Analysis

The performance of the improved MLP model is shown in Figs. 3 and 4. In Fig. 3, the scatter plots show that the predicted and actual values are very close across all the output parameters. This means that the model was capable of learning the complex

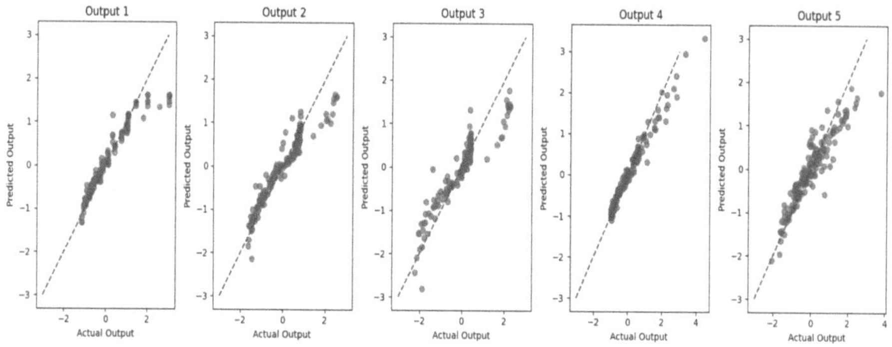

Fig. 3. Scatter plots of predicted vs. actual values for each output parameter using the improved MLP model.

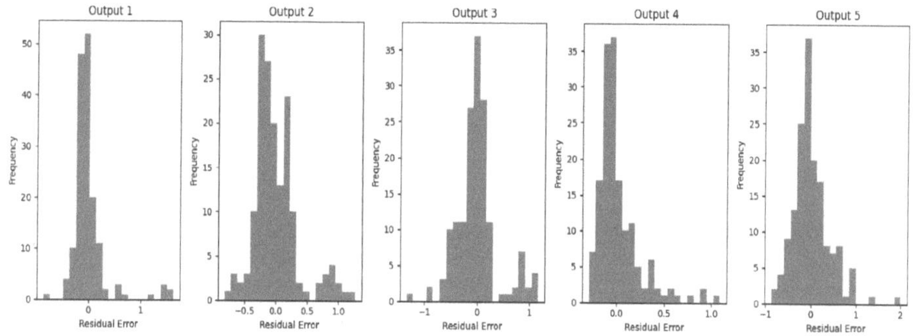

Fig. 4. Residual error histograms for each output parameter using the improved MLP model.

relationships between the FSO link model's inputs and outputs accurately. The residual error histograms in Fig. 4 are mostly centered around zero and less spread out than the initial MLP model, suggesting that the new MLP model was more well-fitted.

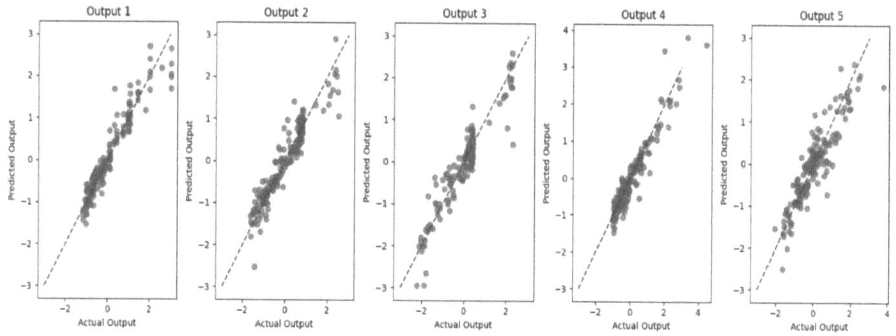

Fig. 5. Scatter plots of predicted vs. actual values for each output parameter using the TabNet model.

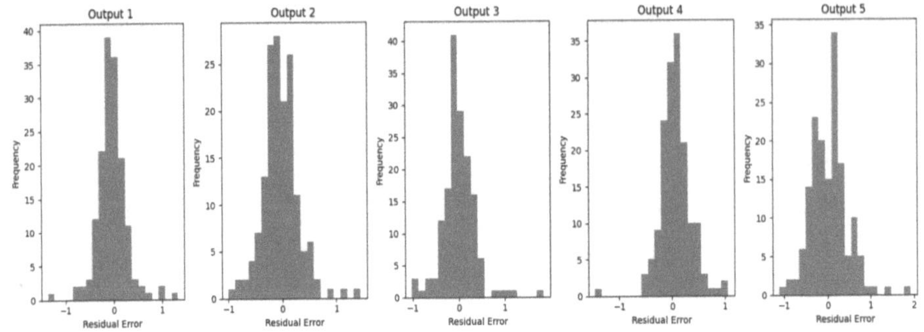

Fig. 6. Residual error histograms for each output parameter using the TabNet model.

The performance of the TabNet model is shown in (Figs. 5 and 6). While the overall performance of the model was still great, the scatter plots in Fig. 5 show more dispersion of the data points around the dashed line compared to the improved MLP. Likewise, the histograms in Fig. 6 also show more spreading around zero than the improved MLP model. Overall, this means that both models achieved great predictive performance, but the MLP model's predictions were slightly more consistent.

Table 1. Measures of variability and central tendency of the optimized data using the improved MLP model.

Parameter	Interquartile Range (IQR)	Standard Deviation	Mean	Median
P_{avg} (W)	6.75E-08	5.09E-08	3.01E-07	2.84E-07
P_{var} (W)	1.18E-14	1.04E-14	5.23E-14	5.50E-14
P_{var}/P_{avg}	4.05E-08	2.58E-08	1.74E-07	1.71E-07
C_n^2 @ surface ($m^{-2/3}$)	2.64E-16	1.62E-16	3.35E-16	3.86E-16
C_n^2 @ next layer ($m^{-2/3}$)	4.80E-17	6.59E-17	1.06E-16	1.04E-16
W_T (m)	0.01615	0.00988	0.05708	0.05742
W_L (m)	2.58022	1.43522	8.7967	8.92716
θ_D (rad)	3.42E-06	2.16E-06	−8.98E-06	−9.53E-06
P_{ref} (W)	1.08E-07	9.52E-08	8.24E-07	8.69E-07
OSNR @ BER = 10^{-3} (dB)	1.653	0.807	18.104	18.271
OSNR @ BER = 10^{-6} (dB)	0.025	0.013	23.992	23.985
a'	0.72444	0.39386	1.76891	1.88266
b'	0.01994	0.01179	0.21064	0.20431

Table 2. Measures of variability and central tendency of the optimized data using the TabNet model.

Parameter	Interquartile Range (IQR)	Standard Deviation	Mean	Median
P_{avg} (W)	7.78E-08	6.48E-08	2.93E-07	2.98E-07
P_{var} (W)	3.37E-14	2.30E-14	5.58E-14	5.57E-14
P_{var}/P_{avg}	1.09E-07	5.85E-08	1.88E-07	1.65E-07
C_n^2 @ surface ($m^{-2/3}$)	1.88E-16	2.01E-16	3.90E-16	3.54E-16
C_n^2 @ next layer ($m^{-2/3}$)	5.27E-17	5.81E-17	9.80E-17	7.79E-17
W_T (m)	0.01257	0.02434	0.07046	0.06264
W_L (m)	2.3854	1.38057	7.89039	7.85765
θ_D (rad)	2.76E-06	1.81E-06	-2.99E-06	-2.95E-06
P_{ref} (W)	2.51E-07	1.72E-07	8.27E-07	8.43E-07
OSNR @ BER = 10^{-3} (dB)	5.0153	4.3661	20.0472	18.678
OSNR @ BER = 10^{-6} (dB)	0.4827	0.7268	24.3562	23.9872
a'	1.22162	0.56068	1.67435	1.53798
b'	0.05361	0.04156	0.22488	0.20425

When used for optimization, both models were able to develop parameter sets that improved the performance of the simulated FSO link. The descriptive statistics (measures of variability and measures of central tendency) of the optimized parameters using both models are listed in (Table 1 and 2). Comparing the two, the MLP model showed less variability, meaning it optimized with more stability. The optimized inputs of the MLP were then re-simulated under low and high turbulence conditions, with the results shown in Table 3. The re-simulation confirmed that the optimized parameters produced results similar to what the model predicted. However, it was observed that the optimization performed better in low turbulence cases and worse in high turbulence cases. This pattern likely happened due to the design of the loss function setup or due to the complexity of the neural network architectures.

Table 3. Re-simulation of optimized data results in low and high turbulence cases using the improved MLP model.

Parameter	First Re-Simulation (Low Turbulence)	Second Re-Simulation (High Turbulence)
P_{avg} (W)	3.46E-07	2.00E-07

(continued)

Table 3. (*continued*)

Parameter	First Re-Simulation (Low Turbulence)	Second Re-Simulation (High Turbulence)
P_{var} (W)	4.98E-14	2.91E-14
P_{var}/P_{avg}	1.44E-07	1.44E-07
C_n^2 @ surface ($m^{-2/3}$)	1.56E-16	7.54E-16
C_n^2 @ next layer ($m^{-2/3}$)	3.93E-17	2.11E-16
W_T (m)	0.0572	0.0572
W_L (m)	8.7838	8.7838
θ_D (rad)	−8.88E-06	−8.88E-06
P_{ref} (W)	8.28E-07	8.28E-07
OSNR @ BER = 10^{-3} (dB)	16.8	28.8
OSNR @ BER = 10^{-6} (dB)	25.1	35.4
a'	2.40197	1.36761
b'	0.17468	0.17711

4 Conclusion

Overall, the findings still show promise that AI and machine learning can be a useful tool in learning and optimizing an FSO link model's performance when using simulated data. Compared to other methods such as AO or hybrid FSO/RF links, the proposed design offers more flexibility and less reliance on physical sensing hardware. While both ANN models performed well, the MLP model had slightly better accuracy and reliability than the TabNet model.

Looking ahead, future work to extend the study includes approaches like investigating other machine learning methods, such as support vector regression (SVR) and Gaussian process regression (GPR), as well as including physical testing with hardware to verify the application of the proposed optimization design outside of computer-simulated environments. To conclude, the study is a step towards autonomous, software-optimized FSO communication links and possibly helps remove the dependency on physical sensing hardware.

References

1. Osborn, J., Townson, M.J., Farley, O.J.D., Reeves, A., Calvo, R.M.: Adaptive optics pre-compensated laser uplink to LEO and GEO. Opt. Express. **29**(4), 6113–6132 (2021)
2. Chen, M., Liu, C., Rui, D., Xian, H.: Experimental results of atmospheric coherent optical communications with adaptive optics. Opt. Commun. **434**, 91–96 (2019)
3. Usman, M., Yang, H.-C., Alouini, M.-S.: Performance analysis of switching based hybrid FSO/RF transmission. In: IEEE 80th Vehicular Technology Conference, pp. 1–5, Vancouver, BC, Canada (2014)

4. Shakir, W.M.R.: Performance evaluation of a selection combining scheme for the hybrid FSO/RF system. IEEE Photonics J. **10**(1), 1–10 (2018)
5. Ibrahim, M., Ahmad, A., Ekin, S., LoPresti, P., Altunc, S., Kegege, O.: Anticipating optical availability in hybrid RF/FSO links using RF beacons and deep learning. IEEE Trans. Mach. Learn. Commun. Netw. **2**(1), 1369–1388 (2024)
6. Loos, G.C., Hogge, C.B.: Turbulence of the upper atmosphere and isoplanatism. Appl. Opt. **18**(15), 2654–2661 (1979)
7. Schmidt, J.D.: Chapter 9. In: Numerical Simulation of Optical Wave Propagation with Examples in MATLAB. SPIE, Bellingham, WA (2010)
8. Hemmati, H.: Chapter 3. In: Near-Earth Laser Communications, 2nd edn. CRC Press, Boca Raton (2020)

Social Network Analysis, Social Media, and Mining

Clique-Based Global Community Hiding Against Overlapping Community Detection

Leyla Tekin[1,2] and Belgin Ergenç Bostanoğlu[2(✉)]

[1] Department of Computer Engineering, İzmir Institute of Technology, İzmir, Turkey
leyla.tekin@usak.edu.tr

[2] Department of Computer Technologies, Karahallı Vocational School, Uşak University, Uşak, Turkey
belginergenc@iyte.edu.tr

Abstract. Although considerable progress has been made in deceiving disjoint community detection algorithms, global concealment of overlapping communities is still an unexplored problem. This study introduces 4clqCH, a new clique-based hiding method that strategically targets edges frequently contained in 4-cliques. This approach is useful in domains like crime detection, since highly cohesive subgraphs may indicate coordinated or illicit activity. The proposed method is evaluated on multiple datasets, demonstrating its effectiveness in deceiving overlapping community detection algorithms, especially those based on cliques and seed expansion. In addition, the study also assesses existing heuristic-based global hiding techniques in the context of overlapping community detection, thus offering a contribution to the field of global community hiding.

Keywords: Overlapping community detection · Community hiding · Community detection attack · Global hiding

1 Introduction

Community detection identifies densely connected communities within a network. These communities often correspond to functional units like social groups, collaborative teams, or thematic clusters. The use of community detection spans numerous real-world applications, such as recommendation systems, fraud detection, and targeted marketing.

As community detection capabilities continue to evolve, the risk of exposing sensitive or private group affiliations has become a growing concern [8]. This situation has led to the emergence of "community hiding", referring to adversarial techniques that seek to prevent the accurate discovery of communities by subtly modifying the network structure. Such modifications can protect individuals or organizations from undesired exposure, especially in scenarios where community membership implies sensitive or strategic relationships. Use cases include

H. R. Arabnia et al. (Eds.): CSCI 2025, CCIS 2968, pp. 481–496, 2027.
https://doi.org/10.1007/978-3-032-28457-0_34

activists or law enforcement agencies concealing cooperate patterns, or malicious actors evading surveillance [16,37]. Moreover, the dual-use nature of community hiding underscores the need for community detection algorithms that are robust to deceptive perturbations.

The problem of community hiding has been formalized across three distinct scales [9,25,35]: target node hiding, target community hiding, and global community hiding. These scales differ in their scope and concealment objectives. Strategies, such as heuristic, optimization-based, and graph learning, have been proposed. While progress has been made in devising attack algorithms against disjoint community detection algorithms, global community hiding in the context of overlapping community structures has not yet been investigated.

This study introduces a new method, 4-clique Community Hiding (4clqCH), which aims to disrupt the output of overlapping community detection algorithms by targeting structural motifs critical to their operation. Specifically, the method identifies and removes high-frequency edges within 4-cliques to degrade clique-based connectivity, thus obfuscating the underlying community structure. Unlike existing approaches that focus on hiding target nodes within overlapping regions [26,38], the 4clqCH addresses the global community hiding, an unexplored direction in the context of overlapping algorithms. The contributions of this study can be listed as follows:

- Addressing the problem of global community hiding targeting overlapping communities.
- Introduction of a new clique-oriented global hiding method, 4clqCH, designed to degrade the effectiveness of overlapping community detection algorithms.
- Evaluation of heuristic-based global community hiding algorithms that do not rely on the output of any specific detection method, against overlapping community detection techniques.
- Empirical assessment of the proposed method on real-world networks, highlighting its ability to conceal community structures at the global scale.

The structure of this paper is as follows. Section 2 provides foundational concepts related to community detection, community hiding, and evaluation measures. Section 3 reviews previous work related to community hiding. The proposed method for addressing the global community hiding problem is detailed in Sect. 4. Then, Sect. 5 presents the performance evaluation. Section 6 discusses the implications of the findings, and finally, Sect. 7 presents the conclusions of the study.

2 Background

This section presents a basic overview of community detection, followed by an introduction to community hiding as its adversarial counterpart, and concludes with performance measures used to assess the effectiveness of hiding methods.

2.1 Community Detection

Complex networks often have an inherent community structure, being considered one of their characteristics. A community within a network is expressed as a subset of nodes that are densely connected with one another, while they have relatively sparse links to nodes outside of this subset. Community detection refers to the task of identifying these cohesive substructures within the network.

Definition 1 (Community Detection). *Consider a graph G with nodes V and edges E. The community detection problem determines the community structure ($\bar{C}$) by partitioning the V into K distinct communities.*

Communities can be categorized into two types: overlapping communities and disjoint (non-overlapping) communities. Overlapping communities enable nodes to belong to more than one groups simultaneously, reflecting the real-world scenarios where individuals often join several groups at the same time. But, non-overlapping communities are strictly partitioned, with each node allocated to a single community, thus creating distinct and mutually exclusive groups.

Research has largely focused on conventional community detection methods, as community hiding aims to deceive them. Nevertheless, the field has advanced notably, particularly through the parallelization of such conventional techniques to enhance their effectiveness and practical usability [15]. Furthermore, deep learning has also shown impressive results in the field of community detection [1].

2.2 Community Hiding

Community detection algorithms can be significantly perturbed when only a small fraction of network links is allowed to be modified, as each link plays a distinct role in preserving the community structure. A community hiding algorithm attempts to deceive community detection algorithms and can be seen as a symmetrical process on the other side of the wall. This process involves strategic adjustments to the network structure, such as edge and/or node update operations, with the objective of fooling the detection algorithms in the identification of the community structure, so that obscuring the correct community affiliations and interactions in the network. Consequently, the hiding algorithms aim to achieve a targeted level of obfuscation by making changes to the network structure according to the number of allowed connections.

Definition 2 (Budget (β)). *It is the upper limit on the number of edge modifications that a community hiding algorithm is allowed.*

Definition 3 (Rewiring). *It is an edge operation in which an edge of a node is deleted and another edge is added to the same node.*

Figure 1 presents an overview of the community hiding process. Initially, the community structure is detected from the original graph through a particular community detection algorithm. Following this, a community hiding algorithm is applied to the graph that executes update operations (i.e., edge operations, such as addition, deletion, and rewiring, and/or node operations, such as addition, deletion, and moving) constrained by a predefined budget. The modified graph is then re-evaluated using the community detection algorithm, which may reveal notable alterations in the resulting community structure. Throughout this study, 'community hiding' and 'community detection attack' are used interchangeably.

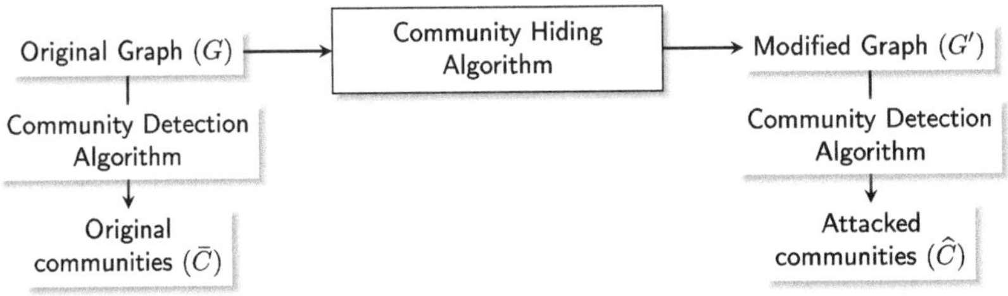

Fig. 1. The process of community hiding.

2.3 Measures for Community Hiding

This section outlines the performance evaluation measures used in this study to assess the effectiveness of community hiding algorithms.

- **Normalized Mutual Information (NMI):** This measure quantifies the similarity between two community partitions (X and Y) through an information-theoretic lens. Here, $H(X)$ represents the entropy associated with partition X, while $I(X, Y)$ denotes the mutual information between the partitions. It is calculated as follows:

$$NMI = \frac{2I(X,Y)}{H(X) + H(Y)}. \tag{1}$$

  This expression is applied to the community structures ($\bar{C}$ and $\widehat{C}$) detected before and after a hiding process [14].
- **Overlapping NMI (ONMI):** It is a measure used to compare overlapping communities [28], It is defined as follows:

$$ONMI = \frac{I(X,Y)}{max(H(X), H(Y))}. \tag{2}$$

  Like NMI, a higher ONMI score indicates greater similarity between the two partitions being compared.

3 Related Work

As community detection becomes increasingly effective in revealing group structures within networks, it simultaneously raises concerns regarding privacy and information exposure. Community hiding has been proposed as a countermeasure, targeting three scales of concealment: target node, target community, and global community structure [9,25,35].

Studies employ some graph learning paradigms—including gradient-based optimization in Graph Convolutional Networks (GCNs) [10], Graph Neural Network (GNN) architectures [22], and deep reinforcement learning [4]—to strategically perturb network structure for concealing target nodes from community detection algorithms. Recently, CH-SNMF [25], designed to address all three community hiding scales, incorporates the clustering strengths of symmetric non-negative matrix factorization (SNMF).

Early research on target community hiding includes the Nagaraja's method [30], the DICE algorithm developed by Waniek et al. [37], and strategies [11,16] grounded in the concept of safeness. Furthermore, Mittal et al. [29] proposed the notion of permanence. Recent studies [4,7,42] have explored the application of reinforcement learning, genetic algorithms, or swarm intelligence to address the problem of community hiding at the mesoscale.

To address the global community hiding problem, Chen et al. [8] present two heuristic strategies, CDA and DBA, which rely on community structure and node degree information, respectively, and the genetic algorithm-based Q-Attack that employs modularity for the fitness function. Liu et al. [27] introduce the REM algorithm to disrupt the community structure by considering community-informed structural entropy. In the GCH algorithm [23], the variational auto-ancoder is used to reconstruct the adversarial graph. The genetic algorithm-based CGN [24] uses NMI as a fitness function and uses initial community structure information to generate the gene pool. Furthermore, Zhao et al. employ genetic algorithms by introducing a self-adaptive framework [41] and a divide-and-conquer approach [40]. Yang et al. [39] propose a heuristic method called LSHA, which focuses on local structures instead of using a community structure. The study [36] explores attack strategies based on edge-importance measures—such as degree, betweenness, and closeness—and subsequently proposes methods to enhance the robustness of community structures.

Most existing efforts have concentrated on devising attack strategies targeting disjoint community detection algorithms. While few techniques [26,38] have emerged to obscure target nodes in overlapping regions, the problem of executing attacks at the global scale against overlapping community detection remains largely unaddressed. Therefore, this study introduces a novel clique-oriented approach to disrupt overlapping community structures at the global scale.

4 Proposed Method

The study [19] points out that several community detection methods based on k-cliques have been developed. In this study, 4-clique Community Hiding, called 4clqCH, is proposed as a global community hiding algorithm. Its aim is to disrupt the most common edges in the 4-cliques, thus changing the community structure relying on the cliques as much as possible.

The 4clqCH algorithm first counts the number of 4-cliques that each edge is a part of. For this, the fast method proposed by the study [32] is used. The counting method employs degree ordering for clique counting. The degree ordering of the graph G is indicated by $\prec$. For vertices i and j, $i \prec j$ means that $deg(i) < deg(j)$ or $deg(i) = deg(j)$ and $i < j$ (comparing their vertex ids). The *degree ordered* DAG (directed acyclic graph) is constructed by orienting all edges of G based on $\prec$.

The pseudocode of the 4clqCH is depicted in Algorithm 1. Line 1 in the pseudocode creates the degree ordered DAG, namely $Gout$, from the input graph. Every 4-clique in the original graph is counted once. The *per-edge* counts for 4-clique pattern are calculated as follows (Lines 3–18); All edges (i, j) are iterated, where $i \prec j$. All the out-wedges containing (i, j) are enumerated, and all triangles containing (i, j) with i as the smallest vertex are found. All pairs of such triangles formed by (i, j) are iterated. For each pair, the missing edge is checked whether it forms a four-clique. Figure 2 shows core patterns for 4-clique counting.

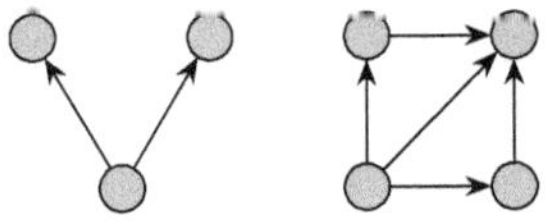

Fig. 2. Core patterns for 4-clique counting (Source: [32]).

After counting, the edges with the highest count is detected from *per-edge* counts (Lines 20–22). The edge that has the minimum degree among the candidate edges is selected as the victim edge (Line 23). Next, the 4-cliques that include the victim edge are determined. The *per-edge* count value of edges involved in these matches is reduced by 1 (Line 28). The victim edge is deleted from the graphs (Line 32).

5 Performance Evaluation

We conduct a comparative analysis to systematically evaluate the results of the 4clqCH algorithm against those produced by Random-Rewire [8], LSHA [39], importance measure-based methods (Degree, Betweenness, and Closeness) [36], and Random-Del. The motivation for choosing these community hiding

Algorithm 1. 4clqCH Algorithm.

Input: Graph G, Budget β
Output: Modified graph G'.

```
Create degree ordered Gout from G
Create arrays tri of size |V| + 1, perEdge of size |E| + 1
for each vertex i ∈ V do
    for each out-nbor j do
        for each next out-nbor k do
            if edge (j, k) exists then
                Insert k into tri
            end if
        end for
        for each k in tri do
            for each next l in tri do
                if edge (k, l) exists then
                    Increase perEdge of edges with {i, j, k, l} by 1
                end if
            end for
        end for
    end for
end for
for t = 0 to β do
    Find maxFreq in perEdge
                                        ▷ if 0, use triangles
    mostFreqs ← Get edges with maxFreq
    (i, j) ← Find edge with min degree in mostFreqs
    comNbrs ← Find common neighbors of i and j (in G)
    for each k in comNbrs do
        for each next l in comNbrs do
            if edge (k, l) exists then
                Decrease perEdge of edges with {i, j, k, l} by 1
            end if
        end for
    end for
    Delete edge (i, j) from G and Gout
end for
G' ← G
```

algorithms is that (i) they can work independently of the result produced by a community detection method during the hiding process, and (ii) they are heuristic approaches.

In our experiments, we employ the Python implementation of the LSHA algorithm provided by its authors and correct it by appropriately setting the local structure index of the deleted edge. As the clique counting method [32], the 4clqCH algorithm is also implemented in C++. Since the code of the other algorithms is not publicly available, we also implement them in C++. The details of the centrality definitions are not given in the study [36], thus edge degree is considered as the sum of node degrees, and edge closeness is calculated according to the definition in the study [6]. The interactive version of these centrality algorithms is considered.

The 4clqCH is compared with the chosen algorithms regarding NMI and ONMI depending on the disjoint and overlapping nature of the community detection algorithm. In addition to the overlapping community detection algorithms, such as Clique Percolation Method (CPM) [31], PercoMCV [20], UMSTMO [3], Demon [13], GREESE [2], the disjoint algorithm Core Expansion [12] is chosen due to its reliance on a seed-based approach. The UMSTMO, Demon and GREESE are chosen to fool although they are not clique-based methods. To find the k-clique communities with CPM, the parallel implementation (COS) [18] is used to solve the scalability issues.

5.1 Community Detection Algorithms

The functionalities of the community detection algorithms are explained below:

CPM [31] is a clique-based algorithm to detect overlapping communities. It characterizes a k-clique community as the union of all k-cliques that are connected through a sequence of adjacent k-cliques, where adjacency means sharing k-1 nodes.

PercoMCV [20] is a hybrid approach that first employs CPM [31] to detect initial communities, then refines the results using eigenvector centrality to reduce unclassified nodes.

Core Expansion [12] first detects the core of each potential community in the network. The core is then expanded iteratively by adding nodes to generate the final communities. It is based on the neighborhood overlap, which is formalized for an edge (u, v) as $c_{uv}/(d_u + d_v - c_{uv} - 2)$, where d_u and d_v are the degrees of nodes u and v, respectively, and c_{uv} is the number of common neighbors of u and v. The initial cores are local maximum nodes, where the weight of a node is the sum of the neighborhood overlap values of its edges.

UMSTMO [3] leverages the union of maximum spanning trees (UMST) as its foundational approach. The method operates in three distinct phases: first, the construction of the UMST T; second, the identification of local communities from T, wherein nodes A, B, and C are considered part of the same community if nodes B and C are connected to A within the UMST T and collectively form a triangle in the original graph G; and third, the integration of preliminary communities based on the importance of their shared nodes.

Demon [13] identifies overlapping communities by analyzing the ego networks (local neighborhoods) of nodes. The local communities for a node are found with a label propagation algorithm, which are then merged according to the similarity threshold to obtain the global communities.

GREESE [2] employs a coupled-seed expansion approach for detecting overlapping communities. A coupled-seed is formed by selecting a node along with its most similar neighboring node, serving as the starting point for expansion. This seed is then iteratively expanded using a fitness function designed to enhance the detection of local communities. Ultimately, local communities with substantial overlap are merged to produce the final set of communities.

5.2 Datasets

Different real world datasets are used from the Network Repository [33] and SNAP [21] platforms, each exhibiting distinct characteristics. Figure 3a summarizes their properties, where Avg. CC is average clustering coefficient and MCS is maximum clique size. Figure 3 illustrates the frequency distribution of k-cliques within the datasets. Specifically, the x-axis represents the size of the cliques (k), while the y-axis indicates the corresponding count of k-cliques present in the dataset. Due to space limitations, results on five out of the seven datasets are reported in this paper.

Graph	Nodes	Edges	Avg. CC	MCS
Karate	34	78	0.571	5
Polbooks	105	441	0.487	6
Football	115	613	0.403	9
Email	1,133	5,451	0.254	12
CA-GrQc	5,242	14,496	0.530	44
Amazon	262,111	1,234,877	0.420	7
Youtube	1,134,890	2,987,624	0.081	17

(a) Graph properties.

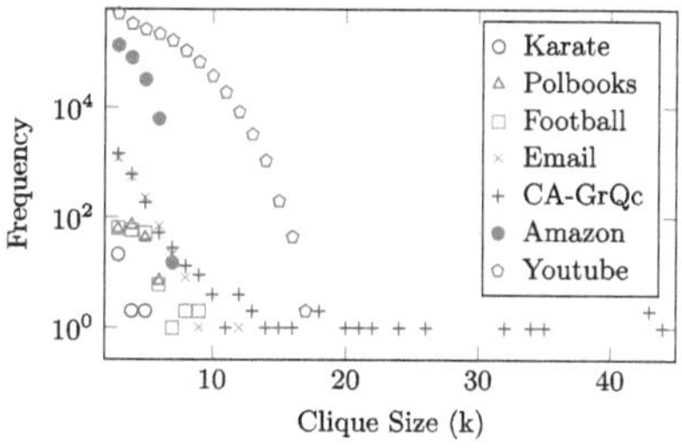

(b) Clique distribution in each graph.

Fig. 3. Overview of dataset characteristics, combining (a) graph properties with (b) the clique distribution observed in each graph.

5.3 Experimental Results

This section presents the discussion regarding the results of the 4clqCH algorithm in comparison to the other algorithms (Figs. 4 and 5), as well as the complexities of the algorithms.

Comparison of the Algorithms: The 4clqCH algorithm produces low ONMI scores across most scenarios when evaluated against the CPM and PercoMCV for the datasets polbooks and email. This is primarily due to its targeted approach of disrupting cliques, particularly 4-cliques, which significantly diminishes the ability of clique-based algorithms, such as CPM and PercoMCV. Additionally, it demonstrates strong performance in deceiving the Core Expansion method across these datasets. Since the Core Expansion relies on neighborhood overlap, the 4clqCH disrupts critical edges in localized regions, thereby reducing neighborhood overlap scores and making it harder for the Core Expansion to identify and expand the cores. Similarly, the Degree [36] algorithm, which focuses on removing high-degree edges, shows some good effectiveness in deceiving the clique-based methods and Core Expansion (on polbooks and email), meaning that high-degree edges are often included within cliques and are important in terms of neighborhood overlap.

The GREESE algorithm operates based on common neighbors and functions similarly to weak clique percolation. The 4clqCH algorithm is highly effective in most scenarios for reducing the ONMI scores of the GREESE algorithm. This effectiveness stems from its strategy of removing edges that are frequently shared among 4-cliques, which play a crucial role in ensuring high neighborhood overlap and supporting the expansion of communities.

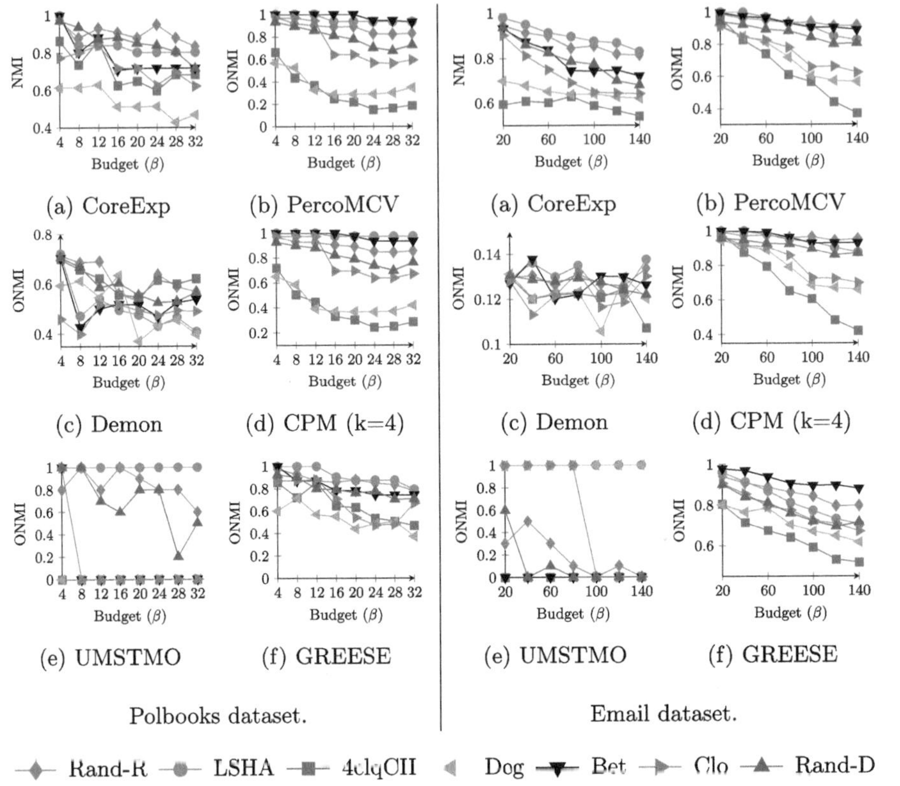

Fig. 4. NMI/ONMI of community hiding algorithms according to the budget on (a) Polbooks and (b) Email datasets.

When the effect of the community hiding algorithms to the Demon algorithm is analyzed on these datasets, it is observed that none of them is superior to the other. Moreover, the hiding algorithms effectively disrupt the functionality of the Demon algorithm on email dataset, indicating that the targeted alterations to edges appear to significantly perturb the structures of the ego neighborhoods and overlapping substructures that the Demon relies upon for its community detection process.

There is no consistency in the success of fooling the UMSTMO algorithm (except for the Betweenness [36] algorithm), that is, the effectiveness of the community hiding algorithms varies according to the characteristics of the datasets. The 4clqCH algorithm performs effectively to deceive the UMSTMO on the polbooks and email datasets. The Betweenness algorithm, which removes edges with the highest betweenness centrality, focuses on edges bridging different communities and maintaining network connectivity. Removing these edges can fragment the network, directly affecting the reliance of the UMSTMO on constructing maximum spanning trees.

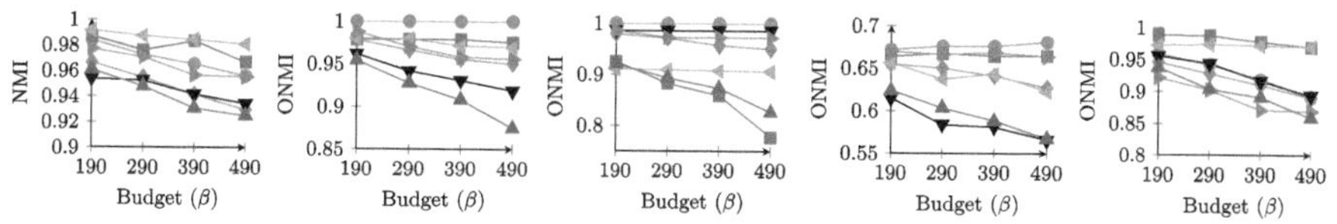

(a) CoreExp (b) CPM (k=4) (c) CPM (k=7) (d) Demon (e) GREESE

Rand-R LSHA 4clqCH Deg Bet Clo Rand-D

Fig. 5. NMI/ONMI of community hiding algorithms according to the budget on CA-GrQc dataset.

For CA-GrQc dataset, the results of some detection algorithms are not shown (due to memory issues in PercoMCV and all results being very close to 1 in UMSTMO). The Random-Del algorithm outperforms the other community hiding methods in terms of NMI/ONMI in most of the scenarios (refer to Fig. 5). The clique distribution within CA-GrQc dataset can be seen from the Fig. 3. The 4clqCH algorithm deletes edges that are most commonly shared across 4-cliques. Since the 4clqCH algorithm spends the budget mostly toward dismantling the larger cliques, it struggles to effectively deceive the CPM algorithm with k = 4. However, when the k-clique communities with the larger k value (k = 7) are identified with the CPM algorithm, it is observed that the 4clqCH algorithm can decrease the ONMI value more.

Table 1. ONMI results to deceive CPM [31] on Amazon and Youtube datasets.

(a) Amazon dataset

k-clique com.	β	4clqCH	Degree	Random-Del
k = 4	3000	**0.9196**	0.9858	0.9902
	4000	**0.9069**	0.9812	0.9868
k = 5	3000	**0.8351**	0.9899	0.9794
	4000	**0.7911**	0.9856	0.9722
k = 6	3000	**0.0336**	0.9889	0.9541
	4000	**0.4609**	0.9851	0.9402
k = 7	3000	**0.0000**	1.0000	0.8889
	4000	**0.2667**	1.0000	0.9111

(b) Youtube dataset

k-clique com.	β	4clqCH	Degree
k = 4	3000	**0.7604**	0.8851
k = 5	3000	**0.4380**	0.8092
k = 10	3000	**0.0862**	0.7449

Table 1 illustrates the ONMI results of the 4clqCH, Degree [36], and Random-Del algorithms to deceive the CPM [31] algorithm on amazon dataset for different values of k ranging from 4 to 7. The 4clqCH algorithm's capability to reduce the ONMI result can be seen even in the lower values of k, showing that it can effectively fool the CPM algorithm. The results of 4clqCH and Degree algorithms against the CPM on youtube dataset for k = 4, 5, and 10 are given in Table 1b. Even though the maximum clique size in youtube dataset is k = 17, the 4clqCH algorithm achieves better results than the Degree algorithm starting from the small k values.

Complexity Analysis: For four-clique counting, the undirected graph G is transformed into a directed acyclic graph $G^{\rightarrow}$, and the four-clique pattern is decomposed into smaller sub-patterns. These sub-patterns are enumerated within $G^{\rightarrow}$, leveraging the directed nature to effectively reduce the combinatorial expansion of the enumeration process. This approach provides a practical counting. The W_{++} and DD represent the counts of out-wedges and directed diamonds, respectively, as illustrated in Fig. 2. The computational cost of determining the number of four-cliques per-vertex and per-edge is $O(W_{++}(G^{\rightarrow}) + DD(G^{\rightarrow}))$, requiring an additional $O(m)$ space, where m is the number of edges in the graph. After counting phase, finding the maximum frequency in per-edge counts, getting the edges with the maximum frequency, and finding the victim edge to be deleted takes $O(m)$ time. Then, the common neighbors of the nodes of the chosen edge are determined in $O(d)$ time (d is the maximum node degree). Decreasing the per-edge count of the edges included in the deleted four-cliques is $O(d^2)$. Hence, the complexity after counting is $O(\beta(m + d^2))$, where β is the budget. So, the total complexity of the 4clqCH algorithm is $O(W_{++}(G^{\rightarrow}) + DD(G^{\rightarrow}) + \beta(m + d^2))$.

The LSHA algorithm identifies local structures within the network and subsequently executes a rewiring attack based on these structures. The complexity of detecting local structures is $O(m + dn)$, where n is the number of nodes in the graph. The time complexity of the attack phase is $O(d(n + m))$. In conclusion, the overall time complexity of the LSHA algorithm is $O(\beta d(n + m))$.

For the Degree algorithm, the time complexity is $O(\beta m)$ and the space complexity is $O(m + n)$.

Calculating the betweenness centrality and closeness centrality is computationally expensive since they need all pairs shortest paths. The well-known Brandes' algorithm [5] operates at a time complexity of $O(nm)$ for unweighted graphs, and a space complexity of $O(m + n)$. When an edge is removed, recalculating the centrality values for the entire graph can become prohibitively costly, especially if this operation is repeated β times. To address this challenge, new algorithms have been developed to update centrality values in dynamic or evolving graphs [34].

The time complexity of Random-Del and Random-Rewire is $O(\beta)$ and $O(\beta n)$, respectively. Their space complexity is $O(m + n)$ for storing the graph.

In summary, since the runtime of 4clqCH is bounded by the numbers of particular patterns, not m and n, its time complexity can be directly compared with that of the Degree and Random algorithms, which have better time complexity. The space complexity of the 4clqCH is $O(m + n)$, being the same as the other community hiding algorithms, except for LSHA, whose space complexity is not reported in the related paper.

6 Discussion on Findings

Numerous overlapping community detection algorithms leverage cliques or localized information to identify community structures. Attacks against these algorithms aim to disrupt the identified community structures, thereby diminishing their performance. The 4clqCH algorithm focuses on targeting edges mostly found within 4-cliques, prioritizing the disruption of larger cliques. This approach aligns with efforts in fields like crime detection and prevention, where reducing the size of the maximum clique is critical. Large cliques often signify cohesive substructures, which can serve as potential sources of coordinated activities, such as terrorist attacks. As such, weakening the cohesiveness of criminal or terrorist networks becomes a crucial objective [17].

The effectiveness of the 4clqCH algorithm has been assessed through experiments on various datasets, demonstrating its success in deceiving overlapping community detection algorithms, particularly those based on cliques or seed-based methodologies. Furthermore, the clique distribution within datasets has been analyzed. According to the given budget, when the frequency of large cliques is high, its performance decreases, as observed in the CA-GrQc dataset; however, it remains effective in modifying communities derived from large k-cliques. Additionally, it deceives CPM algorithm on larger datasets, such as amazon and youtube, across different k-values. While the 4clqCH algorithm matches other hiding methods in space complexity, it is outperformed by Degree and Random algorithms in time complexity.

Although significant research has been dedicated to fooling disjoint community detection algorithms, the domain of overlapping community detection has remained largely unexplored when it comes to systematic hiding strategies at the global scale. This study contributes to the field in two ways: first, by evaluating the efficacy of heuristic-based global community hiding algorithms—independent of detection algorithm results—against overlapping algorithms; and second, by proposing and assessing the performance of a new clique-based method.

7 Conclusion

This study addresses the problem of global community hiding to fool overlapping community detection algorithms. To this end, we introduce a new method called 4clqCH, which specifically distrupts edges that frequently appear within 4-cliques. Experiments on diverse datasets reveal that 4clqCH is effective in misleading a variety of overlapping community detection algorithms, particularly those based on clique-based and seed-expansion paradigms. Moreover, this study also presents a comparative evaluation of existing heuristic-based global hiding algorithms, which have not yet been evaluated for their effectiveness against overlapping community detection. The results show the trade-offs between computational cost and effectiveness in addition to highlighting the usefulness of 4clqCH in practice. Subsequent research could explore the use of alternative graph motifs utilized in overlapping community detection for designing effective hiding strategies. Another research direction may be to develop optimization-based or learning-driven approaches for global community hiding to perturb overlapping community structures.

Acknowledgments. This study builds upon the Ph.D. thesis conducted at zmir Institute of Technology.

Disclosure of Interests. The authors have no competing interests to declare that are relevant to the content of this article.

References

1. Al-Andoli, M.N., Tan, S.C., Cheah, W.P., Tan, S.Y.: A review on community detection in large complex networks from conventional to deep learning methods: a call for the use of parallel meta-heuristic algorithms. IEEE Access **9**, 96501–96527 (2021)
2. Asmi, K., Lotfi, D., Abarda, A.: The greedy coupled-seeds expansion method for the overlapping community detection in social networks. Computing **104**(2), 295–313 (2022)
3. Asmi, K., Lotfi, D., El Marraki, M.: Overlapping community detection based on the union of all maximum spanning trees. Libr. Hi Tech **38**(2), 276–292 (2020)
4. Bernini, A., Silvestri, F., Tolomei, G.: Community membership hiding as counterfactual graph search via deep reinforcement learning. arXiv preprint arXiv:2310.08909 (2023)
5. Brandes, U.: A faster algorithm for betweenness centrality. J. Math. Sociol. **25**(2), 163–177 (2001)
6. Bröhl, T., Lehnertz, K.: A straightforward edge centrality concept derived from generalizing degree and strength. Sci. Rep. **12**(1), 4407 (2022)
7. Chang, Z., Liang, J., Ma, S., Liu, D.: Community hiding: completely escape from community detection. Inf. Sci. 120665 (2024)
8. Chen, J., et al.: GA-based q-attack on community detection. IEEE Trans. Comput. Soc. Syst. **6**(3), 491–503 (2019)
9. Chen, J., Chen, Y., Chen, L., Zhao, M., Xuan, Q.: Multiscale evolutionary perturbation attack on community detection. IEEE Trans. Comput. Soc. Syst. **8**(1), 62–75 (2020)
10. Chen, J., Wu, Y., Xu, X., Chen, Y., Zheng, H., Xuan, Q.: Fast gradient attack on network embedding. arXiv preprint arXiv:1809.02797 (2018)
11. Chen, X., Jiang, Z., Li, H., Ma, J., Philip, S.Y.: Community hiding by link perturbation in social networks. IEEE Trans. Comput. Soc. Syst. **8**(3), 704–715 (2021)
12. Choumane, A., Awada, A., Harkous, A.: Core expansion: a new community detection algorithm based on neighborhood overlap. Soc. Netw. Anal. Min. **10**, 1–11 (2020)
13. Coscia, M., Rossetti, G., Giannotti, F., Pedreschi, D.: Demon: a local-first discovery method for overlapping communities. In: Proceedings of the 18th ACM SIGKDD International Conference on Knowledge Discovery and Data Mining, pp. 615–623 (2012)
14. Danon, L., Diaz-Guilera, A., Duch, J., Arenas, A.: Comparing community structure identification. J. Stat. Mech: Theory Exp. **2005**(09), P09008 (2005)

15. Fazlali, M., Moradi, E., Malazi, H.T.: Adaptive parallel Louvain community detection on a multicore platform. Microprocess. Microsyst. **54**, 26–34 (2017)
16. Fionda, V., Pirro, G.: Community deception or: how to stop fearing community detection algorithms. IEEE Trans. Knowl. Data Eng. **30**(4), 660–673 (2017)
17. Furini, F., Ljubić, I., Martin, S., San Segundo, P.: The maximum clique interdiction problem. Eur. J. Oper. Res. **277**(1), 112–127 (2019)
18. Gregori, E., Lenzini, L., Mainardi, S.: Parallel k-clique community detection on large-scale networks. IEEE Trans. Parallel Distrib. Syst. **24**(8), 1651–1660 (2012)
19. Gupta, S.K., Singh, D.P., Choudhary, J.: A review of clique-based overlapping community detection algorithms. Knowl. Inf. Syst. **64**(8), 2023–2058 (2022)
20. Kasoro, N., Kasereka, S., Mayogha, E., Vinh, H.T., Kinganga, J.: PercoMCV: a hybrid approach of community detection in social networks. Procedia Comput. Sci. **151**, 45–52 (2019)
21. Leskovec, J., Krevl, A.: SNAP datasets: Stanford large network dataset collection (2014). http://snap.stanford.edu/data
22. Li, J., Zhang, H., Han, Z., Rong, Y., Cheng, H., Huang, J.: Adversarial attack on community detection by hiding individuals. In: Proceedings of the Web Conference 2020, pp. 917–927 (2020)
23. Liu, D., Chang, Z., Yang, G., Chen, E.: Community hiding using a graph autoencoder. Knowl.-Based Syst. **253**, 109495 (2022)
24. Liu, D., Chang, Z., Yang, G., Chen, E.: Hiding ourselves from community detection through genetic algorithms. Inf. Sci. **614**, 123–137 (2022)
25. Liu, D., Jia, R., Liu, X., Zhang, W.: A unified framework of community hiding using symmetric nonnegative matrix factorization. Inf. Sci. **663**, 120235 (2024)
26. Liu, D., Yang, G., Wang, Y., Jin, H., Chen, E.: How to protect ourselves from overlapping community detection in social networks. IEEE Trans. Big Data **8**(4), 894–904 (2022)
27. Liu, Y., Liu, J., Zhang, Z., Zhu, L., Li, A.: REM: from structural entropy to community structure deception. Adv. Neural. Inf. Process. Syst. **32** (2019)
28. McDaid, A.F., Greene, D., Hurley, N.: Normalized mutual information to evaluate overlapping community finding algorithms. arXiv (2011)
29. Mittal, S., Sengupta, D., Chakraborty, T.: Hide and seek: outwitting community detection algorithms. IEEE Trans. Comput. Soc. Syst. **8**(4), 799–808 (2021)
30. Nagaraja, S.: The impact of unlinkability on adversarial community detection: effects and countermeasures. In: Atallah, M.J., Hopper, N.J. (eds.) PETS 2010. LNCS, vol. 6205, pp. 253–272. Springer, Heidelberg (2010). https://doi.org/10.1007/978-3-642-14527-8_15
31. Palla, G., Derényi, I., Farkas, I., Vicsek, T.: Uncovering the overlapping community structure of complex networks in nature and society. Nature **435**(7043), 814–818 (2005)
32. Pinar, A., Seshadhri, C., Vishal, V.: Escape: efficiently counting all 5-vertex subgraphs. In: Proceedings of the 26th International Conference on World Wide Web, pp. 1431–1440
33. Rossi, R.A., Ahmed, N.K.: The network data repository with interactive graph analytics and visualization. In: Proceedings of the Twenty-Ninth AAAI Conference on Artificial Intelligence, pp. 4292–4293
34. Shukla, K., Regunta, S.C., Tondomker, S.H., Kothapalli, K.: Efficient parallel algorithms for betweenness-and closeness-centrality in dynamic graphs. In: Proceedings of the 34th ACM International Conference on Supercomputing, pp. 1–12 (2020)
35. Tekin, L., Bostanoğlu, B.E.: A qualitative survey on community detection attack algorithms. Symmetry **16**(10), 1272 (2024)

36. Wang, S., Liu, J.: Constructing robust community structure against edge-based attacks. IEEE Syst. J. **13**(1), 582–592 (2018)
37. Waniek, M., Michalak, T.P., Wooldridge, M.J., Rahwan, T.: Hiding individuals and communities in a social network. Nat. Hum. Behav. **2**(2), 139–147 (2018)
38. Yang, G., Wang, Y., Chang, Z., Liu, D.: Overlapping community hiding method based on multi-level neighborhood information. Symmetry **14**(11), 2328 (2022)
39. Yang, H., Chen, L., Cheng, F., Qiu, J., Zhang, L.: LSHA: a local structure-based community detection attack heuristic approach. IEEE Trans. Comput. Soc. Syst. **11**, 2966–2978 (2023)
40. Zhao, J., Cheong, K.H.: Obfuscating community structure in complex network with evolutionary divide-and-conquer strategy. IEEE Trans. Evol. Comput. (2023)
41. Zhao, J., Wang, Z., Cao, J., Cheong, K.H.: A self-adaptive evolutionary deception framework for community structure. IEEE Trans. Syst. Man Cybern.: Syst. (2023)
42. Zhao, J., Wang, Z., Yu, D., Cao, J., Cheong, K.H.: Swarm intelligence for protecting sensitive identities in complex networks. Chaos, Solitons Fractals **182**, 114831 (2024)

Leveraging the Power of AI and Social Interactions to Restore Trust in Public Polls

Amr Akmal Abouelmagd(✉) and Amr Hilal

Department of Computer Science, Tennessee Technological University, Cookeville, TN, USA
{aabouelma42,ahilal}@tntech.edu

Abstract. The emergence of crowdsourced data has significantly reshaped social science, enabling extensive exploration of collective human actions, viewpoints, and societal dynamics. However, ensuring safe, fair, and reliable participation remains a persistent challenge. Traditional polling methods have seen a notable decline in engagement over recent decades, raising concerns about the credibility of collected data. Meanwhile, social and peer-to-peer networks have become increasingly widespread, but data from these platforms can suffer from credibility issues due to fraudulent or ineligible participation. In this paper, we explore how social interactions can help restore credibility in crowdsourced data collected over social networks. We present an empirical study to detect ineligible participation in a polling task through AI-based graph analysis of social interactions among imperfect participants composed of honest and dishonest actors. Our approach focuses solely on the structure of social interaction graphs, without relying on the content being shared. We simulate different levels and types of dishonest behavior among participants who attempt to propagate the task within their social networks. We conduct experiments on real-world social network datasets, using different eligibility criteria and modeling diverse participation patterns. Although structural differences in social interaction graphs introduce some performance variability, our study achieves promising results in detecting ineligibility across diverse social and behavioral profiles, with accuracy exceeding 90% in some configurations.

Keywords: Social Computing · Centrality Measures · Social Network Analysis · Social Interaction · Public Polls

1 Introduction

Crowdsourced data has transformed social science research by providing large-scale, real-time insights into human behavior, enabling researchers to analyze public opinion, social movements, and cultural trends more efficiently [1]. By leveraging the wisdom of the crowd, social scientists can enhance the diversity

H. R. Arabnia et al. (Eds.): CSCI 2025, CCIS 2968, pp. 497–517, 2027.
https://doi.org/10.1007/978-3-032-28457-0_35

and representativeness of data, offering new opportunities for studying complex social phenomena such as collective decision-making and public policy [2]. Crowdsourced data, like public opinion surveys or public polls, often suffers from nonresponse bias, where certain demographic groups are underrepresented, potentially skewing results and limiting the generalizability of findings [3,4]. Moreover, there is a growing difficulty reaching and persuading potential respondents [3]. For example, the telephone survey response rates declined from 36% in 1997 to 3% in 2023 [4]. On the other hand, there is an observed growth in online opt-in polling compared to conventional methods like phone surveys. While promising, the participation rate in the online polls, and consequently the error rate, are still concerns as it might mislead the public as was evidenced by the significant errors in the 2016 and 2020 US presidential election polls [5,6]. A post-election review of polling by the American Association for Public Opinion Research (AAPOR) cited "polling error of an unusual magnitude" in the 2020 polls [7]. Recurring errors in influential public polls highlight the need for new polling paradigms.

Decentralized peer-to-peer polling can expand participation in surveys and decision-making, but credible outcomes require strict eligibility control. In systems like e-voting [8,9], eligibility is usually enforced by a central authority, which weakens decentralization and adds friction, especially for polling. Removing this dependency simplifies participation and encourages adoption.

To address this problem, we explore how trust in polling data can be restored by leveraging social interactions among participants via social networks. Today, social networks serve as conduits for information, often accompanied by virtual interactions like comments, shares, and likes. We hypothesize that observing social interactions with sufficient dissemination flows can reveal behavioral patterns to infer participants' eligibility without a central authority or relying on people's honesty. We assume honesty as a common operational foundation in our study and investigate the effect of changing its ratio among the participants. This is similar to the assumption that underlies cryptocurrency systems [10,11], which require at least two-thirds of participants to be honest for proper functioning [12]. By combining social behavior with decentralized data-sharing, we aim to develop an AI-based polling mechanism that ensures broad participation and reliability.

In this paper, we explore this concept by addressing the following research questions: (1) How can polling be effectively structured within a practical model based on social dissemination? (2) To what extent can social interactions among participants be observed and analyzed to establish trust in the collected data? (3) What social and technical factors influence the success of this process?

Contributions. To answer these research questions, we present an empirical study aimed at detecting ineligible participation in a polling environment that rely on peer-to-peer communication. We propose a mechanism in which a dissemination graph is constructed using the flow of sharing a polling requests among peers. We leverage centrality measures and machine learning applied to the dissemination graph data to identify ineligible participation. Our contributions can be summarized as follow:

- We propose a data dissemination mechanism for a peer-to-peer polling system that leverages socially induced interactions of poll sharing to learn participants' eligibility. The mechanism encourages both active participation and continued propagation of information.
- We simulate the proposed mechanism using real-world datasets, incorporating different social interaction patterns and eligibility criteria that influence information dissemination in peer-to-peer settings. A dissemination graph is constructed by observing the social dynamics underlying data-sharing activities.
- We employ Graph Neural Networks (GNNs) to analyze the dissemination graph and assess participant eligibility, demonstrating prediction accuracy across various factors that influence the effectiveness of the proposed mechanism.

The remainder of this paper is organized as follows. Section 2 reviews related work. Section 3 introduces the proposed system and key definitions underlying our approach. Section 4 details the system architecture and dynamics, including node types, participation mechanisms, and data forwarding. Section 5 presents the empirical evaluation with the experimental setup and performance analysis. Section 6 reports the main results and observations. Section 7 discusses the study's contributions and limitations. Finally, Sect. 8 concludes and outlines directions for future work.

2 Related Work

Our proposed work intersects three research domains; public opinion polling, electronic voting (e-voting) systems, and blockchain technology. Each of these areas presents unique developments and ongoing challenges. In the following sections, we review relevant literature pertaining to each domain.

2.1 Public Opinion Polling

Public opinion polling has experienced substantial transformations over the past two decades, largely due to technological advancements and changing respondent behaviors. According to a 2023 Pew Research Center study, 61% of U.S. national pollsters altered their polling methodologies between 2016 and 2022, reflecting a shift away from traditional methods [7]. The decline of live telephone polling, which was once the dominant approach, has been driven by increasing costs and declining response rates. In response, many pollsters have turned to online surveys, probability-based panels, and Address-Based Sampling (ABS) to improve the representativeness and reliability of their data. These changes aim to modernize polling, improve data quality, and reduce costs. However, their effectiveness in predicting elections remains uncertain, especially given polling errors in 2016 and 2020 [5]. Ongoing research continues to assess the impact of these changes on the overall accuracy and credibility of public opinion measurements [13].

Cerina et al. [14,15] examine how AI can support public opinion research by combining traditional surveys with large language models (LLMs) that extract structured, survey-like data from social media to improve representativeness, frequency, and cost-efficiency. In contrast, Boelaert [16] argues that LLMs often produce biased, low-variance responses that vary across topics, limiting their usefulness as substitutes for human respondents.

2.2 E-Voting Systems

e-Voting systems have evolved over decades by adopting emerging technologies, with real implementations in countries like Estonia and Switzerland [17,18]. Li et al. [19] address the lack of public traceability in anonymous authentication for e-voting by proposing Traceable Attribute-Based Anonymous Authentication (TABAA), which ensures anonymity, access control, and accountability without trusted third parties. Using zero-knowledge proofs and attribute-based credentials, it enables reusable, unlinkable authentication while preventing double voting through blockchain implementation for decentralized, transparent, and tamper-proof voting. Agrawal et al. [20] introduce a system maintaining voter list privacy while enabling public audits to prevent fraud, allowing voters to verify their listing and auditors to detect fake participation or unfair removals without exposing personal data through cryptographic methods ensuring security and fairness.

2.3 Blockchain Systems

Since its introduction in the Bitcoin white paper [10], blockchain has attracted great interest for building decentralized crowdsourcing systems. Yao et al. [21] propose a decentralized self-tallying e-voting protocol on Ethereum, combining zero-knowledge proofs and homomorphic encryption to ensure secrecy while enabling public ballot verification and result computation.

Muth et al. [22] present Tornado Vote, a DApp for anonymous, fair voting on Ethereum, inspired by Tornado Cash to balance transparency and voter anonymity. Their protocol handles about 10,000 votes in two hours under optimal conditions. However, despite decentralization, current approaches still rely on a central authority for eligibility verification via government-issued IDs and PKI, limiting full decentralization.

2.4 Filling the Gaps

Our goal, in this work, is to address the limitations in the existing polling approaches by combining these three areas. While modern polling approaches, like online surveys, can attract more responses due to ease of use and perhaps privacy preservation [4,7], that opens the door to fraudulent participation leading to wrong or fabricated results. In this work, we address the issue of declining response by harnessing the power of social influence while filtering ineligible responses using AI-assisted graph analysis.

e-Voting systems, including blockchain implementations [11], facilitate secure decentralized voting [9]. Due to high-impact results, e-voting requires strict eligibility verification, necessitating central authorities that undermine decentralization and privacy. Classical e-voting characteristics include anti-coercion and single participation [8]. While not implementing e-voting, we view polling as a relaxed voting version. Our proposed system replaces centralized eligibility checks with AI-assisted filtering, aiming for bounded error margins acceptable in polling domains. We allow multiple participations by design, which fights coercion when time-tracked. Future research will implement a prototype using modified blockchain technology for our application.

3 Proposed System

To address our research questions, we imagine a scenario where a group of people characterized by a simple eligibility criterion is to be addressed by a poll. Instead of following the standard way of asking people to directly participate through mail or online portals, we leverage the power of social connections to disseminate the poll request. Specifically, we propose creating an induced social interaction by prompting the peers of a social network to participate and forward to others from their network of peers in a form similar to snowball sampling [23]. Instead of having a central authority verify the eligibility criterion, we ask people to forward only to those who are eligible to participate where a receiver does not know the forwarder. The forwarding process follows a simple eligibility criterion that's easy to understand and comprehend among average people such as a specific age group, country, or district (similar to how real-world participation frameworks often apply). Using such a basic criterion enables easy and practical implementation while maintaining control over participant selection. It also keeps a low participation barrier by avoiding overly complex constraints that could cause confusion or restrict natural interaction dynamics. Naturally, some participants will commit to the eligibility request and some will not, depending on the level of honest they observe.

We simulate simple honest behavior as it will determine the data flow in the social network. To motivate more honesty, we start the participation flow from credible members of the social network or the society representing key figures in a community (we call them *root nodes* in our experiments and results section). Due to their distinction, these key participants are expected to pass more traffic than an average participant. We assume a certain percentage of honest participants (randomly selected) at the beginning of the simulation where honest participants are more likely to commit to the eligibility criterion than dishonest ones. Like in cryptocurrency systems, we assume there is a way to securely and anonymously identify unique participants, hence allow them to respond to the poll and/or forward to others multiple times as they wish counting only the last participation as the official response. Multi-chance participation/forwarding widens the dissemination scale, increases social interaction, and creates a level of redundancy that we hope can all help differentiate eligible from ineligible participants through their response/forwarding behavior. It also satisfies the anti-coercion requirement in e-voting systems [8]. We use a cap to the length of a forwarding sequence to allow for new flows to stream through the social network for enhanced exploration.

The participants submit their responses to an advisory entity (that could also be decentralized like in blockchain applications [24]), regardless of their decision to forward to others. We assume information about the forwarding flow can be anonymously and securely encoded into the response where a graph of the whole social interactions can be constructed from the submitted responses without revealing the identity of the respondents, called a *a dissemination graph.* This graph is typically a subset of the original social graph but with more information about how it was created. We use machine learning to allow a model to learn the ineligible participants from the social interaction information embedded into the dissemination graph. Figure 1 shows how the dissemination graph is generated over time, with ballot dissemination initiated from root nodes and shaped by network structure and participant behavior.

Modeling social interaction in this experiment was challenging. First, most available social network datasets are highly anonymized, making it difficult to find one suitable for use as a participation criterion. Second, although studies exist on honest social behavior [25], translating their findings into quantifiable parameters for computer simulations proved challenging. Therefore, we had to use probability distributions to model these behaviors as close as we can. We also needed to make more design choices that are plausible from a practical as well as behavioral viewpoint. For example, how many times a user would participate and forward the poll request before feeling bored and how many connections a user would forward to. These and other questions would be better learned from an actual experiment, which we intend to do in a later stage of this research.

4 Key System Configurations

In this section, we define key system terms and describe their configurations. These terms are summarized in Table 1 and we provide more details below.

4.1 Honest Behavior

Unlike structural properties in a network, honesty is inherently person-aware and a self-trait, making it difficult to quantify using simple metrics [26,27]. While people can exhibit a diverse spectrum of honest behavior, we classify participants as either honest or dishonest for our initial investigation. For instance, if the eligibility criterion for participation is a minimum age of 21, an honest participant—whether eligible or ineligible—strictly adheres to the rule and forwards the poll to its connections who are more than 21 years of age. Conversely, a dishonest participant is willing to engage in a malicious or manipulative behavior by forwarding to both eligible and ineligible peers. At the beginning of our simulations, we set an honest ratio and randomly assign honesty to the graph nodes accordingly.

4.2 Root Nodes

In real-world scenarios root nodes should ideally be honest and trusted individuals with high reputation and are well-connected across their larger network. To demonstrate this in our work, we used centrality measures to approximate this selection. Specifically, we use the Betweenness centrality measure [28,29] as our metric for selecting nodes, as it identifies graph vertices that act as critical bridges between different parts of the graph. Nodes with higher Betweenness centrality lie on a larger number of shortest paths, making them influential in facilitating information flow across the network. By leveraging these high-centrality nodes, we enhance the poll propagation, ensuring broader participation by reaching diverse parts of a targeted community. Due to their critical role in the information dissemination process, all of the selected root nodes are chosen from the honest and eligible node sets. We assume this description and behavior of the root nodes matches society key figures or influencers in a real-life

Table 1. Key terms and definitions in the participation dissemination process

Term	Definition
Eligible Node	A participant that meets the predefined criteria for data collection
Ballot Unit	A simulated unit that holds participants' responses. Each root node releases a predefined number of ballot units, which are distributed through the social graph. As nodes forward the poll their connections, a copy of a ballot unit is sent to the next participants. So, the number of these units increases as they propagate down the graph. For safety, each copy of a ballot Unit has a maximum capacity of 7 participations, while the tree-like propagation introduces a level of redundancy
Eligibility Ratio	Represents the proportion of nodes that meet the criteria required to participate in the data collection process
Root Nodes	Participants that serve as starting nodes in the dissemination graph and ignite information flows
Honest Nodes	Participants who follow the voting rules and participate without malicious intent
Eligible Forwarding	Sending the Ballot Unit to an eligible neighbor node so it can participate
Coverage	Percentage of participants that received at least a single ballot unit relative to the total number of nodes in the graph

scenario. A careful selection of these root nodes along with detecting ineligible participants (a main contribution of this paper) helps alleviate possible biases of the snowball-like dissemination process.

4.3 Participation Behavior

We designed our experimental setup to closely mirror real-world scenarios, considering the diverse behaviors of individuals in anonymous participation rounds. In a real world scenario, participants will exhibit varying appetite for filling a poll and/or forwarding to others. To simulate this behavior, we use the node degree centrality measure to set the number of participations for each node separately. We sample this number from an exponential distribution such that 90% of the participation count ($\alpha = 0.9$) lie below 10% of a node's degree ($\beta = 0.1$), as shown in Eq. 1, with a minimum of 1 participation. We compute the rate parameter λ parameterized by the node's degree d, where βd represents 90% of the node's degree. By tying the participation limit to node degree in this way, we aim to mimic the inclination of well-connected people to share more. Additionally, in our setting, root nodes are granted an additional fixed number of participation opportunities to aid their role in the process. This allocation is directly tied to the initial count of ballot units assigned to each root node, facilitating effective dissemination throughout the network. We set these extra participation opportunities to 30.

$$\begin{aligned} &\textit{participation count} \sim \max(\lfloor \mathrm{Exp}(\lambda(d)) \rfloor, 1) \\ &\text{where } \lambda(d) = \frac{-\ln(1-\alpha)}{\beta d}, \alpha = 0.9,\ \beta = 0.1 \end{aligned} \tag{1}$$

4.4 Eligibility Criteria

Participation eligibility is central to developing our system. We adopt a flexible and adaptable criterion that supports diverse use cases across different scenarios. Eligibility can be defined by factors such as age, country of residence, or affiliation with a particular group, reflecting real-world practices where individuals must meet minimum requirements to participate. For instance, in voting processes, citizens are typically required to satisfy conditions like age and residency to qualify.

4.5 Ballot Units Forwarding

Each non-root node that received a ballot unit - whether it participated or not - can forward it to one or more of its social connections. To ensure a realistic and fair selection, we employ the round-robin technique to determine the next connection to receive the ballot unit. This approach ensures balanced distribution and helps reduce bias in the forwarding process.

The forwarding behavior differs between honest and dishonest participants. Regardless of its eligibility status, an honest participant will forward only to

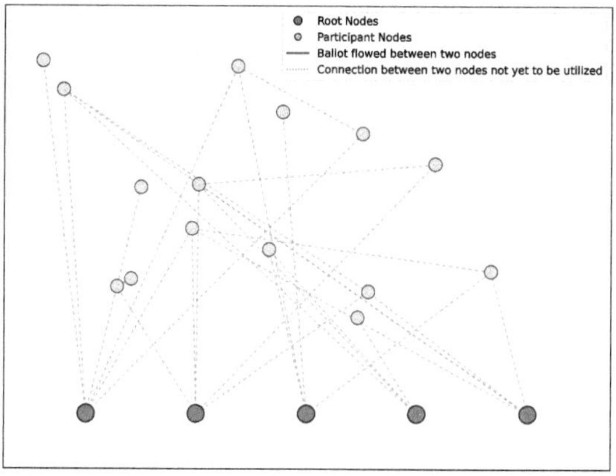

(a)

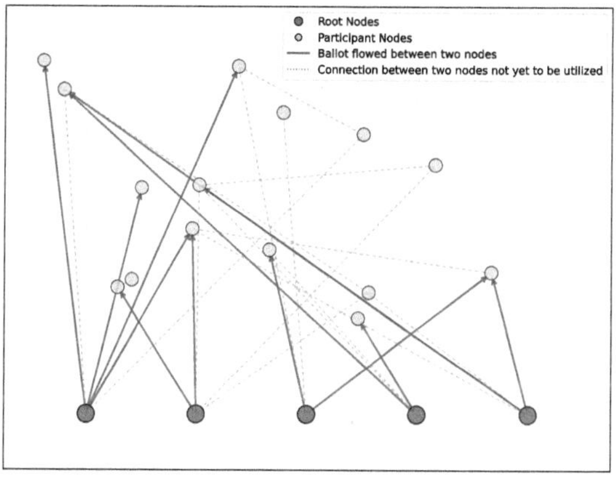

(b)

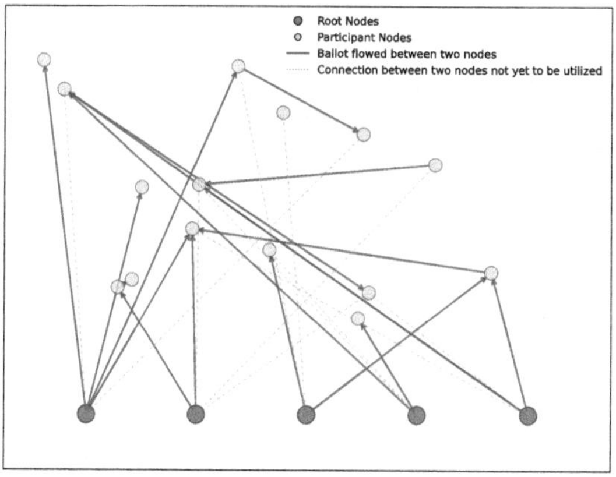

(c)

Fig. 1. Evolution of the dissemination graph over time (a → c)

eligible connections. A dishonest participants, though, may forward to both eligible and ineligible connections. To simulate this behavior we follow a probabilistic decision-making process. Each time a dishonest participant encounters a ballot unit, it samples a value from a uniform distribution. Using a randomized threshold decision rule, the node compares this value against a probability to decide whether to behave legitimately or act maliciously when forwarding. Eligible nodes refrain from sending if they have no eligible neighbors, whereas ineligible nodes send to their eligible neighbors only when they lack ineligible ones.

This approach is justified by previous research [25,30], which highlights key observations about the behavior of malicious participants. Studies suggest that dishonest actors tend to be strategic and cautious, aiming to conceal their fraudulent activities to avoid detection and protect their reputation. Additionally, they often strive to maintain a positive self-image, both publicly and privately. To minimize suspicion, these actors do not engage in continuous fraudulent behavior; instead, they mix malicious actions with legitimate ones. This strategic alternation ensures that their overall conduct appears partially compliant rather than overtly fraudulent, reducing the likelihood of exposure.

Table 2. Summary of datasets used

Dataset	Source	Nodes	Edges
Last.fm Multigraph	UCI Networking Group	465,166	4,120,950
Musae-Twitch (DE)	SNAP	9,498	153,138

5 Experiments

We conducted simulations on multiple datasets to evaluate model performance in predicting participant eligibility under varying community conditions, including the proportions of eligible users, root nodes, and honest participants.

5.1 Datasets

While many datasets exhibit the social-network structure we need for our experiments, we needed datasets that meets the following criteria to fit our empirical study:

1) The dataset has to include individual-based features to be used as eligibility criterion. These features either can be categorical or bear multiple values allowing us to segment the social network based on different values of this feature(s). For example, a feature like age group can be useful, as it enables categorization into predefined ranges (e.g., under 18, 18–30, 31–50, and 50+),

facilitating meaningful analysis. However, descriptive features like objectives, self-reflections would not be effective. It was challenging to find datasets with this criteria because most of the available social networks datasets are highly anonymized.

2) The dataset should contain edges that represent real social connections to allow us to study authentic human behavior rather than artificial or random connections.

We conducted our experiments using two datasets, 1 from UCI Networking Group and 2 datasets from SNAP [31], while exploring diverse domains for the eligibility criteria. These simple characteristics of these datasets are summarized in Table 2, and in details as follows:

1) Last.fm Multigraph Dataset [32]: This dataset is an Internet website for music with social networking features. For this dataset, we used the age attribute of each participant as the primary eligibility criterion. We constructed the edges based solely on the `uname_friends` relationship. The dataset contained a large number of connected components, exceeding 50,000, with the majority being outliers. For our analysis, we selected the largest connected component (465,166 nodes), as it provided the most meaningful structure. The second-largest component, in comparison, had a size of only seven nodes.
2) Musae-Twitch Dataset [33]: This dataset consists of Twitch user-user networks from May 2018, representing gamers who stream in a specific language. Nodes represent users, edges represent friendships, and features include games played, location, and streaming habits. We used the minimum total channel views reported for a user as the eligibility criterion. We used the Germany dataset as it had the highest number of nodes compared to other countries.

5.2 Experiment Setup

The approach we employed to capture the characteristics of social interactions within the network was based on graph node embeddings, which encode the relationships between nodes and the overall graph structure. We generated both static and dynamic node embeddings. Static embeddings were generated using Node2Vec [34], while dynamic embeddings were learned using GraphSAGE [35] with an embedding vector length of 32, which were jointly refined during classification. GraphSAGE was used also as a classifier to predict the node eligibility given the combined static and dynamic node embeddings. Both static and dynamic embeddings were concatenated for the GraphSAGE classifier.

To optimize performance, we tuned the Node2Vec parameters, including embedding vector size, random walk length, and number of walks. Since the two datasets exhibited structural differences, we used dataset-specific configurations for random walk length and number of walks. In both datasets, the embedding dimension was set to 64. For the Twitch dataset, we used a walk length of 32 with 50 walks per node, whereas for the Last.fm dataset, we applied a walk length of 50 with 30 walks per node.

We further experimented with the proportion of root nodes, testing values of 1% and 5%, and capped it at 5% to constrain the number of influencers. Similarly, we varied the proportion of honest nodes, using values of 60% and 80%. In addition, we evaluated four different eligibility ratios per dataset to investigate how model performance varied across configurations. Establishing comparable eligibility ratios across datasets was essential for fair evaluation. However, this proved challenging because the relationship between eligibility criteria and the resulting eligibility ratio is nonlinear. To address this, we tested multiple configurations and selected values that yielded ratios that were close, though not identical, across the two datasets, as summarized in Table 3.

Table 3. Eligibility ratios used across datasets

Dataset	Eligibility 1	Eligibility 2	Eligibility 3	Eligibility 4
Last.FM	86.66%	56.87%	38.67%	26.2%
Twitch	88.67%	55.4%	38.69%	24.86%

6 Results

6.1 Prediction Power

Figures 2 and 3 show the obtained F1-score and accuracy using the graph node embeddings against different levels of eligibility ratios for different levels of root nodes across the two different datasets. While the two datasets exhibit a similar trend, each have a different bracket of performance reflecting the impact of the graph structure that naturally changes across social networks. The figure also

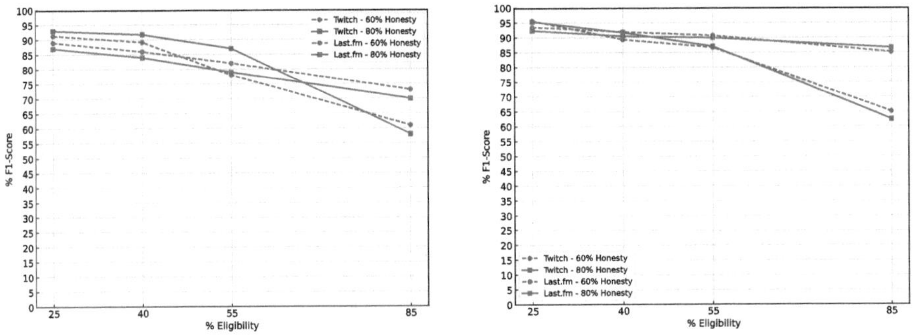

(a) 1% root nodes

(b) 5% root nodes

Fig. 2. Comparison of F1-scores for two datasets under varying eligibility ratios, root coverage, and participant honesty levels.

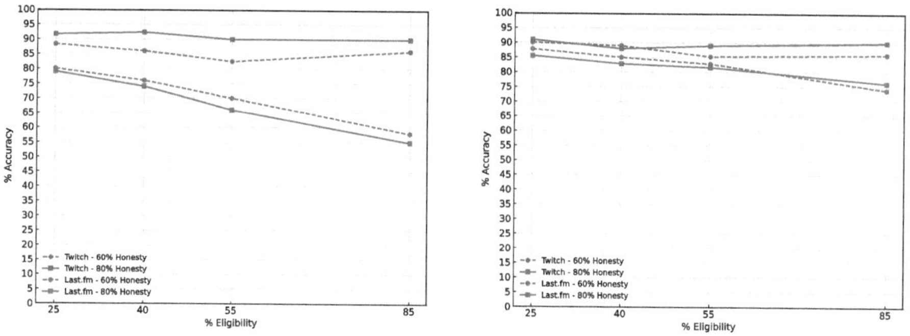

(a) 1% root nodes

(b) 5% root nodes

Fig. 3. Accuracy for the two dataset against different levels of eligibility ratios across different root node levels and different levels of participant honesty.

shows that better predictions can be obtained as the percentage of root nodes increases. This is because root nodes are honest participants who are more likely to forward to eligible participants.

6.2 Coverage

We study the coverage, as defined in Table 1, to inspect how much of the target population have received the poll request. We examined the two cases, 1% root nodes and 5% root nodes, for each dataset under the highest percentage of eligibility across different honesty levels (60%, 80%) as shown in Fig. 4. The low coverage in lower eligibility ratio scenarios is reasonable because ineligible nodes are naturally less likely to receive a participation ballot than in a higher illegibility ratio scenario given the same level of honesty. Figure 4 also shows that the coverage in the Twitch dataset is significantly higher than in the Last.FM dataset. We attribute this to the fact that the nodes in the Twitch dataset have a higher average node degree compared to the Last.fm dataset. Specifically, the Twitch dataset has an average of 32 edges per node, while the Last.FM dataset has 17 edges per node. This affirms again the impact of the graph structure and connectivity on the sought results. We take this into consideration while tuning the system configurations in our ongoing work extension to develop a generalized model.

6.3 Practicality and Applicability

To assess our system's practicality and applicability, we measured each participant's participation frequency which is the number of ballots received and acted upon. Figure 5 (log scale) shows the distribution for 1% and 5% root nodes. Nearly all non-root nodes participated fewer than the 30 additional forwards allowed for roots, with most engaging far less. Although 30 participations is not

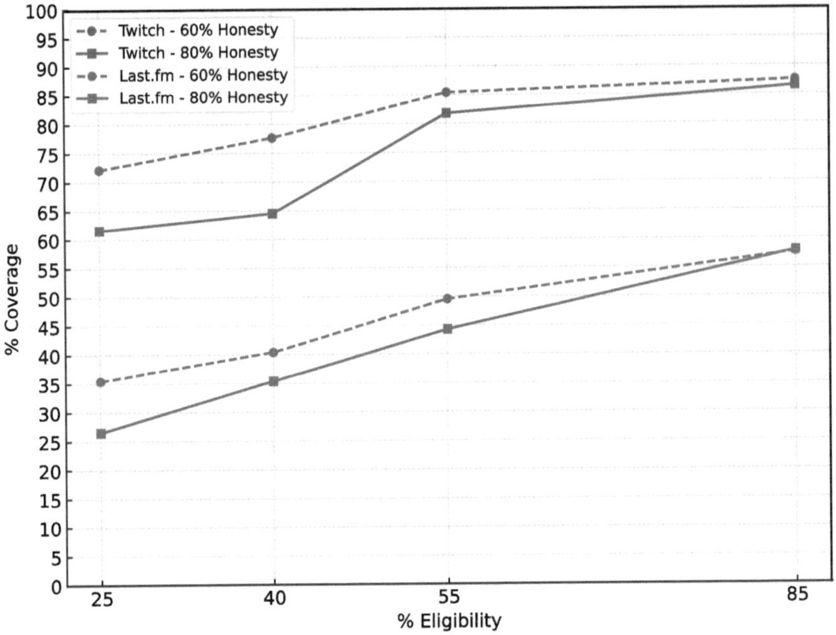

(a) Coverage for 1% root nodes

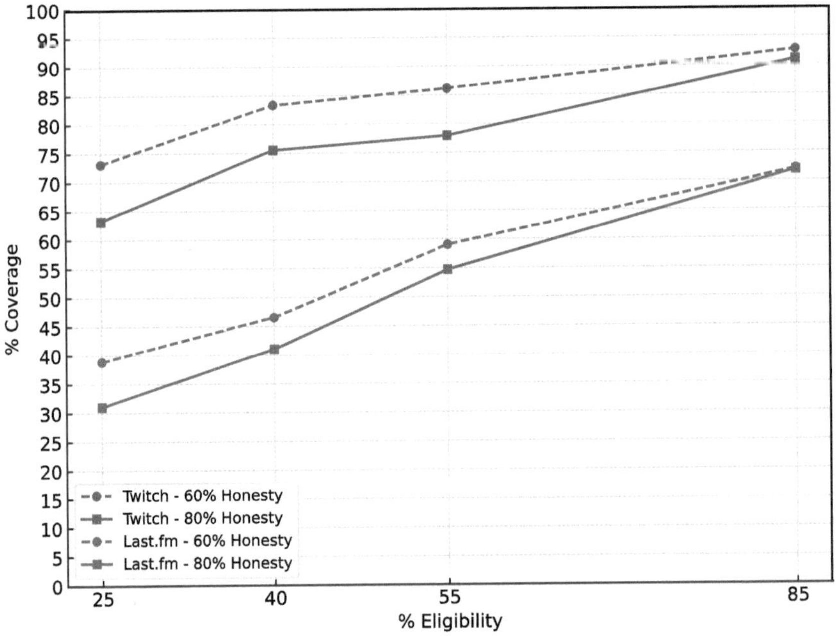

(b) Coverage for 5% root nodes

Fig. 4. Coverage % across 1% and 5% root nodes.

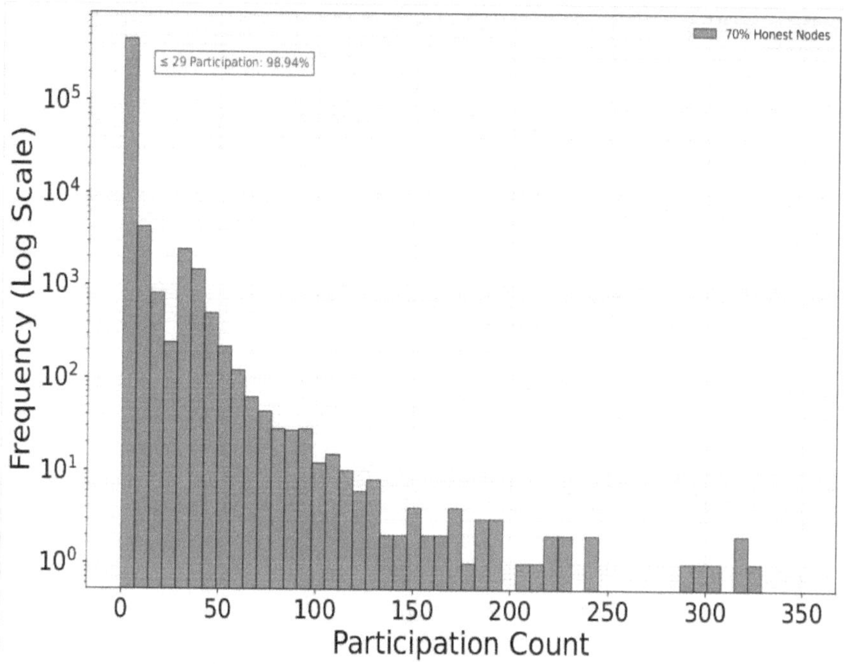

(a) 1% root nodes

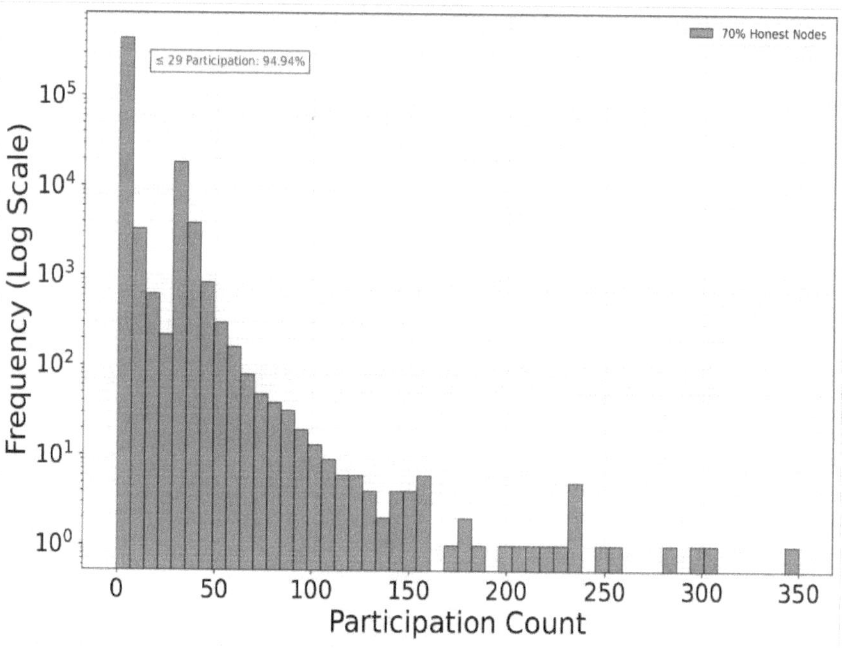

(b) 5% root nodes

Fig. 5. Last.FM Dataset: Histogram for participation count at different root node levels at 70% honest nodes.

excessive, reducing this load remains a goal for future work, and the results suggest our setup is feasible with reasonable interaction levels.

7 Discussion

Our proposed system achieved over 80% F1-score in detecting ineligible participants in the lower eligibility ratio division ($\leq 55\%$), with degrading performance in the upper eligibility ratio. This indicates that the model was able to learn and detect ineligible participants better in the lower eligibility ratio scenarios, which is reasonable as more ineligible participants are present in this lower division. On the other hand, a higher proportion of root nodes enhanced accuracy, reflecting their role as credible and trusted actors. Looking deeper, Fig. 6 shows the values for precision and recall in the 1% root node scenarios. We can observe that the recall was significantly higher than precision. This underscores the model's ability to correctly identifying actual ineligible participants, which we think is more desired in our problem that improving the precision, if we had to choose. However, we aim to improve the precision as well as we further our research in this problem.

A close look at the dissemination graphs structure of the two social datasets reveals fundamental differences that can impact the participants' behavior and consequently the prediction power of our model. We identify and plot three measures to compare the graph structures Fig. 7:

- **Clustering** [36]. Likelihood that a node's neighbors are also connected. Higher means local "cliquishness"
- **Rich club coefficient** [37]: Tendency of high-degree nodes to connect to each other. Higher means presence of elite core of interconnected hubs.
- **Community size** [38]: Size of detected structural groups (via Louvain or label propagation). Higher variation/skew means network has diverse community structure

Figure 7 shows that the Twitch community is more homophilous [39,40] than Last.fm, where higher clustering and rich club coefficients, and lower variation in community structure can be observed in the Twitch network. Therefore, we can expect dishonest behavior to be more odd, hence more observable, in Twitch than in Last.fm. This can explain the difference in prediction power between the two datasets.

For example, the users in the Twitch network tend to form tight-knit groups, like circles of close friends where many people know each other. This is shown by the higher clustering score [36] (~ 0.29) compared to Last.FM dataset. So, unless there is a wide scale collusion within the tightly connected groups, a dishonest behavior (breaking common rules) could be easier to detect. In contrast, Last.fm has very low clustering (~ 0.10). This means users are more loosely connected, more like distant acquaintances. Without strong local connections, it could be harder for a model to learn who's behaving anomalously, so dishonest participations blend in more easily, making it harder for the model to detect

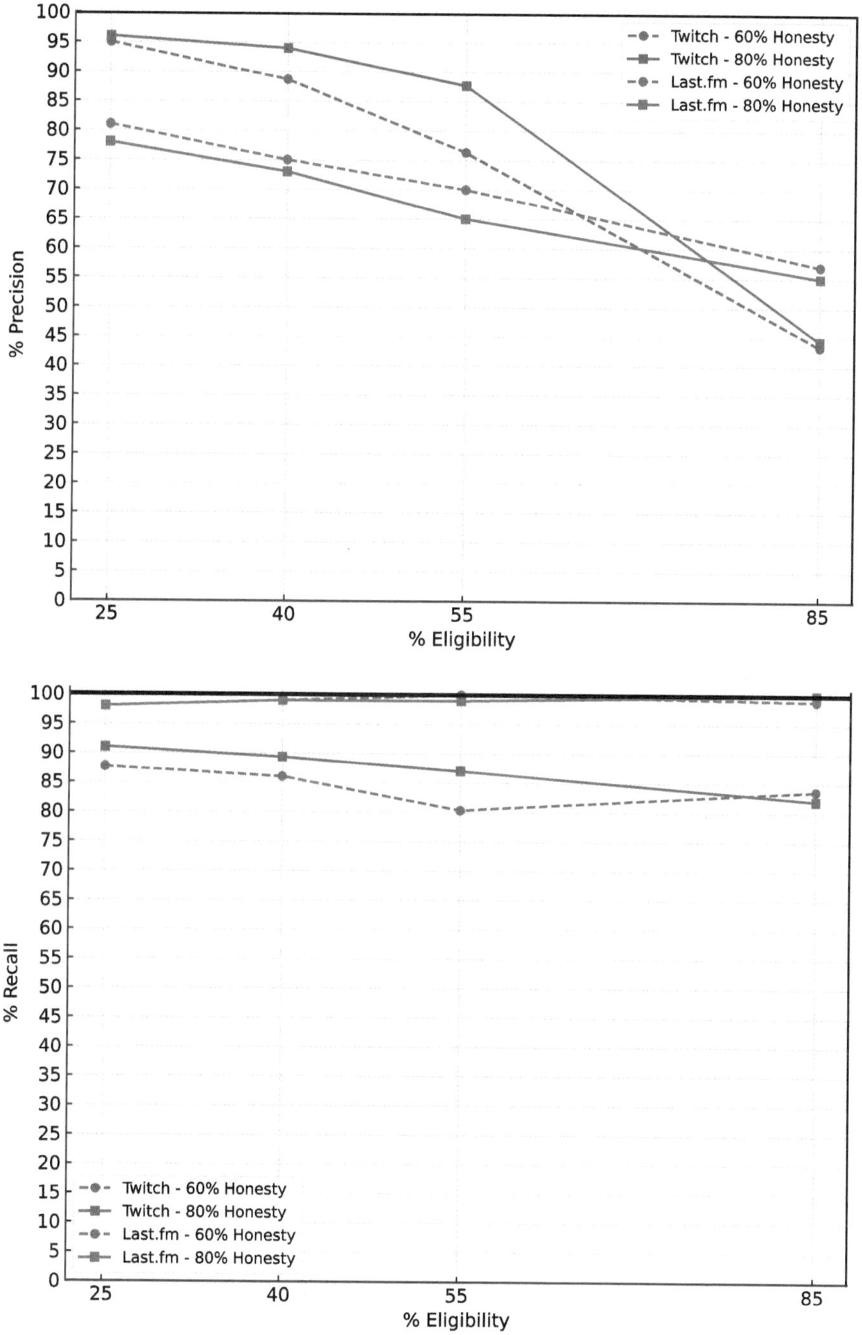

Fig. 6. Precision and recall at 1% root nodes

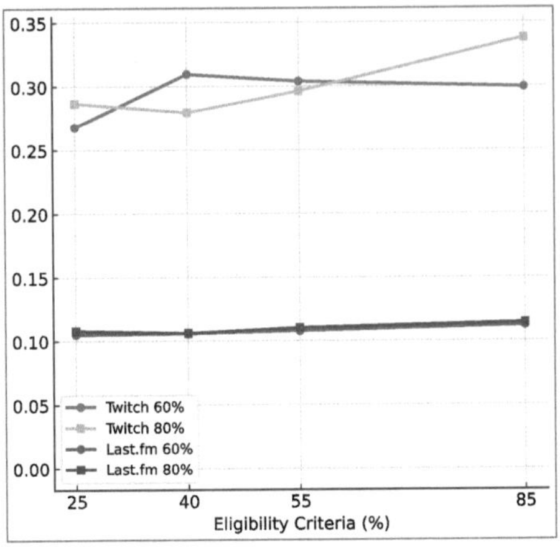

(a) Clustering

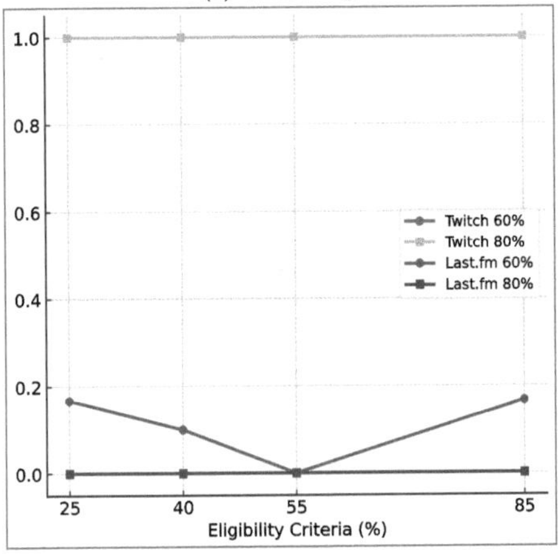

(b) Rich Club Coefficient

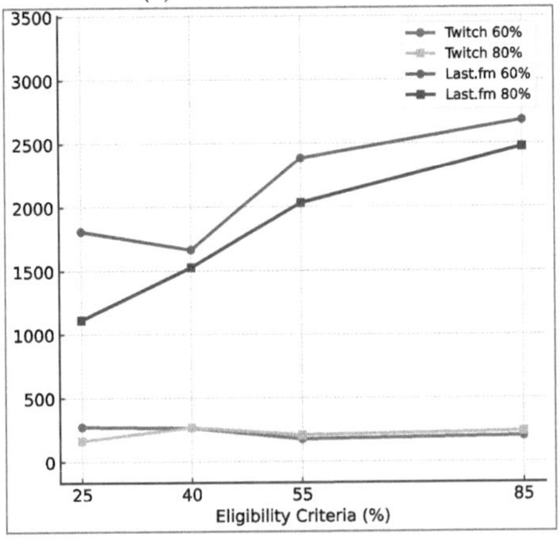

(c) Community Size

Fig. 7. Clustering, rich-club coefficient, and community size reveal complementary aspects of the network's structure

malicious behavior. Contemplating these measures motivates us to improve our prediction models to be more robust against these graph structural differences.

8 Conclusions and Future Work

In this paper, we present an empirical study on how social interactions can enhance the credibility of polls. We use the social interaction to spread poll requests, and apply AI with node embeddings to analyze the resulting dissemination graph, enabling the detection of ineligible participants. Using two real-world datasets, our models achieved promising accuracy, highlighting a new direction for improving polling results with broader participation and improved reliability. Although we approximated human behavior in our simulations, relying on synthetic social data is a limitation since it omits real social and psychological dynamics. We plan to expand this study with real life experiments to refine our findings. We also plan to develop generalizable models that is robust against different graph structures.

References

1. Salganik, M.J.: Bit by Bit: Social Research in the Digital Age. Princeton University Press, Princeton (2019)
2. Howe, J.: Crowdsourcing: Why the Power of the Crowd is Driving the Future of Business. Crown Currency, New York (2009)
3. Kohut, A., Keeter, S., Doherty, C., Dimock, M., Christian, L.: Assessing the representativeness of public opinion surveys. Technical report, Pew Research Center, Washington, DC, USA (2012)
4. Keeter, S.: Public opinion polling basics - a short course explaining how public opinion polling works. Technical report, Pew Research Center, Washington, DC, USA (2023)
5. Kennedy, C., et al.: An evaluation of the 2016 election polls in the united states. Public Opin. Q. **82**(1), 1–33 (2018)
6. Pew Research Center. Key things to know about U.S. election polling in 2024
7. Kennedy, C., Popky, D., Keeter, S.: How public polling has changed in the 21st century (2023)
8. Gibson, J.P., Krimmer, R., Teague, V., Pomares, J.: A review of e-voting: the past, present and future. Ann. Telecommun. **71**, 279–286 (2016)
9. Ohize, H.O., et al.: Blockchain for securing electronic voting systems: a survey of architectures, trends, solutions, and challenges. Clust. Comput. **28**(2), 132 (2024)
10. Nakamoto, S.: A peer-to-peer electronic cash system. Bitcoin **4**(2), 15 (2008). https://bitcoin.org/bitcoin.pdf
11. Al-madani, A.M., Gaikwad, A.T., Mahale, V., Ahmed, Z.A.: Decentralized e-voting system based on smart contract by using blockchain technology. In: 2020 International Conference on Smart Innovations in Design, Environment, Management, Planning and Computing (ICSIDEMPC), pp. 176–180 (2020)
12. Lamport, L., Shostak, R., Pease, M.: The byzantine generals problem. In: Concurrency: The Works of Leslie Lamport, pp. 203–226 (2019)

13. Arletti, A., Tanturri, M.L., Paccagnella, O.: Making online polls more accurate: statistical methods explained. Front. Polit. Sci. **7**, 1592589 (2025)
14. Cerina, R., Duch, R.: Artificially intelligent opinion polling. arXiv preprint arXiv:2309.06029 (2023)
15. Cerina, R.: Possum: a protocol for surveying social-media users with multimodal LLMs. arXiv preprint arXiv:2503.05529 (2025)
16. Boelaert, J., Coavoux, S., Ollion, E., Petev, I., Präg, P.: Machine bias. How do generative language models answer opinion polls? Sociol. Methods Research 00491241251330582 (2025)
17. Gerlach, J., Gasser, U.: Three case studies from switzerland: E-voting. Berkman Cent. Res. Publ. No **3**(2009), 2020–2021 (2009)
18. Mpekoa, N., Van Greunen, D.: E-voting experiences: a case of Namibia and Estonia. In: 2017 IST-Africa Week Conference (IST-Africa), pp. 1–8. IEEE (2017)
19. Li, P., Lai, J., Wu, Y.: Publicly traceable attribute-based anonymous authentication and its application to voting. Secur. Commun. Netw. **2021**(1), 6611518 (2021)
20. Agrawal, P., Jhanwar, M.P., Sharma, S.V., Banerjee, S.: Publicly auditable privacy-preserving electoral rolls. In: Proceedings of the 2024 IEEE 37th Computer Security Foundations Symposium (CSF), pp. 217–232 (2024)
21. Yao, J., Yang, B., Wang, T., Zhang, W.: A distributed self-tallying electronic voting system using the smart contract. Chin. J. Electron. **33**(4), 1063–1076 (2024)
22. Muth, R., Tschorsch, F.: Tornado vote: anonymous blockchain-based voting. In: 2023 IEEE International Conference on Blockchain and Cryptocurrency (ICBC), pp. 1–9. IEEE (2023)
23. Parker, C., Scott, S., Geddes, A.: Snowball sampling. In: SAGE research methods foundations (2019)
24. Peterson, J., Krug, J., Zoltu, M., Williams, A.K., Alexander, S.: Augur: a decentralized oracle and prediction market platform (v2. 0). Whitepaper (2019). https://augur.net/whitepaper.pdf
25. Ayal, S., Gino, F.: Honest rationales for dishonest behavior. American Psychological Association (2012)
26. Ashton, M.C., Lee, K., Vries, R.E.D.: The Hexaco honesty-humility, agreeableness, and emotionality factors: a review of research and theory. Pers. Soc. Psychol. Rev. **18**(2), 139–152 (2014)
27. Verschuere, B., et al.: Registered replication report on Mazar, Amir, and Ariely (2008). Adv. Methods Pract. Psychol. Sci. **1**(3), 299–317 (2018)
28. Freeman, L.C.: A set of measures of centrality based on betweenness. Sociometry 35–41 (1977)
29. Bavelas, A.: A mathematical model for group structures. Hum. Organ. **7**(3), 16–30 (1948)
30. Gino, F., Ayal, S., Ariely, D.: Contagion and differentiation in unethical behavior: the effect of one bad apple on the barrel. Psychol. Sci. **20**(3), 393–398 (2009)
31. Leskovec, J., Krevl, A.: Snap datasets: stanford large network dataset collection. Technical report (2014)
32. Gjoka, M., Butts, C.T., Kurant, M., Markopoulou, A.: Multigraph sampling of online social networks. IEEE J. Sel. Areas Commun. **29**(9) (2011)
33. Rozemberczki, B., Allen, C., Sarkar, R.: Multi-scale attributed node embedding. arXiv preprint arXiv:1909.13021 (2019)
34. Grover, A., Leskovec, J.: node2vec: scalable feature learning for networks. In: Proceedings of the 22nd ACM SIGKDD International Conference on Knowledge Discovery and Data Mining, pp. 855–864 (2016)

35. Hamilton, W., Ying, Z., Leskovec, J.: Inductive representation learning on large graphs. Adv. Neural. Inf. Process. Syst. **30** (2017)
36. Watts, D.J., Strogatz, S.H.: Collective dynamics of 'small-world' networks. Nature **393**(6684), 440–442 (1998)
37. Zhou, S., Mondragón, R.J.: The rich-club phenomenon in the internet topology. IEEE Commun. Lett. **8**(3), 180–182 (2004)
38. Blondel, V.D., Guillaume, J.-L., Lambiotte, R., Lefebvre, E.: Fast unfolding of communities in large networks. J. Stat. Mech: Theory Exp. **2008**(10), P10008 (2008)
39. Khanam, K.Z., Srivastava, G., Mago, V.: The homophily principle in social network analysis: a survey. Multimed. Tools Appl. **82**(6), 8811–8854 (2023)
40. Bisgin, H., Agarwal, N., Xu, X.: Investigating homophily in online social networks. In: 2010 IEEE/WIC/ACM International Conference on Web Intelligence and Intelligent Agent Technology, vol. 1, pp. 533–536. IEEE (2010)

Analyzing Group Polarization and Its Impact on Misinformation Dissemination: A Quantitative Approach

Qiao Zhou and Guang Yu(✉)

School of Management, Harbin Institute of Technology, Harbin, China
22B910018@stu.hit.edu.cn, yug@hit.edu.cn

Abstract. Group polarization triggered by misinformation can significantly undermine legitimate journalism and democratic processes. This study aims to quantitatively analyze the phenomenon of group polarization during the spread of misinformation and its impact on dissemination to provide theoretical support for optimizing strategies to curb misinformation dissemination on social media. The study proposes a model of attitude and behavioral propagation, investigates the polarization evolution process within virtual communities, and quantitatively studies the impact of attitudes on user participation in misinformation dissemination. The results indicate a clear trend of group polarization during the dissemination of misinformation. While the enhancement of attitude polarity does not expedite the information dissemination process, attitude strength significantly promotes dissemination speed, with this effect demonstrating notable heterogeneity across different dissemination periods, user types, and community structures. These findings provide important insights into the dynamic mechanisms of misinformation dissemination and offer guidance for developing more effective strategies for governing misinformation.

Keywords: Misinformation · Group Polarization · Attitude · Communication · Community · CSCI-RTNA

1 Introduction

Recently, the rapid progress of digitalization has positioned social media as a central channel for information dissemination. Social media has established direct pathways between content producers and consumers, significantly reducing the time and space required for information transmission (Kankham and Hou, 2024). However, innovations, while convenient, also facilitate the generation and spread of false information. The online spread of rumors—information and news lacking confirmation or factual basis—can negatively impact real-world society by hindering emergency services and threatening public safety (Wu et al., 2024). Additionally, the proliferation of online misinformation disrupts users' ability to access and evaluate information accurately, often leading to misunderstandings, biases, and extreme emotions within online communities (Chao et al., 2024a). Therefore, it is crucial to explore the motivations behind the sharing of false information.

H. R. Arabnia et al. (Eds.): CSCI 2025, CCIS 2968, pp. 518–533, 2027.
https://doi.org/10.1007/978-3-032-28457-0_36

Driven by the dual forces of irrational online interaction and cognitive differences among the public, the spread of false information is influenced by users' personal preferences and social circles (Luo and Ren, 2022). Preconceived misconceptions deepen users' inherent beliefs, which are further reinforced through communication and repetition within closed systems (Wakefield and Wakefield, 2023). This information imbalance can easily trigger the spread of false information, leading to group polarization, where collective attitudes become more extreme. This not only alters the public perception of major social issues, undermining legitimate news dissemination and democratic processes but also poses significant barriers to peace and stability in the virtual world, potentially affecting political processes in the real world (Yu and Shen, 2024). Therefore, effectively controlling the trend of group polarization during the spread of false information is a key aspect of optimizing platform management strategies (Li and Wang, 2022).

The study of the dissemination of false information on social networks and the phenomenon of group polarization has garnered significant attention within the academic community. The pervasive spread of such information has gradually eroded the boundaries of traditional news channels, making it increasingly challenging to ascertain the legitimacy of diverse news sources and fostering widespread skepticism among readers toward news information (Carlson, 2020). Furthermore, even legitimate news producers may forsake objectivity and neutrality by using sensational headlines to secure higher click-through rates (Zhang et al., 2024). Readers who mistake false information for credible content may develop inaccurate beliefs, exacerbating the trend of ideological polarization (Wakefield and Wakefield, 2023). Researchers are actively investigating group polarization in the digital age through data extracted from texts, images, videos, and other content on social media platforms. It has been confirmed that blog readers exhibit ideological polarization, and a similar trend is observed in news dissemination, where online social networks contribute to an echo chamber effect that further polarizes group opinions (Mall et al., 2024).

However, current research on group polarization in social networks focuses on local factors such as harms and influencing elements (Lu et al., 2022). Recent findings suggest that group polarization results from dynamic evolution, influenced by various factors, including social network structures, interaction patterns between individuals, and external environmental impacts (Xing et al., 2022). Meanwhile, methods such as attitude polarity and intensity scoring face sampling and self-reporting biases, with limited sample capacity and timeliness issues, making them unsuitable for big data research (Ziegler and Fiedler, 2024). Therefore, an in-depth study of group polarization phenomena in online misinformation and its impact on dynamic dissemination in the era of big data is particularly crucial.

This research aims to explore group polarization in the spread of online misinformation through quantitative analysis and assess its impact on dissemination effectiveness, thereby providing theoretical support for optimizing strategies to block the spread of misinformation on social media. Specifically, we will focus on the following research questions:

RQ1: Does group polarization exist in the process of misinformation dissemination on social media?

RQ2: How does the polarity of user attitudes during misinformation dissemination affect their sharing behavior?

2 Research Hypotheses

The research model, illustrated in Fig. 1, is based on the Theory of Reasoned Action (TRA) (Seddig et al., 2022). Like other reasoned action approaches, TRA suggests that behavior is preceded by the intention to perform it. Thus, under similar conditions, the intention to share content on social networks can predict actual behavior. Behavioral intentions are influenced by users' attitudes and subjective norms (Fishbein and Ajzen, 1977). This study evaluates subjective norms using attitude strength, source credibility, and historical post counts, while attitude polarity measures user attitudes. Considering the varying effects of source credibility, community size, and dissemination period, we construct a theoretical model of user attitudes on dissemination behavior. Dissemination effectiveness is then assessed based on speed and scope.

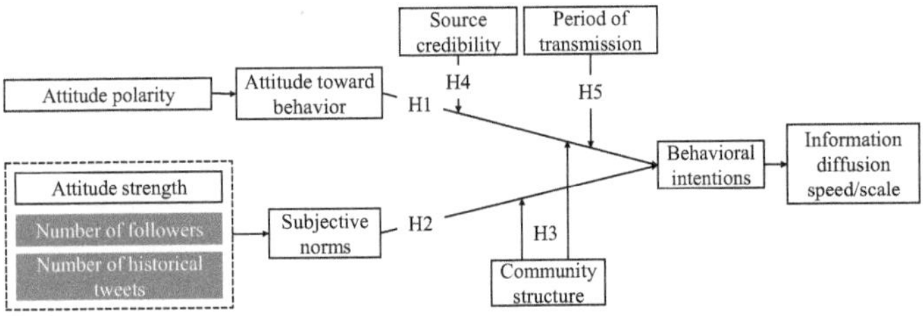

Fig. 1. Model of attitude and behavioral propagation

Recent research on social media has focused on emotions and ideology in group polarization and information dissemination. Users tend to engage more with emotionally charged content, which, whether positive or negative, has strong persuasive power and influences sharing intentions (Song et al., 2023). Negative biases can trigger stronger emotional, behavioral, and cognitive responses (Ziegler and Fiedler, 2024). Content with strong emotions and clear viewpoints significantly impacts individuals, making it more likely to capture attention and be shared (McLaughlin et al., 2020). Ideological opposition increases interaction speed and frequency among homogeneous individuals, intensifying group polarization on social media (Hong and Kim, 2016).

Attitude strength is often a good predictor of behavior, measuring an attitude's persistence and impact. Emotion-based attitudes are generally stronger than cognition-based ones, as emotionally intense information induces physiological responses, making it more likely to be forwarded and disseminated (Rocklage and Luttrell, 2021). Strong attitudes not only influence actions towards the attitude object but also affect related information processing. They persist over time and resist change (Hong and Kim, 2016). Increased attitude strength can stimulate intense emotions and consistent actions, particularly in expressing attitudes, such as stronger attitudes toward political issues predicting more self-reported political actions (Philipp-Muller, 2022). This underscores the importance of attitude strength in information dissemination. Based on these observations, we propose the following hypotheses:

H1 & H2: The enhancement of users' attitude polarity and strength positively influences their willingness to disseminate.

Community size on social media significantly affects the speed and scale of information dissemination. Topological aspects, such as community structure and temporal activity patterns, like burstiness, severely influence the spreading speed in temporal networks (Ma et al., 2024). Community size impacts member influence, susceptibility, and the information cascade process. In large communities, the variety of dissemination paths and member connections increases, enhancing the potential for widespread dissemination and triggering cascade effects (Aral and Walker, 2012). Therefore, the impact of user attitudes on dissemination behavior may vary with community size.

Moreover, users on social networks can express opinions anonymously. When deciding whether to forward a message, recipients consider the credibility of both the source and the message. Studies show that information credibility often correlates positively with source credibility. The reliability of the information source is a key criterion for individuals deciding whether to use information, and the source's identity helps gauge this reliability (Chao et al., 2024b). Platforms like Weibo use verification mechanisms to ensure user authenticity and reliability. Whether a user is verified significantly influences their credibility and is crucial for information dissemination. Recipients are more likely to accept and share arguments from credible sources, such as media accounts. High-credibility users denying rumors can reduce public confidence in those rumors, as seen with debunking from government accounts (Chao et al., 2024b).

Finally, group polarization is a process where members' attitudes change over time. Initial behavior dissemination is influenced by the homogeneity of individuals in the social network, while subsequent processes lead to behavior polarization and differentiation (Koreh and Mandelkern, 2024). González's study on protest recruitment in online social networks found that initially recruited individuals tend to connect with supporters within their networks, while later dissemination involves more interaction with individuals holding different views, triggering intra-group polarization (Centola, 2010). This study distinguishes early and later stages of dissemination networks to explore how group polarity affects user participation during different event dissemination phases, aiding in understanding the evolution of group polarization. Therefore, we further propose:

H3: Community size weakens the impact of attitudes on dissemination speed.
H4: High-credibility user identities enhance the impact of attitudes on dissemination.
H5: Rumor dissemination speed is faster in the early stages of event dissemination.

3 Method

3.1 Data Collection and Community Detection

This study examines misinformation events related to "Haitian Soy Sauce & Food Additives" on Sina Weibo from August 15, 2022, to January 31, 2023, consisting of a total of 66,784 posts. To enhance classification efficiency, we combined manual labeling (20%) with BERT training (80%). The chosen hyperparameters (400–32-2.00e-5-3) and the chinese_macbert_base model achieved a labeling accuracy of 0.815.

A retweet network was constructed using user information flows. Connectivity among users varies from tight to loose, with tight connectivity crucial for emotion contagion and opinion dissemination (Hong and Kim, 2016). Techniques like the Louvain algorithm efficiently identify optimal community partitions. This algorithm uses modularity to partition nodes into distinct communities (Qie et al., 2022). We identified 772 communities in the final retweet network.

3.2 Variable Measurement

Dependent Variable. Information sharing on social media is measured through "reposts." We use propagation speed and scope as dependent variables. Propagation speed is the time difference (in hours) between initial posting and reposting, while propagation scope is the number of reposts.

Independent Variable

Attitude polarity. Attitude, an evaluative response toward an object, influences user behavior. The ABC model delineates attitude into Affect, Behavior Tendency, and Cognition. We measure user attitudes through cognition and affect. Cognition involves understanding, perception, contemplation, and judgment, impacting event comprehension (Sixtus et al., 2023). Emotions are classified as positive, neutral, or negative, and their average defines emotional polarity. A similar approach applies to cognitive polarity (Xing et al., 2022). Attitude polarity is the sum of emotional and cognitive polarity, with community polarity as the average within a community.

Attitude Strength. The HowNet sentiment lexicon measures attitude intensity in texts (Liu et al., 2019*). HowNet considers positive and negative sentiment words, intensity adverbs, and negation words. The resulting equation shows as follows Eq.(1). where* n *represents the number of vocabulary words within the segmented sentence. The symbols* $\mathrm{pos_i}$ *and* $\mathrm{neg_i}$ *denote the emotional attributes of a vocabulary word, with both being either 0 or 1. Similarly,* $\mathrm{adj_i}$ $(\mathrm{adj_i} \geq 1)$ *and* $\mathrm{pri_i}$ $(\mathrm{pri_i} = -1 \text{ or } 1)$ *represent the corresponding intensity adverbs and negation words for each word.*

$$AS = \sum_{i}^{n} pos_i \times adj_i \times pri_i + \sum_{i}^{n} neg_i \times adj_i \times pri_i \tag{1}$$

Community Size. Community size, introduced in 1887 (Tonnies and Loomis, 2017), influences information dissemination and interaction patterns. We categorize community size using the median node count, labeling large communities as 1 and others as 0, creating a binary variable.

User Credibility. Based on message source identity, we classify credibility. Users with organizational backgrounds are deemed more credible, reflecting institutional trust among Chinese users (Chao et al., 2024b). We assign lower credibility to individual users and higher to organizations, media, and governments.

Propagation Stage. Lifecycle theory suggests that events undergo different stages, each with unique patterns. To capture rumor propagation phases, we distinguish between latent and outbreak stages.

Number of Followers and Historical Posts. Social media platforms display basic poster information, like follower count and historical posts. Follower count indicates popularity, reputation, and influence (Chao et al., 2024b), often identifying opinion leaders. Users evaluate poster professionalism based on past behavior, using it as a heuristic cue for responding to current information. User enthusiasm and activity are crucial for engagement; active users are more likely to disseminate information and obtain reposts (Bakshy et al., 2011). Thus, follower count and historical posts are control variables for information sharing.

All analyses were performed using STATA version 17. Feature scaling was applied to variables like follower counts, demonstrating power-law distributions (Robinson and Schumacker, 2009).

4 Results Analysis

4.1 Group Polarization Effects in Social Media

The study plotted daily line graphs of event dissemination volume and polarity changes, as shown in Fig. 2. The dissemination curve indicates that the selected timeframe covers almost all stages of the event's development, making it suitable for analysis. Based on Steven Fink's four-stage model of crisis communication, we divided the dissemination into incubation, outbreak, fluctuation, and resolution phases (Fink, 2013). During the outbreak phase, there is a sharp change in dissemination volume and public attitudes, which dissipates within seven days. In the fluctuation phase, public attitudes stabilize, and dissemination volume fluctuates with secondary issues. In the first two stages, public emotional responses lag behind event intensity and cognition until both gradually subside and approach zero (resolution phase).

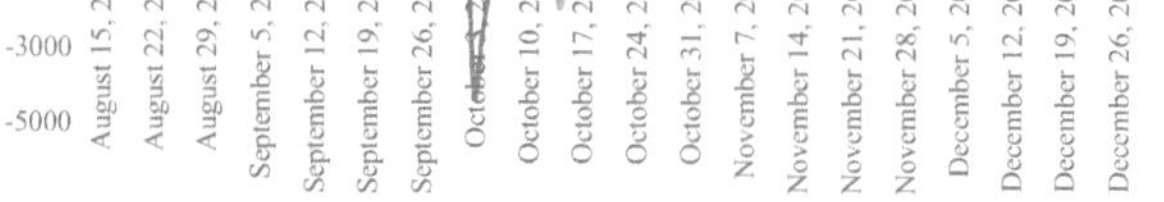

Fig. 2. The spread volume and polarization trends of the event

This study employs the Time-windowed Graph Model to investigate polarization within forwarding networks by establishing temporal nodes (Avrachenkov et al., 2024). Using Gephi, we depict forwarding networks in both early (incubation and outbreak) and later stages (fluctuation and decline), as shown in Fig. 3 a). In this visualization, blue, orange, and red correspond to attitude polarities of 0, 1, and 2, respectively. A comparative analysis reveals significant enhancement in the collective polarity of the final forwarding network compared to earlier phases. During the outbreak phase, there is a surge in nodes joining the network, leading to differentiation and intensification among communities. This is reflected in a V-shaped trajectory of modularity over time and a linear increase in community divisions (Fig. 3 b)).

During concentrated discussions on social media, sentiments, and viewpoints within groups become highly contagious. Group imagination is dynamic and tends to absorb external viewpoints with minimal contemplation, potentially leading to extremism (Fishbein, 1979). The emergence of similar voices incites irrational behavior within groups, causing fluctuations in node polarity and collective cognition that mirror information dissemination trends. As secondary opinions proliferate alongside ambiguous official announcements, group emotions surge during later stages. However, due to media agenda-setting, platform management interventions, and shifts in public attention, group emotions stabilize following this rapid ascent. The anonymity in new media communication significantly contributes to this phenomenon (Deng et al., 2021).

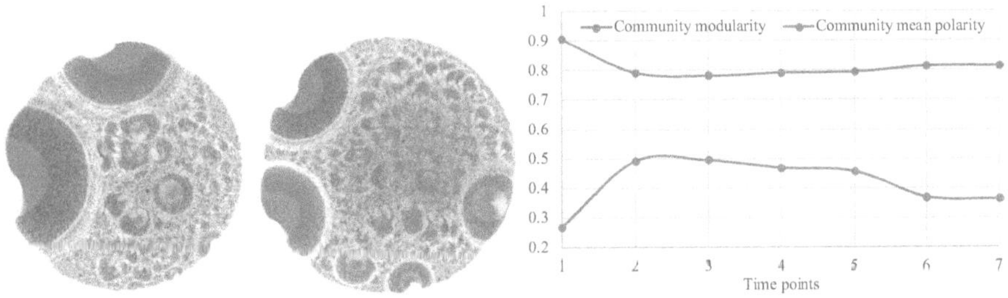

a) b)

Fig. 3. Social media retweet network, community modularity and average polarity

4.2 The Influence of Attitude on User Engagement in Propagation

We used a Negative Binomial Regression Model to explore the relationship between information propagation and user attitudes (Cameron and Trivedi, 2013). We analyzed the moderating effects of source credibility, community structure, and propagation periods. The results were divided into two stages: first, with independent variables, and second, examining interactions. Tables 1 and 2 summarize the regression results for independent variables and interaction terms. Model I analyzes the moderating effects of source credibility and community structure, while Model II examines propagation period heterogeneity. Variance Inflation Factor (VIF) (Tables 3 and 4) was calculated for all variables, showing no multicollinearity concerns (Hair, 2009).

Table 1. Regression results of dependent variable: dissemination speed

Dependent variable: speed	Model I		Model II	
	No interactive term	Interactive term	No interactive term	Interactive term
Attitude strength	6.5939***	−4.4430	10.9979***	−11.8695***
	(1.2144)	(2.8847)	(1.2817)	(2.7813)
1. Attitude polarity	100.4230***	−22.4762**	98.7463***	105.5900***
	(3.8334)	(9.1722)	(3.8285)	(4.7477)
2. Attitude polarity	14.3171*	−89.8884***	18.9187**	68.8097***
	(6.7247)	(16.0904)	(6.7102)	(9.1547)
Source credibility	−86.0314***	−92.8796***	−106.0460***	−107.9879***
	(4.3713)	(6.7224)	(4.5453)	(4.7315)
Community structure	182.6090***	95.1485***	186.8429***	182.5809***
	(3.6517)	(5.0314)	(3.6659)	(3.6856)
Number of followers	3.5515***	3.0886***	3.4327***	3.5176***
	(0.2967)	(0.2947)	(0.2953)	(0.2952)
Number of historical tweets	−8.5020***	−8.0399***	−8.3346***	−8.4677***
	(0.3659)	(0.3594)	(0.3643)	(0.3634)
1. Time			−46.6283***	−28.7502***
			(3.6202)	(4.8794)
1. Attitude polarity × 1. Community structure		212.4547***		
		(7.6852)		
2. Attitude polarity × 1. Community structure		−31.9983**		
		(15.0750)		
1. Attitude polarity × 1. Source credibility		24.0652***		
		(8.8816)		
2. Attitude polarity × 1. Source credibility		129.6282***		
		(17.1273)		
Attitude strength × 1. Community structure		26.2227***		
		(3.1751)		
1. Attitude polarity × 1. Time				−13.5457
				(7.6911)
2. Attitude polarity × 1. Time				−122.6099***
				(13.0301)
Attitude strength × 1. Time				28.3897***
				(3.0494)
_cons	453.8415***	486.9930***	485.3607***	477.8598***
	(4.7348)	(6.5162)	(5.2793)	(5.3280)
N	66784	66784	66784	66784
adj. R^2	0.1013	0.1139	0.1033	0.1057

Note: Significance levels: * p < .1, ** p < .05, *** p < .01

Table 2. Regression results of dependent variable: dissemination scale

Dependent variable: scale	Model I		Model II	
	No interactive term	Interactive term	No interactive term	Interactive term
Attitude strength	0.0588***	−0.1455***	0.0263***	0.0181*
	(0.0039)	(0.0151)	(0 .0043)	(0.0073)
1. Attitude polarity	−0.3233***	−0.5854***	−0.2704***	−0.2230***
	(0.0165)	(0.0395)	(0.0171)	(0.0219)
2. Attitude polarity	−1.1651***	−1.7662***	−1.2151***	−1.2077***
	(0.0309)	(0.1019)	(0.0304)	(0.0410)
Source credibility	−0.9139***	−0.7793***	−0.7023***	−0.7341***
	(0.0149)	(0.0358)	(0.0199)	(0.0229)
Community structure	1.5795***	1.2961***	1.5707***	1.5735***
	(0.0205)	(0.0322)	(0.0200)	(0.0206)
Number of followers	0.1066***	0.1064***	0.1070***	0.1077***
	(0.0021)	(0.0026)	(0.0021)	(0.0021)
Number of historical tweets	−0.0862***	−0.0861***	−0.0832***	−0.0823***
	(0.0040)	(0.0049)	(0.0039)	(0.0039)
1. Time			0.3146***	0.3658***
			(0.0175)	(0.0264)
1. Attitude polarity × 1. Community structure		0.7752***		
		(0.0343)		
2. Attitude polarity × 1. Community structure		−0.5172***		
		(0.0847)		
1. Attitude polarity × 1. Source credibility		−0.0831*		
		(0.0375)		
2. Attitude polarity × 1. Source credibility		0.7747***		
		(0.1185)		
Attitude strength × 1. Community structure		0.2937***		
		(0.0160)		
1. Attitude polarity × 1. Time				−0.1378***
				(0.0350)
2. Attitude polarity × 1. Time				−0.0173
				(0.0596)
Attitude strength × 1. Time				0.0051
				(0.0088)
_cons	9.1161***	9.0168***	8.8112***	8.8160***
	(0.0199)	(0.0310)	(0.0288)	(0.0293)
lnalpha	0.8267***	0.8085***	0.8223***	0.8221***
	(0.0055)	(0.0056)	(0.0056)	(0.0056)
N	66784	66784	66784	66784
adj. R^2	0.0212	0.0226	0.0215	0.0216

Note: Significance levels: * p < .1, ** p < .05, *** p < .01

Table 3. Variance inflation factor of Model I

Variable	No interactive term		Interactive term	
	VIF	1/ VIF	VIF	1/ VIF
Attitude strength	1.21	0.8232	5.93	0.1687
Attitude polarity				
1	1.32	0.7575	7.36	0.1359
2	1.10	0.9114	7.01	0.1426
Source credibility	1.15	0.8669	3.07	0.3258
Community structure	1.21	0.8254	2.43	0.4116
Number of followers	1.62	0.6177	1.64	0.6086
Number of historical tweets	1.71	0.5855	1.71	0.5834
Attitude polarity × Community structure				
1 1			4.31	0.2318
2 1			1.55	0.6447
Attitude polarity × Source credibility				
1 1			5.63	0.1775
2 1			6.67	0.1499
Attitude strength × 1. Community structure			5.98	0.1673
Mean VIF	1.33		4.44	

Table 4. Variance inflation factor of Model II

Variable	No interactive term		Interactive term	
	VIF	1/ VIF	VIF	1/ VIF
Attitude strength	1.29	0.7722	4.88	0.2050
Attitude polarity				
1	1.32	0.7565	2.01	0.4974
2	1.10	0.9086	1.87	0.5346
Source credibility	1.32	0.7549	1.82	0.5508
Community structure	1.22	0.8182	1.22	0.8201
Number of followers	1.62	0.6169	1.82	0.5499
Number of historical tweets	1.71	0.5848	1.85	0.5403
1. Time	1.24	0.8093	2.31	0.4322
Attitude polarity × Time				
1 1			2.84	0.3525
2 1			1.96	0.5111
Attitude strength × 1. Time			4.95	0.2020
Mean VIF	1.35		2.50	

The results indicate that, without interaction terms, Model I shows that reinforcing attitude polarity does not speed up information propagation (Attitude polarity 1: $\beta = 100.423$, $p < 0.01$; Attitude polarity 2: $\beta = 14.317$, $p < 0.1$). It slightly reduces the propagation scale (Attitude polarity 1: $\beta = -0.323$, $p < 0.01$; Attitude polarity 2: $\beta = -1.165$, $p < 0.01$). However, attitude strength significantly and positively influences both propagation speed and scale ($\beta = 6.594$, $p < 0.01$, and $\beta = 0.059$, $p < 0.01$). Therefore, H1 is not supported, while H2 is confirmed.

In small-scale communities, attitude polarity negatively affects propagation speed (β = −22.476, p = 0.014). In large communities, when attitude polarity is 2, it positively moderates propagation speed compared to smaller communities (β = −31.998, p = 0.034). When attitude polarity is 1, large communities positively influence the propagation scale compared to small ones (β = 0.775, p < 0.001). Additionally, attitude strength positively impacts the propagation scale in large communities (β = 0.294, p < 0.001). In analyzing source credibility, the effect of attitude polarity on propagation speed is significantly reduced for texts from celebrity, media, and government accounts compared to ordinary users (Attitude polarity 1: β = 24.065, p < 0.001; Attitude polarity 2: β = 129.628, p < 0.001). The influence on the propagation scale is positive when attitude polarity is 2 (β = 0.775, p < 0.001) and negative when polarity is 1 (β = −0.083, p = 0.027). Thus, H3 and H4 is partially supported.

Finally, early-stage moderation analysis shows that higher attitude polarity's positive impact on propagation speed is more pronounced in early stages than in later stages (β = −122.670, p < 0.001). Thus, H5 is supported.

To ensure reliability, we used hierarchical regression and variable substitution methods. Initially, regression results demonstrated consistent signs and magnitude before introducing interaction terms. Upon inclusion, the analysis revealed heterogeneity in categorical variables. Additionally, substituting interaction volume for engagement volume as the dependent variable confirmed the robustness of our findings. Detailed results of these checks are presented in Table 5.

Table 5. Results of robustness checks.

Variable	Information Diffusion: speed			Information diffusion: retweets
	Master variable	Add regulatory variable	Add interactive item	
Attitude strength	22.2992***	6.5939***	−4.4430	0.0000
	(1.1530)	(1.2144)	(2.8847)	(0.0000)
1. Attitude polarity	186.2422***	100.4230***	−22.4762*	0.0031
	(3.7482)	(3.8334)	(9.1722)	(0.0662)
2. Attitude polarity	2.3634	14.3171*	−89.8884***	−0.9519***
	(6.5882)	(6.7247)	(16.0904)	(0.1764)
Source credibility		−86.0314***	−92.8796***	−0.8232***
		(4.3713)	(6.7224)	(0.0648)
Community structure		182.6090***	95.1485***	1.2458***
		(3.6517)	(5.0314)	(0.0287)

(continued)

Table 5. (*continued*)

Variable	Information Diffusion: speed			Information diffusion: retweets
	Master variable	Add regulatory variable	Add interactive item	
Number of followers		3.5515***	3.0886***	0.0742***
		(0.2967)	(0.2947)	(0.0025)
Number of historical tweets		−8.5020***	−8.0399***	−0.0827***
		(0.3659)	(0.3594)	(0.0044)
1. Attitude polarity × 1. Community structure			212.4547***	1.6954***
			(7.6852)	(0.0383)
2. Attitude polarity × 1. Community structure			−31.9983*	−0.3714**
			(15.0750)	(0.1367)
1. Attitude polarity × 1. Source credibility			24.0652**	−0.1163
			(8.8816)	(0.0651)
2. Attitude polarity × 1. Source credibility			129.6282***	0.3678
			(17.1273)	(0.1934)
Attitude strength × 1. Community structure			26.2227***	0.0005***
			(3.1751)	(0.0000)
Constant	443.6196***	453.8415***	486.9930***	6.6594***
	(2.3182)	(4.7348)	(6.5162)	(0.0613)
N	66784	66784	66784	66784
adj. R^2	0.0402	0.1012	0.1137	

5 Discussion

The findings indicate that attitude strength, rather than polarity, significantly enhances the speed of information dissemination. During the initial stages, attitude polarity positively influences transmission speed; more extreme public attitudes lead to faster dissemination. Early in an event, public emotions and negative sentiments surge, then stabilize due to official statements and shifts in rhetoric from opinion leaders. Early engagement from groups like professional self-media outlets and loyal enterprise users further accelerates dissemination, creating asynchronous development in speed and scale. Additionally, Chinese social media platforms have strengthened policies against online rumors, categorizing false information with a certain number of forwards as illegal. The public's cautious approach to unknown events has markedly improved. Increased psychological pressure from policies compels individuals to protect themselves by curtailing potentially illegal behaviors, reflecting the effectiveness of current governance measures against social media rumors.

Furthermore, attitude polarity has varied effects on dissemination across community structures. Moderate incentives facilitate information flow in large communities; however, strong attitude polarity can inhibit overall scale and reduce speed—a trend seen in all community types. This may be due to risk aversion: higher polarity correlates with stronger emotions and confrontational stances, leading individuals to subconsciously avoid participation, though some may react swiftly under emotional influence.

In smaller communities, attitude strength positively impacts speed, while in larger ones, it enhances scale. Therefore, dynamic monitoring of community size during rumor propagation is essential. Platforms should use big data analytics to assess group polarization and community sizes on specific topics, adjusting strategies for targeted notifications to mitigate overemphasis on negative emotions or confrontational positions, thus reducing secondary public opinion backlash or online group violence.

The moderating effect of source credibility shows that high attitude polarity significantly increases the dissemination scale for accounts with organizational backgrounds, while ordinary user accounts benefit from both attitude polarity and strength-enhancing transmission speeds. In accounts with high credibility, primarily attitude strength, positively affects dispersion scales. Given these characteristics, media entities can boost their credibility on social platforms by promptly disclosing information sources and verification statuses when debunking rumors.

Finally, in the early stages of false information emergence, there is a notable lag in responses from government, media, or celebrity accounts compared to ordinary users, whose forwarding rates exceed those of highly credible sources (see Fig. 4). This discrepancy arises from differences in timeliness and efficacy of rumor-debunking efforts. Policymakers should prioritize addressing online public sentiment and regulate unverified claims' popularity, monitoring group polarization around various topics to mitigate real-world harm caused by online public opinion.

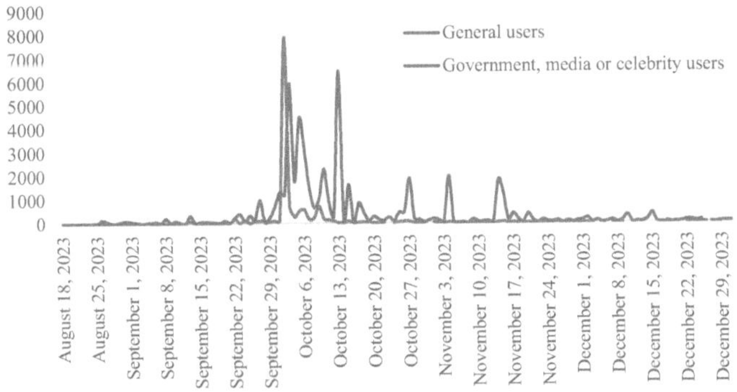

Fig. 4. Dissemination volume on Sina Weibo

5.1 Theoretical and Practical Implications

The theoretical significance of this study is twofold. First, research on group polarization often relies on classical theories and rarely explores its evolution using real social media data. Our dataset covers misinformation events related to public health and safety in 2022, expanding quantitative studies on group polarization with disclosed social media data. Second, unlike previous static studies, this research uses a time-window graph model to dynamically analyze the evolution of group polarization from the perspectives of communities and users. These findings describe group attitudes and polarization patterns in social media networks, enhancing the literature on misinformation dissemination.

From a practical perspective, our analysis reveals that increased attitude strength significantly accelerates dissemination speed. Moreover, attitude polarity influences information dissemination across different community structures, periods, and account types. By leveraging these characteristics and strategically managing the timing of outbreaks and misinformation verification, it is possible to effectively mitigate polarization trends, reducing their scale. This has significant implications for future misinformation governance and debunking efforts. Additionally, the findings reflect the effectiveness of current social media platform measures for managing misinformation.

5.2 Limitations and Future Directions

Our study has limitations. First, we only investigated group polarization on Sina Weibo. Polarization patterns may vary across different countries, platforms, and social events. Second, we observed distinct trends in the speed and scale of misinformation dissemination. However, the relationship between them is influenced by factors such as the nature of the information, dissemination channels, and audience characteristics. Whether this is common in similar communication processes needs further exploration. Lastly, to broadly describe group polarization, we did not differentiate the direction of stance and emotion. Future research could address this with more granularity. Extending the analysis to different countries, platforms, and event types is a future direction. Additionally, refining the division of time window nodes to investigate substantive changes in public opinion and secondary sentiment accompanying misinformation spread will be pursued.

Funding. This work was supported by the National Natural Science Foundation of China, Grant No. 72374056.

References

Aral, S., Walker, D.: Identifying influential and susceptible members of social networks. Science. **337**(6092), 337–341 (2012). https://doi.org/10.1126/science.1215842

Avrachenkov, K., Dreveton, M., Leskela, L.: Recovering static and time-varying communities using persistent edges. IEEE Trans Netw Sci Eng. **11**(2), 2087–2099 (2024). https://doi.org/10.1109/TNSE.2023.3337281

Bakshy, E., Hofman, J.M., Mason, W.A., Watts, D.J.: Everyone's an influencer: quantifying influence on twitter. In: Proceedings of the 4th International Conference on Web Search and Data Mining, pp. 65–74 (2011). https://doi.org/10.1145/1935826.1935845

Cameron, A.C., Trivedi, P.K.: Regression Analysis of Count Data. Cambridge University Press (2013)

Carlson, M.: Fake news as an informational moral panic: the symbolic deviancy of social media during the 2016 US presidential election. Inf. Commun. Soc. **23**(3), 374–388 (2020). https://doi.org/10.1080/1369118X.2018.1505934

Centola, D.: The spread of behavior in an online social network experiment. Science. **329**(5996), 1194–1197 (2010). https://doi.org/10.1126/science.1185231

Chao, F., Wang, X., Yu, G.: Determinants of debunking information sharing behavior in social media users: perspective of persuasive cues. Internet Res. **34**(5), 1545–1576 (2024a). https://doi.org/10.1108/INTR-07-2022-0497

Chao, F., Zhou, Q., Zhao, J., et al.: Trustworthiness matters: effect of source credibility on sharing debunking information across different rumor types. Inf. Process. Manag. **61**(4), 103747 (2024b). https://doi.org/10.1016/j.ipm.2023.103747

Deng, L.F., Sun, W.J., Xu, D.P., Ye, Q.: Impact of anonymity on consumers' online reviews. Psychol. Mark. **38**(12), 2259–2270 (2021). https://doi.org/10.1002/mar.21565

Fink, S.: Crisis Communications: the Definitive Guide to Managing the Message. McGraw Hill (2013)

Fishbein, M.: A theory of reasoned action: some applications and implications. In: Nebraska Symposium on Motivation, vol. 27, pp. 65–116 (1979)

Fishbein, M., Ajzen, I.: Belief, attitude, intention, and behavior: an introduction to theory and research. Philos. Rhetor. **10**(2), 130–132 (1977) http://www.jstor.org/stable/40237022

Gentzkow, M., Shapiro, J.M.: Ideological segregation online and offline. Q. J. Econ. **126**(4), 1799–1839 (2011). https://doi.org/10.1093/qje/qjr044

Hair, J.F.: Exploratory factor analysis. In: Multivariate data analysis (2009)

Hong, S., Kim, S.H.: Political polarization on twitter: implications for the use of social media in digital governments. Gov. Inf. Q. **33**(4), 777–782 (2016). https://doi.org/10.1016/j.giq.2016.04.007

Kankham, S., Hou, J.R.: Community notes vs. related articles: assessing real-world integrated counter-rumor features in response to different rumor types on social media. Int. J. Hum.–Comput. Interact., 1–15 (2024). https://doi.org/10.1080/10447318.2024.2400389

Koreh, M., Mandelkern, R.: Trade-offs of government credibility institutions: market credibility vs. social credibility. J. Eur. Public Policy. **31**(11), 3539–3560 (2024). https://doi.org/10.1080/13501763.2023.2243295

Liu, Y.Q., Duan, Z.K., Iop.: Chinese movie comment sentiment analysis based on HowNet and user likes. J. Phys. Conf. Ser. **1229**(1), 012018 (2019). https://doi.org/10.1088/1742-6596/1229/1/012018

Li, Y., Wang, J.: Cross-network propagation model of public opinion information and its control in coupled double-layer online social networks. Aslib J. Inf. Manag. **74**(2), 354–376 (2022). https://doi.org/10.1108/AJIM-04-2021-0126

Lu, Y.D., Wu, J.J., Tan, Y., Chen, J.: Microblogging replies and opinion polarization: a natural experiment. MIS Q. **46**(4), 1901–1936 (2022). https://doi.org/10.25300/MISQ/2022/15455

Luo, H., Ren, Y.: Analysis of COVID-19 collective irrationalities based on epidemic psychology. Front. Psychol. **13**(4), 825452 (2022). https://doi.org/10.3389/fpsyg.2022.825452

Ma, J.L., Wang, P., Lu, Z.J.: Epidemic spreading on community networks with heterogeneity considering higher-order interactions. Int. J. Mod. Phys. C. (2024). https://doi.org/10.1142/S0129183124501870

Mall, R., Nagpal, M., Salminen, J., et al.: Politics on YouTube: detecting online group polarization based on news videos' comments. SAGE Open. **14**(2) (2024). https://doi.org/10.1177/21582440241256438

McLaughlin, B., Holland, D., Thompson, B.A., Koenig, A.: Emotions and affective polarization: how enthusiasm and anxiety about presidential candidates affect interparty attitudes. Am. Polit. Res. **48**(2), 308–316 (2020). https://doi.org/10.1177/1532673X19891423

Philipp-Muller, A.: Attitude embeddedness and attitude consistency: Major factors in conceptualizing attitude properties, Doctoral dissertation, The Ohio State University (2022)

Qie, H., Li, S.J., Dou, Y., et al.: Isolate sets partition benefits community detection of parallel Louvain method. Sci. Rep. **12**(1) (2022). https://doi.org/10.1038/s41598-022-11987-y

Robinson, C., Schumacker, R.E.: Interaction effects: Centering, variance inflation factor, and interpretation issues. Mult. Linear Regress. Viewpoints. **35**(1), 6–11 (2009)

Rocklage, M.D., Luttrell, A.: Attitudes based on feelings: fixed or fleeting? Psychol. Sci. **32**(3), 364–380 (2021). https://doi.org/10.1177/0956797620965532

Seddig, D., Maskileyson, D., Davidov, E., et al.: Correlates of COVID-19 vaccination intentions: attitudes, institutional trust, fear, conspiracy beliefs, and vaccine skepticism. Soc. Sci. Med. **302** (2022). https://doi.org/10.1016/j.socscimed.2022.114981

Sixtus, E., Krause, F., Lindemann, O., Fischer, M.H.: A sensorimotor perspective on numerical cognition. Trends Cogn. Sci. **27**(4), 367–378 (2023). https://doi.org/10.1016/j.tics.2023.01.002

Song, H., So, J., Shim, M., et al.: What message features influence the intention to share misinformation about COVID-19 on social media? The role of efficacy and novelty. Comput. Hum. Behav. **138**(9) (2023). https://doi.org/10.1016/j.chb.2022.107439

Tonnies, F., Loomis, C.P.: Community and Society. Routledge (2017)

Wakefield, R.L., Wakefield, K.: The antecedents and consequences of intergroup affective polarization on social media. Inf. Syst. J. **33**(3), 640–668 (2023). https://doi.org/10.1111/isj.12419

Wu, Y., Wang, D., Ma, F.: A study on the competitive dissemination of disinformation and knowledge on social media. Aslib J. Inf. Manag. **76**(1), 153–169 (2024). https://doi.org/10.1108/AJIM-03-2022-0163

Xing, Y., Wang, X., Qiu, C., et al.: Research on opinion polarization by big data analytics capabilities in online social networks. Technol. Soc. **68** (2022). https://doi.org/10.1016/j.techsoc.2022.101902

Yu, W.T., Shen, F.: Mapping verification behaviors in the post-truth era: a systematic review. New Media Soc. **26**(3), 1703–1727 (2024). https://doi.org/10.1177/14614448231191138

Zhang, M., Wu, H.T., Huang, Y., et al.: Negative news headlines are more attractive: negativity bias in online news reading and sharing. Curr. Psychol. **43**, 30156–30169 (2024). https://doi.org/10.1007/s12144-024-06646-6

Ziegler, J., Fiedler, K.: Small sample size and group homogeneity: a crucial ingredient to intergroup bias. Personal. Soc. Psychol. Bull. (2024). https://doi.org/10.1177/01461672231223335

Customer Churn Prediction Based on Sentiment Score

Shadha H. Al-Safi(✉), Ahmed Rafea, and Mona Farouk

Computer Science and Engineering Department, American University in Cairo, New Cairo, Egypt

{shadahamed,rafea}@aucegypt.edu, Mona_farouk@eng.cu.edu.eg

Abstract. Customer churn prediction is a critical task in the telecommunications industry, where retaining existing customers is far more cost-effective than acquiring new ones. This study introduces a predictive framework that integrates sentiment analysis with demographic and service-related data to enhance churn prediction accuracy. Sentiment scores are derived using a Random Forest classifier at different granularity levels: binary (positive/negative), three-class (positive/neutral/negative), and five-class (strongly positive to strongly negative). These sentiment features are then combined with traditional customer attributes and processed through an ensemble stacking model that leverages K-Nearest Neighbor (KNN), Support Vector Machine (SVM), and Random Forest (RF) as base learners, with a Multilayer Perceptron (MLP) as the meta learner. The proposed model demonstrates that three-class sentiment integration provides the strongest improvement, achieving 98.86% accuracy, 99.47% AUC, and 97.77% F1-score on the IBM Telco dataset [1]. Results also highlight the benefits of class weighting in addressing imbalance. Beyond accuracy, the framework emphasizes interpretability, computational efficiency, and practical applicability, offering insights for data-driven customer retention strategies. The findings confirm that sentiment-aware ensemble learning can significantly enhance predictive performance, while limitations related to dataset scope and sentiment modeling suggest opportunities for future work using real-world textual feedback and cross-dataset validation.

Keywords: Churn prediction. Sentiment classification. Ensemble model. Class weighted component

1 Introduction

Customer churn, the decision of a customer to discontinue a service, is a major challenge in the telecommunications industry. Retaining existing customers is significantly more cost-effective than acquiring new ones, and therefore the ability to predict churn before it occurs is of high value to service providers. Companies increasingly rely on advanced machine learning techniques to model churn behavior and develop proactive retention strategies.

Most traditional churn prediction approaches focus on demographic, billing, and service usage data. While these features provide useful insights, they often fail to capture

H. R. Arabnia et al. (Eds.): CSCI 2025, CCIS 2968, pp. 534–542, 2027.
https://doi.org/10.1007/978-3-032-28457-0_37

the customer's subjective experience and emotional state, which are critical in understanding churn behavior. Sentiment analysis, which quantifies customer attitudes and opinions, has emerged as a valuable tool to capture these behavioral signals. Integrating sentiment features into churn prediction remains relatively underexplored, especially in ensemble learning contexts.

In this study, we propose a churn prediction framework that augments conventional customer attributes with sentiment scores. The sentiment features are derived using a Random Forest classifier at multiple granularity levels (binary, three-class, and five-class). These are then integrated into a stacking ensemble consisting of KNN, SVM, and RF as base classifiers, with MLP as the meta learner.This design allows the model not only to achieve high predictive accuracy but also to address issues of interpretability and computational efficiency, both of which are critical for real-world deployment.

The contributions of this work are threefold:

- Introducing sentiment analysis as an additional predictive dimension in churn modeling through ensemble learning.
- Evaluating the comparative performance of different sentiment granularities (2-class, 3-class, 5-class) in churn prediction.
- Demonstrating that the integration of sentiment features significantly improves predictive performance, particularly with three-class sentiment scoring.

Finally, this work acknowledges the importance of extending churn models to account for temporal dynamics, since customer dissatisfaction often develops gradually. While the present study uses a static dataset, future work will incorporate time-series features to better capture churn progression.

The paper is structured as follows: Section 2 reviews the literature, Sect. 3 discusses the methodology, Sect. 4 presents experiments and results, and Sect. 5 concludes the study.

2 Literature Review

In customer churn prediction (CCP), machine learning (ML) approaches are widely adopted to identify at-risk customers [2]. Studies often focus on refining prediction models and comparing techniques. For example, Edwine et al. [3] compared advanced ML methods like Random Forest (RF), Support Vector Machines (SVM), and K-Nearest Neighbors (KNN), highlighting the importance of hyperparameter optimization and under sampling for imbalanced data. Beyond traditional ML, deep neural networks have been explored: Chouiekh and El Haj [4] demonstrated that deep convolutional neural networks (DCNNs) outperform traditional ML algorithms such as SVM and RF. Similarly, Usman et al. [5] found that CNNs and RNNs surpass popular classifiers like decision trees and Naïve Bayes. In contrast, Siddika et al. [6] reported that RF outperforms CNN and MLP, whereas Dalli [7] showed that hyperparameters such as activation functions and batch size critically affect neural network performance.

Recent trends favor **ensemble and hybrid methods** over single classifiers. Tavassoli and Koosha [8] employed boosted and bagged ensembles, while Shabankareh et al. [9] proposed a stacking model combining SVM with CHAID decision trees to analyze churn causes. Usman-Hamza et al. [10] developed ensemble decision forest models (LMT, RF, FT) that outperformed base classifiers. Xu et al. [11] introduced a two-level stacking model integrating XGBoost and Logistic Regression, enhancing performance through feature grouping and engineering. Khoh et al. [12] similarly demonstrated that stacking ensembles can outperform single learners in telecom churn, validating ensemble approaches as a dominant trend.

Sentiment **analysis** has also been increasingly linked to churn prediction. Gulati and Raheja [14] found a strong correlation between sentiment growth rates and churn rates using Pearson correlation. Lappeman et al. [13] analyzed tweets to identify churn-related terms in negative sentiment posts, while Kilimci [14] linked customer loyalty to sentiment using contextualized word embeddings. Galitsky et al. [15] applied binomial logistic regression to identify key churn factors, including complaints and service failures. Almuqren et al. [16] incorporated satisfaction scores from tweets into churn prediction, showing that combining sentiment with historical data improves performance. Gulati and Raheja [17] found a strong correlation between sentiment growth rates and churn rates using Pearson correlation

Despite advances, challenges remain. Most existing studies emphasize accuracy while neglecting interpretability and computational efficiency. Ribeiro et al. [18] (LIME) and Lundberg and Lee [19] (SHAP) have shown that interpretable AI techniques can explain model predictions, but these methods remain underutilized in churn prediction. Moreover, many studies rely on a single dataset, raising concerns about generalizability [20]. This motivates our approach, which integrates sentiment signals at multiple granularities (binary, three-class, five-class) into a stacking ensemble, and explicitly considers interpretability, efficiency, and practical deployment

3 Methodology

3.1 Dataset

This study uses the IBM Telco Customer Churn dataset, which contains 7043 records with 33 features covering demographic information, account details, service usage, and churn status. The dataset provides a realistic benchmark for churn prediction tasks, widely adopted in prior literature.

3.2 Sentiment Score Derivation

To capture customer attitude and satisfaction, we derive sentiment scores from survey-related features such as satisfaction ratings, complaints, and service quality indicators. Sentiment labels were generated using a Random Forest classifier trained on these attributes, with outputs categorized into three levels of granularity: (1) binary (positive vs. negative), (2) three-class (positive, neutral, negative), and (3) five-class (strongly positive → strongly negative). The Random Forest was selected due to its robustness to feature noise and interpretability. Features were normalized, and categorical variables were one-hot encoded prior to sentiment classification.

3.3 Feature Selection

We apply **Mutual Information (MI)** to rank features by relevance to the churn label. Features with scores above the dataset mean are retained for model training. The MI between feature X and target Y is computed as:

$$MI(X;Y) = sum_{x\ in\ X} sum_{y\ in\ Y} p(x,y) \log\left(\frac{p(x,y)}{p(x)p(y)}\right) \quad (1)$$

This ensures that both sentiment-based and demographic/service-based features contribute effectively to the ensemble model while reducing dimensionality and overfitting risk.

3.4 Base Learners

The ensemble employs three base learners:

- **KNN**: k = 5, Euclidean distance metric, with probability estimates derived from class frequencies among neighbors.
- **SVM**: radial basis function (RBF) kernel, probability calibration enabled using Platt scaling.
- **Random Forest**: 200 estimators, maximum depth = 12, bootstrap sampling enabled.

These settings were optimized using grid search with five-fold cross-validation to balance accuracy and computational efficiency.

3.5 Meta Learner

The predictions from base learners are used as inputs to a Multilayer Perceptron (MLP) meta learner. The MLP architecture includes an input layer of three neurons (corresponding to the three base learner outputs), one hidden layer with 20 neurons using ReLU activation, and a sigmoid output layer for binary churn classification. The model was trained using Adam optimizer, learning rate = 0.001, batch size = 32, and early stopping after 10 epochs without improvement.

3.6 Class Imbalance Handling

The churn dataset exhibits imbalance, with churned customers forming a minority. To address this, class-weighted loss was applied in both base learners and the meta learner, ensuring that the minority class received higher weight during training. This approach was preferred over oversampling to reduce risk of overfitting.

3.7 Evaluation Metrics

We report performance using: Accuracy, Precision, Recall, F1-score, and AUC-ROC. In addition to these, training time was recorded to assess computational efficiency, and Random Forest feature importances were analyzed for interpretability. These metrics provide a broader view of model applicability beyond accuracy alone.

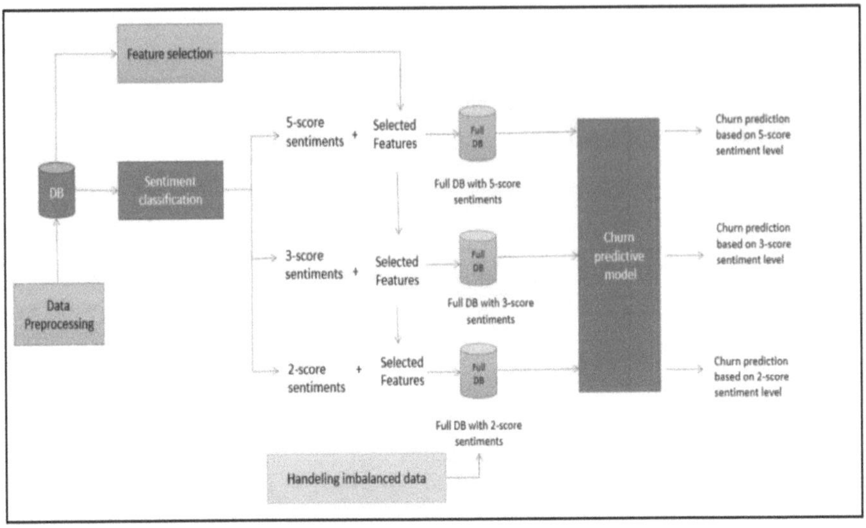

Fig. 1. Ensemble model Architecture

4 Experiments and Results

Python packages such as Scikit-learn [28], Matplotlib [29], and Seaborn [30] were utilized for data preprocessing, model construction, and visualization. Figure 1 shows the architecture of the ensemble model

4.1 Experiment 1: Building the Baseline System

The baseline system was developed in several phases:

Feature Selection: Mutual Information (MI) was used to identify features for churn prediction. The average MI score across all features was 0.015, and features exceeding this threshold were selected. A total of 17 features of various types (binary, categorical, and numeric) were retained.
Hyperparameter Tuning: Grid search with five-fold cross-validation [31, 32, 33] was employed to identify optimal hyperparameters, focusing on maximizing F1-score. This ensured robustness and avoided overfitting.
Model Training: Four classifiers—KNN, SVM, RF, and MLP—were trained on the preprocessed dataset using a 70/30 train-test split. An ensemble was then built using KNN, SVM, and RF as base learners and MLP as a meta-learner. The base learners were trained with five-fold cross-validation, and their probability outputs were aggregated by the meta-learner. Figure 2 illustrates the ensemble architecture performance distribution
Performance Evaluation: The ensemble model achieved the highest F1-score (59.37%), outperforming individual classifiers (RF: 58.14%, MLP: 53.66%, SVM: 44.56%, KNN: 46.48%). As F1-score balances precision and recall, it was chosen as the primary metric. Consequently, the ensemble was selected as the baseline system for further experiments

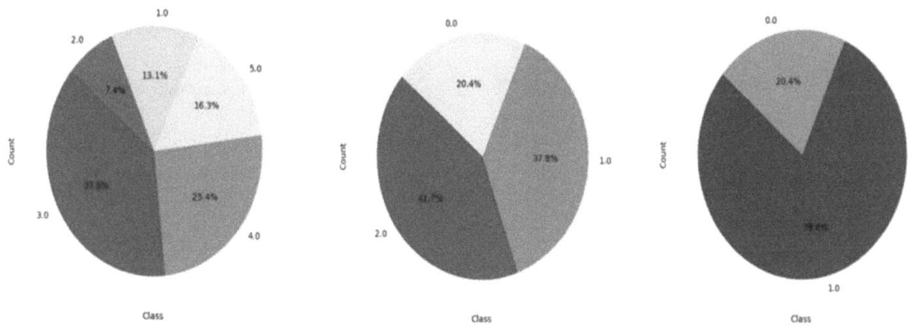

(a) 5-score sentiment (b) 3-score sentiment (c) 2-score sentiment

Fig. 2. Sentiment distributionon

4.2 Experiment 2: Building a Churn Prediction System Using Sentiment

Sentiment classification was performed using Random Forest into three levels:

5-score sentiment: Strongly positive, positive, neutral, negative, strongly negative.
3-score sentiment: Positive, neutral, negative.
2-score sentiment: Positive, negative.

The distributions are shown in Fig. 2, and classification performance is reported in Table 1.

Table 1. Sentiment classification Results

Sentiment level	Accuracy (%)	AUC (%)	Precision (%)	Recall (%)	F1-score (%)
5-score sentiment	43.95	79.51	41.76	43.95	41.87
3-score sentiment	64.85	79.50	63.32	64.85	63.35
2-score sentiment	93.61	96.59	98.51	93.45	95.91

After applying sentiment classifier on the dataset, the derived sentiment labels were appended to customer records, creating three augmented datasets. Table 2 shows the results of Churn classification after adding sentiment results to the Churn data set.

The ensemble classifier was retrained on each dataset. Results show that incorporating the sentiment, the sentiment level yielded the best results: 98.86% accuracy, 99.47% when using AUC, and 97.77% F1-score. The 2-score sentiment also improved the baseline, while the 5-score performed slightly worse than 2- or 3-score sentiment.

Figure 3 shows the ROC curves for ensemble predictions with and without sentiment. Models using sentiment achieved higher AUCs (~98–99%), confirming the strong contribution of sentiment feature. Key findings from Experiment 2: The 3-score sentiment achieved the best overall results, and incorporating sentiment consistently improved churn prediction performance. ROC analysis confirmed that sentiment-based ensembles were better at identifying churners with fewer false positives.

Table 2. Churn classification using Sentiment classification Results

	Accuracy (%)	AUC (%)	F1-score(%)
No-sentiment	80.88	85.76	59.36
5-score sentiment	96.88	97.74	93.66
3-score sentiment	98.86	99.47	97.77
2-score sentiment	98.20	97.56	96.42

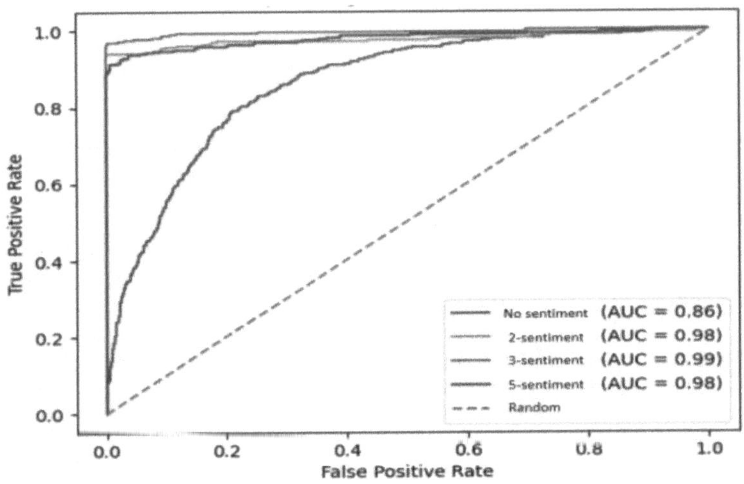

Fig. 3. AUC curve for ensemble predictions

4.3 Experiment 3: Handling Imbalanced Data in 2-Score Sentiment Classification

The 2-score sentiment distribution was imbalanced, favoring positive cases. To address this, class weighting was applied. Grid search optimization determined near-balanced weights for the two classes (Tables 3 and 4).

Table 3. Effect oF Class Weighting on Sentiment Classification)

	Accuracy (%)	AUC (%)	Precision (%)	Recall (%)	F1-Score (%)
2-score sentiment	93.1	96.59	98.51	93.45	95.91
2-score sentiment. + Class weight	93.64	96.70	98.73	93.27	95.92

These results confirm earlier findings that Random Forest is robust to imbalance [33] but adding class weights further improved minority class sensitivity.

Table 4. Ensemble Results with 2-Score Sentiment (Class-Weighted)

F-Score(%)	AUC(%)	Accuracy(%)	F1-Score(%)
96.42	97.56	96.42	96.42
96.81	98.00	96.81	96.81

Key findings from Experiment 3: Class weighting had only marginal impact on overall accuracy but slightly improved minority churn detection, RF's robustness made it a strong choice for sentiment derivation in imbalanced settings.

5 Conclusion and Future Work

This study presented a customer churn prediction framework that integrates sentiment analysis with traditional demographic and service-related features. By employing a stacking ensemble of KNN, SVM, and Random Forest base learners with an MLP meta learner, the model achieved substantial improvements over baseline approaches. Among the evaluated sentiment granularities, the three-class representation proved most effective, yielding 98.86% accuracy, 99.47% AUC, and 97.77% F1-score. These results demonstrate that incorporating sentiment provides significant added value in predicting customer churn.

The findings highlight three main contributions. First, they confirm that sentiment signals are an underutilized yet powerful dimension in churn modeling. Second, the comparative evaluation of 2-class, 3-class, and 5-class sentiment levels clarifies how granularity impacts performance, with three-class sentiment offering the best balance. Third, the stacking ensemble design improves predictive robustness compared to single classifiers.

Despite these strengths, limitations remain. The use of a single dataset constrains generalizability, and the sentiment scores were derived from structured attributes rather than natural customer feedback. Additionally, while the ensemble is computationally efficient at inference, training times are higher compared to individual models.

Future work will focus on applying the framework to real-world textual sentiment sources such as customer reviews, call transcripts, and social media data. Expanding experiments across multiple datasets will test generalizability, and incorporating time-series features will enable modeling of temporal churn dynamics. Further, exploring interpretable AI techniques such as SHAP and LIME can improve transparency and trust in decision-making for telecom operators.

In conclusion, integrating sentiment into churn prediction via ensemble learning offers a promising path toward more accurate, explainable, and practical models for customer retention.

References

1. IBM: Telco customer churn (2022). https://www.ibm.com/docs/en/cognos-analytics/11.1.0?topic=samples-telco-customer-churn
2. Lalwani, P., Mishra, M.K., Chadha, J.S., Sethi, P.: Customer churn prediction system: a machine learning approach. Computing. **104**(2), 271–294 (2022)
3. Edwine, N., Wang, W., Song, W., Ssebuggwawo, D.: Detecting the risk of customer churn in telecom sector: a comparative study. Math. Probl. Eng. **2022**, 1–16 (2022)
4. Chouiekh, A., El Haj, E.H.I.: Deep convolutional neural networks for customer churn prediction analysis. Int. J. Cogn. Inform. Nat. Intell. **14**(1), 1–16 (2020)
5. Usman, M., Ahmad, W., Fong, A.: Design and implementation of a system for comparative analysis of learning architectures for churn prediction. IEEE Commun. Mag. **59**(9), 86–90 (2021)
6. Siddika, A., Faruque, A., Masum, A.K.M.: Comparative analysis of churn predictive models and factor identification in telecom industry. In: 2021 24th International Conference on Computer and Information Technology (ICCIT), pp. 1–6 (2021)
7. Dalli, A.: Impact of hyperparameters on deep learning model for customer churn prediction in telecommunication sector. Math. Probl. Eng. **2022**, 1–11 (2022)
8. Tavassoli, S., Koosha, H.: Hybrid ensemble learning approaches to customer churn prediction. Kybernetes. **51**(3), 1062–1088 (2022)
9. Shabankareh, M.J., Shabankareh, M.A., Nazarian, A., Ranjbaran, A., Seyyedamiri, N.: A stacking-based data mining solution to customer churn prediction. J. Relatsh. Mark. **21**(2), 124–147 (2022)
10. Usman-Hamza, F.E., et al.: Intelligent decision forest models for customer churn prediction. Appl. Sci. **12**(16), 8270 (2022)
11. Xu, T., Ma, Y., Kim, K.: Telecom churn prediction system based on ensemble learning using feature grouping. Appl. Sci. **11**(11), 4742 (2021)
12. Khoh, W.H., Pang, Y.H., Ooi, S.Y., Wang, L.-Y.-K., Poh, Q.W.: Predictive churn modeling for sustainable business in the telecommunication industry: optimized weighted ensemble machine learning. Sustainability. **15**(11), 8631 (2023)
13. Lappeman, J., Franco, M., Warner, V., Sierra-Rubia, L.: What social media sentiment tells us about why customers churn. J. Consum. Mark. **39**(5) (2022)
14. Kilimci, Z.H.: Prediction of user loyalty in mobile applications using deep contextualized word representations. J. Inf. Telecommun. **6**(1), 43–62 (2022)
15. Galitsky, B., De La Rosa, J.-L., Kovalerchuk, B.: Discovering common outcomes of agents communicative actions in various domains, actions in various domains. Knowl.-Based Syst. **24**(2), 210–229 (2011)
16. Almuqren, L., Alrayes, F.S., Cristea, A.I.: An empirical study on customer churn behaviours prediction using Arabic twitter mining approach. Future Internet. **13**(7), 175 (2021)
17. Gulati, V., Raheja, N.: Efficiency enhancement of machine learning approaches through the impact of preprocessing techniques. In: 2021 6th International Conference on Signal Processing, Computing and Control (ISPCC), pp. 191–196 (2021)
18. Ribeiro, M.T., Singh, S., Guestrin, C.: Why should I trust you?: explaining the predictions of any classifier. In: Proceedings of the 22nd ACM SIGKDD International Conference on Knowledge Discovery and Data Mining, pp. 1135–1144 (2016)
19. Lundberg, S.M., Lee, S.-I.: A unified approach to interpreting model predictions. In: Proceedings of the 31st International Conference on Neural Information Processing Systems (NeurIPS), pp. 4768–4777 (2017)
20. Suthaharan, S.: Machine Learning Models and Algorithms for Big Data Classification: Thinking with Examples for Effective Learning, vol. 36. Springer US (2016). https://doi.org/10.1007/978-1-4899-7641-3

An Multimodal Analytics Framework for Facial Expression and User Comments Analysis

Fu-Shing Sun[1](✉) and Miao Guo[2]

[1] Ball State University, Muncie, Indiana, USA
fsun@bsu.edu
[2] University of Connecticut, Storrs, CT, USA

Abstract. Analyzing user comments on social media platforms is essential for understanding user responses and preferences, as well as for enhancing future media production. In this preliminary research, we proposed an ensemble sentiment scoring system that integrates three subsystems: a dictionary-based system, a large language model (LLM)-based system, and a local classification system. By aggregating votes from each subsystem and applying dynamic weighting criteria, a final sentiment score is reached. We conducted a small experiment to evaluate the effectiveness and accuracy of the system. Based on our current design and findings, we discussed several potential directions for future research.

Keywords: social media · facial expression · user comment · sentiment analysis

1 Introduction

In the contemporary digital era, video has emerged as a primary medium for content delivery across various domains, including social networking, entertainment, and education. In such contexts, content creators or presenters convey their messages through recorded or live-streamed videos. Prominent examples include online debates hosted on platforms such as YouTube and educational content delivered via platforms like Coursera. Audiences can actively engage with these materials by posting comments, suggestions, and critiques. These user-generated comments are frequently analyzed to gain insights into audience engagement and to inform the improvement of future content.

In this preliminary study, we aim to conduct a comprehensive analysis of facial expressions exhibited in video content alongside user comments that explicitly reference or evaluate these expressions. The integration of multimodal data—visual cues derived from facial expressions and textual information extracted from user feedback—enables a deeper understanding of audience perception and engagement. By examining how viewers interpret and respond to the emotional and expressive behaviors of the presenter, this research seeks to uncover patterns that reflect the effectiveness of nonverbal communication in video-based interactions. The findings of these analyses can serve as valuable indicators of how the video content is received by its audience and can provide

H. R. Arabnia et al. (Eds.): CSCI 2025, CCIS 2968, pp. 543–551, 2027.
https://doi.org/10.1007/978-3-032-28457-0_38

meaningful feedback to content creators, educators, or communicators for refining presentation styles, improving emotional expressiveness, and enhancing the overall impact of future video productions.

To address these research challenges, we propose a comprehensive analytics framework designed to systematically collect, process, and analyze multimodal data comprising video content and associated user comments. The framework integrates multiple stages of analysis, beginning with data acquisition and preprocessing, followed by the extraction of meaningful features from both visual and textual modalities. Subsequent analytical steps involve the comparison, contrast, and correlation of findings across modalities to uncover relationships between presenters' facial expressions and audience responses. The ultimate goal of this framework is to generate data-driven insights that inform evidence-based conclusions and actionable recommendations for enhancing communication effectiveness and audience engagement in video-based media. In the following sections, we present the methodological design in detail, describe preliminary experiments conducted on a representative YouTube video and its corresponding user comments, and discuss future directions for extending this research.

2 The System Framework

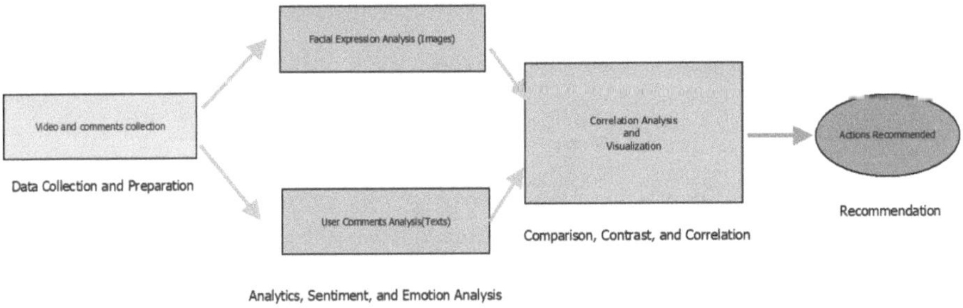

Fig. 1. A system framework for video and user comments sentiment analysis

Figure 1 depicts the overall system framework, which consists of four major modules. We explain each module as follows.

2.1 Data Collection and Preparation

The primary objectives of the data collection and preparation module are to systematically acquire target videos and their corresponding user comments from online platforms. This is accomplished through the use of official data access interfaces, such as the YouTube Data API [3], or through third-party software tools that facilitate large-scale data retrieval. For each video clip, representative image snapshots capturing facial expressions are extracted at regular intervals, with each frame labeled according to its corresponding timestamp. Images that do not clearly depict facial expressions are subsequently discarded to ensure that only meaningful visual data are retained for analysis.

Regarding user comments, this stage is critical for constructing a comprehensive dataset of user-generated textual content, which forms the foundation for subsequent sentiment and discourse analyses. Once the raw data are successfully retrieved, they undergo a series of systematic preprocessing procedures designed to improve their quality, consistency, and analytical suitability. These procedures include cleaning operations to remove unwanted characters and artifacts, normalization of textual elements to achieve consistent formatting, and the exclusion of emojis or other symbols that do not contribute substantively to semantic interpretation. Furthermore, stop words and other non-informative tokens are filtered out to minimize noise and enhance the precision of subsequent analytical tasks. Collectively, these data preparation steps ensure the integrity, clarity, and reliability of the multimodal dataset for downstream analysis.

2.2 Analytics, Sentiment, and Emotion Analyses

In the analytics, sentiment, and emotion analyses stage, the objective is to extract meaningful affective and linguistic insights from both the visual and textual modalities. For the sentiment analysis component, we adopt the conventional three-way polarity classification scheme—positive, negative, and neutral—which provides a coarse-grained understanding of users' overall attitudes toward the video content. This categorization allows for the aggregation of audience opinions and facilitates subsequent quantitative comparisons between visual and textual signals.

For the emotion analysis, we employ a more fine-grained categorization framework consisting of seven discrete emotional states: *anger, disgust, fear, happiness, sadness, surprise,* and *neutral*. These emotion classes are derived from established psychological models of basic emotions, such as Ekman's theory of universal facial expressions. This component aims to identify not only the prevailing emotional tone of user responses but also the nuanced affective expressions displayed by the presenter within the video frames. The framework is designed with flexibility and modularity, enabling the use of either an in-house trained classifier—developed and fine-tuned on domain-specific data—or pretrained large language model (LLM)-based classifiers that leverage transfer learning for enhanced generalization and contextual understanding. This dual approach ensures adaptability to different datasets and allows for empirical comparison between traditional machine learning and state-of-the-art deep learning models in multimodal emotion and sentiment recognition tasks. For the textual analysis of user comments, the system also performs a frequency-based linguistic analysis at multiple levels of granularity, including unigram, bigram, sentence, and document (whole-comment) levels. This analysis provides insights into recurring lexical patterns, common collocations, and frequently co-occurring terms that may indicate dominant discussion themes or emotional triggers. In combination, these analytic procedures enable a comprehensive examination of both expressive cues in facial imagery and affective language in user-generated text, supporting the overall goal of understanding how emotional and attitudinal signals are conveyed and perceived across modalities.

2.3 Comparison, Contrast, and Correlation

The comparison, contrast, and correlation module serves as an integrative analytical stage that synthesizes the findings derived from multiple modalities—namely, facial expressions extracted from video data and textual feedback obtained from user comments. The primary objective of this module is to uncover meaningful relationships and discrepancies between these distinct yet complementary data sources. In the comparison phase, the analysis focuses on identifying similarities and differences in expressive and affective patterns across modalities. For example, it examines whether the emotional tone conveyed by the presenter's facial expressions aligns with the prevailing sentiment observed in audience responses. This step provides insight into the degree of emotional congruence between content delivery and audience perception. The contrast analysis, on the other hand, emphasizes divergences or opposing viewpoints, particularly those expressed by the same user or within the same comment thread. This allows for the detection of mixed or ambivalent attitudes, revealing cases where users may acknowledge certain positive aspects of the content while simultaneously expressing criticism or disapproval of others. Finally, the correlation stage seeks to establish statistical or conceptual relationships between sentiment and emotion signals identified across modalities. By correlating the presenter's emotional expressions with audience sentiment distributions, this step aims to determine whether specific emotional cues (e.g., happiness, sadness, or surprise) tend to elicit consistent or contrasting reactions from viewers. Collectively, these comparative, contrastive, and correlative analyses contribute to a deeper multimodal understanding of audience engagement dynamics and emotional communication effectiveness in video-based interactions.

2.4 Recommendation

Following the comparison, contrast, and correlation stage, the recommendation module is designed to translate analytical insights into actionable guidance for content creators. This module focuses on three primary objectives: encouragement, improvement, and future investigation. In the encouragement component, the system identifies positive patterns in audience responses. For instance, if a particular type of facial expression consistently elicits favorable reactions, such as approval, engagement, or positive sentiment, the module highlights these trends and recommends that the presenter continue incorporating similar expressive behaviors in future content. This serves to reinforce effective communication strategies and maximize audience engagement. The improvement component addresses areas where audience feedback indicates potential shortcomings. For example, if certain facial expressions or presentation styles are consistently associated with negative sentiment or criticism, the module flags these behaviors and suggests modifications or alternative approaches to better align with audience expectations. This feedback mechanism supports iterative refinement and enhances the overall effectiveness of content delivery. Finally, the future investigation component targets cases where audience responses are inconsistent or contradictory. When reactions to a specific facial expression or behavior vary significantly across users or contexts, the module highlights these instances as candidates for deeper exploration. Such investigations aim to uncover underlying factors—such as cultural differences, context-specific interpretations, or individual user preferences—that may explain the observed variability, ultimately informing more nuanced and adaptive content strategies.

Collectively, this recommendation framework leverages multimodal analysis to provide evidence-based guidance, enabling content creators to enhance audience engagement, refine expressive communication, and identify areas for continued research and experimentation. The following section will describe a case study that followed the framework design, from data collection to recommendation.

3 An Experiment

A proof-of-concept experiment was conducted to demonstrate the applicability of the proposed analytical framework, following the sequential stages described in the earlier sections. This experiment illustrates the feasibility of integrating multimodal data collection, processing, and analysis to derive meaningful insights from real-world video content and associated user feedback.

For this study, we selected a 2024 U.S. presidential debate video, publicly available on YouTube through several major news channels. Figure 2 presents a screenshot from a YouTube post published by *The Wall Street Journal*. Figure 3 features a screenshot of representative recent user comments [6]. Using the YouTube Data API [3], approximately 60,000 user comments were collected from six different YouTube postings of the same debate event. Figures. 4 and 5 display headshots extracted from the debate video, which has a total duration of approximately 1 h and 45 min. Image frames were generated at varying time intervals—ranging from 1 s to 60 s—to enable analysis at different temporal granularities. After applying a filtering process to retain only facial-expression-related comments, a subset of several thousand comments was preserved for further analysis.

Fig. 2. A screenshot of the YouTube presidential debate post by the Wall Street Journal

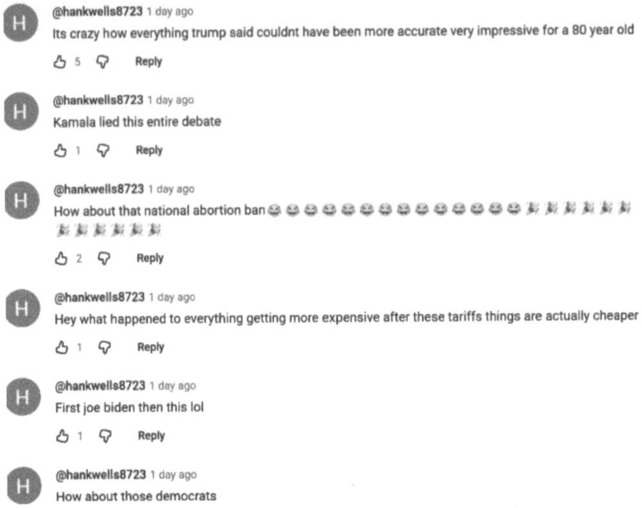

Fig. 3. A screenshot of user comments of the YouTube presidential debate post

Fig. 4. One candidate headshot taken from the presidential debate video recording

For the facial emotion analysis, we employed the DeepFace framework, which produces a probability distribution across the seven emotional categories defined earlier (i.e., *anger, disgust, fear, happiness, sadness, surprise,* and *neutral*). Table 1 summarizes the emotion classification results generated by DeepFace [6] for the first ten facial expressions extracted from the debate video.

For the text-based sentiment analysis, we utilized an LLM-based sentiment classifier available through the Hugging Face platform [7]. Figure 6 provides an example screenshot showing classification results for ten representative comments. To isolate comments specifically referencing facial expressions, we applied an in-house dictionary-based filtering method, as shown in Table 2.

Fig. 5. The other candidate headshot taken from the presidential debate video recording

Table 1. Facial expression emotions classified

Image	Emotion
candidate_2_0.jpg	sad
candidate_2_2.jpg	surprise
candidate_2_4.jpg	sad
candidate_2_6.jpg	neutral
candidate_2_8.jpg	fear
candidate_2_10.jpg	sad
candidate_2_12.jpg	fear
candidate_2_14.jpg	fear
candidate_2_16.jpg	sad
candidate_2_18.jpg	fear

Table 2. Terms used for facial expression

Facial Expression Terms Used for Filtering Comments
Smile, Smirk, Frown, Grimace, Simper, Squint, Glare

Figure 7 presents a visualization of sentiment results closely associated with facial expressions, generated using the SHAP interpretability framework [8]. This visualization highlights the contribution of specific textual elements to the overall sentiment classification. Based on the combined visual and textual analyses, the system generated a recommendation suggesting that the speaker maintain a smiling expression, even during moments of tension or unfavorable discourse. This finding exemplifies how multimodal

```
0
[{'label': 'POSITIVE', 'score': 0.9959543943405151}]
1
[{'label': 'NEGATIVE', 'score': 0.9985212683677673}]
2
[{'label': 'NEGATIVE', 'score': 0.9984055161476135}]
3
[{'label': 'NEGATIVE', 'score': 0.9894289374351501}]
4
[{'label': 'NEGATIVE', 'score': 0.9992617964744568}]
5
[{'label': 'POSITIVE', 'score': 0.9906802773475647}]
6
[{'label': 'POSITIVE', 'score': 0.9953649044036865}]
7
[{'label': 'NEGATIVE', 'score': 0.9910717606544495}]
8
[{'label': 'POSITIVE', 'score': 0.9987867474555969}]
9
[{'label': 'NEGATIVE', 'score': 0.8198723196983337}]
```

Fig. 6. Text sentiment classification using LLM-based classifier

analytics can yield actionable insights to inform communicative strategy and audience engagement.

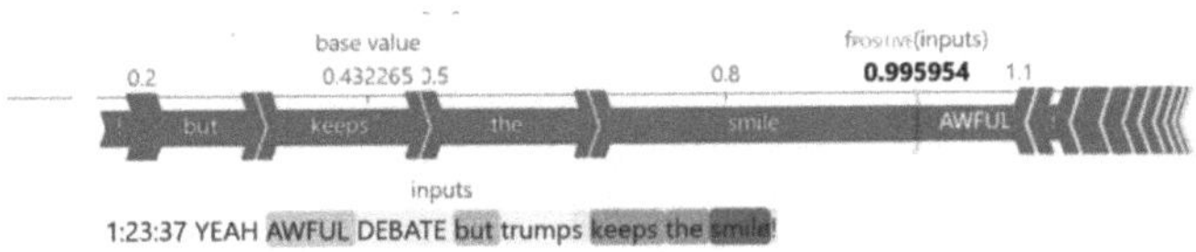

Fig. 7. Sentiment visualization

4 Conclusion and Future Work

In this study, we proposed an analytics framework for facial expression and user comment analysis, designed to integrate multimodal data from video and textual sources. The framework consists of four key modules: (1) data collection and preparation, (2) analytics, sentiment, and emotion analysis, (3) comparison, contrast, and correlation, and (4) recommendation. A proof-of-concept experiment was conducted using a 2024 U.S. presidential debate video retrieved from YouTube. From the collected data, facial expressions were extracted and categorized into multiple emotion classes, while user comments were analyzed to determine their sentiment polarity and emotional tone. The system was able to identify both consistencies and discrepancies between the emotional content conveyed by the candidates' facial expressions and the sentiments reflected in user-generated comments.

For future work, we plan to further examine cases of mismatched sentiment-emotion relationships to better understand their underlying causes. This process may initially require manual verification to ensure the accuracy of classification outputs and to eliminate errors or outliers. In addition, we aim to explore the integration of Retrieval-Augmented Generation (RAG) and LLM-based models within the proposed framework to enhance its analytical capability and contextual understanding in multimodal data environments. Such advancements are expected to strengthen the framework's applicability in broader domains, including social media analytics, education, and human–computer interaction.

References

1. Guo, M., Sun, F.: Local news on Facebook: how television broadcasters use Facebook to enhance social media news engagement. Journal. Pract. **18**, 959–977 (2022). https://doi.org/10.1080/17512786.2022.2074519
2. Guo, M., Sun, F.: Like, comment, or share? Exploring the effects of local television news Facebook posts on news engagement. J. Broadcast. Electron. Media. **64**(5), 736–755 (2020)
3. YouTube Data API: https://developers.google.com/youtube/v3
4. Bird, S., Klein, E., Loper, E.: Natural Language Processing with Python. O'Reilly (2009)
5. https://www.youtube.com/watch?v=VgsC_aBquUE
6. DeepFace: https://github.com/serengil/deepface
7. Hugging Face: https://huggingface.co/
8. SHAP: https://shap.readthedocs.io/en/latest/

Developing the SMART-ACE Framework: Social Media Analytics and Research Translation to Advance Children's Environmental Health

Andrew Larkin(✉), Dixie Jackson, Megan MacDonald, Megan McClelland, Molly L. Kile, and Perry Hystad

College of Health, Oregon State University, Corvallis, OR 97331, USA
larkinan@oregonstate.edu

Abstract. Social media platforms offer novel, real-time datasets that can provide insights into the public's perception of children's environmental health. Systematic analyses of these data could inform targeted research translation and public health actions. We developed SMART-ACE (Social Media Analytics and Research Translation to Advance Children's Environmental Health), an open-source software framework comprising modules that: 1) screen posts for relevance, 2) apply AI to identify and characterize discussion topics, 3) georeference posts, and 4) identify influential social media communities. To evaluate the framework, we analyzed 652 million posts from the social media platform Twitter/X.

SMART-ACE identified over 28 million posts about children and 16 million about schools. In validation datasets (n = 5,000), deep learning model accuracies ranged from 76–98%. Fifty-four percent of posts were georeferenced, including 58% and 68% of posts related to children and schools, respectively. Analysis of 19,348 communities (each with >1,000 child-related posts) showed distinct variations in topics and interactions by community size. Environmental topics captured by topic labeling included wildfires and drinking water. SMART-ACE enables systematic, large-scale analyses of social media to capture perceptions, concerns, and trends in children's environmental health, providing actionable insights for epidemiology, public health research, outreach, and intervention strategies.

Keywords: children · environmental health · social media · AI · GIS · network

1 Introduction

It is important to understand how children are exposed to environmental pollutants and stressors during fetal development and across postnatal critical windows of development [1]. There is a growing body of evidence that exposure to environmental toxicants like heavy metals, organic chemicals, and air pollution adversely impact cognitive [2] and respiratory processes [3, 4] and can increase the risk of severe chronic conditions such as certain cancers [5] in children. These effects are often observed at exposure levels significantly lower than those associated with harm in adults [6], underscoring children's heightened vulnerability during critical windows of development.

H. R. Arabnia et al. (Eds.): CSCI 2025, CCIS 2968, pp. 552–565, 2027.
https://doi.org/10.1007/978-3-032-28457-0_39

Social media platforms represent an underutilized source of real-time data that could significantly enhance understanding of children's environmental health. They have rapidly become a major source of information and conversation, where parents and child caretakers go to share experiences, seek advice, and express concerns. Forty-seven percent of US adults use social media as an information source, including 17% that rely on X (formerly Twitter) for news [7]. Among those who use X, only 31% get their news from mainstream news sources, while 36% acquire their news from politicians, celebrities, and influencers [8]. Young adult trust in social media information is nearly the same as from national news outlets (50% vs. 56%) [9].

Social media provides dynamic, geographically tagged, and temporally immediate insights into public discourse. This information can reveal what issues are gaining attention, how people are responding, and where information gaps may emerge [10]. In other areas of public health, such as infectious disease surveillance and vaccine sentiment tracking, researchers have successfully leveraged social media to capture community attitudes and monitor emerging trends [11]. However, harnessing this vast amount of unstructured data to generate actionable insights into children's environmental health requires sophisticated analytical methods.

To address these opportunities and challenges, we developed the Social Media Analytics and Research Translation to Advance Children's Environmental Health (SMART-ACE) framework. SMART-ACE leverages advanced AI methodologies to systematically extract and analyze children's environmental health information from large-scale social media data, integrating these insights spatially and temporally. This framework aims to provide researchers, policymakers, and public health practitioners with timely, actionable information that can inform effective interventions and improve children's environmental health outcomes.

2 Materials and Methods

SMART-ACE consists of four data analysis modules (Fig. 1), along with a graph database for storing social media records, social network interactions, and analysis products.

2.1 Data Availability

Materials needed to reproduce and implement the SMART-ACE framework including scripts, custom classes, deep learning models, and Neo4j database design are available at the GitHub repository https://github.com/larkinandy/ChildrensHealthSocialMediaASP3IRE. Supplemental Materials referenced in this manuscript are also available at the GitHub repo. Social media posts, metadata, and non-aggregated details extracted from posts (e.g. identified people, places, or topics) cannot be shared due to X Developer User Agreement restrictions.

2.2 Social Media Data Collection

Six hundred and eighty-five million posts were downloaded from the social media platform X by querying the X Developer API.

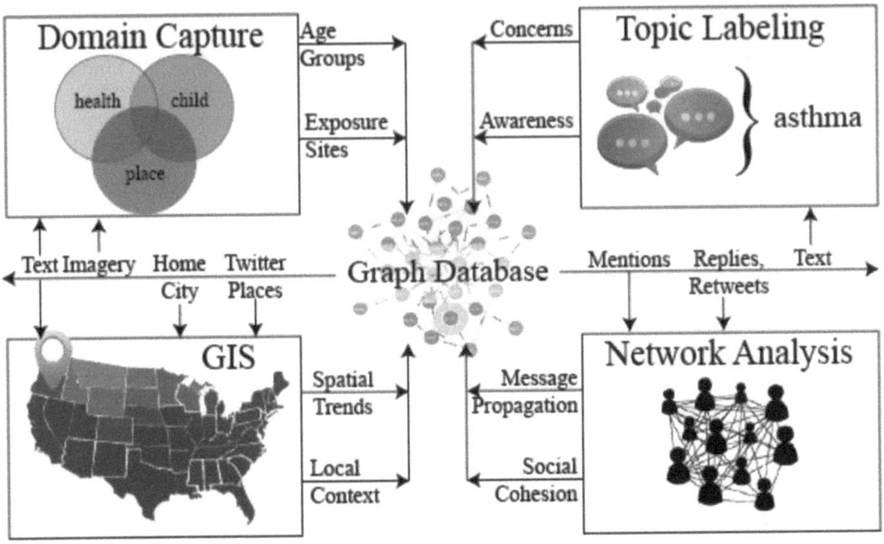

Fig. 1. The Four Primary Components or 'Modules' of the SMART-ACE Framework.

2.3 Module 1: Domain Capture

The domain capture module consisted of a suite of six deep learning models to perform multi-label classification for three high priority domains relevant to children's health:

1. Child age groups. Labels: baby/infant, toddler, elementary school, middle school, and/or high school
2. Places. Labels: home, school, playground/park, neighborhood, and/or daycare center, indoor and/or outdoor
3. Health symptoms or outcomes. Labels: physical, emotional/social, and/or cognitive, positive and/or negative

Developing deep learning models consisted of creating three training datasets, one for each domain, followed by training and evaluating deep learning models.

2.3.1 Creating Training Datasets for Child, Place, and Health Deep Learning Models

For each high priority domain (age, place, and health), 50,000 posts containing keywords relevant to each model (e.g. the words 'baby', 'preemie' were related to child age, Supplemental Table 2) were randomly sampled from downloaded posts. Student workers viewed imagery and text from selected posts and labeled the posts for deep learning model classifications (Supplemental Fig. 1). Workers were trained and certified prior to labeling and were evaluated monthly using quality assurance (QA) posts randomly interspersed throughout the training dataset. Labeled records from students with less than 75% monthly or cumulative QA accuracy, stratified by model, were excluded.

2.3.2 Deep Learning Model Architecture

Architectures differed based on the domain (age, place, health) and whether a social media post included an image (text vs. multimodal, Supplemental Fig. 1).

2.3.3 Model Training and Evaluation

Prior to model training, 10,000 records were removed from each domain training dataset for model tuning (n = 5000) and validation (n = 5000). Tuned hyperparameters included learning rate, dropout rate, number of fixed layers in pre-trained bidirectional encoder representations from transformers (BERT) [12] and residual networks [13] models, and number of nodes in fully connected layers. Early stopping was activated when tuning dataset losses (calculated using binary cross entropy at the end of each epoch) were greater than the lowest recorded losses for three sequential epochs. Models were evaluated by calculating confusion matrices, accuracy, sensitivity, and selectivity for each model and label using the validation dataset.

2.4 Module 2:Topic Labeling

The topic labeling module consisted of a customized topic vector model (n topics = 5000) derived from 5,000,000 children's environmental health social media posts. The topic vector model could be used to assign topics to any text-based social media post based on the cosine distance between document vectors (created from each post using a BERT model) and topic vectors.

2.4.1 Creating the Topic Model

Five million posts which passed module 1 screening were randomly sampled from the downloaded dataset. Text within sampled posts were converted to joint document-word embeddings using a sentence transformer (SBERT) [14] model (multilingual base model, cased text). Embeddings were then used to create a topic vector model using the Top2Vec algorithm [15]. Top2Vec uses Uniform Manifold Approximation and Projection (UMAP) for reducing document-word embedding dimensions and Hierarchical Density-Based Spatial Clustering of Applications with Noise for identifying document-word embedding clusters (i.e. topics) and reducing the number of clusters to a predetermined size (n = 5000 topics for SMART-ACE) [16].

Topic model performance was evaluated by identifying topics from social media posts screened by module 1, calculating the frequency of posts per topic, and comparing posts from similar topics (e.g. school, education, and class).

2.5 Module 3: GIS

The GIS module linked posts to OpenStreetMap (OSM) using three sources of geographic information: 1) self-reported home-town/city from publicly viewable user profiles or 2) opt-in geofences (approximate locations represented by polygons) or coordinates within posts.

2.5.1 Georeferencing Self-Reported Home-Town/City

When opting in to self-report their home town/city, X users had great flexibility. For example, a user who lived in Portland, Oregon might have listed their home town as Portland, PDX (airport code), RipCity, Oregon, or 'cloudy skies and rain all day'. The GIS module attempted to match users' self-reported home towns/cities to one or more of 29,550 cities and towns with populations greater than or equal to 5000 in the 2020 US Census Places dataset. The process entailed a hierarchical series of fuzzy name searches, starting with searches that prioritized high quality matches which equivocally related to one and only town or city. First, the module searched for matches which explicitly designated city and state, or matches which exactly matched city airport codes (e.g. PDX, all caps with no other characters). If an unequivocal match could not be found, the module then searched for matches to large city names whose 2020 US Census population was greater than 500,000 (Supplemental Table 4). If the module was not able to find a match for a town or city, it then attempted to match the user home town/city to a state (e.g. Oregon, or 'OR' in caps with no other characters). Once a match was found, match quality and OSM id corresponding to the US Census Place was added to the user profile in the graph database.

2.5.2 Linking Geofences and Coordinates

When posting, X users could opt-in to share their geographic location as either a geofence (a polygon which can range in extent from a neighborhood to country) or a set of coordinates, called a point of interest (POI), at a public location. Geofence metadata included the variable 'place_type' which indicated the extent of the polygon. Geofence place_types were either 'neighborhood', 'city', 'admin', or 'country'. Geofences with 'neighborhood' spatial extents were spatially joined to the nearest 2020 US census place. Geofences with 'admin' or 'city' spatial extents were joined using a combination of fuzzy name matching and spatial joining: geofence names were compared to names of all census designated places within 100 km and classified as a positive match if the geofence shortname (i.e. without state) perfectly matched to the census place shortname (case insensitive). POIs were joined to OSM by spatially joining POI coordinates to the closest OSM location.

2.6 Module 4: Network Analysis

Social media users are often engaged in children's environmental health discussions with a small but active subset of their entire social media network. Further, these subnetworks are often oriented around a specific domain of children's environmental health, such as a specific age group, common place or activity, or health condition. The network analysis module identified subnetworks within the larger network by developing a children's environmental health profile for each user and then identifying subnetworks with high levels of interaction between users with similar profiles (e.g. similar levels of interest in babies or schools).

2.6.1 Characterizing User Posts, Interactions, and User Networks

User profiles were stored as nodes in the Neo4j graph database. User node properties included a 14-element user profile array corresponding to the percentage of user posts with positive classifications for each of the 14 labels in module 1 deep learning models (Supplemental Table 5). For example, the second and third elements in the profile array corresponded to the percentage of posts in user history relevant to babies and toddlers, respectively. An undirected edge between two user nodes represented one or more interactions between the two nodes (replies, mentions, quotes, and/or retweets). Edge properties included the number of interactions between nodes and cosine similarity between user profile arrays.

2.6.2 Identifying children's Environmental Health Subnetworks Networks

Children's environmental health subnetworks were identified using Louvain Community detection [17], with edges weighted by the product of number of interactions between users and cosine similarity between user profiles (Sect. 2.6.1). Weighting by cosine similarity encouraged Louvain detection to find networks where users were interested in similar domain classifications (e.g. school). It also encouraged the algorithm to place users with low percentages of pertinent interactions on the outer edges of children's environmental health subnetworks (i.e. low centrality).

3 Results

3.1 Data Collection

Summary statistics of downloaded posts are in Supplemental Table 1. The downloaded dataset contained 652 million posts, including 16 million, 19 million, and 109,000 posts that contained the keywords baby, school, and asthma, respectively. Mentions and replies (n = 545 million) were slightly lower than number of posts, and number of unique images (n = 65 million) was an order of magnitude lower than number of posts. The dataset contained on average 19.6 posts per user. The temporal distribution of downloaded posts cannot be shared due to X Developer User Agreement restrictions.

3.2 Domain Capture Module

3.2.1 Training Dataset Quality and Characteristics

After QA screening age, place, and health domain training datasets contained 88,386, 81,062, and 82764 labeled posts, respectively. More than 21% of labeled posts included imagery. Correlations between keywords and relevance to children's environmental health varied significantly by keyword (Supplemental Table 6). For example, only 5.9% of posts with the keyword 'girl' were referring to a child, 52.4% of posts with the keyword 'infant' were referring to babies, and 12.4% of posts with the keyword 'asthma' were referring to a health symptom or condition.

3.2.2 Age, Place, and Health Model Performances

Model accuracies, sensitivities, and specificities in validation datasets are shown in Fig. 2. Additional performance metrics are available in Supplemental Tables 10–12. Overall, age and place model accuracies were above 90% (90.7 to 95.4%). Specificity was the primary driver of high accuracy, with overall specificities between 92.5 to 97.0%. Accuracies and specificities in health models were lower. Average accuracies for the multimodal and text health models were 61.3 and 87.4%, and average specificities were 84.2 and 90.2%, respectively. The multimodal health model accuracies were notably low for physical health and positive health impacts (17.3% and 35.1%).

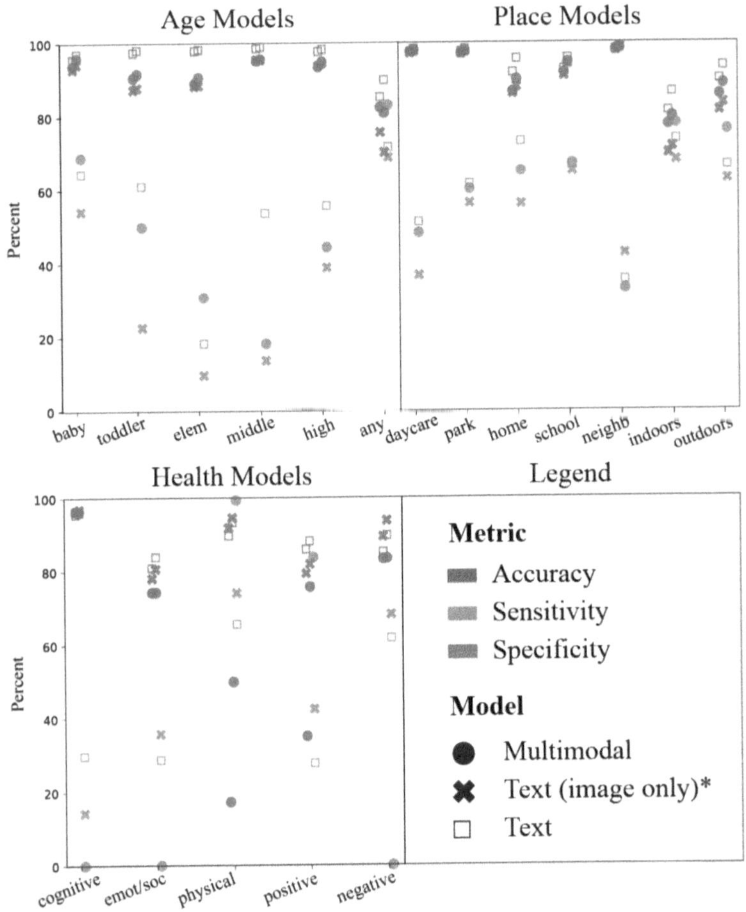

Fig. 2. Models Performance Metrics in Validation Datasets. Abbreviations: **elem** – elementary school age, **middle** – middle school age, **high** – high school age, **nieghb** – neighborhood, **emot/soc** – emotional/social. *Text model performance when restricted to training data with imagery, for the purposes of comparing multimodal and text models performances.

Table 1 shows the number of social media posts with positive labels after applying deep learning models to downloaded posts. Age domain models identified more than 28 million posts with a positive label for mentioning children. Posts per age group ranged from 0.6 million (middle school age) to 2.3 million (high school age). The place domain deep learning models identified 16.1 million posts related to schools, and 3.4 million posts related to home. The number of posts labeled as related to daycares was notably small (0.1 million). The number of posts that mentioned an emotional/social health outcome was notably high (n posts identified by health models = 75 million). Health symptoms and/or outcomes identified by the health models were on average slightly more negative than positive (negative: positive ratio = 1.03).

Table 1. Number of posts with positive labels in the downloaded dataset (n = 652 million)

Age		Place		Health	
label	n	label	n	label	n
baby	1562181	daycare	129819	cognitive	548854
toddler	761563	park	4854563	emotional/social	75101153
elementary school	1419172	home	34262948	physical	16083814
middle school	587119	school	16108475	positive	41831416
high school	2262810	neighborhood	1365019	negative	43070831
child (general)	28102670	indoor	76029440		
		outdoor	27166483		

3.3 Topic Labeling Module

Figure 3 shows a two-dimensional UMAP representation of the topic vector space. Figure 3 also highlights topic vectors (n = 4 per keyword) most semantically like the keyword ‘baby’ (used in social media in numerous contexts), ‘smoke’, and ‘wildfire’ (which can refer to one of several environmental exposures). Vectors related to these keywords were present in multiple quadrants of the vector space, representing their polysemous meaning and usage. For example, a large percentage of posts assigned to a wildfire-related vector in the northeast quadrant (near vectors semantically similar to asthma and smoke) mentioned wildfire air quality conditions, while posts assigned to wildfire related vectors in the southwest quadrant referred to fire danger and economic impacts.

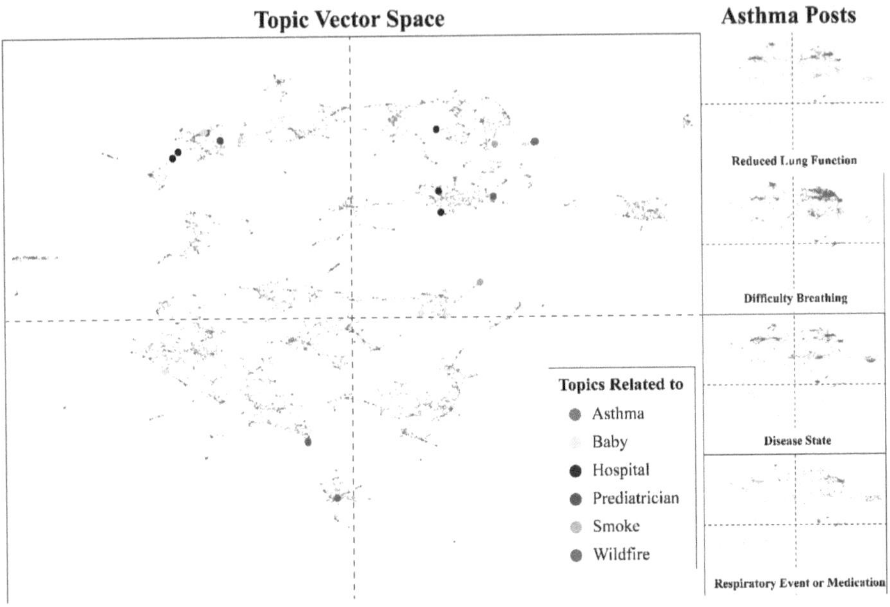

Fig. 3. Topic Vectors and Asthma-Related Posts Represented by a Uniform Manifold Approximation and Projection (UMAP). Left: Vectors semantically similar to polysemous keywords (e.g. baby) were spread across the vector space. Right: Topic vectors captured multiple aspects of living with respiratory diseases, including reduced lung function and difficulty breathing.

3.4 GIS Module

The percent of authors and posts georeferenced to a city or finer spatial extent are shown in Table 2, stratified by georeferenced type (additional metrics are available in Supplemental Table 13). Collectively, 54% of posts were georeferenced, including 58, 68, and 63% of posts that mentioned children, school, and physical health symptoms/outcomes, respectively. Percentages of georeferenced posts were notably high for outdoor (74.2%) and home (71%) and notably low for cognitive health symptoms/outcomes (50%). The overwhelming majority of posts were georeferenced using geofences (51% out of 54%. Note that the 'city full' and 'geofences' georeferencing methods had significant overlap). An additional 52.6 million (8.1%) posts were georeferenced to author state.

Table 2. Percent Georeferenced Posts, Stratified by Georeferencing Method

	Georeferencing method					
	all*	city full	city short	airport	geofences	pois
authors (%)	22.5	19.5	2.8	0.1	NA	NA
posts (%)	53.7	25.3	4.2	0.2	51.2	0.4
age groups						

(*continued*)

Table 2. *(continued)*

	Georeferencing method					
	all*	city full	city short	airport	geofences	pois
child (%)	58.2	28.4	4.1	0.2	48.4	0.9
baby (%)	59.0	26.6	4.1	0.2	50.9	0.4
toddler (%)	60.1	30.1	5.0	0.2	50.4	0.5
elementary (%)	57.5	31.6	4.2	0.1	45.2	1.8
middle school (%)	61.6	32.1	4.5	0.2	50.2	1.6
high school (%)	63.4	31.4	4.2	0.2	52.8	1.3
places						
home (%)	70.9	30.1	4.6	0.3	62.9	0.3
school (%)	67.5	30.6	3.8	0.3	58.1	1.2
daycare (%)	56.4	27.3	4.2	0.2	49.1	0.7
park (%)	67.7	35.1	7.1	0.2	54.0	1.4
neighborhood(%)	66.3	33.0	9.0	0.2	57.7	0.4
outdoors (%)	74.2	42.3	8.4	0.1	60.2	1.4
indoors (%)	71.3	32.2	5.1	0.3	62.3	1.0
health						
emotional/social (%)	67.3	26.4	3.7	0.3	60.1	0.2
cognitive (%)	50.1	25.8	4.2	0.1	41.4	0.4
physical (%)	63.3	29.8	4.8	0.2	54.3	0.3
positive (%)	67.4	27.0	3.7	0.3	59.9	0.3
negative (%)	66.1	27.5	4.2	0.3	58.3	0.1

* all posts that were georeferenced to a city spatial extent of finer. Does not include posts georeferenced at the state level.

3.5 Network Analysis Module

Table 3 shows the summary statistics of communities with 100 or more posts about children. Community sizes ranged from 2 to 48416. The average number of interactions per user was negatively proportional to community size, suggesting that members in smaller communities were on average more interactive and engaged in a greater number of bidirectional communications than larger communities. Smaller communities also had a greater ratio of negative: positive health outcomes and a greater emphasis on physical and smaller emphasis on emotional/social health.

Table 3. Community Summary Statistics, Stratified by Community Size

	Community Size (n users)									
	all (n = 19346)		1 to 10 (n = 630)		11 to 100 (n = 11466)		101 to 1000 (n = 6089)		1000+ (n = 1161)	
	mean	sd	mean	sd	mean	sd	mean	sd	mean	sd
posts (n)	19624.2	70960.5	1917.2	2345.4	3185.2	4023.5	16396.7	18515.4	208510.1	208343.0
users (n)	275.8	825.1	7.6	2.0	47.8	23.5	296.4	211.2	2565.4	2304.8
interactions (n)	5521.0	16840.9	271.2	244.8	1281.0	963.0	6055.4	6342.9	47440.2	50562.0
interactions/user (n)	25.8	17.9	36.8	33.0	28.1	18.0	21.7	14.8	18.4	11.0
age groups										
child (%)	7.2	4.3	12.6	9.0	7.5	4.3	6.3	3.1	5.5	3.6
baby (%)	0.4	0.5	0.8	1.8	0.4	0.5	0.3	0.2	0.3	0.2
toddler (%)	0.2	0.4	0.4	1.6	0.2	0.3	0.2	0.2	0.1	0.1
elementary (%)	0.4	0.9	0.6	2.1	0.4	0.9	0.3	0.5	0.3	0.5
middle school (%)	0.1	0.2	0.2	0.5	0.1	0.3	0.1	0.1	0.1	0.1
high school (%)	0.4	0.5	0.6	1.5	0.4	0.5	0.4	0.3	0.5	0.4
places										
home (%)	4.1	3.8	4.0	4.6	3.6	3.7	4.7	3.9	6.2	3.0
school (%)	2.4	3.8	3.6	9.0	2.2	3.9	2.4	2.7	3.5	3.0
daycare (%)	<0.1	0.2	0.1	0.5	<0.1	0.2	<0.1	<0.1	<0.1	<0.1
park (%)	0.6	1.2	0.6	1.8	0.5	1.1	0.7	1.3	0.9	1.0
neighborhood (%)	0.1	0.3	0.1	0.3	0.1	0.3	0.1	0.4	0.2	0.4
outdoors (%)	2.7	4.2	2.9	5.0	2.4	3.6	3.0	4.4	5.1	6.6
indoors (%)	9.9	8.4	11.0	13.5	8.9	8.4	10.7	7.5	14.6	7.5
health symptoms/outcomes										
emotional/social (%)	9.8	8.3	9.0	8.7	8.5	7.8	11.3	8.6	14.0	8.5
cognitive (%)	0.2	0.4	0.2	0.6	0.2	0.4	0.1	0.2	0.1	0.2
physical (%)	3.9	2.5	4.7	3.4	4.0	2.6	3.6	2.1	3.1	2.1
positive (%)	4.9	5.3	3.9	5.4	4.1	4.9	6.1	5.7	8.0	5.0
negative (%)	8.4	3.5	9.7	5.4	8.4	3.6	8.3	2.8	8.0	3.4

Summary statistics were restricted to communities with 100 or more posts that mentioned children.

4 Discussion

Social media platforms are used by millions of US adults to discuss children's environmental health. For example, in our downloaded dataset of 652 million posts from a single social media platform (X), we 28 million posts about children and 16 million posts about schools. Given the preference among many adults for alternative platforms, it is likely the number of relevant health posts across all social media channels is markedly higher. Social media therefore represents a novel, underutilized data source in epidemiology, providing rich, real-time insights into public perceptions, concerns, and misinformation about children's environmental health. Making open-source tools like SMART-ACE is critical for timely research translation and targeted public health interventions.

SMART-ACE is the first open-source framework optimized for analyzing social media data related to children's environmental health. A key strength of the SMART-ACE framework is its use of advanced AI methods, for example, deep learning models fine-tuned for children's environmental health social media discussions. By automating tasks such as domain capture (module 1) and topic labeling (module 2), SMART-ACE substantially reduces the human resource burden associated with traditional data management and classification tasks. This innovation enables researchers and public health professionals to shift their focus toward interpreting results, designing targeted interventions, and engaging in proactive outreach. This paper represents the foundational demonstration of SMART-ACE's capabilities and establishes a methodological baseline for future work.

The deep learning models within SMART-ACE demonstrated high specificity (approximately 95% removal of unrelated posts) but moderate sensitivity (around 50% capture of relevant posts). Anecdotally, many false negatives were conversational replies, dependent on previous context, highlighting an important limitation. To address this, future AI model iterations could analyze entire conversational threads rather than isolated posts. Additionally, incorporating author-specific and community-level metrics, such as prior engagement or posting history about children, may further improve sensitivity.

Network analysis within SMART-ACE identified over 19,000 distinct online communities discussing children's environmental health, organized by thematic similarity in their posted content. Simplifying networks based on contextual relevance to children's health significantly reduces computational complexity, minimizes irrelevant noise, and enhances the identification of key influencers responsible for health message propagation. However, non-health interactions inevitably influence social network dynamics, potentially affecting trust, reach, and message credibility. A comprehensive, optimal network-analysis approach would integrate interactions across the entire social media ecosystem and strategically weigh community interactions based on specific epidemiological research objectives. The computational resources and comprehensive, free data access required for such analyses remain significant barriers.

In conclusion, SMART-ACE demonstrates that AI methods can effectively transform noisy, massive, and real-time social media datasets into practical epidemiological data and tools. This capability enhances the speed, granularity, and relevance of public health responses by systematically capturing public perceptions, concerns, and misinformation related to children's environmental health. The identified methodological limitations, such as moderate model sensitivity and restrictions on social media data access, outline critical areas for continued research and methodological improvement, including advanced multimodal analyses and more comprehensive contextual modeling. Finally, practical and ethical considerations, particularly maintaining stable data access and addressing potential AI biases, must remain central as these innovative methods are increasingly integrated into epidemiological research and public health practice.

Acknowledgements. Support for this research was provided by a grant from the National Institute of Environmental Health Sciences, National Institutes of Health [P2C ES033432]. The funder had no role in study design, data collection and analysis, decision to publish, or preparation of the manuscript. Computational resources were provided by the NVIDIA Applied Research Accelerator Program.

Disclosure of Interests. The authors have no competing interests to declare that are relevant to the content of this article.

References

1. del Navarrete-Meneses, M., P., Salas-Labadía, C., Gómez-Chávez, F., Pérez-Vera, P.: Environmental pollution and risk of childhood cancer: a scoping review of evidence from the last decade. Int. J. Mol. Sci. **25**(6), 3284 (2024)
2. Lanphear, B.P., Dietrich, K., Auinger, P., Cox, C.: Cognitive deficits associated with blood Lead concentrations <10 microg/dL in US children and adolescents. Public Health Rep. **115**(6), 521–529 (2000)
3. Zhang, Y. et al.: The short-term association between air pollution and childhood asthma hospital admissions in urban areas of Hefei City in China: a time-series study. Environ. Res. **169**, 510–516 (2019)
4. Deryugina, T., Heutel, G., Miller, N.H., Molitor, D., Reif, J.: The mortality and medical costs of air pollution: evidence from changes in wind direction. Am. Econ. Rev. **109**(12), 4178–4219 (2019)
5. Landrigan, P.J., Schechter, C.B., Lipton, J.M., Fahs, M.C., Schwartz, J.: Environmental pollutants and disease in American children: estimates of morbidity, mortality, and costs for lead poisoning, asthma, cancer, and developmental disabilities. I. Environ. Health Perspect. **110**(7), 721 (2002)
6. ATSDR: Agency for toxic substances and /disease registry, exposure dose guidance for body weight. Atlanta, Ga: US. Department of health and human services, public health service, Jan 31 (2023).
7. Reuters Institute Digital News Report.: https://reutersinstitute.politics.ox.ac.uk/sites/default/files/2020-06/DNR_2020_FINAL.pdf (2026)
8. Reuters Institute Digital News Report.: https://reutersinstitute.politics.ox.ac.uk/digital-news-report/2021 (2021)
9. How Americans' Trust in Information from News Organizations and Social Media Sites Has Changed Over Time: https://www.pewresearch.org/short-reads/2022/10/27/u-s-adults-under-30-now-trust-information-from-social-media-almost-as-much-as-from-national-news-outlets (2026)
10. Reveihac, M., Steinmetz, S., Morselli, D.: A systematic literature review of how and whether social media data can complement traditional survey data to study public opinion. Multimed. Tools Appl. **81**(7), 10107–10142 (2022)
11. Charles-Smith, L.E. et al.: Using social Media for Actionable Disease Surveillance and Outbreak Management: a systematic literature review. PLoS One. **10**(10), e0139701 (2015)
12. Devlin, J., Chang, M.-W., Lee, K., Toutanova, K.: BERT: pre-training of deep bidirectional transformers for language understanding. In: Proceedings of the 2019 Conference of the North American Chapter of the Association for Computational Linguistics: Human Language Technologies, vol. 1, pp. 4171–4186 (2019)
13. He, K., Zhang, X., Ren, S., Sun, J.: Deep residual learning for image recognition. In: 2016 IEEE Conference on Computer Vision and Pattern Recognition (CVPR), vol. 1, pp. 770–778 (2019)
14. Reimers, N., Gurevych, I.: Sentence-BERT: Sentence Embeddings Using Siamese BERT-Networks. arXiv preprint (2019). https://arxiv.org/abs/1908.10084

15. Angelov, D.: Top2Vec: Distributed Representations of Topics. arXiv preprint (2020). https://arxiv.org/abs/2008.09470
16. McInnes, L., Healy, J., Melville, J.: UMAP: Uniform Manifold Approximation and Projection for Dimension Reduction. arXiv preprint (2018). https://arxiv.org/abs/1802.03426
17. De Meo, P., Ferrara, E., Fiumara, G., Provetti, A.: Generalized Louvain method for community detection in large networks. In: 2011 11th International Conference on Intelligent Systems Design and Applications, vol. 1, pp. 88–93 (2011)

Visualization of Opinion Trends to Support Understanding Discussions Among Communities of Video Viewers

Ryota Matsumoto[1(✉)], Hiroki Nakayama[1], Ryo Onuma[1], Hiroaki Kaminaga[1], Youzou Miyadera[2], and Shoichi Nakamura[1]

[1] Fukushima University, Kanayagawa 1, Fukushima 960-1296, Japan
matsumoto@cs.sss.fukushima-u.ac.jp, {hnakayama,onuma,kami,nakamura}@sss.fukushima-u.ac.jp
[2] Tokyo Gakugei University, 4-1-1, Nukuikita, Koganei, Tokyo 184-8501, Japan
miyadera@u-gakugei.ac.jp

Abstract. In recent years, the user base of video-sharing platforms has become increasingly diverse, and opportunities for discussion and opinion exchange on these platforms have been growing. In order to effectively utilize this type of discussion, it is important to understand the issues and background from multiple perspectives by grasping the characteristics of the involved viewer communities and their opinion trends. However, this is often difficult because viewers' comments are scattered across numerous videos and the discussions are unstructured. In this research, we aim to realize discussion visualization that emphasizes the understanding of viewer communities and their opinion trends, targeting discussions composed of comments on video-sharing sites. This paper describes a method for extracting viewer communities on the basis of sentiment analysis. We also describe a method for visualizing opinion trends, including the structuring of comments and the suggestion of noteworthy comments. Finally, we describe an experiment on extraction of viewer communities and discuss the features of the proposed methods based on its results.

Keywords: Community of video viewers · Visualization of opinion trend · Extraction of noteworthy comment · Discussion comprehension support

1 Introduction

In recent years, the Internet has increasingly served as a platform for sharing and obtaining information. Among these, video-sharing sites are widely used as a means of information dissemination and gathering [7, 8]. For instance, the usage rate among teenagers reaches 90%, exceeding that of other social media platforms. In addition, as a recent trend, active communication among viewers takes place through comments posted on videos. Different viewer communities with diverse opinions exist in the comment sections of various videos, where positive, negative, and neutral opinions often coexist. Although it is important to objectively understand these, it is not easy.

H. R. Arabnia et al. (Eds.): CSCI 2025, CCIS 2968, pp. 566–572, 2027.
https://doi.org/10.1007/978-3-032-28457-0_40

To address this, methods for performing sentiment analysis on text in comment sections have been reported [1]. However, these methods are limited to the verification of individual comments, and do not lead to an understanding of viewer communities or groups. Next, a study has been reported that analyzes comment data in videos to extract video summaries [2]. However, although the method for extracting comments from videos is useful, the analysis of communities is not considered [3–6]. Furthermore, several studies have proposed methods for community extraction [9–11] and discussion structuring [12, 13]. However, these methods did not support the visualization of trends in emotions or opinions across multiple videos.

In this research, we have aimed to develop a method to support the understanding of two aspects of discussions: emotional trends and reply relationships, using community extraction and discussion structuring.

Accordingly, this method supports users in discovering the focal points of discussion regarding video content by enabling them to grasp the opinion trends for each community extracted from multiple videos, as well as the logical progression within the discussion.

2 Overview of Support System

Figure 1 overviews the discussion support process. In this research, our method first retrieves multiple similar videos and their comment data, starting from a single video of interest on YouTube. Then, it classifies the discussions into three features—positive, negative, and neutral—using community extraction based on sentiment analysis. Next, the system performs discussion structuring considering reply relationships and extracts noteworthy comments from all similar videos. Finally, we integrate these extraction results and visually present the discussion development by community and the noteworthy comments to users in an easy-to-understand format. This enables users to objectively understand the discussion's focal points and main opinions by overviewing opinion trends and discussion developments over a wider range, starting from a single video of interest on YouTube.

3 Extraction of Communities of Video Viewers Based on Sentiment Analysis

3.1 Extraction of Communities Sharing Same Sentiments

This method extracts and classifies viewer communities based on the analysis of the video title, description, and viewer comments of the video of interest, following the steps below.

Step 1: Collection of Similar Videos

First, the system extracts frequent terms from the title and description of the single video of interest specified by the user and defines them as keywords representing the features of that video. Furthermore, it searches for videos on YouTube using the extracted keywords as queries. From the search results, if a video contains those keywords in its title or description, it is collected as a "similar video."

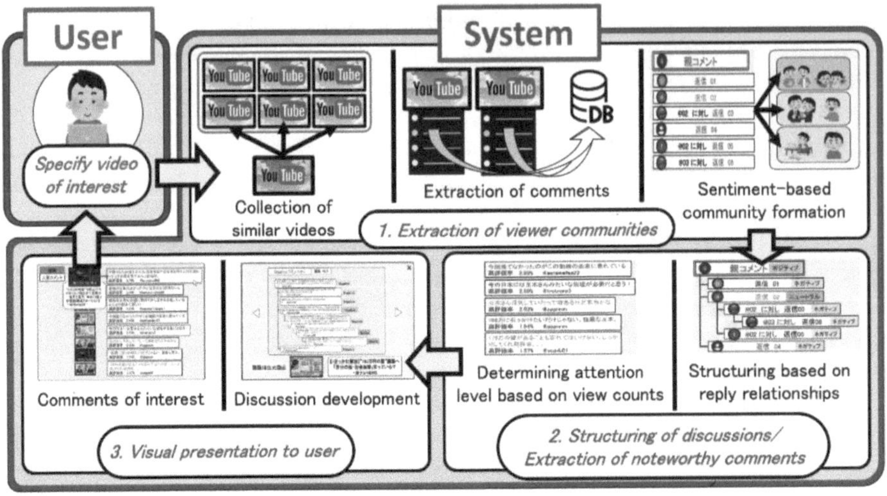

Fig. 1. Support Overview.

Step 2: Extraction of Viewers Commenting on Videos

Using an extraction tool linked with the YouTube Data API, the system extracts comment data, such as countless User_IDs and comment bodies, from the collected similar videos.

Step 3: Morphological Analysis of Comments

The system extracts nouns and verbs from the collected comments using the morphological analysis engine MeCab. In this research, we define a "discussion" as a (parent) comment that has received multiple replies (child comments).

Step 4: Extraction of Communities

The system performs community classification focusing only on these "discussions." First, using the Japanese Sentiment Polarity Dictionary (for verbs and adjectives), it conducts a positive/negative determination on the morphologically analyzed comment sentences based on their positive and negative scores (Fig. 2). Next, based on these scores, the system divides the viewers within the discussions into three communities: "a viewer group with many users who left positive comments", "a viewer group with many users who left neutral comments," and "viewer group with many users who left negative comments" (Fig. 2).

3.2 Structuring Comments that Compose the Discussions

As preparation for visualizing the "discussions" within the community, the system structures and classifies the comments within those discussions. It structures the flow of the discussion from the parent comment to its replies, as well as displaying the sentiment (positive/negative) for each comment, which was determined using the previously

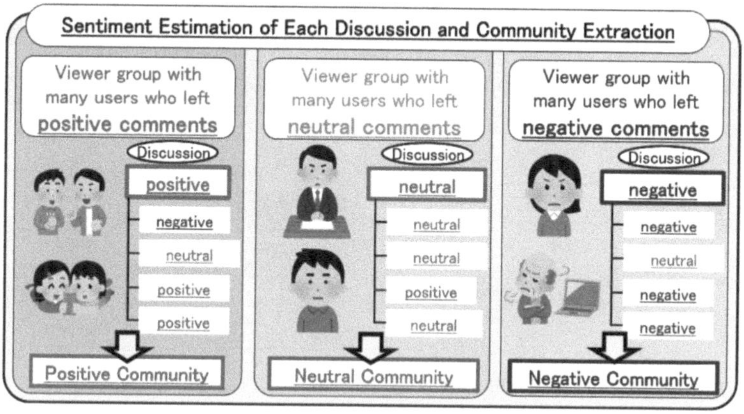

Fig. 2. Community Extraction Based on Sentiment Analysis.

described method. This method promotes the understanding of topic shifts within the actual discussions and counter-arguments to the parent comments (Fig. 3).

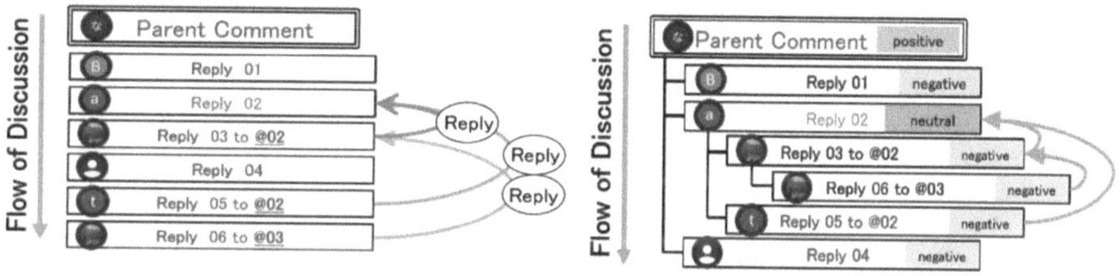

Fig. 3. Structuring of Comments in Discussions.

3.3 Extraction of Noteworthy Comments for Similar Videos

It is difficult to objectively grasp the support or attention for comments within the similar videos used in this research, as their view counts differ. Therefore, this method extracts noteworthy comments for similar videos by sorting the comments based on their degree of attention, using their "like counts." The method is summarized as follows.

First, the system utilizes the "comment like count" and the "view count" of the original video, which are acquired when extracting comments using the YouTube Data API. Next, based on the API data, the system obtains the ratio "comment like count / original video view count" for each comment. The system uses this obtained ratio as the "noteworthiness score" for the comment and extracts comments with a score above a certain level as the noteworthy comments for similar videos (Fig. 4).

3.4 Visualization of Discussion and Opinion Trends

When presenting to the user, we classify and display the items in two categories: "Discussions" and "Noteworthy Comments." Users can switch between these views using the button displayed in the upper left.

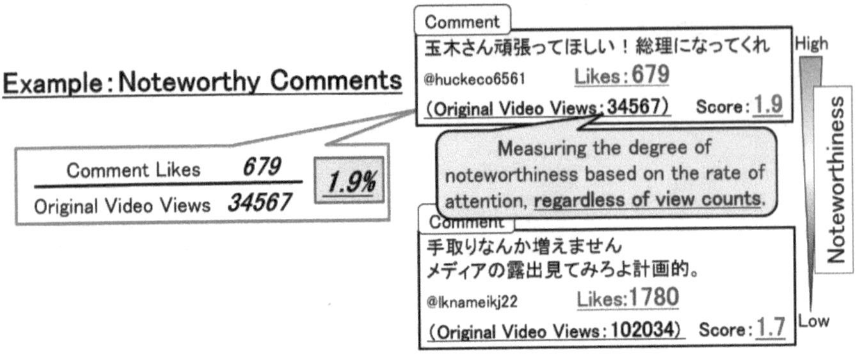

Fig. 4. Evaluation of Noteworthiness of Comments.

In the first visualization, the "Discussion Visualization," the system displays the discussions structured as described in Sect. 3.2, along with the original videos from which the comments were extracted. This view enables users to select the community and discussion they wish to browse.

In the second visualization, the "Noteworthy Comment Visualization," the system displays the noteworthy comments for similar videos, as described in Sect. 3.3, along with their respective noteworthiness scores and the original videos from which the comments were extracted. Furthermore, the system also displays the sentiment (positive/negative) of the comments to make it easier for users to judge them. The visualized interface is shown in Fig. 5.

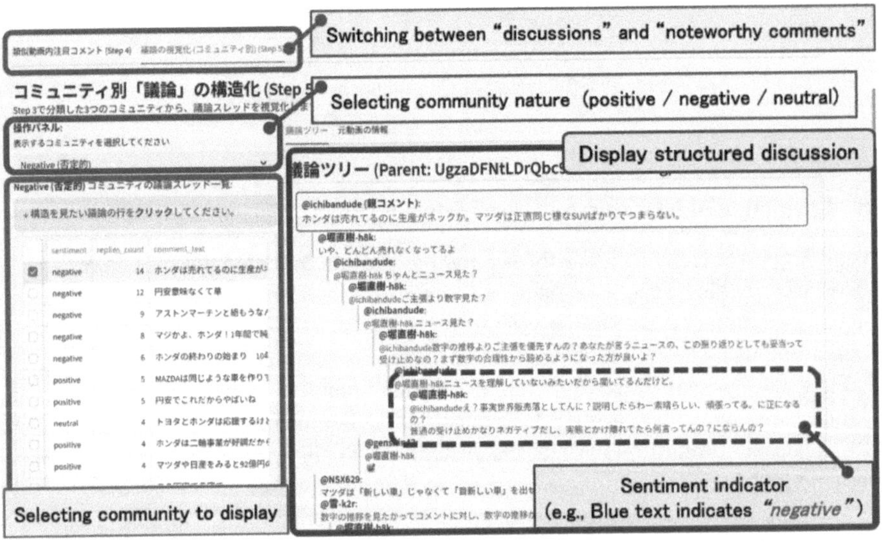

Fig. 5. Visual Presentation of a Structured Discussion and its Elements.

4 Experiment and Discussion

4.1 Experiment Overview

We conducted an experiment to verify the effectiveness of the community extraction method on the basis of sentiment analysis. First, we extracted communities on "five news program videos with different contents" using the proposed method.

Next, we prepared the correct answer data by visually inspecting all discussions and performing community extraction. Finally, we calculated the estimation accuracy by comparing the extraction result by the proposed method with the correct answer data.

4.2 Results and Discussion

We confirmed that the series of processes, from sentiment analysis of comment sentences to sentiment scoring for each comment and extraction of viewer communities, could be performed as intended. The results of the community extraction are shown in Table 1. Considering that the subject of this experiment was news programs, where emotions are less likely to be expressed explicitly in text, it can be said that the accuracy was reasonable.

On the other hand, there were many cases where if the topic of the original video (from which the discussion was extracted) was itself negative, the discussion was classified into the negative community regardless of its actual content. Therefore, it is necessary to consider measures such as changing the weighting of the comment sentiment analysis based on the sentiment (positive/negative) of the original video's topic.

Table 1. Results of Viewer Community Extraction

Accuracy	Precision	Recall	F-value
0.61	0.61	0.67	0.64

5 Conclusion

In this paper, we propose a visualization method for opinion trends based on comment extraction and discussions within specific communities. We described a development method for a visualization system that enables users to objectively understand discussions and opinion trends by grasping an overview of viewer communities from multiple videos. We also described methods for extracting viewer communities across videos, their opinion trends, and determining discussion properties.

Future work includes further verification and improvements of the proposed methods through case studies.

References

1. Novendri, R., Callista, A.S., Pratama, D.N., Puspita, C.E.: Sentiment analysis of YouTube movie trailer comments using Naïve bayes. Bull. Comput. Sci. Electr. Eng. **1**(1), 26–32 (2020)
2. Nawaz, S., Rizwan, M., Rafiq, M.: Recommendation of effectiveness of YouTube video contents by qualitative sentiment analysis of its comments and replies. Pak. J. Sci. **71**(4), 91–97 (2019)
3. Zhao, J., Chevalier, F., Collins, C., Balakrishnan, R.: Facilitating discourse analysis with interactive visualization. IEEE Trans. Vis. Comput. Graph. **18**(12), 2639–2648 (2012)
4. Hoque, E., Carenini, G.: ConVis: a visual text analytic system for exploring blog conversations. Comput. Graph. Forum. **33**(3), 221–230 (2014)
5. Liu, Y., Yu, S.: A visual analytics system for comment moderation. Int. J. Semant. Web Inf. Syst. **20**(1) (2024)
6. Lee, S.-C., Song, J., Ko, E.-Y., Park, S., Kim, J., Kim, J.: SolutionChat: real-time moderator support for chat-based structured discussion. In: Proceedings of the 2020 CHI Conference on Human Factors in Computing Systems, pp. 1–12 (2020)
7. Lange, P.G.: Publicly private and privately public: social networking on YouTube. J. Comput. Mediat. Commun. **13**(1), 361–380 (2007)
8. Burgess, J., Green, J.: Agency and controversy in the YouTube community. In: Handbuch Soziale Praktiken und Digitale Alltagswelten, pp. 105–116 (2020)
9. Matsuo, Y. et al.: POLYPHONET: an advanced social network extraction system from the web. J. Web Semant. **5**(4), 262–278 (2007)
10. Hamasaki, M., Takeda, H., Hope, T.: Network analysis of an emergent massively collaborative creation community. In: Proceedings of the 3rd International AAAI Conference on Web and Social Media (ICWSM 2009), pp. 222–225 (2009)
11. Matsumura, N., Ohsawa, Y., Ishizuka, M.: Discovery of emerging topics by co-citation graph on the web. In: Proceedings 5th International Conference on Knowledge-Based Intelligent Engineering Systems & Allied Technologies (KES 2001), pp. 168–172 (2001)
12. Tomobe, H., Nagao, K.: Discussion ontology: knowledge discovery from human activities in meetings. In: New Frontiers in Artificial Intelligence: JSAI 2006 Conference and Workshops Lecture Notes in Artificial Intelligence (LNAI), vol. 4384, pp. 33–41. Springer (2007)
13. Ito, T., Imi, Y., Ito, T., Hideshima, E.: COLLAGREE: a facilitator-mediated large-scale consensus support system. In: Proceedings of the 2nd Collective Intelligence Conference (2014)

Visualization of Others' Judgment Tendencies in Identifying Useful Disaster Alerts on Social Media

Sho Abe[1](✉), Hiroki Nakayama[1], Ryo Onuma[1], Hiroaki Kaminaga[1], Youzou Miyadera[2], and Shoichi Nakamura[1]

[1] Fukushima University, 1 Kanayagawa, Fukushima, Japan
sho@cs.sss.fukushima-u.ac.jp, {hnakayama,onuma,kami, nakamura}@sss.fukushima-u.ac.jp
[2] Tokyo Gakugei University, 4-1-1 Nukuikita, Koganei, Tokyo, Japan
miyadera@u-gakugei.ac.jp

Abstract. In the event of diverse natural disasters such as heavy rains, landslides, and earthquakes, posts on social media by individuals which warn the imminent danger or the need to evacuate (disaster alerts) have often aided evacuation decisions. However, the vast volume of posts includes information of low usefulness, making it essential to identify. Although such identification ability requires acquisition through experience, accumulating meaningful experience is difficult. This research has been aiming at developing a mechanism that supports the accumulation of experience in identifying disaster alerts on social media. This paper describes an observation support interface to help inexperienced users understand their own and others' judgment tendencies by observing the cases of identification. Specifically, we describe the structure of interface and a way for visualizing both the user's own and others' judgment results along with the elements they referenced. Experimental results using a prototype system suggested that reviewing past cases of identifying useful disaster alerts while comparing them to others' judgments may encourage users to reconsider the elements they reference.

Keywords: Disaster Alert · Identification of Useful Posts · Understanding of Judgment Tendencies · Actions for Damage Mitigation · Social Media

1 Introduction

In recent years, the sources of anxiety among citizens have diversified, including earthquakes, heavy rains, and the resulting landslides and mudslides. While information provided by the Japan Meteorological Agency and local governments is undoubtedly essential for evacuation and damage mitigation, there are also numerous cases where individuals have been saved by utilizing posts on social media that warn of imminent danger or the need to evacuate—hereinafter referred to as disaster alerts. However, a massive amount of information is circulated on social media, requiring users to discern and evaluate the credibility of what they encounter. In particular, during disasters, the

H. R. Arabnia et al. (Eds.): CSCI 2025, CCIS 2968, pp. 573–579, 2027.
https://doi.org/10.1007/978-3-032-28457-0_41

situation changes rapidly, and users must quickly identify reliable information and make appropriate decisions. Nevertheless, identifying which information deserves attention requires experience. Many users tend to take information at face value, failing to assess it critically, and as a result, they cannot make full use of valuable individual posts.

To address these issues, there have been methods for collecting information posted on social media to assist in grasping disaster situations [1–4, 10–15], analyses of hoax diffusion during disasters [5, 9], and methods for automatically extracting useful information from social media to promote behavioral change during disasters [6–8]. However, these existing methods focused primarily on collecting and analyzing useful information and did not take into account promoting the inexperienced gain the experience in identifying valuable information on social media.

This research aims to address these issues by developing a mechanism that facilitates the understanding of others' judgment tendencies when identifying disaster alerts on social media. The proposed method directs users' attention to potentially valuable information during disasters and promotes the experience in selecting useful information.

2 Overview of Support System

Figure 1 shows an overview of the system for promoting the experiences in identifying disaster alerts on social media. This research targets articles written in Japanese posted on Twitter (currently X), a social media platform known for its high real-time nature and potential to provide useful information.

First, based on observations of articles containing disaster information, we define word groups: those representing each disaster type and those indicating imminent danger or the necessity of evacuation. Next, using the Twitter API, the system collects articles containing these words as preliminary candidate articles for each disaster type. Attribute information such as posting date and time, post ID, text, location data, impressions, retweets, and likes is then extracted and stored in a database.

To select disaster alerts from the accumulated preliminary candidate articles, two analytical perspectives are adopted. The first is that the system evaluates changes in the level of attention a certain period after an article is posted, selecting articles that show particularly rapid increases in attention as disaster alerts. The second is that it extracts characteristic conditions and processes involved in discovering useful articles during search behavior, defined as "successful contexts." When a user's search behavior matches these contexts, the system selects corresponding articles as disaster alerts. The selected disaster alerts are visually presented to users, who then judge whether the articles warrant attention.

To more effectively promote users' identification experience, two presentation strategies are introduced. The first strategy involves gradually adjusting the content of the articles presented based on the user's level of experience. During the early stages of system use, the system presents only articles that are highly likely to be disaster alerts, encouraging the development of an identifying mindset. As users accumulate experience—evaluated based on the number of identifications—the system gradually begins to include articles that may not be disaster alerts, thereby fostering more practical experience. Another strategy enables comparison of others' judgments and the factors referenced during identification. When identifying, each user selects the factors they prioritized in addition to their own judgment. Meanwhile, the system also displays the factors prioritized by others during their judgments. This mechanism allows users to objectively understand their own judgment tendencies and characteristics, and to identify their strengths and weaknesses through comparison with others.

Through these presentations, users are expected to deepen their identification experience while reflecting on their judgment tendencies.

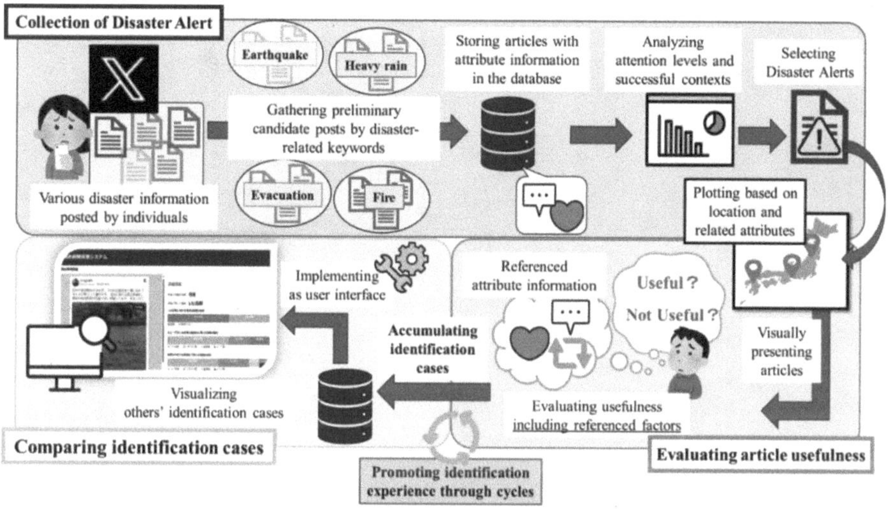

Fig. 1. Overview of the Support System

3 A User Interface for Supporting the Observation of Judgment Tendencies in Identifying Useful Disaster Alerts

Figure 2 shows the interface of the proposed system that supports users in reviewing past cases of disaster alert identification. After users have conducted identifications using the system, it displays the image of each identified article along with the disaster type and the factors referenced during the identification process, enabling users to easily review their past identification cases. Additionally, to allow intuitive browsing of past cases, the system displays articles in a scrollable list format, enabling users to swipe through them sequentially. When focusing on a single article's identification case, that article is visually highlighted in red to attract attention.

When a user selects an article, the system displays detailed information about the corresponding identification case (Fig. 3). In addition to the user's own judgment and the factors referenced during decision-making, the system presents details such as the article's text, attached images, posting date and time, number of likes, and number of reposts. Furthermore, it supports the user's understanding of their own judgment tendencies by displaying the distribution of identification results among others, the proportion of factors referenced by others when identifying the same article, and the proportion of factors the user has referred to in their own previous identifications.

Through these visual comparisons, users can objectively reflect on their judgment tendencies and improve their identification awareness.

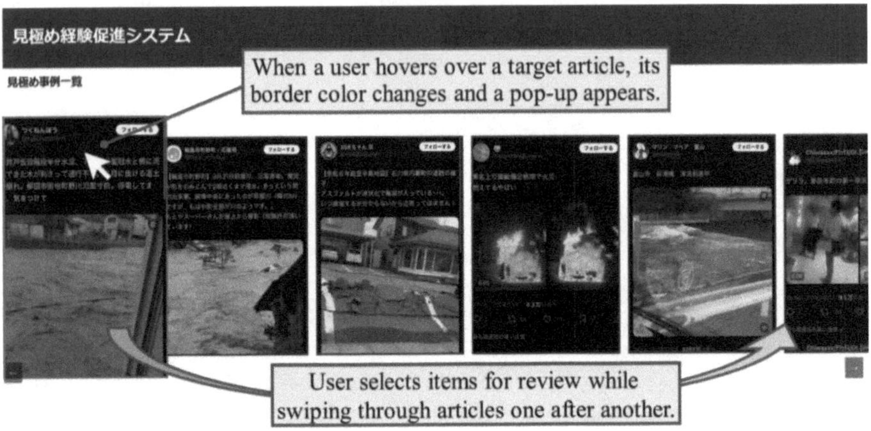

Fig. 2. List of Past Identification Cases

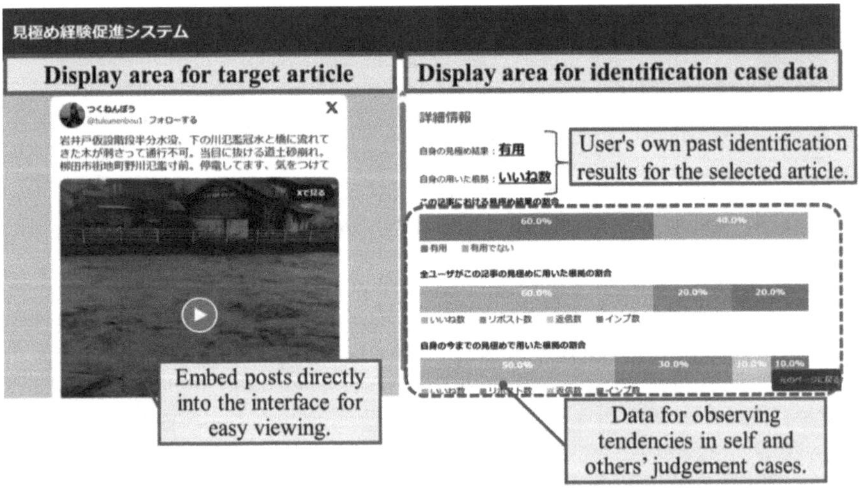

Fig. 3. Presentation of Identification Results and Others' Judgment Tendencies

4 Experiments and Discussion

4.1 Experimental Overview

In this experiment, the authors selected ten actual posts containing disaster alert information from Twitter. The experiment was conducted in two stages.

In the first stage, evaluation files (for Stage 1) were distributed to four undergraduate participants. Each file contained the target articles for identification and input fields for responses. Participants individually evaluated each article on their own PCs, determining whether the article was "useful for disaster response" and which elements they referenced when making that judgment. No information regarding others' judgments or reasoning was provided, ensuring that each participant evaluated the articles solely based on their own judgment. After the evaluations, the completed files were collected and the responses were entered into a database on author's PC running a XAMPP environment.

In the second stage, the same ten articles were presented again. Using the observation support interface that synchronized with the data collected in Stage 1, participants could simultaneously view their own and others' previous identification results along with the factors referenced during those identifications. Under these conditions, they were asked to re-evaluate whether each article was "useful or not," using evaluation files prepared for Stage 2.

By comparing the results of the two stages, we examined the impact of the observation support interface on both the usefulness judgments and the factors emphasized during those judgments.

4.2 Results and Discussion

The results of the experiment are summarized as follows. The average number of cases per participant in which the "mainly referenced factor" changed between Stage 1 and Stage 2 was 5.5. Meanwhile, the average number of cases where the identification result itself changed between the two stages was 0.25 per participant. These findings suggest that, by observing how others made identifications, participants reconsidered which factors deserved attention, leading to positive behavioral outcomes.

Furthermore, a questionnaire survey using a five-point scale was conducted to assess participants' understanding and comprehension of the information presented by the system. The survey results are as shown in Table 1.

These results indicate that the survey outcomes were favorable and consistent with the experimental findings.

On the other hand, post-experiment interviews revealed that participants tended to focus more on the article text itself than on others' reactions. Therefore, it will be necessary to conduct further experiments with a larger number of articles and over a longer period to carefully validate the effectiveness of the proposed method.

Table 1. Results of the questionnaire evaluating the observation support interface.

Question Content	Average Scores
Did you find it easy to operate the system?	4.75
Were you able to easily find the identification cases you wanted to review?	4.00
Overall, were you able to grasp your past identification cases effectively?	4.25
Overall, were you able to grasp others' identification tendencies effectively?	3.50
Did comparing your identifications with others' help you understand your own judgment tendencies?	3.75
Were you able to effectively grasp the distribution of identification results for a given article?	4.75
Were you able to effectively grasp the trends of factors referenced in the identification of a given article?	4.25
Were you able to effectively grasp the factors you have emphasized in your past identifications?	4.25
Do you think supporting the understanding of your own and others' identification tendencies is important?	4.75

5 Conclusion

This paper presented a method to support the understanding of users' judgment tendencies when identifying disaster alerts on social media. Based on experiments conducted using the developed the observation support interface, we discussed the characteristics and potential of the proposed method.

Future work includes conducting more practical and large-scale experiments to further verify the effectiveness of the proposed approach and advance its improvement.

References

1. Urakawa, G., Shiota, A., Kurihara, R.: Building information processing to extract social media information to support creating common operational picture. J. Disaster Inf. Stud. **2**(16), 179–189 (2018)
2. Cu, Q. et al.: An attempt to grasp the disaster situation of "the 2018 Hokkaido eastern Iburi earthquake" using SNS information. J. Disaster Res. **14**(9), 1170–1184 (2019)
3. Miura, A., Toriumi, F., Komori, M., Matsumura, N., Hiraishi, K.: Relationship between emotion and diffusion of disaster information on social media: case study on 2011 Tohoku earthquake. Trans. Jpn. Soc. Artif. Intell. **31**(1), NFC-A_1-9 (2016)
4. Yamaguchi, Y., Ikawa, Y., Amagasa, T., Kitagawa, H.: User location inference using local events in social media. Trans. Inf. Process. Soc. Jpn. **6**(5), 23–37 (2013)
5. Mendoza, M., Poblete, B., Castillo, C.: Twitter under crisis: can we trust what we RT? In: Proceedings of the First Workshop on Social Media Analytics (SOMA '10), pp. 71–79 (2010)

6. Rokuse, T., Uchida, O.: Extracting useful information for disaster response by paraphrasing Tweeets using sequence-to-sequence model. In: Proceedings of the 37th Annual Conference of the Japanese Society for Artificial Intelligence, pp. 1–4 (2023)
7. Kitajima, R., Kamimura, R., Uchida, O., Toriumi, F.: A study on tweet classification by neural networks. In: Proceedings of the 29th Annual Conference of the Japanese Society for Artificial Intelligence, p. 2BS-NFC-02a-2 (2015)
8. Kitajima, R., Kamimura, R., Uchida, O., Toriumi, F.: Potential information maximization: potentially-driven information maximization and its application to tweets classification and interpretation. Int. J. Comput. Inf. Syst. Ind. Manag. Appl. **8**, 42–51 (2016)
9. Nadamoto, A., Miyabe, M., Aramaki, E.: Analysis of microblog Rumors and correction texts for disaster situation. In: Proceedings of the International Conference on Information Integration and Web-Based Applications & Services, pp. 44–52 (2013)
10. Rogstadius, J., Vukovic, M., Teixeira, C., Kostakos, V., Karapanos, E., Laredo, J.: Crisis-Tracker: crowdsourced social media curation for disaster awareness. IBM J. Res. Dev. **57**(5), 1–13 (2013)
11. Ngamassi, L., Zhang, J., Malik, A., Edbert, D.: Social media visual analytic toolkits for disaster management: a review of the literature. In: Proceedings of the 14th International Conference on Information Systems for Crisis Response and Management (ISCRAM 2017), pp. 592–603 (2017)
12. Arvandi, A., Rokne, J., Alhajj, R.: Twitter analysis in emergency management: recent research and trends. Soc. Network Anal. Min. **14**(1), 1–22 (2024)
13. Vieweg, S., Hughes, A., Starbird, K., Palen, L.: Microblogging during two natural hazards events: what twitter may contribute to situational awareness. In: Proceedings of the SIGCHI Conference on Human Factors in Computing Systems (CHI'10), pp. 1079–1088 (2010)
14. Bruns, A., Burgess, J.: Social media use during the 2011 Japan earthquake: how twitter transforms the locus of crisis communication. Media Int. Aust. **149**, 28–40 (2013)
15. Peary, B., Shaw, R., Takeuchi, Y.: Utilization of social Media in the East Japan Earthquake and Tsunami and its effectiveness. J. Nat. Disaster Sci. **34**(1), 3–11 (2012)

Author Index

H. R. Arabnia et al. (Eds.): CSCI 2025, CCIS 2968, pp. 581–582, 2027.
https://doi.org/10.1007/978-3-032-28457-0

FSC
www.fsc.org
MIX
Papier | Fördert gute Waldnutzung
FSC® C083411

Kolibri 360 GmbH
Ferdinand-Jühlke-Straße 7
99095 Erfurt, Deutschland
produktsicherheit@kolibri360.de